ABOUT THE COVER

The "Skydeck Ledge" in the Willis Tower (formerly the Sears Tower) is a cantilevered, all-glass viewing booth that extends outward from the building at a height of 1450 ft above the streets of Chicago. The clear floor of this skydeck consists of a 1-1/2-inch-thick structural composite made up of three 1/2-inch glass layers bonded together by plastic sheets called *interlayers*. Since the skydeck opened in July 2009, thousands of visitors each day have enjoyed the thrilling view from high atop the world's third-tallest building. To ensure their safety on the skydeck, the structure is designed to bear five tons of weight, as well as the loads imposed by those gusty winds for which Chicago is so well known. *(Courtesy of DuPont*[TM] *SentryGlas®, Wilmington, Delaware, www.sentryglas.com)*

D0151030

STATICS AND STRENGTH
OF MATERIALS

STATICS AND STRENGTH OF MATERIALS

Seventh Edition

H. W. Morrow

Professor Emeritus
Nassau Community College
Garden City, New York

Robert P. Kokernak

Professor Emeritus
Fitchburg State College
Fitchburg, Massachusetts

Prentice Hall

Boston   Columbus   Indianapolis   New York   San Francisco   Upper Saddle River
Amsterdam   Cape Town   Dubai   London   Madrid   Milan   Munich   Paris   Montreal   Toronto
Delhi   Mexico City   São Paulo   Sydney   Hong Kong   Seoul   Singapore   Taipei   Tokyo

Vice President and Editor in Chief: Vernon R. Anthony

Acquisitions Editor: David Ploskonka

Editorial Assistant: Nancy Kesterson

Director of Marketing: David Gesell

Executive Marketing Manager: Derril Trakalo

Senior Marketing Coordinator: Alicia Wozniak

Marketing Assistant: Les Roberts

Project Manager: Maren L. Miller

Senior Managing Editor: JoEllen Gohr

Associate Managing Editor: Alexandrina Benedicto Wolf

Senior Operations Supervisor: Pat Tonneman

Operations Specialist: Laura Weaver

Art Director: Jayne Conte

Cover Designer: Axell Designs

Cover Image: *DuPont™ SentryGlas®*

AV Project Manager: Janet Portisch

Full-Service Project Management: Ravi Bhatt/Aptara®, Inc.

Composition: Aptara®, Inc.

Printer/Binder: LSC Communications/Kendallville

Cover Printer: LSC Communications/Kendallville

Text Font: Minion

Credits and acknowledgments borrowed from other sources and reproduced, with permission, in this textbook appear on the appropriate page within the text. Unless otherwise stated, all artwork has been provided by the author.

Library of Congress Cataloging-in-Publication Data

Morrow, H. W. (Harold W.)

    Statics and strength of materials / H. W. Morrow, Robert P. Kokernak. —7th ed.

      p. cm.

    Includes index.

    ISBN-13: 978-0-13-503452-1

    ISBN-10: 0-13-503452-3

    1. Strength of materials.   2. Statics.   3. Strength of materials—Problems, exercises, etc.   4. Statics—Problems, exercises, etc.   5. Trigonometry.   I. Kokernak, Robert P. II. Title.

TA405.M877 2011

620.1'12—dc22

                                        2009031922

Prentice Hall
is an imprint of

www.pearsonhighered.com

ISBN 10:   0-13-503452-3

ISBN 13: 978-0-13-503452-1

13 2019

For James Kokernak and Christine Smith

You don't choose your family.
They are God's gift to you,
as you are to them.

—Bishop Desmond Tutu

PREFACE

The objective of this seventh edition is to cover statics and strength of materials at an elementary level, where calculus is not required. However, for instructors who use the text to teach in accredited programs in the technologies, sections requiring calculus are included. Those sections relate to centroids and moments of inertia of plane areas and deflection of beams by integration. Marked with an asterisk in the table of contents, they can be omitted without a loss of continuity.

Statics and Strength of Materials is written for students enrolled in the industrial technology or engineering technology curriculum and in university-level courses for nonengineering majors, such as architecture. It is also useful for self-study and can serve as a reference for courses in materials, materials testing, machine design, and structural design.

NEW TO THIS EDITION

As with past editions, we have tried to keep this text as *reality based* as possible. Toward that end, the following features have been added to this edition of the book.

- The introductory section (Section 1.1) has been expanded to include examples of *catastrophic failures*. It demonstrates to readers how important statics and strength of materials are in our everyday lives.
- The number of one-page *Application Sidebars,* which have proven so popular with students and faculty alike, has been increased from 19 to 25. Each sidebar is heavily illustrated and describes the real-life application of the material being discussed.
- A short subsection describing some *recent developments* in materials technology has been added to Section 11.8. This creates an awareness of the wide variety of research worldwide, which results in enhanced material properties and unique applications for existing materials.
- A new section (Section 6.8) on *cable analysis* has been added to this edition. Many textbooks, particularly those that are not calculus based, ignore this topic, but an increasing use of cables on structural projects demands that students have at least a basic understanding of the behavior and analysis of these elements.

We have also made several format changes to this seventh edition, each aimed at making the book easier for students to use.

- The book is now accompanied by an *animated CD,* which shows worked examples from various topics in the text. References to the CD at appropriate points are indicated by CD icons in the margins of the text as shown at the bottom left of this page.
- Although there are still more than 975 practice problems to accompany this text, approximately 800 appear here in printed form, with the remainder moved to the Companion Website. This was done to accommodate the new material described earlier, while keeping the book at a manageable size for students. These problems are at various levels of difficulty and offers a good balance in the use of U.S. customary units and the international system of units (SI).
- The new two-color design highlights features throughout the book and increases the clarity of figures and line drawings. We have also increased the number of photographs to help readers develop a better understanding of the relationship between theory and practice.

ORGANIZATION OF TEXT

Care has been taken to present the various topics clearly in a simple and direct fashion and to avoid information overload. To that end, more than 200 examples illustrate the principles involved.

Chapters 1 through 9 focus on statics and begin with a review of basic mathematics. Trigonometric formulas and the component method are employed to solve concurrent force problems. A discussion of the resultant and equilibrium of nonconcurrent forces follows, with special emphasis on the theorem of moments. Then the force analysis of structures and machines, and concurrent and nonconcurrent force systems in space, are presented. The chapters on statics conclude with friction, centers of gravity, centroids, and moment of inertia of areas.

Strength of materials is covered in Chapters 10 through 18. The chapters begin with the study of stress and strain in axially loaded members. This is followed by discussions of shear stresses and strains in torsion members, bending and deflection of beams, combined stress using Mohr's circle, columns, and structural connections.

The majority of the material in this book was originally written by H. W. Morrow, who prepared two preliminary editions for use at Nassau (NY) Community College from 1976 to 1979. These were followed by three editions, published in 1981, 1993, and 1998 by Prentice Hall. Subsequent editions represent a joint collaboration between Mr. Morrow and R. P. Kokernak of Fitchburg (MA) State College.

SUPPLEMENTS

New! Companion Website

Prentice Hall's Companion Website provides support resources and an interactive learning environment for students. Tap into this robust site at **http://www.pearsonhighered .com/morrow** to enrich your learning experience. The Companion Website is organized by chapter and includes Chapter Objectives and approximately 175 Practice Problems.

For the Instructor

The Online Instructor's Resource Manual (ISBN-10: 0-13-245434-3) contains valuable information to assist instructors in the classroom.

ACKNOWLEDGMENTS

We want to thank the following individuals for their help in providing illustrations and/or technical information used in this edition: Colin Daviau of Momentum PR & Branding; Robert Donahue of GE Transportation; John Hillman of HC Bridge Company; Janine LaMaie of BERGER/ABAM Engineers; Richard Piasecki of Piasecki Steel Construction Corp.; Charles Popenoe of Stress Indicators, Inc.; Chip Fogg of DuPont Performance Materials; and Rick Scott, Rich Peabody, and Brian Wasielewski of Rodney Hunt Company. Additional thanks to Dick Moisan for both his photographic expertise and suggestions pertaining to graphic design and to Gene and Jamie Fleck of Gene's Service Center for allowing use of their facilities to prepare several of the illustrations used in this edition. Finally, thanks to our family members for their continued support and inspiration, especially Jean, Amelia, and Charlotte Kokernak, and Abigail and Rachel Smith.

We also want to acknowledge the reviewers of this text: Zarjon Baha, Purdue University; Daniel Hoch, University of North Carolina at Charlotte; and Roushdy Nakhla, Morrisville State College.

Users of the text are encouraged to write the authors with suggestions for improvements in future editions. Such material may be sent via e-mail to *rkokernak@comcast.net* or mailed directly to this address: Robert P. Kokernak, PO Box 1038, Ashburnham, MA 01430.

H. W. Morrow
Ft. Lauderdale, FL

R. P. Kokernak
Ashburnham, MA

CONTENTS

*Sections denoted by an asterisk indicate material that can be omitted without loss of continuity.

APPLICATION SIDEBARS

BASIC CONCEPTS

CHAPTER OBJECTIVES

This chapter reviews the basic mathematical skills necessary to the study of statics and strength of materials. Before proceeding to Chapter 2, you should be able to

■ Perform numerical computations and conversions using both U.S. customary units and the International System of Units (SI).

■ Use standard trigonometric formulas in the analysis of right and oblique (nonright) triangles.

■ Solve groups of two and three simultaneous linear equations.

1.1 INTRODUCTION: CATASTROPHIC FAILURES

Statics is that branch of mechanics involving the study of forces and the effect of forces on physical systems that are in equilibrium. This topic is discussed in Chapters 1 through 9. One reason for the study of statics is to determine all the forces, or loads, that act on a system and the reactions that develop in response to those loads. It is easy to imagine, for example, the tremendous forces that are applied to large buildings (Fig. 1.1). These include such things as the weight of the structure itself, wind and snow loads, extraordinary forces caused by earthquakes, and the everyday effects of people and equipment that enter or leave the building on a regular basis. However, many manufacturing processes (Figs. 1.2 and 1.3) also utilize large forces that require careful analysis and management both for production efficiency and safety.

Statics is one of the oldest branches of science. Its origins date back to the Egyptians and Babylonians, who used statics in the building of pyramids and temples. Among the earliest written records are the theories developed by Archimedes (287–212 B.C.), who explained the equilibrium of the lever and the law of buoyancy in hydrostatics. However, modern statics dates from about A.D. 1600 with the use by Simon Stevinus of the principle of the parallelogram of forces.

Strength of materials, or mechanics of materials, establishes the connection between the external forces applied to a physical system, the resulting deflections or deformations of that system, and the intensity of the internal forces (stress) in the system. We study strength of materials to learn methods for the analysis and design of various load-bearing structures and machines. This topic is considered in Chapters 10 through 18.

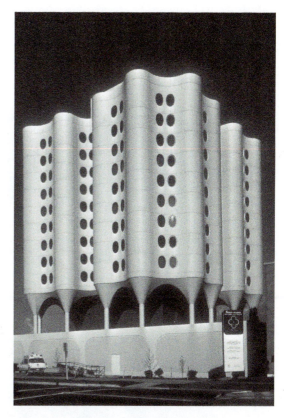

FIGURE 1.1. In addition to the tremendous forces due to their own weights, large buildings must be able to sustain extraordinary loads from other sources. This structure, the St. Joseph Medical Center in Tacoma, Washington, was completed in 1974 and features a unique earthquake-resisting system between the base and tower. Even after the 6.8-magnitude Nisqually Earthquake that shook the Puget Sound area in 2001, the facility remains undamaged.

(Courtesy of BERGER/ABAM Engineers, Inc., Federal Way, Washington, www.abam.com)

FIGURE 1.2. Heavy industries often require significant lifting capacity at some point on their production lines. This facility, which manufactures about 900 train locomotives each year, utilizes a crane and supporting structure that is capable of lifting the 400,000-lb weight of a completely assembled locomotive and moving the unit from one location to another within the plant. The empty cab shown here weighs only 15,000 lbs and is being moved into position for painting.

(Courtesy of GE Transportation RWD, Erie, Pennsylvania, www.getransportation.com)

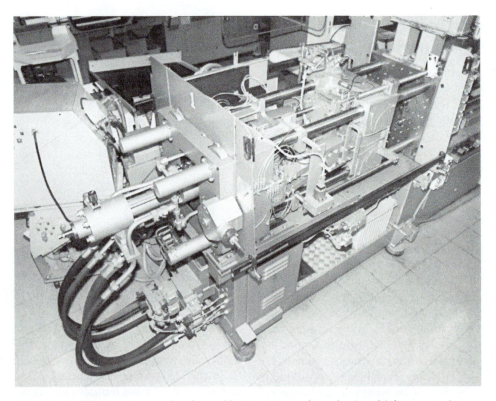

FIGURE 1.3. Plastic parts are often formed by injecting molten plastic at high pressure into a mold consisting of two or more mating sections. Molding machines such as the one shown here contain hydraulic pistons that exert clamping forces to prevent the mold from opening during the injection process. The forces provided by these pistons are multiplied mechanically by a toggle mechanism; although clamping force varies with mold size (90 tons for the machine pictured), total forces in excess of two million pounds are not unusual.

Although not as old as statics, strength of materials dates back to the time of the Renaissance. Galileo, in his book *Two New Sciences,* published in 1638, made reference to the properties of structural materials and discussed the strength of beams. As early as 1678, English physicist and mathematician Robert Hooke published the experimental relationship between force and displacement that is now called Hooke's law. In 1802, this work was quantified by Thomas Young, who defined and measured the material property known as *modulus of elasticity.* However, much of what is now called strength of materials was developed by French investigators in the late 1700s and the early 1800s. Most notable among them were Coulomb and Navier.

So why are statics and strength of materials important? The answer is that any object whose purpose is to support or transmit some combination of loads, and that is made from real materials, is subject to the types of analyses presented in this book. These objects may be as large as bridges and skyscrapers; of moderate size such as planes, trains, or automobiles; or small, such as screwdrivers, hinges, springs, and other mechanical components found in virtually all the consumer products that we use on a daily basis.

If the design of products remained unchanged, then there would be little innovation, and the demand for architects, engineers, or technologists who analyze the feasibility of new products would be small. However, there will always be a steady stream of improvements in materials, manufacturing processes, and construction techniques, so that new products inspired by style, safety, or cost will naturally follow.

A good example of structural innovation is the Grand Canyon Skywalk shown in Fig. 1.4. A cantilevered, horseshoe-shaped walkway that opened to tourism in 2007, the structure extends almost 70 ft from the canyon wall and is 4000 ft above the Colorado River. Because safety is such an obvious design factor here, the walkway was engineered to withstand 100 mph winds and the extreme conditions of a magnitude 8.0 earthquake. Although this structure can support 800 people, each weighing 175 lbs, actual pedestrian traffic is limited to a maximum of 120 persons on the skywalk at one time.

Unfortunately, innovation always involves risk, until a new design, material, or technique has proven itself safe and reliable (see Application Sidebar 1). However, even when well-established technologies yield a catastrophic event, the result is often a firestorm of media coverage and bad publicity. Typical examples include plane crashes and structural failures such as the deadly collapse of Minnesota's I-35W highway bridge into the Mississippi River (see Fig. 1.5).

Many failures occur because of poor mechanical design, such as those caused by overloads, insufficient strength, or excessive deflections. However, a number of other sources

(a)

(b)

FIGURE 1.4. (a) Owned by the Hualapai Indian Tribe, this 500-ton walkway is anchored in place by 94 steel rods, each 46 ft long, driven down into the limestone cliffs. *(Courtesy of DuPont™ SentryGlas®, Wilmington, Delaware, www.sentryglas.com)* (b) Tourists can view the 4000-ft drop beneath them through a 5-layer tempered-glass floor that is 2.8 in. thick. The top layer is removable and can be replaced when scratches and weathering degrade visibility. *(Courtesy of Grand Canyon Skywalk at Grand Canyon West, Arizona, www. destinationgrandcanyon.com)*

COMPRESSION
REINFORCEMENT
(Concrete Arch)

BEAM SHELL
(Fiber-Reinforced
Plastic)

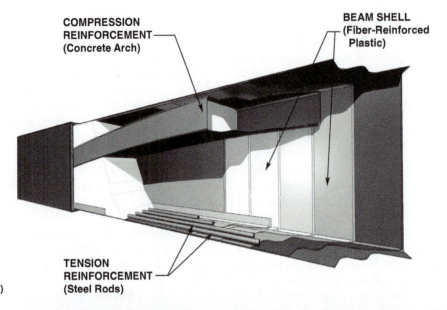

TENSION
REINFORCEMENT
(Steel Rods)

SB.1(a)

SB.1(b)

Another good example of innovation in the use of building materials and techniques is the recent development of a *hybrid-composite beam* for use on short- and intermediate-span bridges. Conventional beams of steel or reinforced concrete are heavy, expensive, and prone to corrosion, factors that become significant for cities and towns faced with an aging infrastructure in need of repair or replacement.

The hybrid-composite beam shown in SB.1(a) consists of three main elements made of three different materials: load-carrying capability of the beam is provided by a *concrete* arch cast in the shape of a parabolic curve; the bottom ends of this arch are tied together by strands of high-strength *steel*; and the concrete and steel are encased in a protective shell made of *plastic* resin reinforced by carbon or glass fibers. The beam is essentially a tied-arch in a fiberglass box with all the unnecessary concrete removed. As a result, the weight of these beams is approximately one-tenth that of a steel or precast concrete beam having the same span. This lighter weight allows for reduced shipping and installation costs, and the fiber-reinforced plastic shell increases resistance to the corrosion experienced by many beams, especially those located in harsh climates.

The first bridge constructed using hybrid-composite beams was a 57-ft single-span bridge completed in Lockport Township, Illinois, in 2008. (Installed as shown in SB.1(b), the beams support a standard 8-in. thick deck made of reinforced concrete.) Although this new beam design makes efficient use of material properties and offers many potential improvements over conventional techniques, only time will tell if the technology is valid and whether it will become an accepted part of the construction industry. (*Courtesy of HC Bridge Co., LLC, Wilmette, Illinois, www.hcbridge.net*)

FIGURE 1.5. Catastrophic failures generally receive extraordinary amounts of media coverage. A recent example was the collapse of Minnesota's I-35W bridge on August 1, 2007, in which 13 people died when the structure plunged into the Mississippi River. Although other factors are often responsible for such failures, public perception is usually that poor engineering design is to blame, primarily because of inadequate strength or overloading of critical members.

(Photo supplied by the Minnesota Department of Transportation, dot.state.mn.us)

can lead to dramatic failures. Some of these contributing factors are poor maintenance practices; material deterioration due to cracking or corrosion; unforeseen events such as fires, floods, accidents, and sabotage; and economic considerations that can result in shoddy materials or negligence in the manufacturing or construction processes.

Here are a few of the more famous "failures" involving statics and/or strength of materials that have occurred in the past 50 years, some as single events, others taking place over an extended period of time, but virtually all with dire consequences.

- While on a training voyage in 1963, the submarine USS *Thresher* broke apart and sank at a depth of approximately 1600 feet. Investigators determined that this disaster began with the *material failure* of a casting, pipe, or weld in the boat's high-pressure water system, resulting in the deaths of all 129 people on board.

- In 1981, two skyway balconies in the Kansas City Hyatt Regency Hotel collapsed onto a dance floor below, killing 114 people. *Poor design* of the suspension rod-washer-nut assembly used to support the balconies caused this accident, which is considered to be the worst nonterrorist structural disaster in U.S. history.

- One of the world's worst high-speed train crashes took place in June 1998, near Eschede, Germany, involving an intercity express train traveling at a speed of 125 mph. The cause of this crash was a *faulty wheel rim* that broke

and embedded itself in the floor of a passenger car, causing the train to derail and slamming the cars into a bridge abutment. There were 101 fatalities.

- Early in the morning of June 28, 1983, a 100-ft by 3-lane section of the Mianus River Bridge on I-95 in Greenwich, Connecticut, collapsed. Although this highway normally carried 100,000 vehicles per day, the morning traffic was light, and only three deaths occurred. Collapse of the 25-year-old bridge was traced to *metal corrosion* and *fatigue* in the bridge's supporting structure, conditions that had gone unnoticed due to *deferred maintenance*.

- Beginning in 1979, more than 86,000 Bjork-Shiley Convexo-Concave mechanical heart valves were implanted. In 1986, the devices were recalled because of *metal fractures* that would cause a patient's heart to contract and thus require immediate emergency surgery. By December 2003, 633 valve failures had been reported, and about two-thirds of those patients who had a defective valve died, many within several minutes.

- Just hours after 5000 basketball fans had left the Hartford (CT) Civic Center Coliseum on January 18, 1978, a two and one-half acre section (about 100,000 square feet) of the building's roof collapsed because of the weight of snow and ice. Because of the timing, there were no injuries. Investigators found that the roof's supporting framework was *poorly designed*, and that the steel

FIGURE 1.6. In any major city worldwide, the skyline is dotted with construction cranes. The size of these structures, and the magnitudes of their loads and counterbalances, makes them susceptible to both tipping and collapse.

(Courtesy of the Portland Cement Association, Skokie, Illinois, www.concrete.org)

members and bolts used to fabricate the framework were of *inadequate strength.*

- In 2000, Bridgestone-Firestone recalled 6.5 million of its tires because of reports of tread separation resulting in fatal crashes of certain sport utility vehicles manufactured by the Ford Motor Company. This recall, reputed to cost Ford $3 billion, was also accompanied by numerous lawsuits over the 271 deaths and 700 serious injuries blamed on these tires. Although investigation revealed that several factors contributed to the tread-separation problem, a specific *manufacturing process* used to produce the tires was found to be the main culprit.

- On November 12, 2001, American Airlines Flight 587 crashed into a residential neighborhood near Queens, New York, killing 265 people. It is believed that a series of quick rudder swings, made in response to turbulence, produced *excessive stresses* on the plane's framework, causing the vertical stabilizer and rudder to fall from the aircraft.

- Seen via television by millions of viewers, the space shuttle *Challenger* was launched on January 28, 1986. Only 73 seconds into its flight, at an altitude of 46,000 ft and a speed of Mach 2, *mechanical failure* of a simple O-ring seal allowed a huge fireball to engulf the spacecraft. This led to structural damage, which caused the shuttle to disintegrate, killing all seven crew members in what was arguably the most dramatic and widely viewed disaster of our time.

- It is estimated that approximately 125,000 construction cranes, both free standing and building mounted

(see Fig. 1.6), are now in use in the United States alone. Some of these structures are 20 to 30 stories high, and all are susceptible to tipping or collapse because of *unbalanced forces.* In the decade from 1997 to 2006, there were 818 crane-related fatalities in the United States. From 2003 to 2006, the highest number of deaths occurred in Texas (42), Florida (27), California (25), and Louisiana (17).

Ultimately, each of the preceding examples involved a material failure caused by an object's inability to carry the loads applied to it at a particular instant in time. In just these ten selected cases, over 1300 people lost their lives, and the total value of property that was damaged or destroyed is staggering. Hopefully, the need for some level of knowledge in the areas of statics and strength of materials is now apparent to you. We begin our study, then, with a review of the basic mathematical tools needed to quantify loads and their effects on real materials.

1.2 FUNDAMENTAL QUANTITIES: UNITS

The two fundamental quantities used in the solution of statics problems are force and length.

Force may be defined as the action of one body on another that tends to change the shape or state of motion, or both, of the other body. Aside from a simple push or pull that we can exert on a body with our hands, the force we are most familiar with is gravitational force. The gravitational force or attraction exerted by the earth on a body is the weight of the body. Other familiar forces are electrical or magnetic attraction, wind forces on a surface, automobile tire traction on the pavement, and so on.

Length is a measure of size or relative position. It is used to describe the size of a body and to locate the position of the forces that act on a body.

To describe a force, we need to specify three things: magnitude, direction, and point of application. The magnitude of a force is given by a certain number of force units. The direction may be given by the angle the force makes with a selected reference axis. The point of application is the point at which the force is applied.

A length can be described by magnitude only, and the magnitude is given by a certain number of length units. The magnitude of both force and length are defined by arbitrarily chosen units.

U.S. Customary Units

Engineers in the United States have commonly used the pound (lb) as the unit of force and the foot (ft) as the unit of length. The pound force is defined as the weight of a specific platinum cylinder placed at sea level and at a latitude of 45°. Other commonly used units of force are the kilopound (kip), equal to 1000 lb, and the ton, equal to 2000 lb. The foot is defined as 0.3048 meter. (See the following paragraph for the definition of meter.) Other units of length based on the foot are the mile (mi), equal to 5280 ft; the inch (in.), equal to 1/12 ft; and the yard (yd), equal to 3 ft.

International System of Units (SI)

A modernized version of the metric system, called the International System of Units (SI) has been adopted throughout the world. The United States has been slow to convert to the SI system. Therefore, engineers, technicians, and technologists should be familiar with both the SI and U.S. customary systems of units.

In the SI system, the meter (m) is the unit of length, and the newton (N) is the unit of force. The meter was at one time defined as the distance between parallel lines marked on a bar kept at standard temperature and pressure and located near Paris, France. The latest definition is based on the wavelength of a color line in the spectrum of the gas krypton. The advantage of such a definition is that it can be reproduced at any location. The meter is 3.2808 ft. The unit of force, the newton, is a derived unit. It is based on the change in the state of motion of a standard body on which the force acts. The newton (N) is approximately two-tenths of a pound of force or, more precisely, 1 N = 0.2248 lb. (The pound force is approximately 4-1/2 N or, more precisely, 1 lb = 4.4482 N.)

Weight and Mass

The *weight* or *force of gravity* on a body is determined from the *mass* of the body. The unit of mass is a kilogram (approximately equal to the mass of 0.001 m^3 of water) and is defined by the mass of a small platinum-iridium bar locked in an airtight vault near Paris, France.[†]

To find the weight W of a body in newtons (N) from the mass m in kilograms (kg), we use the equation

$$W = mg \qquad (1.1)$$

where g is the acceleration due to gravity in meters per second squared (m/s^2). The acceleration varies only slightly from place to place on earth. We will use the approximation 9.81 m/s^2. For example, to find the weight of a body with a mass of 2.5 kg, we multiply by 9.81. $W = 2.5(9.81) = 24.5$ kg $(m/s^2) = 24.5$ N.

Multiples and submultiples of length and force commonly used are the kilometer (km), equal to 1000 m; the millimeter (mm), equal to 0.001 m; the kilonewton (kN), equal to 1000 N; and the meganewton (MN), equal to 1,000,000 N. Thus, we see that the prefix milli means 0.001, or 10^{-3}, the prefix kilo means 1000, or 10^3, and the prefix mega means 1,000,000, or 10^6. Except for the U.S. customary kilopound (kip) mentioned earlier, as well as the kilopound per square inch (ksi), use of prefixes is generally confined to the SI system of units. Other common prefixes may be found inside the back cover of this book.

1.3 SI STYLE AND USAGE

Precise rules of style and usage have been established in the SI system. Several of the rules follow.

A centered dot is used to separate units that are multiplied together. For example, for a newton · meter, the moment of force, we write N · m. This helps avoid confusion with the millinewton, which would be written mN.

Except for the prefix kilo (k) in kilograms, prefixes should be avoided in the denominator of compound units that are answers. For example, as a measure of stress we use N/m^2 rather than N/mm^2, to avoid having the prefix milli (m) appear in the denominator.

Numerical values are written using standard mathematical notation: Numbers having four digits to the left of the decimal point are written without commas (9876); numbers having five or more digits to the left of the decimal point are written with commas (98,760 and 30,000,000); decimal values less than 1 are preceded by a zero to the left of the decimal (0.543); and repeating decimal values are indicated by a bar located over the portion of the decimal that repeats (11/6 = 1.833333... = 1.8$\bar{3}$, 81/33 = 2.454545... = 2.$\overline{45}$).

1.4 CONVERSION OF UNITS

As a general rule, we will solve problems in the same units as those used to give data for the problem. In special cases, it may be necessary to convert from one system of units to another.

[†]The kilogram is the last scientific unit of measure defined by a physical contrivance, a single bar, rather than a constant of nature. Ongoing research is directed at defining the kilogram in terms of standards that can be duplicated anywhere. A possible definition may be based on Avogadro's number—a measure of the number of molecules present in the volume of a mole of gas (0.02241 m^3) at a fixed temperature and pressure. Researchers say that with luck the change may come within the decade.

TABLE 1.1	Equivalents of Customary and SI Units

U.S. Customary Units	SI Units
Length	
1 in. = 25.40 mm	1 mm = 0.03937 in.
1 in. = 0.02540 m	1 m = 39.37 in.
1 ft = 0.3048 m	1 m = 3.281 ft
Force	
1 lb = 4.448 N	1 N = 0.2248 lb
1 kip = 4.448 kN	1 kN = 0.2248 kip

To convert a length or force in one set of units to another, we multiply by the appropriate conversion factor. For example, to convert a length L of 12.5 in. into m, we must replace inches by meters. From Table 1.1, we see that

$$1\,\text{m} = 39.37\,\text{in.} \quad \text{or} \quad \frac{1\,\text{m}}{39.37\,\text{in.}} = 1$$

Because this ratio has the value of unity, the value of L will not be changed if we write

$$L = 12.5\,\text{in.} \left(\frac{1\,\text{m}}{39.37\,\text{in.}} \right)$$

Performing the numerical calculations and canceling units that appear in the numerator and denominator, we obtain the desired result:

$$L = 12.5(0.0254) = 0.318\,\text{m}$$

The number 0.0254 was used to convert length in inches to length in meters. It is called a *conversion factor*.

To convert a stress of 18,540 lb/in.^{2} to units of N/m^2, we replace pounds by newtons and inches by meters. From Table 1.1, we have

$$1\,\text{N} = 0.2248\,\text{lb} \quad \text{or} \quad \frac{1\,\text{N}}{0.2248\,\text{lb}} = 1$$

and

$$1\,\text{m} = 39.37\,\text{in.} \quad \text{or} \quad \frac{39.37\,\text{in.}}{1\,\text{m}} = 1$$

Because each ratio is equal to unity, the value of stress is not changed if we write

$$\text{stress} = 18,540\,\frac{\text{lb}}{\text{in.}^2} \left(\frac{1\,\text{N}}{0.2248\,\text{lb}} \right)\left(\frac{39.37\,\text{in.}}{1\,\text{m}} \right)\left(\frac{39.37\,\text{in.}}{1\,\text{m}} \right)$$

Calculating and canceling units, we have

$$\text{stress} = 18,540(6895) = 127.8 \times 10^6 \frac{\text{N}}{\text{m}^2}$$

The number 6895 is a conversion factor from lb/in.^{2} to N/m^2. From tables inside the Back Cover we see that 10^6 newtons is equal to a meganewton (MN); therefore,

$$\text{stress} = 127.8 \frac{\text{MN}}{\text{m}^2}$$

Conversion factors for several other quantities may also be found inside the back cover of this book.

1.5 NUMERICAL COMPUTATIONS

Accuracy

The accuracy of a numerical value is often expressed in terms of the number of *significant digits* that the value contains. What are significant digits? Any nonzero digit is considered significant; zeroes that appear to the left or right of a digit sequence are used to locate the decimal point and are not considered significant. Thus, the numbers 0.00345, 3.45, 3450, and 3,450,000 all contain three significant digits represented by the sequence 3–4–5. Zeroes bounded on both sides by nonzero digits are also significant; 0.0005067, 5.067, 50.67, and 506,700 each contain four significant digits, as represented by the numerical sequence 5–0–6–7.

The accuracy of a solution can be no greater than the accuracy of the data on which the solution is based. For example, the length of one side of a right triangle may be given as 20 ft. Without knowing the possible error in the length measurement, it is impossible to determine the error in the answer obtained from it. We will usually assume that the data are known with an accuracy of 0.2 percent. The possible error in the 20-ft length would therefore be 0.04 ft.

To maintain an accuracy of approximately 0.2 percent in our calculations, we will use the following practical rule: use four digits to record numbers beginning with 1 and three digits to record numbers beginning with 2 through 9. Thus, a length of 19 ft becomes 19.00 ft, a length of 20 ft becomes 20.0 ft, and a length of 43 ft becomes 43.0 ft.

You will notice one exception to this rule throughout the text: Values of the trigonometric functions are traditionally written to four decimal places, and that practice will be followed here, not for increased accuracy, but to clarify the computations used in worked examples.

Rounding Off Numbers[†]

If the data are given with greater accuracy than we want to maintain (see Fig. 1.7), the following rules may be used to round off their values:

1. When the digit dropped is greater than 5, increase the digit to the left by 1. *Example:* 23.56 ft becomes 23.6 ft.

2. When the digit dropped is less than 5, drop it without changing the digit to the left. *Example:* 23.34 ft becomes 23.3 ft.

3. When the digit dropped is 5 followed only by zeros, increase the digit to the left by 1 only if it becomes even. If the digit to the left becomes odd, drop the 5 without changing the digit to the left. *Example:* 23.5500 ft rounded to three numbers becomes 23.6 ft, and 23.4500 ft becomes 23.4 ft. (This practice is often referred to as the *round-even rule*.)

[†]American Society of Mechanical Engineers (ASME) *Orientation and Guide for Use of SI (Metric) Units,* 9th edition, 1982, p 11. By increasing the digit to the left for a final 5 followed by zeroes only if the digit becomes even, we are dividing the rounding process evenly between increasing the digit to the left and leaving the digit to the left unchanged.

FIGURE 1.7. This sign, which greets visitors to a midwestern city, implies that every man, woman, and child in the population there has been counted. Do you believe that the number shown is 100% accurate? Was it ever accurate to five significant digits? What number of significant digits do you think might be more realistic?

SIGNIFICANT DIGITS ▼

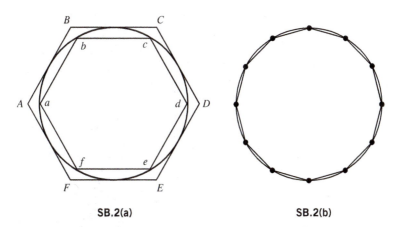

SB.2(a)   SB.2(b)

A classic example of *significant digits* involves the numerical value of *pi* (π), the constant used in computing areas and circumferences of circles. Archimedes, a Greek mathematician, engineer, and physicist, who died in 212 B.C., developed a method of calculating pi that was used for more than 1800 years. His technique, which was based on the definition of pi as the ratio of a circle's circumference to its diameter, used straight-sided figures called *polygons* that were *inscribed* (drawn inside) and *circumscribed* (drawn outside) a circle whose diameter was one unit. In SB.2(a), the perimeter of the inscribed polygon (distance a–b–c–d–e–f–a) is always less than the circumference of the circle, whereas the perimeter of the circumscribed polygon (distance A–B–C–D–E–F–A) is always greater than the circumference. As the number of sides for both polygons increases [demonstrated by the inscribed polygon of SB.2(b)], the two perimeters approach, or *converge on*, the actual circumference of the circle. By the late nineteenth century, manual computations such as these had pushed the known value of pi to 528 digits. By the mid-twentieth century, helped by electronic calculators, that accuracy had been extended to 2038 digits, and by the beginning of the twenty-first century, aided now by computers, the value of pi is known to more than 1.24 *trillion* digits! Although the full value of pi has little application in most areas of technology, it is commonly used today for testing new computers and for training programmers. The many calculations required to determine pi to multimillion-digit accuracies allows computer engineers to check the accuracy of a machine and to verify the integrity of software and memory. (By the way, the value of pi, to forty *decimal places*, is 3.1415926535897932384626433832795028841971.)

Very Large and Very Small Numbers

Calculations for the strengths of materials frequently involve very large and very small numbers. For example, a material property of steel known as its modulus of elasticity has a numerical value of 30,000,000 psi, while its coefficient of thermal expansion has a value of 0.0000065/°F. Both of these quantities are often used in the same formula or equation. For this reason, you should develop some degree of proficiency with the exponential representation of numbers (scientific notation). Numerical values in common use are expanding in both directions: Nanotechnology is now being used to describe material behavior at the atomic level, where distances are measured in *nanometers* (0.000000001 m or 10^{-9} m), while government budgets are no longer presented in millions or billions of dollars, but instead in trillions (and soon quadrillions) of dollars. Some of these named large numbers in common use are listed inside the back cover of this book.

Calculators

Electronic calculators and computers are widely available for use in engineering. Their speed and accuracy make it possible to do difficult numerical computations in a routine manner. However, because of the large number of digits appearing in solutions, their accuracy is often misleading. As pointed out previously, the accuracy of the solution can be no greater than the accuracy of the data on which the solution is based. Care should be taken to retain sufficient digits in the intermediate steps of the calculations to ensure the required accuracy of the final answer. Answers with more significant digits than are reasonable should not be recorded as the final answer. An accuracy greater than 0.2 percent is rarely justified.

PROBLEMS

1.1 Convert the following lengths to millimeters and meters: (a) 17.9 in., (b) 6.4 ft, (c) 13.8 in., and (d) 95.2 ft.

1.2 Convert the following lengths to inches and feet: (a) 10.2 m, (b) 45.0 m, (c) 204 mm, and (d) 4600 mm.

1.3 Convert 60 mi/hr to an equivalent number of ft/s.

1.4 Electric power is generally measured in watts (W), while mechanical power is often measured in horsepower (hp). If 1 hp = 745.7 W:
   a. Find the horsepower of a 60-W lightbulb.
   b. How many watts are equivalent to a 210-hp automobile engine?

1.5 Convert the following forces to newtons and kilonewtons: (a) 23.5 lb, (b) 5.8 kip, (c) 250 lb, and (d) 15.9 kip.

1.6 Convert the following forces to pounds and kilopounds: (a) 52.9 N, (b) 6.85 kN, (c) 1200 N, and (d) 20.8 kN.

1.7 Determine the weight of the following masses in newtons and kilonewtons: (a) 250 kg, (b) 45.0 Mg, (c) 375 kg, (d) 25.0 Mg, and (e) 140.0 kg.

1.8 Determine the mass of the following bodies in kilograms if their weights are (a) 2000 N, (b) 3.50 kN, (c) 1200 N, (d) 4.40 kN.

1.9 Water has the specific weight of 62.428 lb/ft^3 at 42.8°F. (a) What is the specific weight expressed in SI units (N/m^3 and kN/m^3)? (b) If one gallon contains 231 $in.^3$, how much does a 55-gal drum full of water weigh?

1.10 Reinforced concrete weighs 150 lb/ft^3. What is the weight in lb per linear ft of a concrete beam with a 10-in. × 22-in. cross section? What is the weight of the beam in kN/m and the mass in kg/m?

1.11 In the SI system pressure may be expressed in pascals (Pa). One Pa = 1 N/m^2. What is the pressure in pascals of 1 $lb/in.^2$? The atmospheric pressure at sea level is 14.7 $lb/in.^2$. What is the atmospheric pressure in pascals?

1.12 The stress at yield for a steel is 250 MN/m^2 and for an aluminum alloy it is 55 MN/m^2. What is the stress at yield for steel and aluminum in $lb/in.^2$ and $kip/in.^2$?

1.13 A measure of land area, the acre, has an area of 43,560 ft^2. What is the area of an acre in m^2?

1.14 The ultimate tensile strength (stress) for steel is 400 MN/m^2 and for an aluminum alloy it is 70 MN/m^2. What is the ultimate tensile strength for steel and aluminum in $lb/in.^2$ and $kip/in.^2$?

1.6 TRIGONOMETRIC FUNCTIONS

We will find trigonometric functions and trigonometric formulas very useful in solving statics problems. A brief review of trigonometry follows.

Consider the following right triangle [Fig. 1.8(a)], a three-sided closed figure where one of the angles is a right angle (90°). Sides b and c of the triangle form angle A, and sides a and c form angle B. The right angle is angle C.

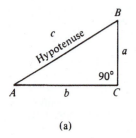

(a)

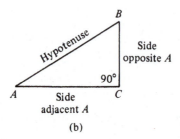

(b)

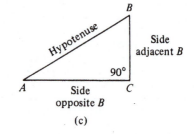

(c)

FIGURE 1.8

The side *c* opposite the right angle is called the *hypotenuse*. The other two sides are named with reference to either angle *A* or angle *B*. When referring to angle *A*, the side opposite is *a* and the adjacent side is *b* [Fig. 1.8(b)]. Similarly, with angle *B*, the side opposite is *b* and the adjacent side is *a* [Fig. 1.8(c)].

Trigonometric functions are ratios of the length of the sides of a right triangle. The three trigonometric functions of angle *A* that are of interest are known as the tangent, sine, and cosine of the angle *A*. They are abbreviated as tan *A*, sin *A*, and cos *A* but are always read in full. For example, tan *A* is read tangent of *A* or tangent of angle *A*. The functions are defined as follows:

$$\tan A = \frac{\text{length of side opposite angle}}{\text{length of side adjacent angle}} = \frac{a}{b} \quad (1.2)$$

$$\sin A = \frac{\text{length of side opposite angle}}{\text{length of the hypotenuse}} = \frac{a}{c} \quad (1.3)$$

$$\cos A = \frac{\text{length of side adjacent angle}}{\text{length of the hypotenuse}} = \frac{b}{c} \quad (1.4)$$

They are not independent but are related in the following way:

$$\frac{\sin A}{\cos A} = \frac{a/c}{b/c} = \frac{a}{b} = \tan A$$

For a given angle *A*, the trigonometric function is a dimensionless constant. Their values are usually irrational numbers represented by nonrepeating, nonterminating decimals. We will use an electronic calculator for the values of the trigonometric functions. Values of trigonometric functions

TABLE 1.2 Trigonometric Functions

Angle	Sine	Cosine	Tangent
0°	0.0	1.0000	0.0
30°	0.5000	0.8660	0.5774
45°	0.7071	0.7071	1.0000
60°	0.8660	0.5000	1.7320
90°	1.0000	0.0	∞

for some common angles are tabulated in Table 1.2 for illustration purposes only.

Two other relationships are helpful in the analysis of right triangles. The first of these applies to *any* triangle and states that the sum of the three angles in a triangle is *always* equal to 180°. In other words,

$$A + B + C = 180° \quad (1.5)$$

For a right triangle, *C* = 90°, so Eq. (1.5) becomes

$$A + B = 90° \quad (1.6)$$

The second relationship, known as the *Pythagorean theorem*, only applies to right triangles and states that *the square of the hypotenuse is equal to the sum of the squares of the remaining two sides*. In equation form,

$$a^2 + b^2 = c^2 \quad (1.7)$$

Eq. (1.7) represents a convenient method for checking numerical results obtained using the trigonometric ratios of Eqs. (1.2) through (1.4).

Example 1.1   The shadow of a flagpole falls on the ground as shown in Fig. 1.9(a). The length of the shadow on the ground measures 20 ft, and the angle that the sun's rays make with the ground is 60°. Find the height, *H*, of this flagpole.

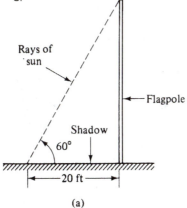

(a)

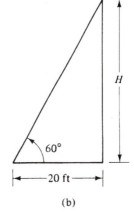

(b)

FIGURE 1.9

Solution: In Fig. 1.9(b), we show the right triangle required for solution. The side adjacent to the 60° angle is 20 ft, and the side opposite the 60° angle is the height of the flagpole, *H*. From Eq. (1.2),

$$\tan 60° = \frac{\text{side opposite 60° angle}}{\text{side adjacent 60° angle}} = \frac{H}{20}$$

$$H = 20 \tan 60° = 20(1.7320) = 34.64$$

Height of flagpole = 34.6 ft   **Answer**

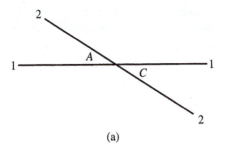
(a)

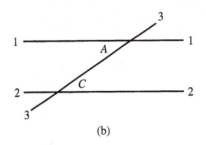
(b)

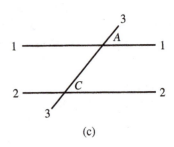
(c)

FIGURE 1.10

Real problems often contain several triangles. To help recognize similarities between figures, a few basic equalities are helpful. Consider the three cases shown in Fig. 1.10. When two straight lines intersect [Fig. 1.10(a)], angles A and C, called *vertical angles*, are equal. When parallel lines 1–1 and 2–2 are both intersected by a third straight line 3–3 [Fig. 1.10(b)], angles A and C, known as *alternate interior angles*, are equal. If two parallel lines are intersected by a third line [Fig. 1.10(c)], angles A and C, called *corresponding angles*, are equal. As the following example demonstrates, problems containing multiple triangles often rely on one or more of these relationships for solution.

Example 1.2   Stair supports, called *stringers*, are to span the vertical and horizontal distances shown in Fig. 1.11. The shaded sections of each stringer are to be cut and removed. Calculate length, *L*, of board required, as well as dimensions "a" and "b," which define the angles of cut at top and bottom of the stringer. Assume that the original boards have right angles at all four corners.

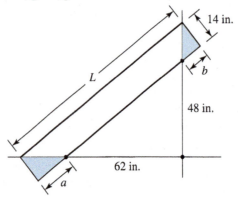

FIGURE 1.11

Solution: The three related triangles are redrawn in Fig. 1.12. (Note that, for clarity, the figure is not drawn to scale.) Using the given side lengths in triangle "1," we may compute the following quantities:

$$\tan A = \frac{48 \text{ in.}}{62 \text{ in.}} = 0.7742$$

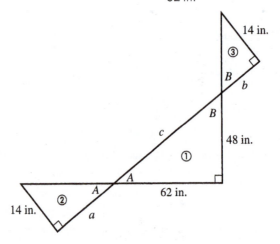

FIGURE 1.12

Then, $\tan^{-1}(0.7742) = A = 37.7°$

From Eq. (1.6), $B = 90° - A = 52.3°$

Using Eq. (1.7),

$$c^2 = (48\text{ in.})^2 + (62\text{ in.})^2$$

or $c = 78.4\text{ in.}$

Moving to triangle "2," vertical angle A is located as shown.

Then,

$$\frac{14\text{ in.}}{a} = \tan A = 0.7742$$

and $a = \dfrac{14\text{ in.}}{0.7742} = 18.1\text{ in.}$ **Answer**

Similarly, in triangle "3,"

$$\frac{14\text{ in.}}{b} = \tan B = \tan 52.3° = 1.2938$$

and $b = \dfrac{14\text{ in.}}{1.2938} = 10.8\text{ in.}$ **Answer**

Finally,

$$L = a + b + c = 18.1\text{ in.} + 10.8\text{ in.} + 78.4\text{ in.}$$
$$L = 107\text{ in.}$$ **Answer**

PROBLEMS

1.15 For the triangle shown, determine the third side c and the $\sin\theta$ and $\cos\theta$ if (a) $b = 950\text{ mm}$, $h = 475\text{ mm}$; (b) $b = 8\text{ ft}$, $h = 6\text{ ft}$.

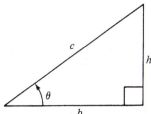

PROB. 1.15

1.16 In the figure, an 8-m ladder is leaning against a house at an angle of 70° with the ground. Determine (a) the

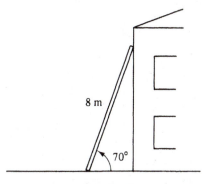

PROB. 1.16

distance of the bottom of the ladder from the house, and (b) the distance of the top of the ladder from the ground.

1.17 In the figure, $d = 300\text{ m}$, $\theta_1 = 45°$, and $\theta_2 = 35°$. Determine distances b and c.

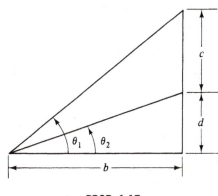

PROB. 1.17

The angle the line of sight forms with the horizontal is called the *angle of elevation* if the line of sight is above the horizontal and the *angle of depression* if it is below the horizontal.

1.18 A man 5 ft 8 in. tall casts a shadow 10 ft long. Determine the angle of elevation of the sun above the ground.

1.19   From a point 40 m away from the base of a flagpole, the angle of elevation of the top is 28°. Determine the height of the flagpole.

1.20   The angle of depression of a ship from the top of a 288-ft-high lighthouse is 22°. Determine how far the ship is from the lighthouse.

1.21   Two buildings are on opposite sides of a 30-m-wide street. The taller building is known to be 124 m high. If the angle of elevation of the roof of the taller building from the roof of the lower building is 42°, determine the height of the lower building.

Problems in structural analysis and surveying often contain multiple triangles or else require the construction of imaginary lines to form appropriate right triangles.

1.22   A building lot has the dimensions shown in the figure. Find the amount of road frontage (distance *AD*) for this lot. (*Hint:* First construct an imaginary line from point *A* perpendicular to lot line *CD*.)

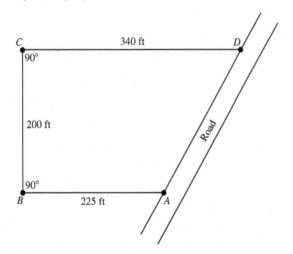

PROB. 1.22

1.23   Member *AB* in the figure is 6.0 m long and is to be held in the position shown by cable *AC*, whose length is 8.0 m. Find distance *x* to the point at which this cable should be fastened. (*Hint:* From point *A*, draw a vertical line downward that is perpendicular to the extension of line *CB*; this creates a pair of "nested" right triangles, one within the other.)

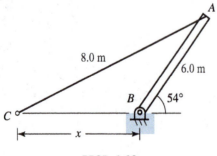

PROB. 1.23

1.24   The 14-in. square concrete column shown in the figure is hoisted using a chain that is 80 in. long. If the lifting

hook is placed at the midpoint, *B*, of section *ABC* of this chain, find the angle formed at *B* between *AB* and *BC*.

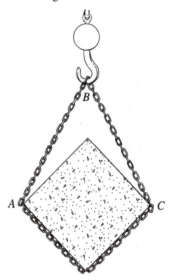

PROB. 1.24

1.25   A force of 40 N is applied to an L-shaped bracket as shown in the figure. To compute the twisting effect of this force about point *A*, it is necessary to determine perpendicular distance *AB* from point *A* to the line of action of the force. Find *AB*.

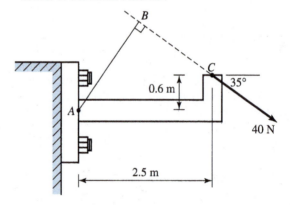

PROB. 1.25

1.26   The roof truss for a residential structure is shown. This truss is symmetric about its centerline (member *CG*), and angles *AHB*, *HGC*, and *GFD* are each 90°:

   a.   Find each angle within the truss.

   b.   How many linear feet of lumber are required to fabricate this truss?

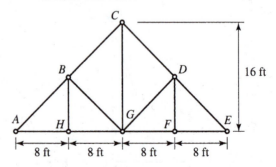

PROB. 1.26

1.7 TRIGONOMETRIC FORMULAS

There are many applications in statics where the triangle to be considered does not involve a right angle. For such applications, the law of cosines and the law of sines may prove useful. However, you should note that both of these relationships apply to *any triangle,* whether that triangle is a right triangle or an *oblique* (nonright) triangle.

Law of Cosines

The square of any side of a triangle is equal to the sum of the squares of the other two sides minus twice the product of these sides and the cosine of their included angle.

The triangle in Fig. 1.13 has sides of length a, b, and c. Angle A is opposite side a, angle B is opposite side b, and angle C is opposite side c. Applying the cosine law, we have formulas for the length of a, b, and c as follows:

$$a^2 = b^2 + c^2 - 2bc\cos A$$
$$b^2 = a^2 + c^2 - 2ac\cos B \qquad (1.8)$$
$$c^2 = a^2 + b^2 - 2ab\cos C$$

When applied to a right triangle for which angle $C = 90°$, the law of cosines yields the Pythagorean theorem. Also, for angles between 90° and 180°, all cosine values are *negative.* If such a value appears in any law of cosines equation, all three terms on the right side of the equation are *added* together. Failure to take into account this consolidation of negative signs in the third, or cosine, term is one of the most frequent computational errors made using the law of cosines.

Law of Sines

The ratio of the side of any triangle to the sine of the opposite angle is a constant. Applying the sine law to the triangle (Fig. 1.13), we have

$$\frac{a}{\sin A} = \frac{b}{\sin B} = \frac{c}{\sin C} \qquad (1.9)$$

One pitfall exists when applying the law of sines. Any two positive angles whose sum is 180° (called *supplementary angles*) have identical sine values. Using your hand calculator, you can verify that

$$\sin 20° = \sin 160° = 0.3420$$
$$\sin 50° = \sin 130° = 0.7660$$
$$\sin 85° = \sin 95° = 0.9962$$

However, under the arcsine or inverse sine function, calculators are only able to display one angular value, and most are programmed to yield the smaller of two angles. Thus,

$$\arcsin(0.9962) = \sin^{-1}(0.9962) = 85°$$

This situation has some basis in reality. Consider, for example, the oblique triangle ABC in Fig. 1.14(a). If side a, side b, and angle A are known, then Eq. (1.9) yields

$$\sin B = \frac{b}{a}\sin A$$

However, by swinging side b to the new position shown in Fig. 1.14(b), we can create a *different* angle, B', for which the same numerical value of sine is obtained. Because two solutions are possible, this condition is known as the *ambiguous case* of the law of sines. It is good practice, then, to check your final computed results using the law of cosines.

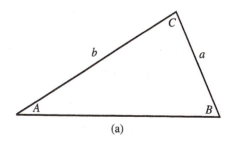

(a)

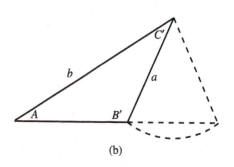

(b)

FIGURE 1.14

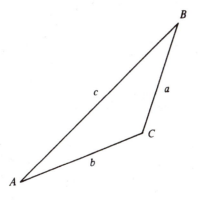

FIGURE 1.13

Example 1.3 A triangle [Fig. 1.15(a)] has sides of 3 ft and 4 ft and the included angle equal to 135°. Find the third side and the angle it forms with line *AE.*

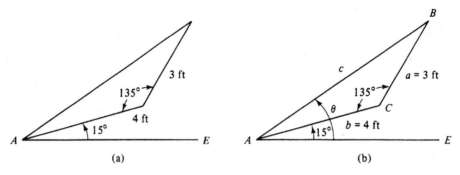

FIGURE 1.15

Solution: In Fig. 1.15(b), sides and angles of the triangle are identified with those shown in Fig. 1.13. Side a = 3 ft, side b = 4 ft, and angle C = 135°. We may solve for side c by applying the law of cosines.

$$c^2 = a^2 + b^2 - 2ab\cos C$$

$$\cos 135° = -0.7071$$

$$c^2 = (3)^2 + (4)^2 - 2(3)(4)(-0.7071) = 41.97$$

$$c = 6.48 \text{ ft} \qquad\qquad \textbf{Answer}$$

To find angle A, we apply the law of sines.

$$\frac{a}{\sin A} = \frac{c}{\sin C} \qquad \frac{3}{\sin A} = \frac{6.48}{\sin 135°}$$

Solving for $\sin A$, we have

$$\sin A = \frac{3\sin 135°}{6.48} = \frac{3(0.7071)}{6.48} = 0.3274$$

$$A = \arcsin(0.3274) = 19.1°$$

The angle that side c of the triangle forms with line AE is

$$\theta = A + 15° = 19.1° + 15° = 34.1° \qquad\qquad \textbf{Answer}$$

Example 1.4 A triangle has sides of 6 m and 12 m, and one of the angles is 35°, as shown in Fig. 1.16(a). Find the third side and the other two angles.

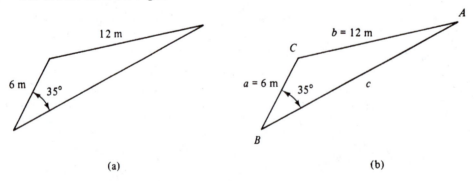

FIGURE 1.16

Solution: The sides and angles of the triangle are identified in Fig. 1.16(b) with those shown in Fig. 1.13. Because angle C between a and b is unknown, we cannot apply the law of cosines directly. Applying the law of sines to find angle A, we have

$$\frac{b}{\sin B} = \frac{a}{\sin A} \qquad \frac{12}{\sin 35°} = \frac{6}{\sin A}$$

$$\sin A = \frac{6(\sin 35°)}{12} = \frac{6(0.5736)}{12} = 0.2868$$

$$A = \arcsin(0.2868) = 16.7° \qquad\qquad \textbf{Answer}$$

Angles $A + B + C = 180°$; therefore, $C = 128.3°$. We may now solve for the third side of the triangle by applying the law of cosines.

$$c^2 = a^2 + b^2 - 2ab\cos C$$
$$\cos 128.3° = -0.6198$$
$$c^2 = (6)^2 + (12)^2 - 2(6)(12)(-0.6198) = 269$$
$$c = 16.4\,\text{m}$$

Answer

PROBLEMS

1.27 From a point on the ground, a balloon was observed to have an angle of elevation of 37°. After the balloon ascended 400 m, it was observed to have an angle of elevation of 56°. Determine the height of the balloon above the ground on the first observation.

1.28 Determine, for the triangle shown, the missing sides and angles with the following sides and angles given.
   a. $a = 7$ in., $B = 40°$, and $C = 30°$
   b. $a = 3$ m, $b = 6$ m, and $C = 48°$
   c. $a = 8$ ft, $b = 7$ ft, and $A = 60°$
   d. $a = 4$ m, $b = 7$ m, and $c = 9$ m

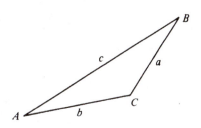

PROB. 1.28

1.29 Determine the lengths of the diagonals AC and BD of the parallelogram shown with the following sides and angle.
   a.   8 in., 12.5 in., and 65°
   b.   550 mm, 320 mm, and 55°
   c.   10.3 ft, 12.5 ft, and 45°
   d.   5 m, 12 m, and 125°

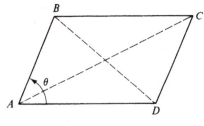

PROB. 1.29

1.30 A boom is supported by a cable as shown. Determine the length of the cable from A to B and the angle the cable makes with the horizontal.

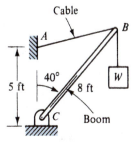

PROB. 1.30

1.31 A boom is supported by two cables, as shown. Cable BD is 28.3 ft long, and cable CD is 44.7 ft long. Determine the length of the boom and the angles cable BD and cable CD make with the horizontal.

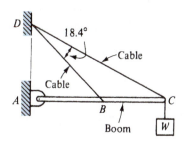

PROB. 1.31

1.32 For the crankshaft, connecting rod, and piston shown, determine the distance the piston travels as the position of the crankshaft given by the angle θ moves from $\theta = 35°$ to $\theta = 90°$.

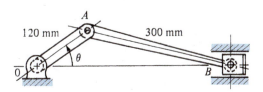

PROB. 1.32

1.33 The truss supports loads as shown. Determine the following angles: (a) $\angle GCB$, (b) $\angle AHJ$.

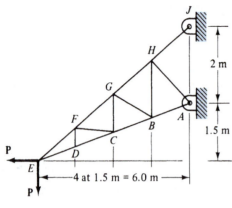

PROB. 1.33

1.34 For the pin-connected tower shown, determine the following angles: (a) $\angle BAH$, (b) $\angle BCG$, (c) $\angle DEF$.

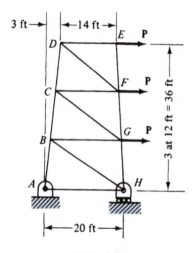

PROB. 1.34

1.35 A telephone pole of height "h" is installed vertically on the side of a hill that makes an angle of 25° with the horizontal. On the uphill side of this pole, a 9.15-m guy wire is anchored to the ground at a point 5.50 m from the base of the pole.
  a. Find height, h, of the pole.
  b. Find distance, d, on the downhill side to the point at which a 13.7-m guy wire must be anchored.

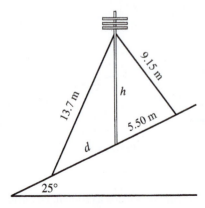

PROB. 1.35

1.36 A homeowner plans to cut down a dead pine tree. Because of neighboring houses, the tree must fall toward the man's own residence, which is 85 ft from the base of the tree. Although it appears that the falling tree will just clear this structure, he measures the angle of elevation to the top of the tree as 28.0°, then moves 75 ft directly toward the tree and finds the new angle of elevation to be 42.0°. Will the tree miss his house?

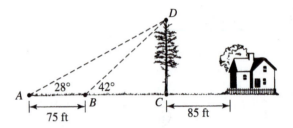

PROB. 1.36

1.37 A landowner wants to excavate a gravel embankment located on his property. The embankment slopes upward for 225 ft at an angle of 20° above the horizontal and then levels off onto a flat wooded area as shown in the figure. In granting a permit to excavate, the local conservation commission has specified that (1) the existing boundary of the wooded area (point B) must remain undisturbed, and (2) a graded embankment (line BD), to be created after gravel has been removed, must make an angle of no more than 50° with the horizontal. (a) to what horizontal distance, w, can the landowner excavate his embankment and satisfy these conditions? (b) If the landowner plans to excavate this embankment for 1/8 mi. along its length, how many cubic yards of gravel will be removed? (c) If the average dump truck can carry 20 cubic yards of gravel, how many such trucks must visit the site?

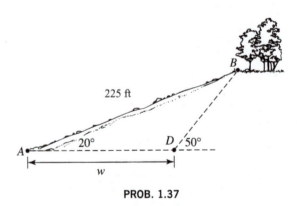

PROB. 1.37

1.38 To accommodate solar collectors, the rear of a 32-ft-wide house is to be framed using 18-ft rafters set at an angle of 41.5°. Find the rafter length, R, and angle B required for construction on the front of this house.

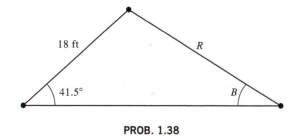

PROB. 1.38

1.39　A coupling contains three bolts equally spaced on a *bolt circle* of diameter d. If the center-to-center distance between bolts is 14 cm, find d.

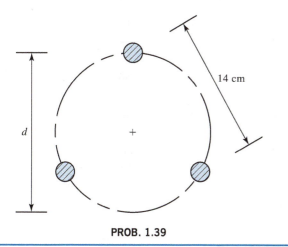

PROB. 1.39

1.8　LINEAR EQUATIONS AND DETERMINANTS

Systems of linear equations are used to solve for the resultant and equilibrium of forces in a plane and in space. We give some of the algebraic methods for solving such equations. Included is the application of determinants to their solution.

Solution by Substitution

Consider the following linear equations:

$$-0.325P + 0.500Q = 0 \qquad \text{(a)}$$

$$0.890P - 0.600Q = 5.00 \qquad \text{(b)}$$

From Eq. (a),

$$Q = \frac{0.325P}{0.500} = 0.650P \qquad \text{(c)}$$

Substituting the value of Q from Eq. (c) in Eq. (b), we have

$$0.890P - 0.600(0.650P) = 5.00$$

or　　　　　　　　　　　　　　$P = 10$　　**Answer**

Substituting the value of P into Eq. (c), we write

$$Q = 0.650(10) = 6.50 \qquad \textbf{Answer}$$

Solution by Addition or Subtraction

We multiply or divide the equations by numbers that are chosen so that the coefficient of either P or Q terms are of equal magnitude.

Consider again the previous two linear equations:

$$-0.325P + 0.500Q = 0 \qquad \text{(a)}$$

$$0.890P - 0.600Q = 5.00 \qquad \text{(b)}$$

If we divide Eq. (a) by 0.5 and Eq. (b) by 0.6, we have the following:

$$-0.650P + Q = 0 \qquad \text{(d)}$$

$$1.483P - Q = 8.33 \qquad \text{(e)}$$

Adding Eqs. (d) and (e), we obtain

$$0.833P = 8.33 \quad \text{or} \quad P = 10 \qquad \textbf{Answer}$$

Substituting $P = 10$ into Eq. (d), we have

$$Q = 0.650(10) = 6.5 \qquad \textbf{Answer}$$

The methods of substitution and of addition and subtraction are also applicable to linear equations of more than two unknowns. You should note, however, that for any set of simultaneous equations, a solution can be obtained only if the number of unknowns is equal to the number of equations.

Solution by Determinants

A determinant consists of a square array of elements bounded by two vertical straight lines. The arrangement is said to be square when the number of rows in the array is equal to the number of columns. It is conventional to number the rows, R, and columns, C, beginning with the top left-hand element. Thus, a determinant containing two rows and two columns (called a *second-order* determinant) might be labeled as shown below at left; in practice, however, such labels are rarely used, and determinants typically appear in the form shown at the right.

$$
\begin{array}{cc}
 & \text{C1 \ C2} \\
\text{R1} & \begin{vmatrix} a & b \\ c & d \end{vmatrix} \\
\text{R2} &
\end{array}
\qquad
\begin{vmatrix} 2 & 3 \\ 7 & 5 \end{vmatrix}
$$

Any group of elements contained in a determinant must be combined using a specific sequence of operations. The three steps used to evaluate a second-order determinant are these:

1. *Multiply* the element in the *upper left* corner of the array (R1–C1) with its *diagonal* counterpart in the *lower right* corner (R2–C2). This upper-left-to-lower-right diagonal is known as the *principal* diagonal.

2. *Multiply* elements on the *secondary* diagonal from the *lower left* corner (R2–C1) to the *upper right* corner (R1–C2).

3. *Subtract* the product obtained in step 2 from the product obtained in step 1.

The entire process may be summarized as follows:

$$\begin{vmatrix} a & b \\ c & d \end{vmatrix} = (a)(d) - (c)(b)$$

Determinants containing only numerical elements may be reduced to a single number. For example,

$$\begin{vmatrix} 3 & 2 \\ 4 & 5 \end{vmatrix} = (3)(5) - (4)(2) = 15 - 8 = 7$$

Note that the subtraction defined in step 3 is independent of any negative signs attached to elements within the array:

$$\begin{vmatrix} 3 & -2 \\ 4 & 5 \end{vmatrix} = (3)(5) - (4)(-2) = 15 + 8 = 23$$

Example 1.5   Determine the value for each of the following second-order determinants.

$$\begin{vmatrix} -0.65 & 1.0 \\ 1.483 & -1.0 \end{vmatrix} \quad \begin{vmatrix} 0 & 1.0 \\ 8.33 & -1.0 \end{vmatrix} \quad \begin{vmatrix} -0.65 & 0 \\ 1.483 & 8.33 \end{vmatrix}$$

Solution:

$$\begin{vmatrix} -0.65 & 1.0 \\ 1.483 & -1.0 \end{vmatrix} = (-0.65)(-1.0) - (1.0)(1.483) = -0.833 \qquad \textbf{Answer}$$

$$\begin{vmatrix} 0 & 1.0 \\ 8.33 & -1.0 \end{vmatrix} = (0)(-1.0) - (1.0)(8.33) = -8.33 \qquad \textbf{Answer}$$

$$\begin{vmatrix} -0.65 & 0 \\ 1.483 & 8.33 \end{vmatrix} = (-0.65)(8.33) - (0)(1.483) = -5.41 \qquad \textbf{Answer}$$

Now consider two linear equations expressed in general terms with letters for coefficients as follows:

$$a_1 x + b_1 y = c_1 \qquad \textbf{(f)}$$
$$a_2 x + b_2 y = c_2 \qquad \textbf{(g)}$$

To eliminate y, multiply Eq. (f) by b_2 and Eq. (g) by b_1. Thus,

$$a_1 b_2 x + b_1 b_2 y = b_2 c_1 \qquad \textbf{(h)}$$
$$a_2 b_1 x + b_1 b_2 y = b_1 c_2 \qquad \textbf{(i)}$$

Subtracting Eq. (i) from Eq. (h),

$$(a_1 b_2 - a_2 b_1)x = b_2 c_1 - b_1 c_2$$

or

$$x = \frac{(b_2 c_1 - b_1 c_2)}{(a_1 b_2 - a_2 b_1)} \qquad \textbf{(1.10)}$$

Using similar techniques to eliminate x, we obtain

$$y = \frac{(a_1 c_2 - a_2 c_1)}{(a_1 b_2 - a_2 b_1)} \qquad \textbf{(1.11)}$$

Both equations look complicated, but each can actually be expressed as the ratio of two second-order determinants. This relationship, known as Cramer's rule, is stated as

$$x = \frac{\begin{vmatrix} c_1 & b_1 \\ c_2 & b_2 \end{vmatrix}}{\begin{vmatrix} a_1 & b_1 \\ a_2 & b_2 \end{vmatrix}} \qquad \textbf{(1.12)}$$

and

$$y = \frac{\begin{vmatrix} a_1 & c_1 \\ a_2 & c_2 \end{vmatrix}}{\begin{vmatrix} a_1 & b_1 \\ a_2 & b_2 \end{vmatrix}} \qquad \textbf{(1.13)}$$

Note that expanding these determinants yields Eqs. (1.10) and (1.11) directly.

The denominator of both ratios is formed by the coefficients of x and y and is known as the *determinant of coefficients*. The numerator of the solution for x has the coefficients multiplying x (a_1 and a_2) replaced by the constant terms (c_1 and c_2). In like manner, the numerator of the solution for y has the coefficients multiplying y replaced by the constant terms.

Care should be taken if any of the determinants in Eqs. (1.12) and (1.13) reduce to a value of zero. If the numerator is zero but the denominator has some real value, A, then the equations are said to be consistent or independent, and zero is one of the solution values. However, division by zero is undefined, so if the determinant of coefficients is zero, the equations are either inconsistent (represent physically incompatible conditions) or dependent (the same equation written in two different forms). In other words,

$$x = \frac{0}{A} = 0 \qquad \text{(Real solution value)}$$

$$x = \frac{A}{0} = \infty \qquad \text{(Inconsistent equations)}$$

$$x = \frac{0}{0} = \infty \qquad \text{(Dependent equations)}$$

Example 1.6 Solve the following sets of linear equations using determinants and Cramer's rule.

a. $5x - 6y - 3 = 0$ b. $3x + 4y = 5$
$\quad\quad 4x = 7y$ $\quad\quad\quad\quad 4.5x + 6y = 6$

Solution: a. The given equations must first be aligned. In this example, the constant terms have been "boxed" to emphasize their placement within the determinants.

$$5x - 6y = \boxed{3}$$
$$4x - 7y = \boxed{0}$$

To solve for x, the determinants are arranged according to Cramer's rule and evaluated as

$$x = \frac{\begin{vmatrix} \boxed{3} & -6 \\ \boxed{0} & -7 \end{vmatrix}}{\begin{vmatrix} 5 & -6 \\ 4 & -7 \end{vmatrix}} = \frac{(3)(-7) - (0)(-6)}{(5)(-7) - (4)(-6)} = \frac{-21 - 0}{-35 + 24} = \frac{-21}{-11}$$

$$x = \frac{21}{11}$$ **Answer**

Similarly, for y,

$$y = \frac{\begin{vmatrix} 5 & \boxed{3} \\ 4 & \boxed{0} \end{vmatrix}}{\begin{vmatrix} 5 & -6 \\ 4 & -7 \end{vmatrix}} = \frac{(5)(0) - (4)(3)}{-11} = \frac{0 - 12}{-11} = \frac{12}{11}$$

$$y = \frac{12}{11}$$ **Answer**

b. From Cramer's rule,

$$x = \frac{\begin{vmatrix} 5 & 4 \\ 6 & 6 \end{vmatrix}}{\begin{vmatrix} 3 & 4 \\ 4.5 & 6 \end{vmatrix}} = \frac{(5)(6) - (6)(4)}{(3)(6) - (4.5)(4)} = \frac{6}{0}$$

$$y = \frac{\begin{vmatrix} 3 & 5 \\ 4.5 & 6 \end{vmatrix}}{\begin{vmatrix} 3 & 4 \\ 4.5 & 6 \end{vmatrix}} = \frac{(3)(6) - (4.5)(5)}{(3)(6) - (4.5)(4)} = \frac{-4.5}{0}$$

Because the determinant of coefficients is zero, these equations are inconsistent and have no solution.

Determinants of the Third Order

Cramer's rule also applies to sets of simultaneous linear equations that contain more than two equations in two unknowns. Solutions to the generic three-equation set shown here, for example, are represented by the ratios of determinants given in Eqs. (1.14), (1.15), and (1.16). Note that each denominator contains the determinant of coeffi-cients and that the constant terms have been "boxed" to emphasize the pattern by which they are substituted in each numerator.

$$a_1x + b_1y + c_1z = \boxed{d_1} \quad\quad \text{(j)}$$
$$a_2x + b_2y + c_2z = \boxed{d_2} \quad\quad \text{(k)}$$
$$a_3x + b_3y + c_3z = \boxed{d_3} \quad\quad \text{(l)}$$

$$x = \frac{\begin{vmatrix} d_1 & b_1 & c_1 \\ d_2 & b_2 & c_2 \\ d_3 & b_3 & c_3 \end{vmatrix}}{\begin{vmatrix} a_1 & b_1 & c_1 \\ a_2 & b_2 & c_2 \\ a_3 & b_3 & c_3 \end{vmatrix}} \qquad (1.14)$$

$$y = \frac{\begin{vmatrix} a_1 & d_1 & c_1 \\ a_2 & d_2 & c_2 \\ a_3 & d_3 & c_3 \end{vmatrix}}{\begin{vmatrix} a_1 & b_1 & c_1 \\ a_2 & b_2 & c_2 \\ a_3 & b_3 & c_3 \end{vmatrix}} \qquad (1.15)$$

$$x = \frac{\begin{vmatrix} a_1 & b_1 & d_1 \\ a_2 & b_2 & d_2 \\ a_3 & b_3 & d_3 \end{vmatrix}}{\begin{vmatrix} a_1 & b_1 & c_1 \\ a_2 & b_2 & c_2 \\ a_3 & b_3 & c_3 \end{vmatrix}} \qquad (1.16)$$

To obtain numerical solutions using Cramer's rule, we must be able to evaluate a third-order determinant. Several methods are available, the most common of which is known as *expansion by minors*. With this technique, each element in the determinant is assigned a positive or negative sign depending on its location in the array. If the sum of the element's row number and column number is *even*, a *positive* sign is applied; if the sum is *odd*, a *negative* sign is entered. In the following determinant, these signs are shown in parentheses at the upper left of each element. The (−2) element, for instance, is located in the second row (R2) and the first column (C1); because the sum of this element's row and column numbers is odd (2 + 1 = 3), a negative sign appears in the parentheses.

$$\begin{vmatrix} {}^{(+)}1 & {}^{(-)}4 & {}^{(+)}{-3} \\ {}^{(-)}{-2} & {}^{(+)}{-5} & {}^{(-)}7 \\ {}^{(+)}3 & {}^{(-)}6 & {}^{(+)}{-1} \end{vmatrix}$$

A third-order determinant containing numerical elements may be reduced to a single numerical value by "expanding" the determinant along any row or column in the array. Selection of a particular row or column is arbi-

trary, but those that contain small positive numbers invite fewer errors in the computational process. For the previous determinant, we will expand along the top row, so only the assigned (+) or (−) signs for that row will apply. These signs are shown in the following determinant.

$$\begin{vmatrix} {}^{(+)}1 & {}^{(-)}4 & {}^{(+)}{-3} \\ -2 & -5 & 7 \\ 3 & 6 & -1 \end{vmatrix}$$

The expansion process itself is carried out as follows: Each element in the top row, including its assigned (+) or (−) value, is multiplied by the *minor determinant* obtained when the *row and column containing that element have been deleted* from the original determinant. Starting with the element in the upper left corner of the array (element R1–C1), delete the row and column containing that element. (This row and column are boxed out in the following determinant.) The element is now multiplied by its minor determinant, which contains the remaining elements in the original array.

$$\begin{vmatrix} {}^{(+)}1 & {}^{(-)}4 & {}^{(+)}{-3} \\ -2 & -5 & 7 \\ 3 & 6 & -1 \end{vmatrix} = (+)1 \begin{vmatrix} -5 & 7 \\ 6 & -1 \end{vmatrix}$$

Proceeding to the second element in the top row,

$$\begin{vmatrix} {}^{(+)}1 & {}^{(-)}4 & {}^{(+)}{-3} \\ -2 & -5 & 7 \\ 3 & 6 & -1 \end{vmatrix} = (-)4 \begin{vmatrix} -2 & 7 \\ 3 & -1 \end{vmatrix}$$

Finally, the third element in our top row yields

$$\begin{vmatrix} {}^{(+)}1 & {}^{(-)}4 & {}^{(+)}{-3} \\ -2 & -5 & 7 \\ 3 & 6 & -1 \end{vmatrix} = (+)(-3) \begin{vmatrix} -2 & -5 \\ 3 & 6 \end{vmatrix}$$

Expansion by minors, then, allows us to represent the original third-order determinant as

$$1\begin{vmatrix} -5 & 7 \\ 6 & -1 \end{vmatrix} - 4\begin{vmatrix} -2 & 7 \\ 3 & -1 \end{vmatrix} - 3\begin{vmatrix} -2 & -5 \\ 3 & 6 \end{vmatrix}$$

Performing the indicated operations yields

$$1[(-5)(-1) - (6)(7)] - 4[(-2)(-1) - (3)(7)] - 3[(-2)(6) - (-5)(3)]$$

$$1[5 - 42] - 4[2 - 21] - 3[-12 + 15]$$

$$1[-37] - 4[-19] - 3[+3]$$

$$-37 + 76 - 9$$

$$+30 \qquad \textbf{Answer}$$

Example 1.7   Determine the value of the following determinant by multiplying each element of the first row by its signed minor and adding the three products.

$$\begin{vmatrix} -6 & -2 & 3 \\ 2 & 1 & -2 \\ 12 & 1 & 1 \end{vmatrix}$$

Solution:

$$\begin{vmatrix} -6 & -2 & 3 \\ 2 & 1 & -2 \\ 12 & 1 & 1 \end{vmatrix} = -6\begin{vmatrix} 1 & -2 \\ 1 & 1 \end{vmatrix} - (-2)\begin{vmatrix} 2 & -2 \\ 12 & 1 \end{vmatrix} + 3\begin{vmatrix} 2 & 1 \\ 12 & 1 \end{vmatrix}$$

$$= -6[1(1) - (-2)(1)] + 2[2(1) - (-2)(12)]$$
$$+ 3[2(1) - 1(12)]$$
$$= -6(3) + 2(26) + 3(-10) = 4 \qquad \textbf{Answer}$$

Example 1.8   For the determinant of Example 1.7, multiply each element of the second column by its signed minor and add the three products.

Solution:

$$\begin{vmatrix} -6 & -2 & 3 \\ 2 & 1 & -2 \\ 12 & 1 & 1 \end{vmatrix} = -(-2)\begin{vmatrix} 2 & -2 \\ 12 & 1 \end{vmatrix} + 1\begin{vmatrix} -6 & 3 \\ 12 & 1 \end{vmatrix} - 1\begin{vmatrix} -6 & 3 \\ 2 & -2 \end{vmatrix}$$

$$= 2[2(1) - (-2)(12)] + 1[-6(1) - 3(12)]$$
$$- 1[-6(-2) - 3(2)]$$
$$= 2(26) + 1(-42) - 1(6) = 4 \qquad \textbf{Answer}$$

Expansion of a Determinant Along the Diagonals

Third-order determinants can also be evaluated by expanding along the diagonals. *This method is applicable to second- and third-order determinants only.* We add the first two columns of the determinant to the right-hand side of the determinant as shown here. There are now a total of six diagonals and six products of elements along the diagonals. The three products of elements along the diagonals that run downward from left to right are added, and the three products along the diagonals that run downward from right to left are subtracted to form the value of the determinant.

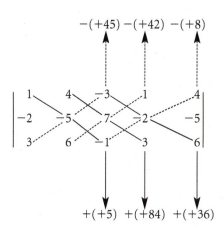

Combining these six values in horizontal form yields the same numerical value obtained previously with expansion by minors:

$$+(5) + (84) + (36) - (45) - (42) - (8) = +30$$

Example 1.9   Evaluate the third-order determinant shown by expanding along the diagonals.

$$\begin{vmatrix} 3 & 1 & 8 \\ 5 & 5 & -3 \\ 11 & 7 & 9 \end{vmatrix}$$

Solution: Adding the first two columns of the determinant to the right-hand side and expanding along the diagonals, we have the following result.

$$\begin{vmatrix} 3 & 1 & 8 \\ 5 & 5 & -3 \\ 11 & 7 & 9 \end{vmatrix}\begin{matrix} 3 & 1 \\ 5 & 5 \\ 11 & 7 \end{matrix} = \begin{matrix} +3(5)(9) + 1(-3)(11) + 8(5)(7) \\ -8(5)(11) - 3(-3)(7) - 1(5)(9) \end{matrix}$$

$$= -40 \qquad \textbf{Answer}$$

Example 1.10   Solve the following linear equations by Cramer's rule. Expand the determinants along the diagonals.

$$3P + 2Q + 2R = 5$$
$$2P - 3Q - R = -3$$
$$P + 2Q + R = 4$$

Solution: Computing values for the determinant of coefficients, D, and the determinants PD, QD, and RD for the numerators of P, Q, and R, respectively, is as follows:

$$D = \begin{vmatrix} 3 & 2 & 2 \\ 2 & -3 & -1 \\ 1 & 2 & 1 \end{vmatrix}\begin{matrix} 3 & 2 \\ 2 & -3 \\ 1 & 2 \end{matrix} = 3(-3)(1) + 2(-1)(1) + 2(2)(2)$$
$$-(2)(-3)(1) - 3(-1)(2) - 2(2)(1) = 5$$

$$PD = \begin{vmatrix} 5 & 2 & 2 \\ -3 & -3 & -1 \\ 4 & 2 & 1 \end{vmatrix}\begin{matrix} 5 & 2 \\ -3 & -3 \\ 4 & 2 \end{matrix} = 5(-3)(1) + 2(-1)(4) + 2(-3)(2)$$
$$-(2)(-3)(4) - 5(-1)(2) - 2(-3)(1) = 5$$

Dividing by D,

$$P = \frac{5}{5} = 1 \qquad \textbf{Answer}$$

$$QD = \begin{vmatrix} 3 & 5 & 2 \\ 2 & -3 & -1 \\ 1 & 4 & 1 \end{vmatrix}\begin{matrix} 3 & 5 \\ 2 & -3 \\ 1 & 4 \end{matrix} = 3(-3)(1) + 5(-1)(1) + 2(2)(4)$$
$$-(2)(-3)(1) - 3(-1)(4) - 5(2)(1) = 10$$

Dividing by D,

$$Q = \frac{10}{5} = 2 \qquad \textbf{Answer}$$

$$RD = \begin{vmatrix} 3 & 2 & 5 \\ 2 & -3 & -3 \\ 1 & 2 & 4 \end{vmatrix}\begin{matrix} 3 & 2 \\ 2 & -3 \\ 1 & 2 \end{matrix} = 3(-3)(4) + 2(-3)(1) + 5(2)(2)$$
$$-5(-3)(1) - 3(-3)(2) - 2(2)(4) = -5$$

Dividing by D,

$$R = \frac{-5}{5} = -1 \qquad \textbf{Answer}$$

The repetitive solution of simultaneous equations is best accomplished using current mathematical technology. Most graphing calculators, and many nongraphing models as well, can solve sets of more than 20 simultaneous equations containing an equal number of unknowns. Generally, the operator simply enters the total number of equations contained in the set and the values for each numerical coefficient. As quickly as this data can be entered, the calculator yields a set of solution values. The following example closely approximates the techniques and successive screens displayed by one popular brand of calculator.

Example 1.11   Use the capabilities of a graphing calculator to solve this set of simultaneous linear equations:

$$2x + 3y = 4$$
$$5x + 6y = 7$$

Screen 1

SIMULTANEOUS EQUATIONS

Number = **2**

Screen 2

$a1,1\ X1 + a1,2\ X2 = b1$

$a1,1 = \mathbf{2}$

$a1,2 = \mathbf{3}$

$b1 = \mathbf{4}$

Screen 3

$a2,1\ X1 + a2,2\ X2 = b2$

$a2,1 = \mathbf{5}$

$a2,2 = \mathbf{6}$

$b2 = \mathbf{7}$

Screen 4

$X1 = -1$

$X2 = 2$

FIGURE 1.17

Solution: Figure 1.17 shows four successive screens displayed by a typical graphing calculator. Screen 1 indicates that the "simultaneous equation" function is being used and asks for the number of equations in the set. Here, the operator enters "2." (This value and all subsequent entered values are shown in boldface type.) Screens 3 and 4 then ask for values of the numerical coefficients. Note that some brands of calculators use equations in the format:

$$(a_{1,1})X1 + (a_{1,2})X2 = b_1$$
$$(a_{2,1})X1 + (a_{2,2})X2 = b_2$$

These "a" subscripts represent row and column numbers for each coefficient of the variables $X1$ and $X2$ rather than x and y. When the appropriate "a" and "b" values are entered, Screen 4 displays the solution:

$$X1(=x) = -1$$
$$X2(=y) = 2$$

Another way to solve simultaneous equations is through the use of simple computer programs written in the language known as BASIC (Beginners All-purpose Symbolic Instruction Code). Virtually all of today's personal computers contain a version of this language and will perform the calculations described by the programs listed in Tables 1.3 and 1.4. Once installed on your computer, these programs allow for the solution of two- and three-equation sets of simultaneous linear equations by the method of Cramer's rule. These programs are written for the notational format used in Eqs. (f), (g), (j), (k), and (l), listed earlier in this section. Example 1.12 demonstrates a typical solution provided by the program shown in Table 1.3.

TABLE 1.3 BASIC Program for Two Simultaneous Linear Equations

```
 10 PRINT "ENTER COEFFICIENTS FROM EQUATIONS OF THE FORM:"
 20 PRINT "A1*X + B1*Y = C1"
 30 PRINT "A2*X + B2*Y = C2"
 40 PRINT
 50 INPUT "A1 = "; A1: INPUT "B1 = "; B1: INPUT "C1 = "; C1:
 60 PRINT
 70 INPUT "A2 = "; A2: INPUT "B2 = "; B2: INPUT "C2 = "; C2:
 80 PRINT
 90 DENOM = (A1 * B2)-(A2 * B1)
100 IF DENOM = 0 THEN 150
110 PRINT
120 PRINT "X = "; (B2 * C1 - B1 * C2) / DENOM
130 PRINT "Y = "; (A1 * C2 - A2 * C1) / DENOM
140 GOTO 160
150 PRINT "DIVISION BY ZERO, NO SOLUTION"
160 PRINT
170 END
```

TABLE 1.4 BASIC Program for Three Simultaneous Linear Equations

```
 10 PRINT "ENTER COEFFICIENTS FOR EQUATIONS OF THE FORM:"
 20 PRINT "A1*X + B1*Y + C1*Z = D1"
 30 PRINT "A2*X + B2*Y + C2*Z = D2"
 40 PRINT "A3*X + B3*Y + C3*Z = D3"
 50 PRINT
 60 INPUT "A1 = "; A1: INPUT "B1 = "; B1
 70 INPUT "C1 = "; C1: INPUT "D1 = "; D1
 80 PRINT
 90 INPUT "A2 = "; A2: INPUT "B2 = "; B2
100 INPUT "C2 = "; C2: INPUT "D2 = "; D2
110 PRINT
120 INPUT "A3 = "; A3: INPUT "B3 = "; B3
130 INPUT "C3 = "; C3: INPUT "D3 = "; D3
140 PRINT
150 DENOM1 = (A1 * B2 * C3) + (A3 * B1 * C2) + (A2 * B3 * C1)
160 DENOM2 = (A3 * B2 * C1) + (A1 * B3 * C2) + (A2 * B1 * C3)
170 DENOM = DENOM1 - DENOM2
180 IF DENOM = 0 THEN 330
190 XNUM1 = (B2 * C3 * D1) + (B1 * C2 * D3) + (B3 * C1 * D2)
200 XNUM2 = (B2 * C1 * D3) + (B3 * C2 * D1) + (B1 * C3 * D2)
210 XNUM = XNUM1 - XNUM2
220 YNUM1 = (A1 * C3 * D2) + (A3 * C2 * D1) + (A2 * C1 * D3)
230 YNUM2 = (A3 * C1 * D2) + (A1 * C2 * D3) + (A2 * C3 * D1)
240 YNUM = YNUM1 - YNUM2
250 ZNUM1 = (A1 * B2 * D3) + (A3 * B1 * D2) + (A2 * B3 * D1)
260 ZNUM2 = (A3 * B2 * D1) + (A1 * B3 * D2) + (A2 * B1 * D3)
270 ZNUM = ZNUM1 - ZNUM2
280 PRINT
290 PRINT "X = "; XNUM / DENOM
300 PRINT "Y = "; YNUM / DENOM
310 PRINT "Z = "; ZNUM / DENOM
320 GOTO 340
330 PRINT "DIVISION BY ZERO, NO SOLUTION"
340 PRINT
350 END
```

Example 1.12   Using the computer program from Table 1.3, solve the following equation set:

$$0.5736x = 0.9659y$$

$$0.8192x = 630 + 0.2588y$$

Solution: Once this program has been loaded onto your machine, press the appropriate key (generally F2) to activate or "run" the program. Lines 10, 20, and 30 will cause the following display to appear on the screen of your monitor:

```
"ENTER COEFFICIENTS FROM EQUATIONS OF THE FORM:
A1*X + B1*Y = C1
A2*X + B2*Y = C2
A1 = ?"
```

This display reminds us that our equations must be put in the required form:

$$0.5736x - 0.9659y = 0$$

$$0.8192x - 0.2588y = 630$$

A blinking cursor will be located just to the right of the last line on the display (A1 = ?). At this point, you should type in "0.5736" and then press the ENTER key, causing the next line to appear: "B1 = ?" The process repeats itself until all six numerical coefficients have been entered. These steps are summarized as follows.

When This Appears on Screen:	You Should Type:	
A1 = ?	0.5736	ENTER
B1 = ?	−0.9659	ENTER
C1 = ?	0	ENTER
A2 = ?	0.8192	ENTER
B2 = ?	−0.2588	ENTER
C2 = ?	630	ENTER

After the last ENTER, your machine calculates a value for the determinant of coefficients (line 90), here called DENOM. If this value is zero, the equations are either inconsistent or dependent, and the machine is instructed (lines 100, 150) to display the message: "DIVISION BY ZERO, NO SOLUTION." If, as in this case, a real solution exists, the machine calculates values for x (line 120) and y (line 130) and instantaneously displays the solution values as follows:

$$\text{"X} = 946.6402$$

$$\text{Y} = 562.1626\text{"}$$ **Answer**

PROBLEMS

Solve the systems of linear equations in Probs. 1.40 through 1.43 by (a) substitution, (b) subtraction or addition, and (c) determinants.

1.40 $-15x + 21y = 12$
$-2x + 3y = 17$

1.41 $19u - 20v = -22$
$20u - 21v = -23$

1.42 $5m - 3n = 9$
$3m - 5n = -9$

1.43 $2A + 2B = 3$
$A - B = -0.9$

1.44   Derive the original equation set that yields the following determinants.

$$x = \frac{\begin{vmatrix} 0 & -5 \\ -4 & 1 \end{vmatrix}}{\begin{vmatrix} 3 & -5 \\ 6 & 1 \end{vmatrix}}$$

1.45   Derive the original equation set that yields the following determinants. Are these equations valid? Solve for x and y.

$$y = \frac{\begin{vmatrix} -7 & 10 \\ 9 & -5 \end{vmatrix}}{\begin{vmatrix} -7 & 10 \\ 9 & -5 \end{vmatrix}}$$

1.46 The rates charged by a local delivery service consist of a fixed cost, F, and a variable rate, r, based on the weight, W, of the package being delivered. Total cost, C, to the customer is determined by the following formula: $C = F + rW$. If it costs \$10.10 to deliver a 12-lb package and \$13.95 to deliver a 19-lb package, find fixed rate, F, and the cost per pound, r. Write two simultaneous equations and solve using Cramer's rule.

1.47 A contractor purchases 350 pieces of 2×4 lumber and 200 pieces of 2×6 lumber at a total cost of \$1077.50. Several days later he buys another 140 of the 2×4s, and 125 more of the 2×6s for a total of \$527.75. What is the cost of a 2×4? Of a 2×6? Write two simultaneous equations and solve using any method.

Solve the system of equations in Probs. 1.48 through 1.51 by Cramer's rule. Evaluate the determinants by minors or by expanding along the diagonals.

1.48
$$2x + 3y - 2z = -7$$
$$x + y + z = 2$$
$$-x - 3y + 2z = 5$$

1.49
$$P + 3Q + 2R = 2$$
$$-2P - 2Q + 3R = 1$$
$$-P + Q + R = -1$$

1.50
$$y + 2z = 1$$
$$-2x + 2y - z = 3$$
$$3x - y + z = 2$$

1.51
$$2A + 3B + 2C = 3$$
$$2A + B - 4C = 4$$
$$A + 2B + C = 2$$

1.52 A machine shop is preparing to ship three iron castings whose total weight is 1867 lb. The difference in weight between the heaviest and lightest pieces is 395 lb, and twice the weight of the lightest casting is 427 lb less than the sum of the two heaviest pieces. Write three simultaneous equations and solve using Cramer's rule.

RESULTANT OF CONCURRENT FORCES IN A PLANE

2.2 GRAPHICAL REPRESENTATION OF FORCES: VECTORS

Physical quantities such as length and temperature require a magnitude for their complete description and are called *scalar* quantities. Forces require both magnitude and direction and are called *vector* quantities.

Vector quantities such as forces can be represented graphically by arrows drawn to an appropriate scale. The length of the arrow represents the magnitude, and the direction in which the arrow points represents the direction. In Fig. 2.2, we show a force of 15 N that forms a positive angle of 20° with the *x* axis.

In print, a vector is commonly represented in boldface (heavy) type (**F**). The same letter in lightface italic type (*F*) represents the magnitude of the same vector. In longhand or typewritten copy, a vector can be indicated by an arrow over the letter ($\vec{F}$) or by underlining the letter (*F*). Underlining the letter has the advantage that it can be done with the typewriter or computer.

CHAPTER OBJECTIVES

This chapter discusses the graphical and algebraic methods by which forces are combined. After completing the chapter, you should be able to

- Depict the graphical representation of a force vector.
- Explain the parallelogram method and triangular method for adding vectors to find their resultant.
- Determine the components of a vector along any specified pair of axes.
- Use graphical techniques and the trigonometry of oblique triangles to add two or more vectors.
- Compute the rectangular components of a force.
- Use the method of rectangular components to find the resultant of two or more vectors.
- Apply the concept of a negative force to the subtraction of vectors.

2.1 INTRODUCTION

A system of several forces is *concurrent* when all forces acting on the body have the same point of application, or the lines of action of the forces intersect at a common point. When all forces act in a single plane, the forces are said to be *coplanar*. Examples of concurrent force systems are shown in Fig. 2.1. The forces in the examples act in the plane of the paper. In this chapter, we consider various methods of finding the resultant of concurrent force systems.

2.3 RESULTANT OF TWO CONCURRENT FORCES: VECTORS

It has been found by *experiment* that a concurrent force system can be replaced by a single force or resultant. The resultant has the same physical effect as the force system it replaces.

Parallelogram Method

The resultant of two concurrent forces can be obtained graphically by constructing a parallelogram, as shown in Fig. 2.3. The forces **P** and **Q** are drawn to scale. The forces

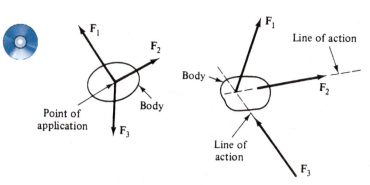

FIGURE 2.1

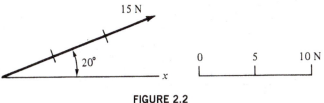

FIGURE 2.2

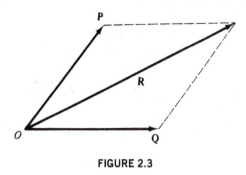

FIGURE 2.3

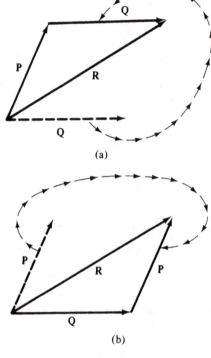

(a)

(b)

FIGURE 2.4

acting at *O* form two sides of the parallelogram. The diagonal that passes through *O* is the resultant **R**. This is known as the *parallelogram law,* which states that two forces whose lines of action intersect can be replaced by a single force, which is the diagonal of a parallelogram that has sides equal to the two forces. The parallelogram law is based on experimental evidence only. It cannot be proved or derived by mathematics.

A force is a *vector* quantity. Vectors may be defined as mathematical quantities that have magnitude and direction and add according to the parallelogram law.

Triangular Method

The parallelogram construction shown in Fig. 2.3 suggests that forces may also be added by arranging vectors in a tip-to-tail fashion. The method as shown in Fig. 2.4(a) and (b) consists of moving either force parallel to itself until the tail coincides with the tip of the fixed force. The closing side of the triangle forms the resultant with the tail of the resultant at the tail of the fixed force and the tip of the resultant at the tip of the moved force. As we can see from the construction, the order in which the forces are combined does not change the resultant.

Example 2.1   The two forces **S** and **T** act as shown in Fig. 2.5(a) at point *O*. Obtain their resultant graphically by the parallelogram method and the triangular method.

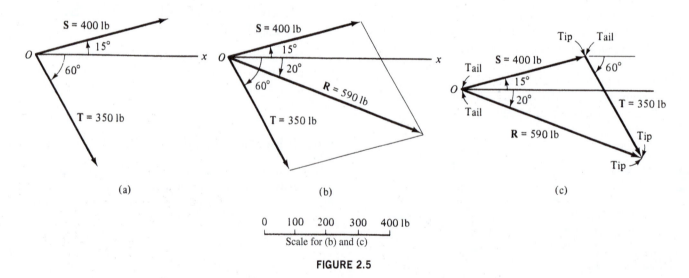

(a)

(b)

(c)

0   100   200   300   400 lb

Scale for (b) and (c)

FIGURE 2.5

Solution: **Graphical Parallelogram Method:** A parallelogram with sides equal to *S* and *T* is drawn to scale [Fig. 2.5(b)]. The diagonal forms the resultant. The magnitude and direction of the resultant are measured and found to be

$$R = 590 \text{ lb} \searrow 20°$$

Answer

Graphical Triangular Method: Draw fixed force **S** at point O as shown in Fig. 2.5(c). Move force **T** parallel to itself until its tail coincides with the tip of force **S**. The resultant forms the closed side of a triangle. The tail of the resultant is at the tail of the fixed force, and the tip of the resultant is at the tip of the moved force. Measurement of the magnitude and direction of the resultant gives

$$R = 590 \,\text{lb} \,\text{⤴} 20°$$ **Answer**

Example 2.2 A ship is pulled by two tugboats, as shown in Fig. 2.6(a). The resultant force is 3 kip parallel to the x axis. Find the forces exerted by each tugboat graphically by the triangular method.

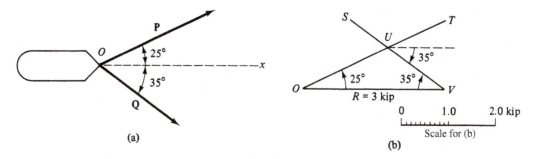

(a)   (b)

FIGURE 2.6

Solution: The resultant of 3 kip is drawn to scale parallel to the x axis from O to V as shown in Fig. 2.6(b). The line OT is drawn through O at an angle of 25° and the line VS through V at an angle of 35°. The two lines intersect at point U. Force **P** is directed from O to U, and force **Q** is directed from U to V. Measuring the magnitude of the forces, we obtain

$$P = 2.0 \,\text{kip} \qquad Q = 1.5 \,\text{kip}$$ **Answer**

One disadvantage of graphical techniques, of course, is the level of accuracy obtainable. Working carefully and to a large scale will normally yield results that are within 1 to 3 percent of theoretical values. By comparison, our stated goal in Sec. 1.4 of the preceding chapter was a computational accuracy of 0.2 percent. However, as the following example demonstrates, use of appropriate trigonometry allows us to algebraically replicate our graphical constructions.

Example 2.3 The two forces **P** and **S** act as shown in Fig. 2.7(a) at point A. Obtain their resultant by trigonometry.

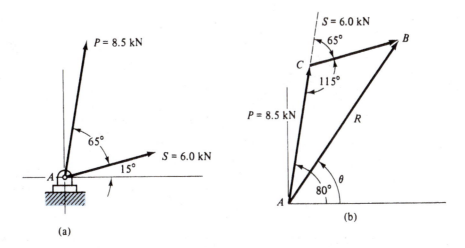

(a)   (b)

FIGURE 2.7

Solution: We use the triangular method, arranging **P** and **S** in a tip-to-tail fashion. The resultant **R** forms the third side of the triangle [Fig. 2.7(b)]. From the construction, angle $C = 115°$. Applying the law of cosines yields

$$R^2 = P^2 + S^2 - 2PS \cos C$$
$$= (8.5)^2 + (6.0)^2 - 2(8.5)(6.0)\cos 115°$$
$$R = 12.30 \text{ kN}$$

From the law of sines, we have

$$\frac{\sin A}{S} = \frac{\sin C}{R}; \qquad \frac{\sin A}{6.0} = \frac{\sin 115°}{12.3}$$

Solving for angle A, we have

$$\sin A = \frac{6.0(\sin 115°)}{12.3} = 0.4421$$
$$A = \arcsin 0.4421 = 26.2°$$

The direction of angle $\theta = 80 - A = 53.8°$.

$$\mathbf{R} = 12.30 \text{ kN} \angle 53.8° \qquad\qquad \textbf{Answer}$$

Example 2.4 Two cables are used to pull a truck as shown in Fig. 2.8(a). The resultant force lies along the x axis. Determine the magnitude of the force **Q** and of the resultant **R** by trigonometry.

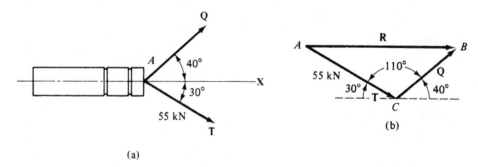

(a)

(b)

FIGURE 2.8

Solution: The force Q at an angle of 40° is added to the force T at an angle of 30° to form the resultant R at an angle of 0° in a tip-to-tail fashion as shown in Fig. 2.8(b). From the construction, angle $B = 40°$ and angle $C = 110°$. Applying the law of sines yields

$$\frac{\sin 30°}{Q} = \frac{\sin 40°}{55} = \frac{\sin 110°}{R}$$

Solving for Q and R, we have

$$Q = \frac{55(\sin 30°)}{\sin 40°} = 42.8 \text{ kN} \qquad R = \frac{55(\sin 110°)}{\sin 40°} = 80.4 \text{ kN} \qquad \textbf{Answer}$$

PROBLEMS

2.1 and 2.2   Determine graphically the resultant force on the screw eye shown by the triangular method.

2.3 and 2.4   The bracket supports two forces as shown. Obtain their resultant by trigonometry.

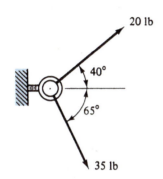

PROB. 2.1

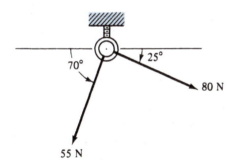

PROB. 2.2

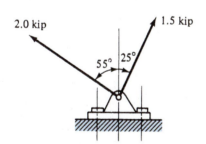

PROB. 2.3

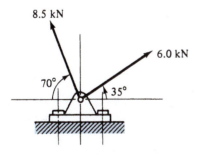

PROB. 2.4

2.5   A barge is pulled by two ropes with tensions **P** and **Q** as shown. If $P = 300$ lb and $Q = 525$ lb, determine the resultant force applied on the barge by trigonometry.

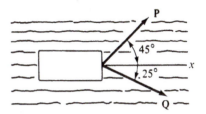

PROB. 2.5 and PROB. 2.6

2.6   A barge is pulled by two ropes with tensions **P** and **Q**. The force $P = 24$ kN, and the resultant of the two forces acts along the x axis. Determine the force Q and the resultant force applied on the barge by trigonometry.

2.7   Two forces **P** and **Q** have the resultant **R** as shown. If $R = 35$ kip, $P = 15$ kip, and $\theta_1 = 30°$, determine the force Q and the angle θ_2 it makes with the vertical by trigonometry.

2.8   Two forces **P** and **Q** have the resultant **R** as shown. The resultant $R = 14$ kN, $\theta_1 = 40°$, and $\theta_2 = 25°$. Determine the forces P and Q by trigonometry.

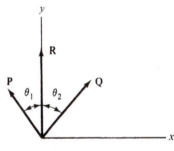

PROB. 2.7 and PROB. 2.8

2.4   RESULTANT OF THREE OR MORE CONCURRENT FORCES

When the resultant of three or more concurrent forces is required, either the parallelogram or triangular method may be used. However, the parallelogram construction becomes awkward, and the triangular method is preferred.

In Fig. 2.9(a) and (b) the resultant of three forces **P, Q,** and **S** is determined by the repeated application of the triangular method. First, we add **P** and **Q** to find the resultant **R**$_{1,2}$. Then, we add **R**$_{1,2}$ and **S** to find the resultant **R** of the three forces.

Algebraic results could have been obtained here by applying the trigonometry of oblique triangles, first to triangle OPQ in Fig. 2.9(b) and then to triangle OQS in the same figure. However, this procedure becomes quite cumbersome as the number of added vectors increases. In Sec. 2.6, we will

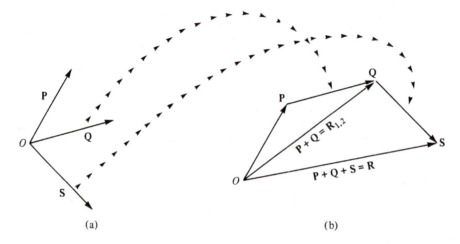

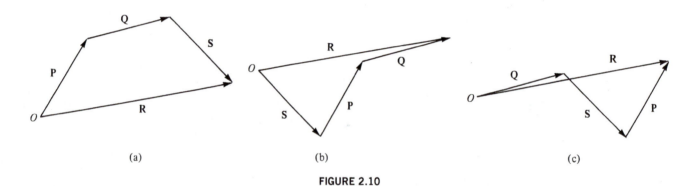

FIGURE 2.9

FIGURE 2.10

discuss a technique that allows any number of vectors to be added in a single step and yields exact results.

A modification of the triangular method known as the *polygon method* can also be used to find the resultant. In the polygon method, the intermediate resultants, in this example $R_{1,2}$, need not be included in the construction.

The polygon method, shown in Fig. 2.10(a), can be described in the following way. Leave the fixed force **P** at point *O*.

Move **Q** parallel to itself until its tail coincides with the tip of **P**. Next, move **S** parallel to itself until its tail coincides with the tip of **Q**. The resultant **R** forms the closing side of the polygon with the tail of the resultant at the tail of the fixed force **P** and the tip of the resultant at the tip of the last moved force **S**.

The order of addition of the forces does not change the resultant, as we see from the constructions shown in Fig. 2.10(b) and (c).

Example 2.5   The four forces [Fig. 2.11(a)] act at point *P*. Obtain their resultant graphically by the polygon method.

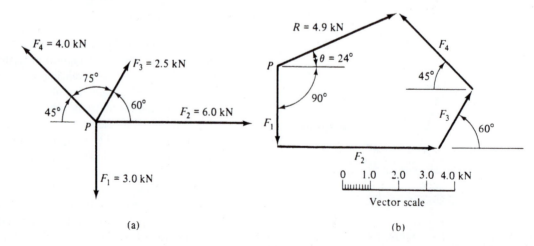

FIGURE 2.11

Solution: The four forces are drawn to scale in a tip-to-tail fashion in Fig. 2.11(b). The resultant acts from the tail of the first force to the tip of the last. The magnitude and direction angle are measured and the resultant is found to be

$$\mathbf{R} = 4.9\,\text{kN} \measuredangle 24°$$

Answer

PROBLEMS

2.9 and 2.10 Obtain the resultant of the concurrent forces shown graphically by the polygon method. For 2.9, check your results algebraically using the trigonometry of oblique triangles. How accurate were your graphical values?

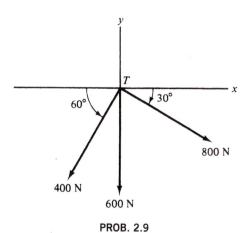

PROB. 2.9

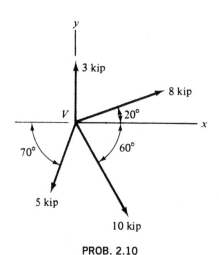

PROB. 2.10

2.11 The gusset plate shown is subjected to four concurrent forces. If $F_1 = 150$ lb, $F_2 = 240$ lb, $F_3 = 180$ lb, and $F_4 = 300$ lb, determine the magnitude and direction of the resultant graphically by the polygon method.

2.12 The forces on the gusset plate shown in the figure are $F_1 = 6.0$ kN, $F_2 = 4.8$ kN, $F_3 = 4.0$ kN, and $F_4 = 8.0$ kN.

Determine the magnitude and direction of the resultant graphically by the polygon method.

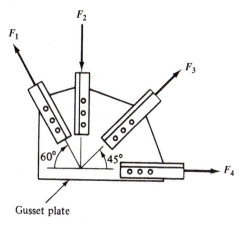

PROB. 2.11 and PROB. 2.12

2.5 COMPONENTS OF A FORCE VECTOR

As we saw in Secs. 2.3 and 2.4, two or more forces may be replaced by their resultant. By reversing the process, any force **F** can be replaced by any number of forces whose resultant is **F**. These replacement forces are called *components* of the force **F**.

Although force **F** can be replaced by an infinite number of different components, two components at right angles to each other are the most useful. They are called *rectangular components* and are usually determined in the horizontal and vertical direction, as shown in Fig. 2.12(a). They may also be found in any two directions at right angles to each other, as shown in Fig. 2.12(b).

A method for finding the rectangular components of a force vector graphically can be described as follows. Let a force **F** be represented by a vector drawn to a convenient scale [Fig. 2.12(a) and (b)]. The tail end of the vector is point O, and the arrow end of the vector is point B. Through point O, draw any two axes at right angles to each other. The angle that the force **F** makes with the x axis is θ (theta). To find the rectangular component of **F** along the x axis, we draw a line parallel to the y axis from the arrow end of **F** at B to point C on the x axis. The vector from O to C represents the component of **F** along the x axis and is represented by $\mathbf{F}_x$. Similarly, by drawing a second line parallel to the x axis from B to point A on the y axis, we find $\mathbf{F}_y$, the component of **F** along the y axis.

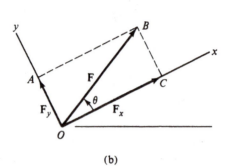

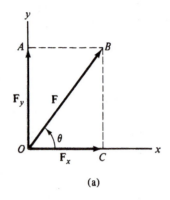

FIGURE 2.12

Rectangular components of a force vector may also be found mathematically by using trigonometric functions. Notice that triangle OBC in Fig. 2.12(a) and (b) is a right triangle, and length OA is equal to length CB. From the definition of the cosine of an angle,

$$\cos\theta = \frac{OC}{OB} = \frac{F_x}{F} \quad \text{or} \quad F_x = F\cos\theta \qquad (2.1)$$

and from the definition of the sine of an angle

$$\sin\theta = \frac{CB}{OB} = \frac{F_y}{F} \quad \text{or} \quad F_y = F\sin\theta \qquad (2.2)$$

The components F_x and F_y are considered either positive or negative, depending on whether they act in the positive or negative direction of the x and y axes.

Example 2.6   Using trigonometry, find the horizontal and vertical components of the forces shown in Fig. 2.13(a).

Solution: Sketch the force **P** = 14 N from the origin to point B at an angle of 55° with the positive x axis and force **Q** = 12.5 N from the origin to point B' at an angle of 65° with the negative x axis as in Fig. 2.13(b).

Next, apply Eqs. (2.1) and (2.2) to force **P** at an angle of $\theta = 55°$ with the positive x axis. The components of **P** are both positive because they act in the positive x and y directions.

$$P_x = P\cos\theta = 14\cos 55° = 8.03\text{ N} \qquad \textbf{Answer}$$
$$P_y = P\sin\theta = 14\sin 55° = 11.47\text{ N} \qquad \textbf{Answer}$$

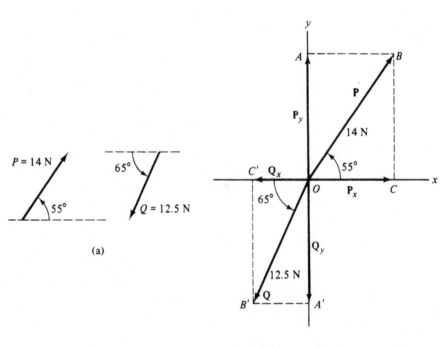

FIGURE 2.13

Proceed similarly for force **Q** at an angle $\theta = 65°$ with the negative x axis. The components of **Q** are both negative because they act in the negative x and y direction.

$$Q_x = -Q\cos\theta = -12.5\cos65° = -5.28\text{ N}$$ **Answer**
$$Q_y = -Q\sin\theta = -12.5\sin65° = -11.33\text{ N}$$ **Answer**

An alternative method for finding the components of **Q** without introducing a negative sign to account for the direction is to use the counterclockwise angle **Q** forms with the positive x axis. Because force **Q** forms a counterclockwise angle of $180° + 65° = 245°$ from the positive x axis, we have

$$Q_x = Q\cos\theta = 12.5\cos245° = -5.28\text{ N}$$
$$Q_y = Q\sin\theta = 12.5\sin245° = -11.33\text{ N}$$

Example 2.7   Determine the rectangular components of the forces shown in Fig. 2.13(a) with respect to axes that form angles of 30° and 120° with the horizontal.

Solution: Rotate the x and y axes to the new position as in Fig. 2.14. Force **P** forms an angle of 25° with the new x axis, and **Q** forms an angle of 35° with the new negative x axis.

Applying Eqs. (2.1) and (2.2) to force **P** at an angle of $\theta = 25°$ with the positive x axis, we obtain

$$P_x = P\cos\theta = 14\cos25° = 12.69\text{ N}$$ **Answer**
$$P_y = P\sin\theta = 14\sin25° = 5.92\text{ N}$$ **Answer**

Proceed similarly for force **Q** at an angle of $\theta = 35°$ with the negative x axis.

$$Q_x = -Q\cos\theta = -12.5\cos35° = -10.24\text{ N}$$ **Answer**
$$Q_y = -Q\sin\theta = -12.5\sin35° = -7.17\text{ N}$$ **Answer**

Or, using the alternative method, force **Q** forms a counterclockwise angle of $180° + 35° = 215°$ from the positive x axis; therefore,

$$Q_x = 12.5\cos215° = -10.24\text{ N}$$
$$Q_y = 12.5\sin215° = -7.17\text{ N}$$

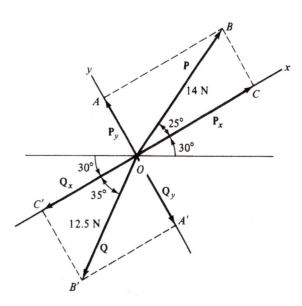

FIGURE 2.14

PROBLEMS

2.13 If $P = 35$ kN and $\theta = 30°$, find the components of **P** parallel and perpendicular to the lines (a) *mn* and (b) *rs*.

2.14 If $P = 2000$ lb and $\theta = 45°$, find the components of **P** parallel and perpendicular to the lines (a) *mn* and (b) *rs*.

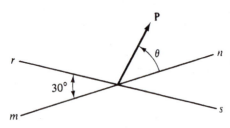

PROB. 2.13 and PROB. 2.14

2.15 Determine components, in the horizontal and vertical direction, of the following forces: (a) 28 kN ⤢$35°$, (b) 1500 lb ⤡$15°$, (c) 10.2 kip ⤣$75°$, and (d) 450 N ⤢$50°$.

2.16 Determine the rectangular components of the forces given in Prob. 2.15 with respect to the x' and y' axes as shown.

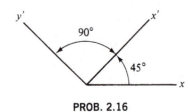

PROB. 2.16

2.17 A packing crate weighing 1150 N is placed on a loading ramp that is inclined at an angle of $34°$ as shown. Find the components of weight, $F1$ and $F2$, which act perpendicular to and parallel to the ramp surface, respectively.

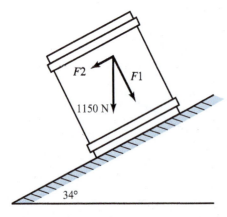

PROB. 2.17

2.18 Boom member *ABC* can be raised or lowered by hydraulic cylinder *DB*. For the position shown, this cylinder pushes on point *B* with a force of 1650 lb in direction *DB* as indicated by the dotted line. Find the components of this force that are parallel to and perpendicular to member *ABC*.

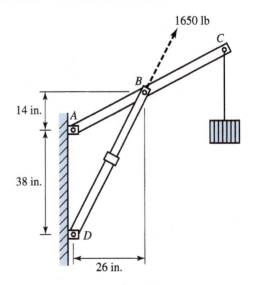

PROB. 2.18

2.6 RESULTANT OF CONCURRENT FORCES BY RECTANGULAR COMPONENTS

The resultant of three or more concurrent forces may be found graphically by the polygon of forces. The usual mathematical solution, however, is based on the method of rectangular components.

To develop the rectangular component method we consider, in Fig. 2.15(a), two forces **P** and **Q**, which act at point A and form angles θ_1 and θ_2 with the x axis. The x and y components of **P** and **Q** can be found graphically by the method described in Sec. 2.5. They may also be found by Eqs. (2.1) and (2.2) as follows:

$$P_x = P\cos\theta_1 \qquad P_y = P\sin\theta_1$$

and

$$Q_x = Q\cos\theta_2 \qquad Q_y = -Q\sin\theta_2$$

Because the components P_x and Q_x lie on the x axis [Fig. 2-15(a)], they may be added algebraically to find the resultant in the x direction, R_x. That is,

$$R_x = \Sigma F_x = P_x + Q_x$$

where the Greek capital letter Σ (sigma) means the algebraic sum and ΣF_x means the algebraic sum of the x components of the forces.

Similarly, the components of P_y and Q_y lie on the y axis [Fig. 2.15(a)] and can be added algebraically to find the resultant in the y direction, R_y. That is,

$$R_y = \Sigma F_y = P_y + Q_y$$

where ΣF_y means the algebraic sum of the y components of the forces.

Adding the resultants in the x and y directions, $\mathbf{R}_x$ and $\mathbf{R}_y$ [Fig. 2.15(b)], we find the resultant **R** of the force system. The magnitude and direction of the resultant can be found

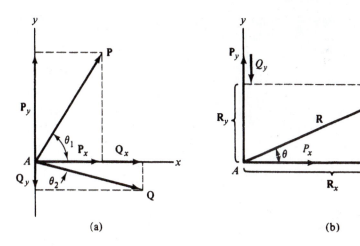

FIGURE 2.15

from the Pythagorean theorem and the definition of the tangent of the angle θ. That is,

$$R^2 = R_x^2 + R_y^2 = (\Sigma F_x)^2 + (\Sigma F_y)^2 \qquad (2.3)$$

$$\tan\theta = \frac{R_y}{R_x} = \frac{\Sigma F_y}{\Sigma F_x} \qquad (2.4)$$

Unlike the algebraic application of our parallelogram and triangular methods, rectangular components can be used to find the resultant of *any* number of forces, exactly and in a single step.

Example 2.8   Find the resultant of the concurrent force system shown in Fig. 2.16(a) by the method of rectangular components.

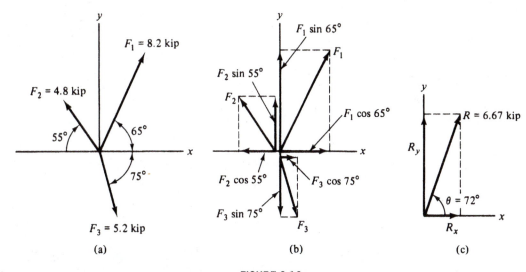

FIGURE 2.16

Solution: The x and y components of each force are determined by trigonometry, as shown in Fig. 2.16(b). Use of a table helps organize the data and minimize computational errors.

Values for each component are shown in Table (A). Recall that x components are positive if they act to the right and negative to the left, and y components are positive if they act upward and negative if they act downward.

The resultants in the x and y directions are given by the sums shown in Table (A).

$$R_x = \Sigma F_x = 2.058 \text{ kip}$$

and

$$R_y = \Sigma F_y = 6.341 \text{ kip}$$

TABLE (A) Example 2.8				
Force	Magnitude (kip)	Angle (deg)	x Component (kip)	y Component (kip)
F_1	8.2	65	3.465	7.432
F_2	4.8	55	−2.753	3.932
F_3	5.2	75	1.346	−5.023
			$\Sigma F_x = 2.058$ kip (R_x)	$\Sigma F_y = 6.341$ kip (R_y)

The magnitude and direction of the resultant is determined from the Pythagorean theorem and the definition of the tangent of an angle. From Fig. 2.16(c), we write

$$R^2 = (2.058)^2 + (6.341)^2 = 44.44$$
$$R = 6.67 \text{ kip}$$

and

$$\tan\theta = \frac{R_y}{R_x} = \frac{6.341}{2.058} = 3.0811 \qquad \theta = 72.0°$$

$$\mathbf{R} = 6.67 \text{ kip at } +72.0° \qquad\qquad \textbf{Answer}$$

Example 2.9 For the concurrent force system shown in Fig. 2.17(a), find the resultant by the rectangular component method.

Solution: An alternative method for finding the components of the force without introducing a negative sign to account for the direction is to measure the angle that the force forms with the positive x axis. Angles measured in the counterclockwise direction are positive, and angles measured in the clockwise direction are negative.

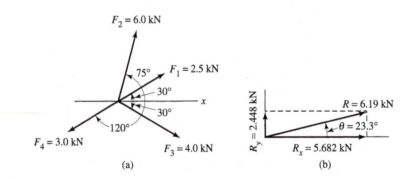

(a)   (b)

FIGURE 2.17

The x and y components of the forces are determined by trigonometry and shown in Table (A).

TABLE (A) Example 2.9				
Force	Magnitude (kN)	Angle θ^a (deg)	$F_x = F\cos\theta$ (kN)	$F_y = F\sin\theta$ (kN)
F_1	2.5	30	2.165	1.25
F_2	6.0	75	1.553	5.796
F_3	4.0	−30	3.464	−2.0
F_4	3.0	−120	−1.5	−2.598
			$\Sigma F_x = 5.682$ kN (R_x)	$\Sigma F_y = 2.448$ kN (R_y)

[a]The angle θ is positive when measured from the positive x axis in a counterclockwise direction and negative when measured in a clockwise direction.

By adding the x components and y components of the forces, we have the resultant in the x and y directions.

$$R_x = 5.682 \text{ kN} \qquad R_y = 2.448 \text{ kN}$$

From the Pythagorean theorem and the definition of the tangent of an angle,

$$R^2 = R_x^2 + R_y^2 = (5.682)^2 + (2.448)^2 = 38.28$$
$$R = 6.19 \text{ kN}$$

and
$$\tan\theta = \frac{R_y}{R_x} = \frac{2.448}{5.682} = 0.4308 \qquad \theta = 23.3°$$

$$\mathbf{R} = 6.19 \text{ kN at} + 23.3° \qquad \qquad \textbf{Answer}$$

This vector is shown in Fig. 2.17(b).

In addition to simplifying computations with vectors, rectangular components allow us a different perspective on the effects produced when forces are applied to real objects.

Example 2.10 Repeat Example 2.4 using rectangular components.

Solution: The vector addition shown in Fig. 2.8(b) is redrawn here as Fig. 2.18. Note that the original vectors **T** and **Q** (shown in dashed lines) have been replaced by their rectangular components. From this diagram, we see that two effects are created simultaneously:

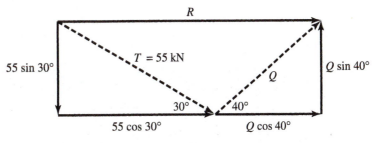

FIGURE 2.18

1. Those force components perpendicular to the truck axis must cancel to keep the truck from being pulled sideways. Stated algebraically,

$$(55 \text{ kN})\sin 30° = Q\sin 40°$$

or
$$Q = 55 \text{ kN} \times \frac{\sin 30°}{\sin 40°} = 42.8 \text{ kN} \qquad \qquad \textbf{Answer}$$

2. Those components parallel to the truck axis combine to yield resultant R. In other words,

$$R = (55 \text{ kN})\cos 30° + Q\cos 40°$$
$$= 47.6 \text{ kN} + (42.8 \text{ kN})\cos 40°$$
$$= 80.4 \text{ kN} \qquad \qquad \textbf{Answer}$$

2.7 DIFFERENCE OF TWO FORCES: VECTOR DIFFERENCES

In this section, we define what is meant by a negative force. The definition must be consistent with what we have already learned about the addition of two forces and will agree with the usual rules of algebra. From the usual rules of algebra, we write

$$\mathbf{F} - \mathbf{F} = 0 \quad \text{or} \quad \mathbf{F} + (-\mathbf{F}) = 0$$

Therefore, the force $-\mathbf{F}$ when added to the force $\mathbf{F}$ has a resultant equal to zero. Such a force must have the same magnitude as force **F** and be directed in the opposite direction. It follows that the force **F₁** may be subtracted from force **F₂** by reversing **F₁** and adding it to force **F₂** by any of the various methods previously discussed.

Example 2.11   For the concurrent force system shown in Fig. 2.19(a), find $\mathbf{R} = \mathbf{F_1} + \mathbf{F_2} - \mathbf{F_3}$ by the rectangular component method.

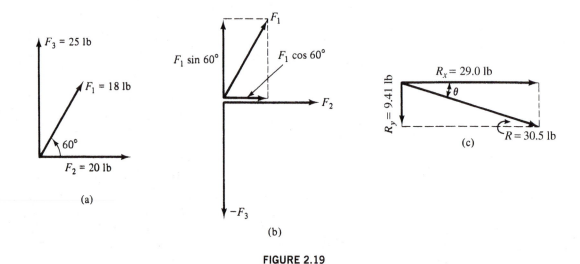

FIGURE 2.19

Solution: Replace **F₃** by a −**F₃** and proceed as in Example 2.9. The x and y components of each force are determined by trigonometry in Fig. 2.19(b). Calculations for the components are shown in Table (A).

TABLE (A) Example 2.11				
Force	Magnitude (lb)	Angle θ^a (deg)	$F_x = F\cos\theta$ (lb)	$F_y = F\sin\theta$ (lb)
F₁	18	60	9.0	15.59
F₂	20	0	20.0	0
−**F₃**	25	−90	0	−25.0
			$\Sigma F_x = 29.0$ lb (R_x)	$\Sigma F_y = -9.41$ lb (R_y)

[a]The angle θ is positive when measured from the positive x axis in a counterclockwise direction and negative when measured in a clockwise direction.

By adding the x components and y components of the forces, we have the resultant in the x and y directions.

$$R_x = \Sigma F_x = 29.0\,\text{lb} \qquad R_y = \Sigma F_y = -9.41\,\text{lb}$$

The magnitude and direction of the resultant is determined from the Pythagorean theorem and the definition of the tangent of an angle. From Fig. 2.19(c), we write

$$R^2 = (29.0)^2 + (-9.41)^2 = 929.5$$
$$R = 30.5\,\text{lb}$$

and
$$\tan\theta = \frac{R_y}{R_x} = \frac{-9.41}{29.0} \qquad \theta = -18.0°$$

$$\mathbf{R} = 30.5\,\text{lb at} -18.0°$$ **Answer**

PROBLEMS

For 2.19 through 2.36, use the method of rectangular components.

2.19 Two forces act on the free end of a cantilever beam, as shown; find the magnitude and direction of the resultant.

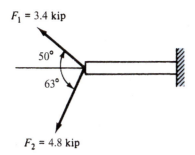

F_1 = 3.4 kip

50°

63°

F_2 = 4.8 kip

PROB. 2.19

2.20 Two forces act at the middle of a simply supported beam, as shown; find the magnitude and direction of the resultant.

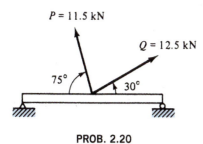

P = 11.5 kN

Q = 12.5 kN

75°

30°

PROB. 2.20

2.21 For the forces shown, P = 2.15 kN, and Q = 3.46 kN. Determine the resultant.

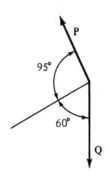

P

95°

60°

Q

PROB. 2.21

2.22 For the forces shown, S = 3.81 kip, T = 4.73 kip, and U = 3.65 kip. Determine the resultant.

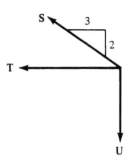

S

3

2

T

U

PROB. 2.22

2.23 Three forces act on the pile, as shown. If the resultant of the three forces is equal to 500 N and is directed vertically, determine the magnitude and direction of **S**.

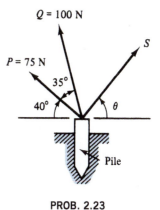

Q = 100 N

S

P = 75 N

35°

40°

θ

Pile

PROB. 2.23

2.24 Three forces act on the pin-connected tower, as shown. If the resultant of the three forces is equal to 4800 lb and is directed horizontally, determine the magnitude and direction of **V**.

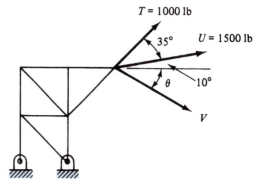

T = 1000 lb

35°

U = 1500 lb

θ

10°

V

PROB. 2.24

2.25 Solve Prob. 2.1.

2.26 Solve Prob. 2.2.

2.27 Solve Prob. 2.3.

2.28 Solve Prob. 2.4.

2.29 Solve Prob. 2.9.

2.30 Solve Prob. 2.10.

2.31 Solve Prob. 2.11.

2.32 Solve Prob. 2.12.

2.33 Using the figure for Prob. 2.19, find $\mathbf{F}_1 - \mathbf{F}_2$.

2.34 Using the figure for Prob. 2.20, find $\mathbf{P} - \mathbf{Q}$.

2.35 Using the figure for Prob. 2.21, find $\mathbf{Q} - \mathbf{P}$.

2.36 Using the figure for Prob. 2.22, find $\mathbf{S} + \mathbf{T} - \mathbf{U}$.

EQUILIBRIUM OF CONCURRENT FORCES IN A PLANE

CHAPTER OBJECTIVES

In this chapter we apply the techniques of Chapters 1 and 2 to solve the simplest type of statics problem: the equilibrium of forces applied to a single point. After studying this material, you should be able to

- Draw a complete and accurate free-body diagram showing the forces acting on a point in equilibrium.

- Explain the physical conditions required for equilibrium of a point, and demonstrate how these conditions are described mathematically.

- Apply the graphical, algebraic, and trigonometric methods of earlier chapters to solve for any unknown forces acting on a point in equilibrium.

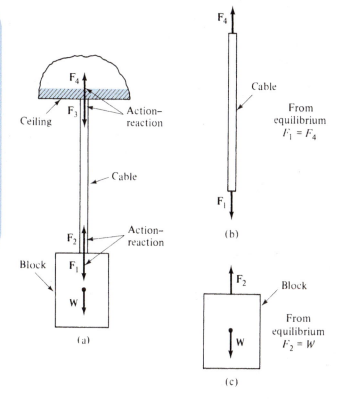

FIGURE 3.1

3.1 CONDITIONS FOR EQUILIBRIUM

When the resultant of a force system acting on a body is zero, the body is in equilibrium. If a body is in equilibrium, the body will either remain at rest, if originally at rest, or in motion, if originally in motion. We are concerned here with the *equilibrium of bodies at rest.*

In Chapter 2, we considered various methods for finding the resultant of concurrent force systems. We make use of those methods to solve problems involving concurrent force systems that act on bodies at rest—force systems that have a zero resultant.

3.2 ACTION AND REACTION

In the process of separating a body from its surroundings, we make use of a principle known as *Newton's third law,* which states that action equals reaction. That is, if a body *A* exerts a force on body *B*, body *B* exerts a force on body *A* equal in magnitude, opposite in direction, and having the same line of action. Newton's third law cannot be proven mathematically. It agrees with intuition, and deduction from the third law agrees with experiment.

Consider a block attached to a cable that is supported by the ceiling. In Fig. 3.1(a), we show the action–reaction pairs of forces. The force $\mathbf{F_1}$ represents the force exerted on the cable by the block. The reaction $\mathbf{F_2}$ is equal and opposite and represents the forces exerted on the block by the cable. The force $\mathbf{F_3}$ represents the force exerted on the ceiling by the cable. Its reaction $\mathbf{F_4}$ is equal and opposite and represents the force exerted on the cable by the ceiling. The forces acting on the cable and on the block are shown in Fig. 3.1(b) and (c). Forces $\mathbf{F_1}$ and $\mathbf{F_4}$ do not constitute an action–reaction pair because they act on the same body—the cable. The same is true for forces $\mathbf{F_2}$ and $\mathbf{W}$ that act on the block. However, both the cable and the block are in equilibrium with $F_4 = F_1$ and $F_2 = W$. Therefore, $\mathbf{F_3} = \mathbf{W}$. Thus, the cable can be thought of as transmitting the weight of the block from the block to the ceiling.

3.3 SPACE DIAGRAM, FREE-BODY DIAGRAM

In statics, we solve problems involving structures, machines, or other physical bodies. A sketch called a *space diagram* may be used to describe the physical problem to be solved.

Statics involves the forces or interactions of bodies on each other. The bodies must be separated from each other so that unknown forces may be determined. To this end, we select and free a body from its surroundings. A diagram showing the forces acting on the body is then drawn. Such a diagram is called a *free-body diagram*. When the free-body diagram involves a concurrent force system, the problems can be solved by the methods of this chapter.

3.4 CONSTRUCTION OF A FREE-BODY DIAGRAM

In this section, we consider various physical bodies that are described by space diagrams. For each example, a free body will be isolated, and all forces acting on the body will be shown.

Among the forces to be shown will be the weight. We will see in Chapter 9 that the weight of the body acts through a point called the *center of gravity* of the body. It is directed from the center of gravity downward toward the center of the earth. For a uniform body, the center of gravity is at the geometric center of the body. For nonuniform bodies, the center of gravity's location is usually designated as "c.g.," or by the symbol ⬕.

It is not possible to overemphasize the importance of clear and accurate free-body diagrams to the analysis of problems in static force equilibrium. The purpose of such diagrams is to isolate all important mathematical data from the pictorial descriptions of these problems. Toward that end, your diagrams should include

1. Any given angles and dimensions.
2. All forces, including weights, whose magnitudes and directions are specified or implied.
3. Any unknown angles and forces. (These should each be labeled and the directions of forces indicated by arrowheads. If you are not sure of the direction in which a force acts, assume a direction. As we will see in the next section, our solutions not only yield numerical results, but also indicate the correct directions of all forces required for equilibrium.)

The following example illustrates both the general technique for construction of free-body diagrams and the way in which these diagrams may be used to determine the forces created by external loads.

Example 3.1   Three packages, A, B, and C, are stacked (at rest) on a table, as shown in Fig. 3.2. If the package weights are 15, 40, and 65 lb, respectively, what contact force do you think exists between packages A and B? Packages B and C? Package C and the table? Draw a free-body diagram of each package and show algebraically what the actual value is for each contact force.

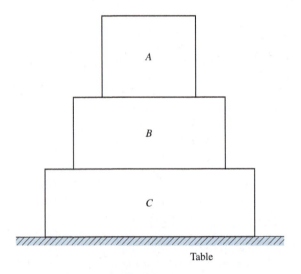

Table

FIGURE 3.2

Solution: We begin by separating A from B, B from C, and C from the table, as indicated in Fig. 3.3. Because each pair of surfaces represents a point of physical contact, Newton's third law tells us that these surfaces exert *equal but opposite* forces on each other. Therefore, we draw the force pairs F_{AB}, F_{BC}, and F_{table} as shown. How do we know in which direction to draw these forces? On more complicated free-body diagrams, we often simply *guess* at the correct directions for each force. However, here our own experience should tell us that the weight of each package pushes downward on the package or table below, whereas the lower packages and the table

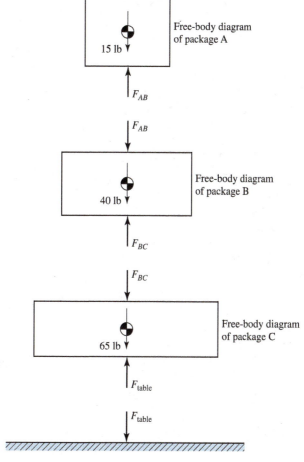

15 lb

Free-body diagram
of package A

F_{AB}

F_{AB}

40 lb

Free-body diagram
of package B

F_{BC}

F_{BC}

65 lb

Free-body diagram
of package C

F_{table}

F_{table}

FIGURE 3.3

itself must push upward to support the cumulative weights above. Our forces, then, are drawn *downward* on the *top* surfaces of B, C, and the table, and *upward* on the *bottom* surfaces of A, B, and C. The weights of A, B, and C, assumed to act downward at each package's geometric center, are also drawn on the appropriate package. In its free-body diagram, then, each package is now *disconnected from any other object*. It is "free in space" (an isolated, or "free body"), but the *effect* of every adjacent object is retained in the form of a contact force. For the packages to physically remain in *equilibrium* (at rest), the vertical forces that act on A, B, and C must balance. In other words, on each package, the sum of the forces up must equal the sum of the forces down. By inspection of the free-body diagram for A,

$$F_{AB} = 15\,\text{lb}$$ **Answer**

For the free-body diagram of B,

$$F_{BC} = F_{AB} + 40\,\text{lb} = 15\,\text{lb} + 40\,\text{lb} = 55\,\text{lb}$$ **Answer**

For the free-body diagram of C,

$$F_{table} = F_{BC} + 65\,\text{lb} = 55\,\text{lb} + 65\,\text{lb} = 120\,\text{lb}$$ **Answer**

As we might expect, each surface supports the combined weights of all packages located above that surface.

The next few examples illustrate the construction of free-body diagrams for various groups of connected objects, including blocks, cables, weightless bars (called *links*), and smooth surfaces. The physical basis for each type of contact force will be developed later in this chapter and in Chapter 4. These contact forces, also known as *reactive* forces because they develop *in reaction to* the external loads on a system, are then summarized at the beginning of Chapter 5. For the types of problems to be analyzed and solved in this chapter, your completed free-body diagrams should contain a total of no more than *two* unknown forces and/or angles.

Example 3.2   The 12.5-kN block is supported by cables as shown in Fig. 3.4(a). Draw the free-body diagrams of the block and point *B*.

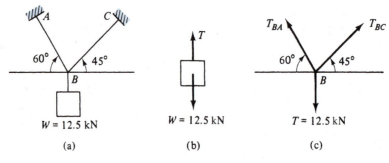

(a)   (b)   (c)

FIGURE 3.4

Solution: Consider the block as a free body [Fig. 3.4(b)]. The force *T* of the cable acts upward along the cable, away from the body, and the 12.5-kN weight of the body acts downward toward the center of the earth. The two forces act along the same straight line. They form a collinear force system. Because the block is at rest, the forces are in equilibrium, and the resultant is equal to zero. Therefore, $T - W = 0$ or $T = W = 12.5\,\text{kN}$. The vertical cable can be thought of as transmitting the weight of the block from the block to point *B*.

A cable can support a tensile force; therefore, in the free body of point *B* [Fig. 3.4(c)], each cable is in tension and acts away from point *B*.

Example 3.3   The 2.55-kg block in Fig. 3.5(a) is supported by a cable that passes over a frictionless pulley. Draw the free-body diagram for the pulley.

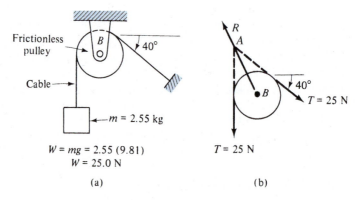

(a)   (b)

FIGURE 3.5

Solution: From Eq. (1.1), the block weighs

$$W = mg = 2.55(9.81) = 25.0\,\text{N}$$

In Example 3.2, we saw that the tension in the vertical cable was equal to the weight of the body supported by the cable. Therefore, the tension in the cable is $T = 25.0\,\text{N}$.

We now draw a free-body diagram of the frictionless pulley [Fig. 3.5(b)]. The tensile forces in the cable on each side of the frictionless pulley are the same. This can be proved by the methods of Chapter 5. Thus, the tensile force on the left and right of the pulley is equal to 25 N. Both tensile forces are directed away from the pulley, and their lines of action intersect at point *A*. For equilibrium, the reactive force *R* of the axle at *B* on the pulley must also act through point *A*. Ropes, strings, and cords are analyzed in the same way as the cable in a free-body diagram.

Example 3.4　　The links *AB* and *BC* support a rope that is attached to the 5-kip load, as shown in Fig. 3.6(a). Draw a free-body diagram of point *B*.

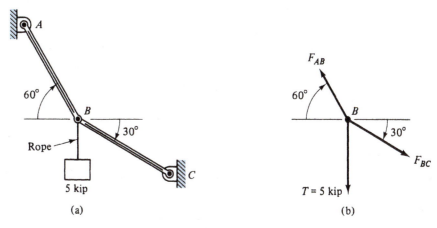

FIGURE 3.6

Solution:　　In Example 3.2, we saw that the tension in the vertical cable was equal to the weight of the body supported by the cable. Therefore, the tension in the rope is $T = 5$ kip. The links *AB* and *BC* are weightless bodies joined to the supports at *A* and *C* and together at *B* by frictionless pins. Thus, each link or body is subject to two forces—one at each end. As such, they are special cases of two-force bodies. The equilibrium of two-force bodies is discussed in Chapter 5, where we show that the force in a link acts in the direction of the link.

Consider the frictionless pin at *B* as a free body [Fig. 3.6(b)]. The tensile force *T* acts down—away from *B*—and the forces in the link F_{AB} and F_{BC} act in the direction of the links. The forces in the links can act either toward *B* or away from *B*. The correct directions must be found from the conditions of equilibrium.

Example 3.5　　Consider a 100-lb block that is supported by a cord and smooth plane, as shown in Fig. 3.7(a). Draw the free-body diagram of the block.

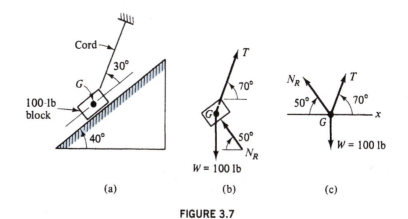

FIGURE 3.7

Solution:　　We isolate the block as the free body [Fig. 3.7(b)]. Three forces are acting on it. The tensile force *T* on the cord acts along the cord and away from the body. The reactive force N_R of the plane on the block acts normal to the plane at the surface of contact and toward the body. The 100-lb weight of the body acts from the center of gravity *G* downward toward the center of the earth. The lines of action of the three forces intersect at point *G* and thus represent a concurrent force system. In Fig. 3.7(c), we move force N_R along its line of action so that all the forces are directed away from point *G*.

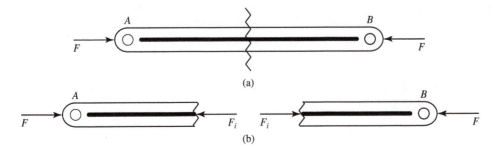

FIGURE 3.8

Free-body diagrams can also be used to find the *internal* forces carried by an object. Consider the rigid weightless link *AB* in equilibrium, as shown in Fig. 3.8(a). Applied as shown, the two equal-but-opposite forces, **F**, tend to *crush* this link, a condition known as *compression*. If we make an imaginary cut through the link at any cross section along its length [Fig. 3.8(b)], the *internal forces exerted by each piece of the link on the other* appear as an equal-but-opposite pair, **F**$_i$. For equilibrium of both pieces, these forces must be in the directions shown, and **F**$_i$ must equal **F**. From this, we see that *AB* carries the constant load, **F**, everywhere along its length and is in compression at all points.

Similarly, flexible members such as cables, ropes, cords, and strings can support forces [Fig. 3.9(a)] that tend to *stretch* the member, creating a condition known as *tension*. Cutting the member at any point [Fig. 3.9(b)] reveals the constant internal tensile forces, **F**$_i$.

As we will see in later chapters, this use of free-body diagrams can help us analyze the forces at any point in a system. For instance, weight **W** of Fig. 3.10(a) is suspended by a cable that is supported on a frictionless pulley. Knowing that the internal forces, **T**, in the cable are constant, we may cut the member at points such as *A*, *B*, and *C* to begin construction of multiple free-body diagrams [Fig. 3.10(b)] that may aid in our analysis.

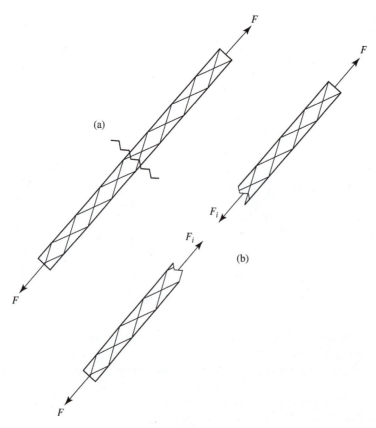

FIGURE 3.9

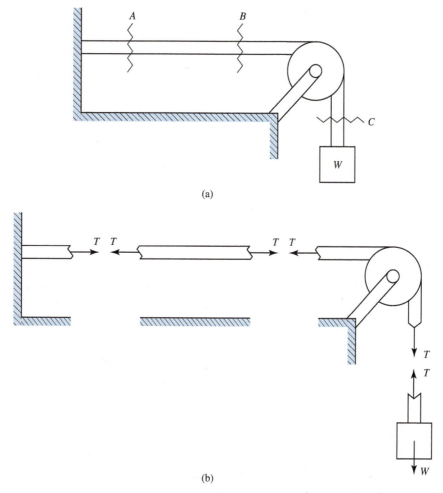

(a)

(b)

FIGURE 3.10

PROBLEMS

Assume pulleys frictionless, struts and links weightless, and inclined planes smooth.

3.1 through 3.8. Draw a free-body diagram of point *B*.

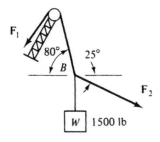

PROB. 3.1

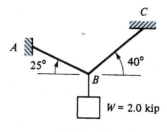

PROB. 3.2

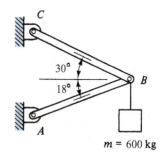

$m = 600$ kg

PROB. 3.3

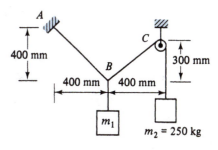

$m_2 = 250$ kg

PROB. 3.4

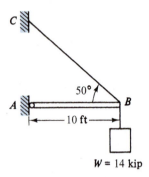

PROB. 3.5

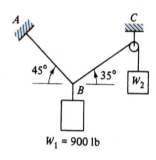

PROB. 3.6

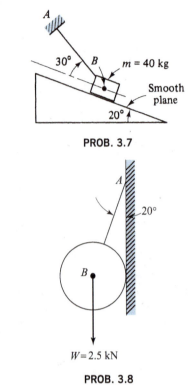

PROB. 3.7

PROB. 3.8

3.5 THREE CONCURRENT FORCES IN EQUILIBRIUM

When a body is in equilibrium under the action of three concurrent forces, the problem can be solved by drawing a free-body diagram of the point through which all three forces pass, and then constructing a force triangle using the methods presented in Chapter 2. This completed force triangle should satisfy the two conditions defined by our rule for vector addition:

1. Vectors must be added in a tail-to-tip fashion.
2. For equilibrium to exist, the resultant must equal zero. This can only be true if the vector addition begins and ends at the same point. Stated another way, our force triangle must form a closed figure.

We illustrate the method with the following examples.

Example 3.6   A crate weighing 6000 N is suspended from two cables as shown in Fig. 3.11(a).

   a. Draw a free-body diagram of point *B*.
   b. Construct a force triangle that satisfies the conditions for equilibrium of point *B*.
   c. Use your force triangle to find the tensions required in cables *AB* and *BC*.

Solution: a. The appropriate free-body diagram is shown in Fig. 3.11(b). Note that F_{AB} and F_{BC} are *assumed* to be in the directions shown; their actual directions will be verified by the force triangle of part (b).

   b. We begin our force triangle by drawing the known weight of 6000 N in a downward direction. Through one end of this vector, we construct a line that is in the same direction as F_{AB}, extending the line indefinitely in both directions. Through the other end of our known vector, we draw a line in the same direction as F_{BC}, extending this line indefinitely in both directions. These constructions are shown in Fig. 3.12(a). Note that our weight vector and the two drawn lines form a closed figure, thus ensuring a zero resultant. By drawing arrowheads in the required tail-to-tip fashion of Fig. 3.12(b), we can verify that the directions of F_{AB} (upward to the left) and F_{BC} (upward to the right) agree with the directions assumed on our free-body diagram. As shown in Fig. 3.13(a) and (b), these same results would have been obtained if the lines parallel to F_{AB} and F_{BC} had been drawn at opposite ends of the weight vector.

   c. From Fig. 3.12(b) or 3.13(b), the law of sines yields

$$\frac{6000 \text{ N}}{\sin 60°} = \frac{F_{AB}}{\sin 70°}$$

$$F_{AB} = 6510 \text{ N} \qquad\qquad \textbf{Answer}$$

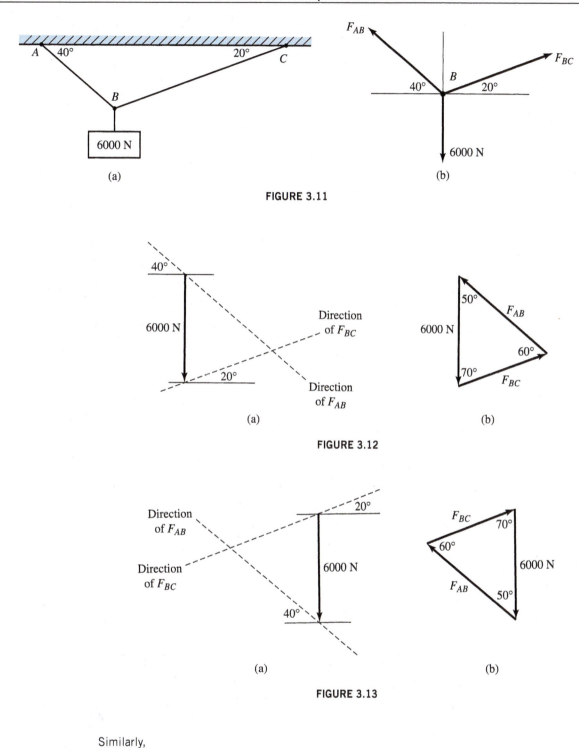

FIGURE 3.11

FIGURE 3.12

FIGURE 3.13

Similarly,

$$\frac{6000 \text{ N}}{\sin 60°} = \frac{F_{BC}}{\sin 50°}$$

$$F_{BC} = 5307 \text{ N} \qquad \qquad \textbf{Answer}$$

Note that this problem contained two unknowns, F_{AB} and F_{BC}. If a third supporting cable, BD, were added as shown in Fig. 3.14(a), a third unknown, F_{BD}, would be introduced. The resulting vector addition would yield a force *polygon* similar to that of Fig. 3.14(b). Although this construction shows a specific combination of values for F_{AB}, F_{BC}, and F_{BD}, which form the closed figure *abcd*, different force combinations forming the polygons *abc′d′* or *abc″d″* would be equally valid. No matter which combination is selected, however, there is insufficient information to solve for any numerical values of force. Such a problem contains too many unknowns and is said to be *statically indeterminate*.

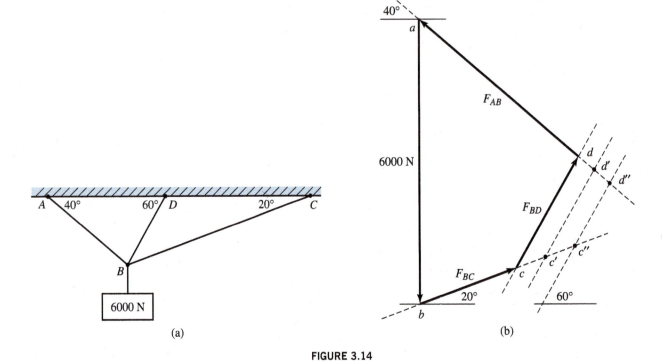

FIGURE 3.14

Example 3.7   Three concurrent forces in equilibrium acting at point *B* are shown in the free-body diagram of Fig. 3.6(b). The direction of the forces in links *AB* and *BC* was assumed to act away from *B*. Find the magnitude of the forces $\mathbf{F_{AB}}$ and $\mathbf{F_{BC}}$.

Solution: Draw the known force $T = 5$ kip. Through the end of **T**, draw a line parallel to $\mathbf{F_{BC}}$, and through the other end of **T**, draw a line parallel to $\mathbf{F_{AB}}$. Two possibilities exist as shown in Fig. 3.15(a) and (b). In either case, the lines of action intersect to form a triangle. The force triangle can now be constructed around either of these triangles [Fig. 3.16(a) and (b)].

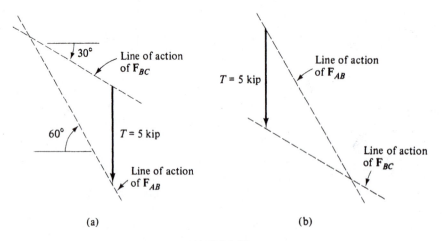

FIGURE 3.15

Because the three forces are in equilibrium and by construction form a closed triangle,

$$\mathbf{T} + \mathbf{F_{AB}} + \mathbf{F_{BC}} = \mathbf{0}$$

Applying the law of sines to either triangle, we have

$$\frac{F_{AB}}{\sin 120°} = \frac{F_{BC}}{\sin 30°} = \frac{5}{\sin 30°}$$

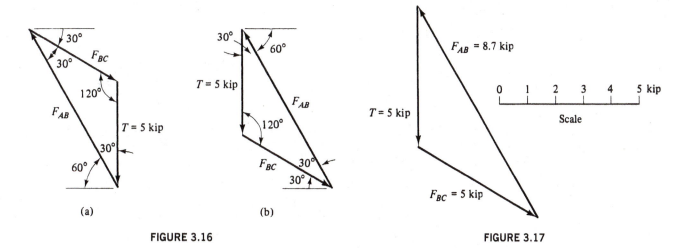

(a)   (b)

FIGURE 3.16   **FIGURE 3.17**

Therefore,

$$F_{AB} = \frac{\sin 120°}{\sin 30°}(5) = 8.66 \text{ kip}$$ **Answer**

and

$$F_{BC} = \frac{\sin 30°}{\sin 30°}(5) = 5.0 \text{ kip}$$ **Answer**

The solution may also be obtained graphically. Select an appropriate scale and draw vector forces $T = 5$ kip, $\mathbf{F_{BC}}$, and $\mathbf{F_{AB}}$ in a tip-to-tail fashion (Fig. 3.17). The magnitudes of the unknown forces are found to be

$$F_{BC} = 5 \text{ kip} \qquad F_{AB} = 8.7 \text{ kip}$$ **Answer**

Example 3.8   The three forces in Fig. 3.18 are in equilibrium. The magnitude of all three forces is known, but the directions of $\mathbf{F_2}$ and $\mathbf{F_3}$ are unknown. Find the directions of $\mathbf{F_2}$ and $\mathbf{F_3}$.

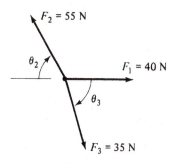

FIGURE 3.18

Solution: **Graphical Method:** Select an appropriate scale. Draw $F_1 = 40$ N to scale from point A to point B [Fig. 3.19(a)]. With a compass set for a radius $r_2 = F_2 = 55$ N, draw an arc from the tip of $\mathbf{F_1}$, point B. With a compass set for radius $r_3 = F_3 = 35$ N, draw a second arc from the tail of $\mathbf{F_1}$, point A. The two arcs intersect at point C. Because the force triangle must close, $\mathbf{F_2}$ acts from B to C, and force $\mathbf{F_3}$ acts from C to A, as shown in the force triangle, Fig. 3.19(b). Measuring the direction of the forces, we have

$$\theta_2 = 39° \quad \text{and} \quad \theta_3 = 87°$$ **Answer**

Mathematical Method: A force triangle is constructed in Fig. 3.19(c). Applying the law of cosines to the triangle, we have

$$F_2^2 = F_1^2 + F_3^2 - 2F_1 F_3 \cos A$$
$$(55)^2 = (40)^2 + (35)^2 - 2(40)(35)\cos A$$

or $$\cos A = -0.0714, \qquad A = 94.1°$$

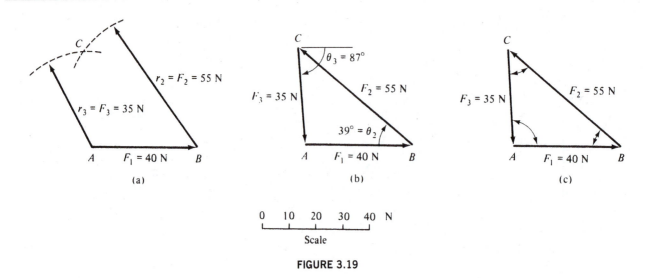

FIGURE 3.19

From the law of sines, we write

$$\frac{\sin B}{F_3} = \frac{\sin A}{F_2} \qquad \frac{\sin B}{35} = \frac{\sin 94.1°}{55}$$

$$\sin B = \frac{35}{55} \sin 94.1° = 0.6347 \qquad B = 39.4°$$

Because $A + B + C = 180°$, $C = 46.5°$. Therefore, the directions of the forces are

$$\theta_2 = B = 39.4° \qquad\qquad\qquad \textbf{Answer}$$

$$\theta_3 = B + C = 39.4° + 46.5° = 85.9° \qquad \textbf{Answer}$$

Example 3.9   The three forces acting at A in Fig. 3.20(a) are in equilibrium. The magnitude and direction of $\mathbf{F_1}$ and $\mathbf{F_2}$ are known. Determine $\mathbf{F_3}$.

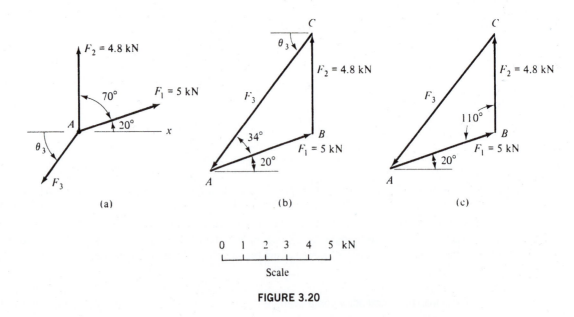

FIGURE 3.20

Solution: **Graphical Method:** Select the scale. Draw $\mathbf{F_1}$ from A to B and $\mathbf{F_2}$ from B to C. Force $\mathbf{F_3}$ must be represented by a vector drawn from C to A [Fig. 3.20(b)]. Measuring the magnitude and direction of $\mathbf{F_3}$, we have

$$\mathbf{F_3} = 8\,\text{kN} \nearrow 54° \qquad\qquad \textbf{Answer}$$

Mathematical Method: A force triangle is constructed in Fig. 3.20(c). Angle $B = 90° + 20° = 110°$. From the law of cosines,

$$F_3^2 = F_1^2 + F_2^2 - 2F_1F_2 \cos B$$
$$= (5)^2 + (4.8)^2 - 2(5)(4.8) \cos 110° = 64.46$$
$$F_3 = 8.03 \text{ kN}$$

Applying the law of sines yields

$$\frac{\sin C}{F_1} = \frac{\sin B}{F_3} \qquad \frac{\sin C}{5} = \frac{\sin 110°}{8.03}$$

Thus,
$$\sin C = \frac{5}{8.03} \sin 110° = 0.5851 \qquad C = 35.8°$$

The angle $\theta_3 = 90° - C = 90° - 35.8° = 54.2°$.

$$\mathbf{F_3} = 8.03 \text{ kN} \nearrow 54.2° \qquad \qquad \textbf{Answer}$$

3.6 FOUR OR MORE FORCES IN EQUILIBRIUM

When a body is in equilibrium under the action of four or more concurrent forces, the problem may be solved *graphically* by drawing the force polygon, provided that we have no more than two unknowns (magnitude and/or direction). The force polygon method is illustrated in the following example.

Example 3.10 The four forces acting at point B, as shown in Fig. 3.21(a), are in equilibrium. The magnitude and direction of $\mathbf{F_1}$ and $\mathbf{F_2}$ are known. The magnitude of $\mathbf{F_4}$ is known, and the direction of $\mathbf{F_3}$ is known. Find the direction of $\mathbf{F_4}$ and the magnitude of $\mathbf{F_3}$ by constructing a force polygon.

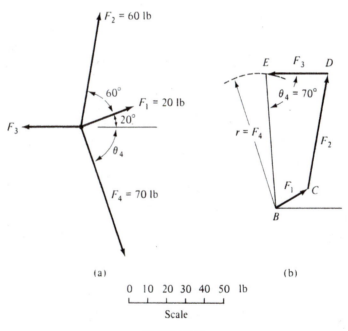

(a)

(b)

0 10 20 30 40 50 lb

Scale

FIGURE 3.21

Solution: As shown in Fig. 3.21(b), we start drawing the polygon with the forces of known direction and magnitude. In tip-to-tail fashion, draw forces $\mathbf{F_1}$ and $\mathbf{F_2}$ from B to C and C to D. Through point D, draw a line parallel to the direction of $\mathbf{F_3}$, and around point B draw an arc of radius $r = F_4 = 70$ lb. The arc and line intersect at point E. Force $\mathbf{F_3}$ acts from D to E, and force $\mathbf{F_4}$ acts from E to B. Measuring, we obtain

$$F_3 = 52 \text{ lb} \quad \text{and} \quad \theta_4 = 70° \qquad \qquad \textbf{Answer}$$

For a mathematical solution, we use the method of components as described in the following section.

(a)   (b)

FIGURE 3.22. (a) This 650-ft television tower is held in place by three sets of guy wires located in vertical planes around the tower. Each four-wire set is attached to a concrete pier (see "A" in photo). (b) In this close-up view of a pier, the resultant of these four concurrent, coplanar guy wire forces is upward to the right and must be counteracted through some combination of forces exerted by the ground on the pier. For a further analysis of tower guy wires, see SB-8(a) and (b) in Chapter 7.

(Courtesy of WAND-TV, Decatur, Illinois, www.wandtv.com)

Fig. 3.22 shows an example of multiple forces in equilibrium.

3.7 EQUILIBRIUM BY RECTANGULAR COMPONENT METHOD

In the *rectangular component method,* each force is replaced by its x and y components. The resultant in the x direction, obtained by adding the x components of the forces, must add to zero, and the resultant in the y direction, obtained by adding the y components of the forces, must also add to zero. That is,

$$R_x = \Sigma F_x = 0 \qquad \textbf{(3.1)}$$

$$R_y = \Sigma F_y = 0 \qquad \textbf{(3.2)}$$

The following examples will illustrate the rectangular component method. These examples use standard x and y axes that are *horizontal* and *vertical,* respectively. However, you should note that this is *not* a requirement of Eqs. (3.1) and (3.2), and that the coordinate system may be rotated to any convenient position as long as the axes themselves remain perpendicular to each other. For instance, problems containing an inclined plane often use axes that are *parallel* and *perpendicular* to the surface of the plane to simplify the trigonometric computations required for a solution.

Example 3.11   Repeat Example 3.6 using rectangular components.

Solution: The original free-body diagram of Fig. 3.11(b) is shown in Fig. 3.23(a) with the rectangular components of F_{AB} and F_{BC} included. In Fig. 3.23(b), these components and the 6000 N load are all applied to point B. From this figure, we see that two physical conditions must be satisfied *simultaneously* for point B to remain in equilibrium:

1. The two horizontal force components must be exactly equal in magnitude and opposite in direction.

2. Our two vertical force components must support the applied load of 6000 N.

These two conditions are represented in notational form as Eqs. (3.1) and (3.2):

$$\Sigma F_x = 0 \qquad \text{and} \qquad \Sigma F_y = 0$$

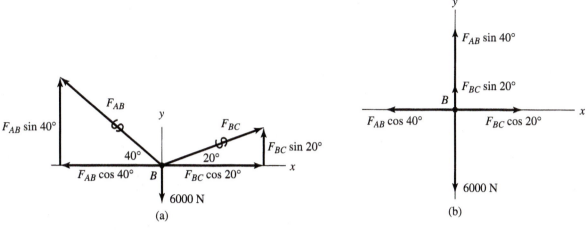

FIGURE 3.23

The actual algebraic equations must be written to agree with those force directions shown on our free-body diagram. If forces to the right and forces to the left are both assigned positive values, then our notational equations become

$$\text{Sum of forces to right} - \text{Sum of forces to left} = 0$$

or

$$\text{Sum of forces to right} = \text{Sum of forces to left}$$

Then, from Fig. 3.23(b),

$$F_{BC} \cos 20° = F_{AB} \cos 40° \qquad \textbf{(a)}$$

Similarly, taking forces both upward and downward as positive

$$\text{Sum of forces upward} - \text{Sum of forces downward} = 0$$

or

$$\text{Sum of forces upward} = \text{Sum of forces downward}$$

Again, from Fig. 3.23(b),

$$F_{BC} \sin 20° + F_{AB} \sin 40° = 6000 \text{ N} \qquad \textbf{(b)}$$

Eqs. (a) and (b) can also be obtained by inspection of the vector addition shown in Fig. 3.24. Because all our forces are horizontal or vertical, the closed figure formed by this tail-to-tip addition is a rectangle. Setting the opposite sides equal to each other yields Eqs. (a) and (b) directly.

Aligning Eqs. (a) and (b) and substituting the appropriate sine and cosine values yields

$$0.9397 \, F_{BC} - 0.7660 \, F_{AB} = 0 \qquad \textbf{(c)}$$

$$0.3420 \, F_{BC} + 0.6428 \, F_{AB} = 6000 \qquad \textbf{(d)}$$

Then, from Cramer's rule,

$$F_{BC} = \frac{\begin{vmatrix} 0 & -0.7660 \\ 6000 & 0.6428 \end{vmatrix}}{\begin{vmatrix} 0.9397 & -0.7660 \\ 0.3420 & 0.6428 \end{vmatrix}} = \frac{4596}{0.8660}$$

$$F_{BC} = 5307 \text{ N} \qquad \textbf{Answer}$$

and

$$F_{AB} = \frac{\begin{vmatrix} 0.9397 & 0 \\ 0.3420 & 6000 \end{vmatrix}}{\begin{vmatrix} 0.9397 & -0.7660 \\ 0.3420 & 0.6428 \end{vmatrix}} = \frac{5638.2}{0.8660}$$

$$F_{AB} = 6510 \text{ N} \qquad \textbf{Answer}$$

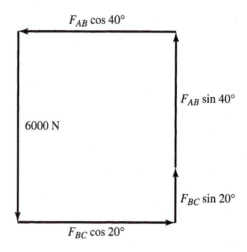

FIGURE 3.24

You should take a few moments to compare the algebra involved in this solution to that of Example 3.6. Note also that the two equations [(3.1) and (3.2)] simply offer an alternative method for describing those conditions under which $R = 0$. By comparison, in Example 3.6, the two conditions necessary for a zero resultant were described using trigonometric relationships defined by the law of sines.

Example 3.12 The three forces in Fig. 3.25(a) are in equilibrium. Determine F_1 and F_3 by the method of rectangular components.

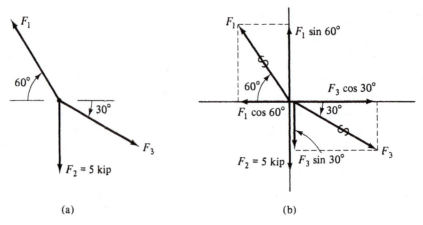

(a)   (b)

FIGURE 3.25

Solution: The x and y components of each force are determined by trigonometry, as shown in Fig. 3.25(b), and tabulated in Table (A). From equilibrium Eqs. (3.1) and (3.2),

$$\Sigma F_x = 0 \qquad -0.5000F_1 + 0.8660F_3 = 0 \tag{a}$$

and

$$\Sigma F_y = 0 \qquad +0.8660F_1 - 5.0 - 0.5000F_3 = 0 \tag{b}$$

From Eq. (a),

$$F_3 = \frac{0.5000}{0.8660}F_1 = 0.577F_1 \tag{c}$$

Substituting the value of F_3 from Eq. (c) in terms of F_1 in Eq. (b), we have

$$0.8660F_1 - 5.0 - 0.5000(0.577F_1) = 0$$

or

$$F_1 = 8.67 \text{ kip} \qquad\qquad \textbf{Answer}$$

TABLE (A)   Example 3.12

Force	Magnitude (kip)	Angle (deg)	x Component (kip)	y Component (kip)
F_1	F_1	60	$-0.5000F_1$	$+0.8660F_1$
F_2	5	90	0	-5.0
F_3	F_3	30	$+0.8660F_3$	$-0.5000F_3$
			$\Sigma F_x = 0$ Eq. (a)	$\Sigma F_y = 0$ Eq. (b)

Substituting the value of F_1 into Eq. (c), we write

$$F_3 = 0.577(8.67) = 5.00 \text{ kip}$$ **Answer**

Check:

$$\Sigma F_x = -0.5000(8.67) + 0.8660(5.00) = 0 \quad \text{OK}$$

Example 3.13 The four forces shown in Fig. 3.26 are in equilibrium. Determine F_3 and F_4 by the method of rectangular components.

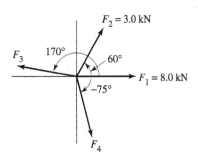

FIGURE 3.26

Solution: The x and y components of each force are determined by trigonometry and shown in Table (A). The direction angles for the forces are measured from the positive x axis. From Table (A) for equilibrium in the x and y directions, we write

$$\Sigma F_x = 0 \quad -0.9848F_3 + 0.2588F_4 + 9.5 = 0 \quad \text{(a)}$$
$$\Sigma F_y = 0 \quad 0.1736F_3 - 0.9659F_4 + 2.60 = 0` \quad \text{(b)}$$

TABLE (A)   Example 3.13

Force	Magnitude (kN)	Angle θ^a (deg)	$F_x = F\cos\theta$ (kN)	$F_y = F\sin\theta$ (kN)
F_1	8.0	0	8.0	0
F_2	3.0	60	1.5	2.60
F_3	F_3	170	$-0.9848F_3$	$0.1736F_3$
F_4	F_4	-75	$0.2588F_4$	$-0.9659F_4$
			$\Sigma F_x = 0$	$\Sigma F_y = 0$

[a]Angle θ is measured from the positive x axis.

Solving Eqs. (a) and (b) for F_3 and F_4 by Cramer's rule, we have

$$D = \begin{vmatrix} -0.9848 & +0.2588 \\ +0.1736 & -0.9659 \end{vmatrix} = -0.9848(-0.9659) - (0.2588)(0.1736) = 0.9065$$

$$DF_3 = \begin{vmatrix} -9.5 & +0.2588 \\ -2.60 & -0.9659 \end{vmatrix} = -9.5(-0.9659) - (0.2588)(-2.60) = 9.850$$

$$DF_4 = \begin{vmatrix} -0.9848 & -9.5 \\ +0.1736 & -2.60 \end{vmatrix} = -0.9848(-2.60) - (-9.5)(0.1736) = 4.210$$

Dividing by D, we have

$$F_3 = \frac{9.850}{0.9065} = 10.87 \text{ kN} \qquad F_4 = \frac{4.210}{0.9065} = 4.64 \text{ kN} \qquad \textbf{Answer}$$

Check:

$$\Sigma F_y = 0.1736(10.87) - 0.9659(4.64) + 2.60 = 0 \qquad \text{OK}$$

PROBLEMS

3.9 through 3.11 Determine the tension in the cables (a) graphically, (b) mathematically from the force triangle, and (c) by the method of rectangular components.

3.12 and 3.13 Determine the tension in the cable and the tension or compression in the strut (a) graphically, (b) mathematically from the force triangle, and (c) by the method of rectangular components.

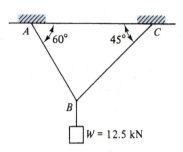

PROB. 3.9

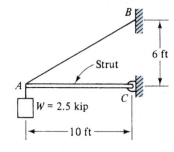

PROB. 3.12

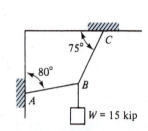

PROB. 3.10

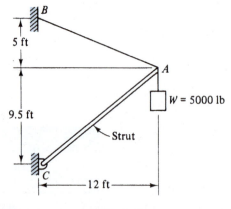

PROB. 3.13

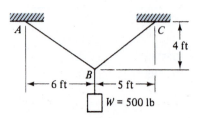

PROB. 3.11

3.14 Determine the force F and the normal reaction of the smooth plane on block A (a) graphically, (b) mathematically from the force triangle, and (c) by the method of rectangular components.

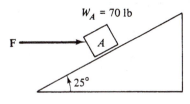

PROB. 3.14

3.15 and 3.16   Determine the tension in the cable and the normal reaction of the smooth plane on block A (a) graphically, (b) mathematically from the force triangle, and (c) by the method of rectangular components.

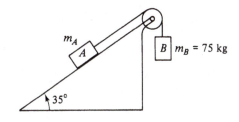

PROB. 3.15

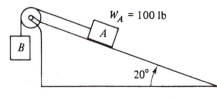

PROB. 3.16

3.17   A uniform crate weighing 15.0 kN is lifted by a sling, ABC, a small pulley, and cable BD. If the tension in the sling is limited to 9.0 kN, determine the minimum length of the sling. Neglect the radius of the pulley.

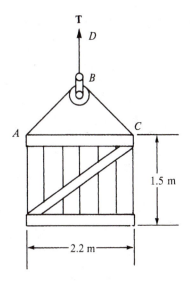

PROB. 3.17

3.18   Use the figure for Prob. 3.2. Determine the tension in the cables by the method of rectangular components.

3.19   Use the figure for Prob. 3.3. Determine the tension or compression in the struts by the method of rectangular components.

3.20   Use the figure for Prob. 3.4. By the method of rectangular components, find the tension in the cables.

3.21   Use the figure for Prob. 3.7. Determine the tension in the cable and the normal reaction of the smooth plane on block B by the method of rectangular components.

3.22   Use the figure for Prob. 3.8. Find the tension in the cable and the force of the wall by the method of rectangular components.

3.23 and 3.24   The four forces are in equilibrium. Determine F_4 and θ_3 (a) graphically from the force polygon, and (b) by the method of rectangular components.

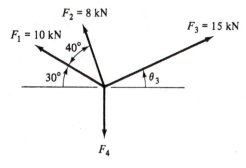

PROB. 3.23

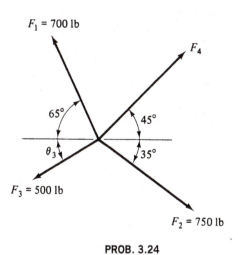

PROB. 3.24

RESULTANT OF NONCONCURRENT FORCES IN A PLANE

CHAPTER OBJECTIVES

In this chapter, we discuss the rotational effects, called *moments,* created by forces applied at different points on an object. After completing this material, you should be able to

- Compute the moment of a force about any selected point, using either the definition of moment or Varignon's theorem.

- Find the magnitude and location of the resultant for a system of parallel or nonparallel forces.

- Describe the combination of forces that form a couple, and define the moment of a couple.

- Resolve a given force into an equivalent force–couple combination.

- Determine the magnitude and location of the resultant for both uniform and triangular distributed loads.

4.1 INTRODUCTION

In Chapters 2 and 3, we studied concurrent force systems. In each case, the free body reduced to a single point or particle on which the forces acted. With nonconcurrent force systems, we consider not only the physical size of the body, but also the various points on the body where the forces are applied.

The bodies considered in statics are usually assumed to be rigid. A rigid body is one that does not deform or change shape. However, bodies such as structures or machines are never completely rigid. They deform or change shape as a result of the forces that act on them. When the deformations or changes in shape are small, they can be neglected. Such changes will become important when we study strength of materials, where we determine the stresses and deflections produced by the loads.

With the free body occupying more than a point, we must consider force systems that are *not concentrated* at a point but *distributed* over a surface or throughout a volume. Examples of such systems are wind pressure, water pressure, and the weight of a body. Distributed force systems are discussed in Sec. 4.9.

4.2 TRANSMISSIBILITY

The *principle of transmissibility* states that a force **F** acting at a point A on a rigid body may be transmitted or moved to any other point B on its line of action without changing the effect of the force on the rigid body. The principle is illustrated in Fig. 4.1(a) and (b).

The principle of transmissibility has limitations. Consider a bar acted on by two equal and opposite forces [Fig. 4.2(a)]. Making use of the principle of transmissibility, we move the force that acts at B to A, as shown in Fig. 4.2(b). From the point of view of rigid-body statics, the systems are both in equilibrium and identical. However, in Fig. 4.2(a), the bar shown is in tension. There is an internal force between A and B equal to F, and the length of the bar will increase. In the bar shown in Fig. 4.2(b), the internal force between A and B is zero, and the length of the bar will remain unchanged. Thus, we see that the principle of transmissibility cannot be applied, except with care, to problems involving internal forces and deformations.

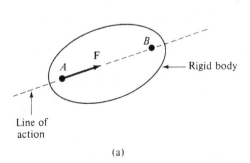

(a)

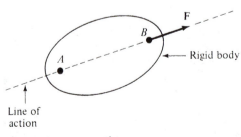

(b)

FIGURE 4.1

 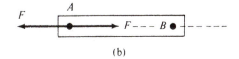

FIGURE 4.2

4.3 MOMENT OF A FORCE

The tendency of a force to produce rotation about an axis is called the *moment* of a force about an axis. Consider a wrench that is applied to a nut on a bolt as shown in Fig. 4.3(a). To obtain the maximum rotation or turning effect on the nut, we know from common experience that the force should be applied perpendicular to the handle as far away from the axis through the center of the bolt at *O* as possible. The following definition is consistent with common experience: *The moment of a force about an axis is defined as the product of the force and the perpendicular distance from the line of action of the force to that axis.* With forces in a plane, the moment of a force about a point is understood to mean the moment about an axis perpendicular to the plane at that point. We define the magnitude of the moment of a force in equation form with the aid of Fig. 4.3(b). The perpendicular distance from the line of action of *F* to point *O* is shown as *BO*. Let *BO* be equal to *d*. Therefore, the moment of the force *F* about *O* is equal to the product of *F* and *d*. In equation form,

$$M_O = Fd \qquad (4.1)$$

The point *O* is the moment center, and the distance *d* is the arm of the force. A moment is also referred to as a *torque*, and the two terms are used interchangeably. (In fact, it is customary to describe the amount of "twist" applied to a threaded fastener as a torque rather than a moment. This is also true when expressing the rotational characteristics of revolving shafts, such as those found on electric motors, internal combustion engines, pumps, generators, and a variety of tools.) In U.S. customary units, moments or torques may be expressed in lb-ft, kip-ft, or lb-in., and in SI units in N · m, kN · m, or N · mm.

From the general definition given by Eq. (4.1), you should note several details related to the moment of a force:

1. The magnitude of a moment may be changed by varying either force *F* or perpendicular distance *d*. In the previous wrench example, for instance, to increase the moment, we can push harder on the wrench, make optimum use of distance *OA* by changing the direction of *F* so that it is perpendicular to the wrench handle, or use a wrench that has a longer handle.

2. The moment of a force about any point has a *direction* (clockwise or counterclockwise) associated with it. Direction is generally indicated by labeling the magnitude of a moment using a curved arrow or through the appropriate use of positive and negative signs.

3. A given force can have different moments (magnitude, direction, or both) about different points. For this reason, moment of a force is always specified relative to a particular point. The point chosen can be anywhere on, *or off*, an actual object.

4. If the line of action of a force passes through a point, then *d* = 0, and the force has no moment about that point. In the previous example, this would be equivalent to throwing away the wrench and trying to tighten the nut by pushing directly on it.

4.4 THEOREM OF MOMENTS

In Fig. 4.3(c), we resolve the force **F** into *x* and *y* components. The component F_x has no moment about *O* because it acts directly through *O*. Thus, the sum of the moments about *O* of the components of the force are given by

$$M_O = F_y r \qquad (a)$$

From the construction, we see that $d = r \sin\theta$ or $r = d/\sin\theta$ and $F_y = F\sin\theta$. Substituting into Eq. (a), we have

$$M_O = F_y r = F\sin\theta \frac{d}{\sin\theta} = Fd$$

The result agrees with the definition of the moment of the force about *O* [Eq. (4.1)]. Thus, the moment of the force **F**

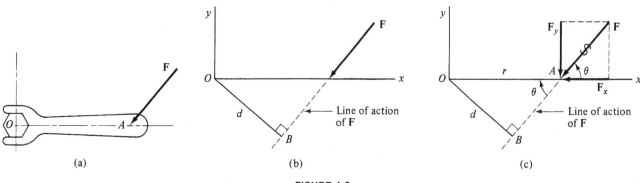

FIGURE 4.3

TORQUE MEASUREMENT AND APPLICATION

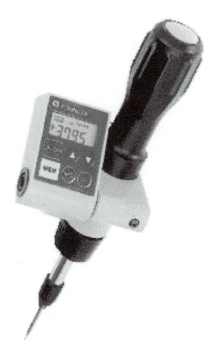

SB.3(a)

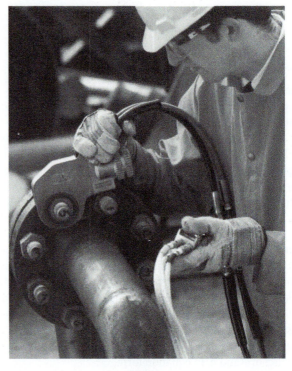

SB.3(b)

For connections that use threaded fasteners, one important consideration is "how tight" the fastener should be. In most instances, the torque applied to a screw, nut, or bolt can provide a quantitative measure of the force developed within the fastener, and thus within the connection itself. For this reason, minimum values of torque are often specified for threaded fasteners to maintain the integrity of a connection or to ensure a desired clamping force between surfaces in contact. A variety of specialty tools are available that measure the exact amount of "twist" applied to a fastener. This screwdriver with digital readout [SB.3(a)] can measure torques as low as 14 oz-in. (0.0729 lb-ft) and is well suited to assembly and production applications. *(Courtesy of Tohnichi America Corp., Northbrook, Illinois, www.tohnichi.com)* The hydraulically driven torque wrench shown here [SB.3(b)] can produce torques as high as 175,000 lb-ft and is able to operate in restricted spaces. *(Courtesy of Hydratight Sweeney, a Dover Company, Englewood, Colorado, www.hydratightsweeney.com)*

about O is equal to the sum of the moments of the components of **F** about O. Our discussion here represents the proof of a special case of the *theorem of moments*, also known as *Varignon's theorem*. The theorem of moments states that *the moment of a force about a point is equal to the sum of the moments of the components of that force about the same point.*

The following examples will be used to illustrate methods for finding moments of forces and application of the theorem of moments.

Example 4.1   Find the moment of the 120-lb force at *C* about points *A*, *B*, *C*, and *D* for the rigid body in Fig. 4.4(a).

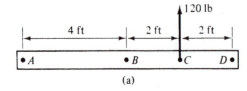

(a)

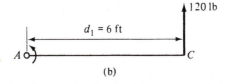

(b)

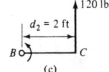

(c)

(d)

FIGURE 4.4

Solution: To help us learn the method for finding the moment of a force, we will draw a sketch that shows the *force,* the *moment center,* and the *arm* of the force.

Moment about A: The force tends to produce rotation about A in the counterclockwise direction, Fig. 4.4(b). By convention, the *counterclockwise* direction will be *positive.* The moment about A is given by

$$M_A = Fd_1 = (120)(6) = 720\,\text{lb-ft}$$
$$= 720\,\text{lb-ft}\,\circlearrowleft \qquad\qquad \textbf{Answer}$$

Moment about B: As shown in Fig. 4.4(c), the force tends to produce counterclockwise rotation about B; thus, the moment about B is

$$M_B = Fd_2 = (120)(2) = 240\,\text{lb-ft}$$
$$= 240\,\text{lb-ft}\,\circlearrowleft \qquad\qquad \textbf{Answer}$$

Moment about C: The moment of the force about C will be zero because the force acts *directly* through C and can have no moment about C.

$$M_C = 0 \qquad\qquad \textbf{Answer}$$

Moment about D: The force tends to produce clockwise rotation about D [Fig. 4.4(d)]. By convention, *clockwise* rotation will be *negative.* The moment about D is given by

$$M_D = Fd_3 = -(120)(2) = -240\,\text{lb-ft}$$
$$= 240\,\text{lb-ft}\,\circlearrowright \qquad\qquad \textbf{Answer}$$

Example 4.2   Determine the moment about C due to the force **F** of magnitude 100 N applied at point A to the bracket shown in Fig. 4.5(a).

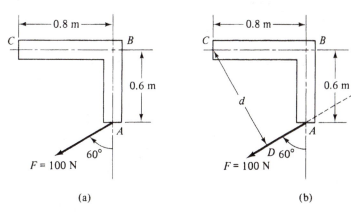

(a)   (b)

FIGURE 4.5

Solution 1:   We construct the diagram shown in Fig. 4.5(b) to determine the arm of the force, d, that is, the perpendicular distance from the force to the moment center at point C. From the right triangle ABE,

$$\tan 30° = \frac{AB}{BE} = \frac{0.6}{BE}$$

$$BE = \frac{0.6}{\tan 30°} = 1.039\,\text{m}$$

From the right triangle CDE,

$$\sin 30° = \frac{d}{CE} = \frac{d}{1.039 + 0.8}$$

$$d = 1.839 \sin 30° = 0.9195\,\text{m}$$

Therefore, because a clockwise rotation by convention is negative,

$$M_C = Fd = -100(0.920) = -92.0\,\text{N·m}$$
$$= 92\,\text{N·m}\,\circlearrowright \qquad\qquad \textbf{Answer}$$

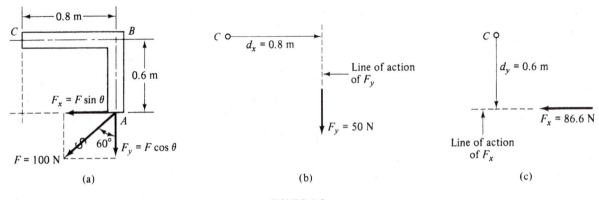

FIGURE 4.6

Solution 2:   We find the x and y components of the 100-N force and use the theorem of moments to find the moment about C.

The x and y components of the force in Fig. 4.6(a) are given by $F_x = F\sin\theta$ and $F_y = F\cos\theta$; therefore,

$$F_x = 100\sin 60° = 86.6\,\text{N}$$
$$F_y = 100\cos 60° = 50.0\,\text{N}$$

To assist us in learning the method, we draw a sketch of each component of the force together with its arm and moment center C [Fig. 4.6(b) and (c)].

The moment about C must be equal to the algebraic sum of the moments of the components about C. Both components produce clockwise or negative rotation; therefore,

$$M_C = -50(0.8) - 86.6(0.6) = -91.96\,\text{N·m}$$
$$= 92\,\text{N·m}\ \circlearrowright \qquad\qquad\text{**Answer**}$$

Solution 3:   One common variation of the theorem of moments uses a different set of rectangular components than the horizontal and vertical pair used in Solution 2. In Fig. 4.7(a), we may draw a line from point C to point A directly. The length of this line and the resulting angle formed at C are 1.0 m and 36.9°, respectively. Then, the angle between force F and line CA becomes 66.9°, and the rectangular components *along* line CA and *perpendicular* to CA are 39.3 N and 92.0 N as shown in Fig. 4.7(b). The 39.3 N component passes through point C and so has no moment about that point; the 92.0 N component has CA as its arm and so creates the moment about C:

$$M_C = 92.0\,\text{N} \times 1.0\,\text{m} = 92\,\text{N·m}\ \circlearrowright \qquad\qquad\text{**Answer**}$$

As expected, the answers obtained by all three solutions are the same. Unless the arm of a force can be obtained by inspection, it is usually simpler to solve the problem by the theorem of moments.

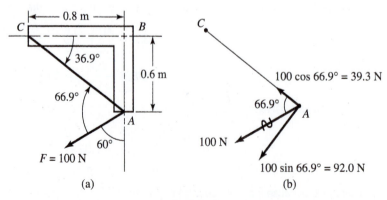

FIGURE 4.7

Example 4.3   Determine the moment of the three forces that act on the truss shown in Fig. 4.8(a) about point A and point D.

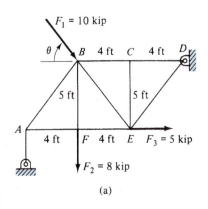

(a)

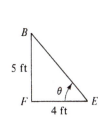

(b)

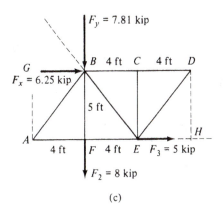

(c)

FIGURE 4.8

Solution: To find the x and y components of the force $F_1 = 10$ kip, we consider the right triangle *BFE* [Fig. 4.8(b)]. From the triangle,

$$\tan \theta = 5/4 \qquad \theta = 51.34°$$

The slope of $F_1 = 10$ kip is the same as member *BE* of the truss, and the x and y components of the force are given by

$$F_x = 10 \cos 51.34° = 6.25 \text{ kip}$$
$$F_y = 10 \sin 51.34° = 7.81 \text{ kip}$$

We draw the truss showing $\mathbf{F_2}$, $\mathbf{F_3}$, and the components of $\mathbf{F_1}$ in Fig. 4.8(c).

Moments about A: We see from Fig. 4.8(c) that the arm of F_x is *GA*, the arm of F_y and F_2 is *FA*, and the arm of F_3 is zero. Therefore, the algebraic sum of the moments about A (counterclockwise positive) is

$$\Sigma M_A = -F_x(GA) - F_y(FA) - F_2(FA) + F_3(0)$$
$$= -6.25(5) - 7.81(4) - 8(4) = -94.49$$
$$= 94.5 \text{ kip-ft } \circlearrowleft \qquad\qquad \textbf{Answer}$$

Moments about D: The arm of F_x is zero, the arm of F_y and F_2 is *BD*, and the arm of F_3 is *HD* [Fig. 4.8(c)]. Therefore, the algebraic sum of the moments about D (counterclockwise positive) is

$$\Sigma M_D = +F_x(0) + F_y(BD) + F_2(BD) + F_3(HD)$$
$$= 7.81(8) + 8(8) + 5(5) = 151.48$$
$$= 151.5 \text{ kip-ft } \circlearrowright \qquad\qquad \textbf{Answer}$$

PROBLEMS

4.1 Determine the moment of the forces about points A, B, C, and D for the beam shown.

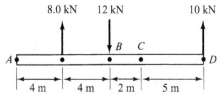

PROB. 4.1

4.2 For the parallel-force system shown, determine the moment of the forces about points O, P, Q, and R.

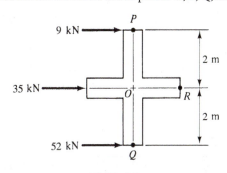

PROB. 4.2

4.3 Find the moment of the 65 lb force about point A:
   a. Using Eq. (4.1) and moment arm AB
   b. Using Varignon's theorem with the horizontal and vertical components of the 65 lb force
   c. Using the theorem of moments with rectangular force components along, and perpendicular to, a line drawn from point A to point C

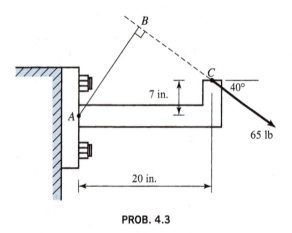

PROB. 4.3

4.4 Determine the moment of the forces about points A, B, and C for the cantilever beam shown.

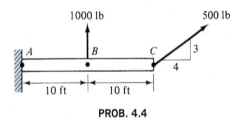

PROB. 4.4

4.5 Determine the moment of the forces shown about points D, E, F, and G.

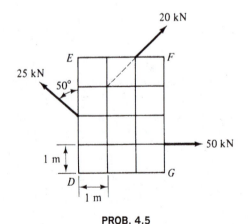

PROB. 4.5

4.6 Determine the moment of the forces shown about points P, Q, R, and S.

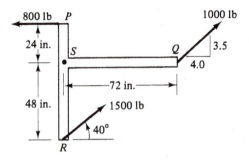

PROB. 4.6

4.7 For the circular disk shown, determine the moment of the forces about points L, M, N, and O.

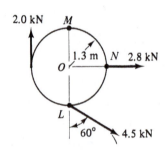

PROB. 4.7

4.5 RESULTANT OF PARALLEL FORCES

The resultant of a parallel system of forces must have the same or equivalent effect as the parallel system. That is, the resultant force must be equal to the sum of the parallel forces, and the sum of the moments of the forces about any point must be equal to the moment of the resultant force about that same point.

Consider the system of parallel forces F_1, F_2, and F_3 with moment arms of r_1, r_2, and r_3 measured from O in Fig. 4.9(a). The resultant of this force system must produce the same effect as the original force system. To produce the same effect, the sum of the forces must be equal to the resultant force, and the sum of the moments of the forces about a point must be equal to the moment of the resultant about that same point. Thus, the force system in Fig. 4.9(b) is the resultant of the force system in Fig. 4.9(a) if

$$\Sigma F = F_1 + F_2 + F_3 = R \tag{a}$$

and $$\Sigma M_O = F_1 r_1 + F_2 r_2 + F_3 r_3 = R\bar{r}$$

or $$\bar{r} = \frac{F_1 r_1 + F_2 r_2 + F_3 r_3}{F_1 + F_2 + F_3} \tag{b}$$

The magnitude of the resultant R is given by Eq. (a), and the location of the resultant from point O is given by Eq. (b).

Not all parallel-force systems have a simple force as the resultant. A system of two parallel forces of equal magnitude acting in opposite directions with different lines of action is called a *couple*. The couple does not have a single force as a resultant. The couple is discussed in Sec. 4.7.

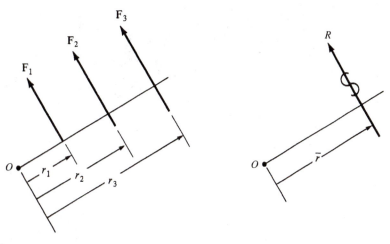

(a) Parallel-force system   (b) Resultant or equivalent force

FIGURE 4.9

Example 4.4   For the parallel-force system acting on the rod in Fig. 4.10(a), determine the resultant or equivalent force.

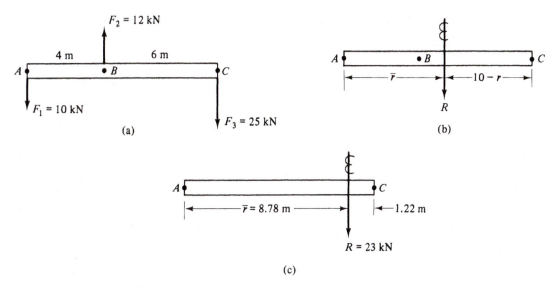

FIGURE 4.10

Solution: The resultant or equivalent force acting on the rod is shown in Fig. 4.10(b). To be equivalent,

$$R = \Sigma F_y = -F_1 + F_2 - F_3 = -10 + 12 - 25 = -23 \text{ kN}$$

or $R = 23 \text{ kN} \downarrow$

and $\Sigma M_A = F_2(\overline{AB}) - F_3(\overline{AC}) = -R\bar{r}$

$$12(4) - 25(10) = -23\bar{r}$$

or $\bar{r} = \dfrac{-202}{-23} = 8.78 \text{ m}$

We check the results by calculating moments with respect to C.

$$\Sigma M_C = R(10 - \bar{r})$$

$$10(10) - 12(6) = 23(10 - 8.78)$$

$$28 = 28.06 \quad \text{OK}$$

The magnitude and location of the resultant is shown in Fig. 4.10(c).

PROBLEMS

4.8 For the wheel loads shown, determine the resultant. Locate the resultant with respect to A and B.

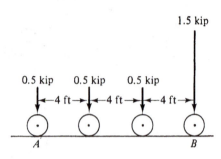

PROB. 4.8

4.9 Determine the resultant of the parallel-force system shown. Locate the resultant with respect to A and D.

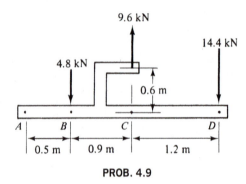

PROB. 4.9

4.10 The pin-connected truss is acted on by forces as shown. Determine the resultant and locate with respect to points A and B.

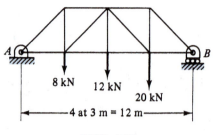

PROB. 4.10

4.11 Determine the resultant of the parallel-force system shown. Locate the resultant with respect to points A and D.

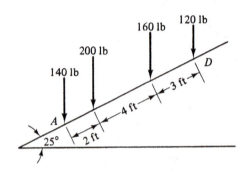

PROB. 4.11

4.6 RESULTANT OF NONPARALLEL FORCES

The resultant of a nonparallel system of forces is found in essentially the same way as the resultant of a parallel-force system. By resolving each force into rectangular components, we have two sets of parallel forces at right angles to each other. We illustrate the procedure in the following example.

Example 4.5 Determine the resultant of the force system shown in Fig. 4.11(a). Locate the resultant with respect to point A.

Solution: The x and y components of the forces are calculated as follows:

$$F_{x_1} = -F_1 \cos 60° = -12(0.5000) = -6.0 \text{ kip}$$
$$F_{x_2} = +F_2 \cos 45° = +20(0.7071) = 14.14 \text{ kip}$$
$$F_{x_3} = +F_3 \cos 15° = +25(0.9659) = 24.15 \text{ kip}$$

and

$$F_{y_1} = +F_1 \sin 60° = +12(0.8660) = 10.39 \text{ kip}$$
$$F_{y_2} = +F_2 \sin 45° = +20(0.7071) = 14.14 \text{ kip}$$
$$F_{y_3} = -F_3 \sin 15° = -25(0.2588) = -6.47 \text{ kip}$$

In the free-body diagram [Fig. 4.11(b)], each force has been replaced by its x and y components. Adding forces in the x and y directions, we have

$$\Sigma F_x = R_x = -6.0 + 14.14 + 24.15 = 32.29 \text{ kip}$$
$$\Sigma F_y = R_y = 10.39 + 14.14 - 6.47 = 18.06 \text{ kip}$$

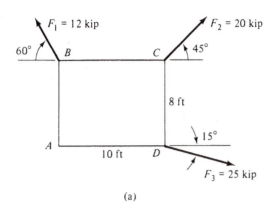

(a)

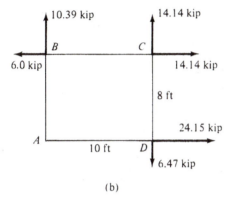

(b)

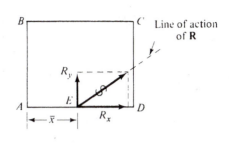

(c) Resultant or
equivalent force

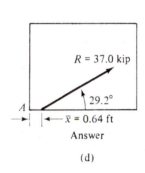

Answer

(d)

FIGURE 4.11

The resultant is given by

$$R^2 = R_x^2 + R_y^2 = (32.29)^2 + (18.06)^2 = 1369$$

$$R = 37.0 \, \text{kip}$$

$$\tan\theta = \frac{18.06}{32.29} = 0.5593 \qquad \theta = 29.2°$$

$$\mathbf{R} = 37.0 \, \text{kip} \, \angle 29.2° \qquad\qquad \textbf{Answer}$$

Therefore, the resultant is directed upward and to the right. To find the position of the resultant with respect to A, we assume that the line of action of **R** is as shown in Fig. 4.11(c). From the principle of transmissibility, the resultant **R** can be moved to any point on its line of action. We select point E for convenience because the x component of the resultant will have no moment about A.

To conclude our solution, we sum the moments about A of the forces shown in Fig. 4.11(b) and equate them to the moments about A of the components of the resultant as shown in Fig. 4.10(c). Thus,

$$\Sigma M_A = R_y \bar{x} + R_x(0)$$

$$\Sigma M_A = 6(8) + 14.14(10) - 14.14(8) - 6.47(10)$$

$$= 11.58 \, \text{kip-ft}$$

$$R_y \bar{x} = 18.06 \bar{x}$$

$$18.06 \bar{x} = 11.58$$

$$\bar{x} = \frac{11.58}{18.06} = 0.641 \, \text{ft}$$

$$\mathbf{R} = 37.0 \, \text{kip} \, \angle 29.2°$$

Located 0.64 ft to the right of A   **Answer**

The answer is shown in Fig. 4.11(d).

PROBLEMS

4.12 A cantilever truss is acted on by forces as shown. Find the resultant and locate the resultant with respect to A and C.

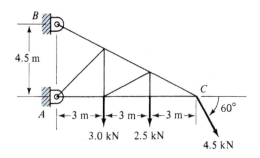

PROB. 4.12

4.13 Determine the resultant of the forces acting on the hook-shaped member shown. Locate the resultant with respect to B and C.

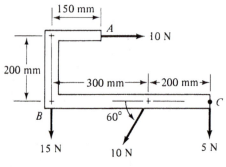

PROB. 4.13

4.14 Determine the resultant of the force system shown. Locate the resultant with respect to points A and D.

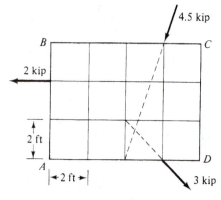

PROB. 4.14

4.15 A pin-connected truss is acted on by forces as shown. Find the resultant and locate with respect to points A and D.

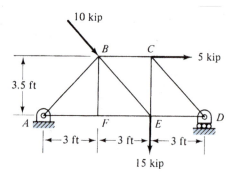

PROB. 4.15

4.16 Determine the resultant of the forces acting on the pin-connected tower shown. Locate the resultant with respect to the right side of the tower—FJ.

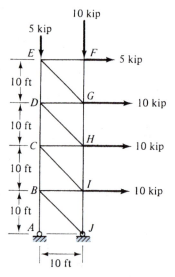

PROB. 4.16

4.17 For the force system acting on the member shown, find the resultant and locate with respect to M and N.

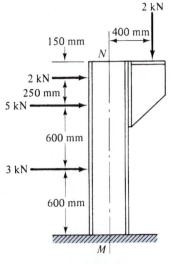

PROB. 4.17

4.7 MOMENT OF A COUPLE

We consider a special parallel-force system consisting of two parallel forces of equal magnitude acting in opposite directions, as shown in Fig. 4.12. Such a system is called a *couple*. The plane in which the forces act is called the plane of the couple, and the perpendicular distance d, between the lines of action of the forces, is called the *arm of the couple*.

Adding the forces of the couple to find the resultant force, we have $R = F - F = 0$. Thus, the resultant force is equal to zero. Next, we sum the moments of the forces about a point O to find the moment of the couple about O.

$$\Sigma M_O = -Fx + F(x + d) = Fd \,\circlearrowright \qquad (4.2)$$

Therefore, we see that the moment of a couple depends only on the forces in the couple and the arm of the couple. That is, the moment of a couple is the same about any point in the plane of the couple. A couple produces a pure moment—a turning effect only. In Fig. 4.13, we show several equivalent

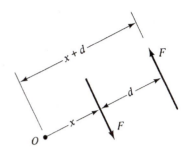

FIGURE 4.12

couples. The couples are equivalent because their moments are equal. From Fig. 4.13(a)–(c),

$$M = Fd = 15(6) = 12(7.5) = 10(9) = 90\,\text{lb-ft}$$
$$= 90\,\text{lb-ft.} \,\circlearrowright$$

To add two or more couples, we simply calculate the algebraic sum of their moments.

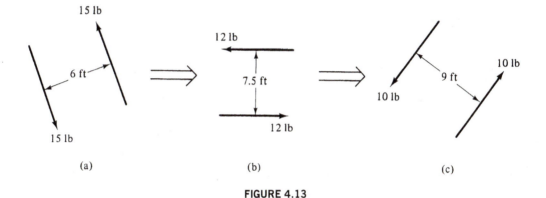

(a)   (b)   (c)

FIGURE 4.13

Example 4.6   Find the resultant moment of the couple that acts on the plate shown in Fig. 4.14(a).

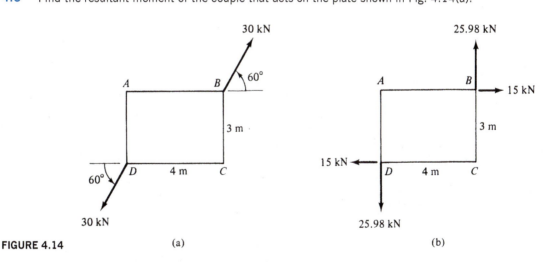

FIGURE 4.14   (a)   (b)

Solution: The x and y components of the 30-kN forces are calculated and displayed on the free-body diagram of the plate shown in Fig. 4.14(b). The 15-kN forces have an arm of 3 m and form a clockwise couple. The 25.98-kN forces have an arm of 4 m and form a counterclockwise couple. Therefore,

$$M = 25.98(4) - 15(3) = 58.92\,\text{kN·m}$$
$$= 58.9\,\text{kN·m} \,\circlearrowright$$ **Answer**

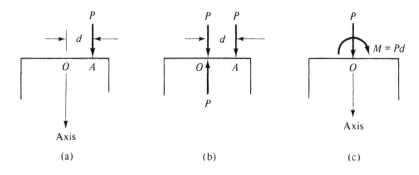

FIGURE 4.15

4.8 RESOLUTION OF A FORCE INTO A FORCE AND COUPLE

Consider a force **P** acting on a body at point A, which is located a distance d away from point O, as shown in Fig. 4.15(a). In certain physical applications, it is convenient to replace the force **P** at A by a force **P** at O and a couple. The replacement or resolution may be described as follows. Introduce two forces **P**

and −**P** at O [Fig. 4.15(b)]. The force system is unchanged because the two forces at O add to zero. However, the force directed up at O and the force directed down at A form a couple with the clockwise moment $M = Pd$. Thus, we are left with a force downward at O and a clockwise couple with moment equal to Pd, as shown in Fig. 4.15(c). The resolution is complete. We have replaced the force at A by a force at O and a couple.

Example 4.7   Replace the force at B by a force at C and a couple [Fig. 4.16(a)].

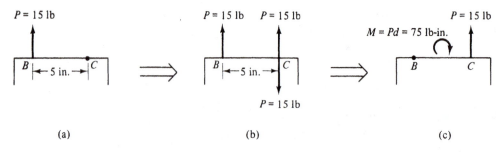

FIGURE 4.16

Solution: We introduce two 15-lb forces at C, one up and the other down, as shown in Fig. 4.16(b). The force up at B and down at C form a clockwise couple with a moment $M = Pd = 15(5) = 75$ lb-in. The resolution is shown in Fig. 4.16(c), where $M = 75$ lb-in. ↺ and $P_{atC} = 15$ lb ↑.

PROBLEMS

4.18 A couple acts at A as shown. Replace the couple by (a) two vertical forces at B and C, and (b) two horizontal forces at D and E.

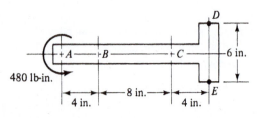

PROB. 4.18

4.19 The three couples shown are equivalent. Determine (a) the distance d for the second couple and (b) the force F for the third couple.

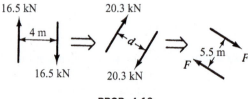

PROB. 4.19

4.20 Find the resultant moment for the couples that act on the plate shown.

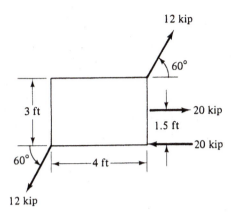

12 kip

60°

3 ft

20 kip

1.5 ft

20 kip

60°

4 ft

12 kip

PROB. 4.20

4.21 Replace the force at C by (a) a force at A and a couple, and (b) a force at B and a couple.

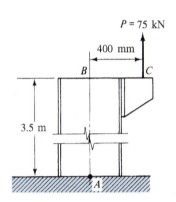

$P = 75$ kN

400 mm

B C

3.5 m

A

PROB. 4.21

4.22 Replace the force at A by (a) a force at B and a couple, and (b) a force at C and a couple.

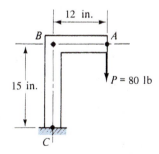

12 in.

B A

15 in.

$P = 80$ lb

C

PROB. 4.22

4.23 The bracket is attached to a plate by two bolts as shown. (a) Replace the force at D by a force at C and a couple. (b) Replace the couple from part (a) by two vertical forces at A and B.

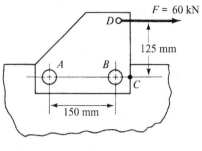

$F = 60$ kN

D

125 mm

A B

C

150 mm

PROB. 4.23

4.24 The disk is acted on by forces as shown. Replace the forces by a force at O and a couple. (*Hint:* Replace each force by a force at O and a couple. Add the forces at O and the couples to find a single force at O and a single couple.)

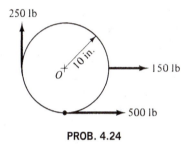

250 lb

O 10 in.

150 lb

500 lb

PROB. 4.24

4.25 The grid is acted on by forces as shown. Replace the forces by a force at A and a couple. (*Hint:* Replace each force by a force at A and a couple. Add the forces at A and the couples to find a single force at A and a single couple.)

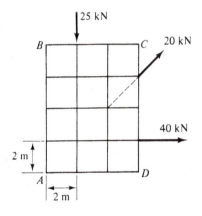

25 kN

B C 20 kN

40 kN

2 m

A D

2 m

PROB. 4.25

4.26 A section cut from an offset link has dimensions and a load as shown. Replace the force at A by a force at B and a couple. (*Hint:* Find the components of the load. Replace each component by a force at B and a couple. Add the forces and couples to find a single force at B and a single couple.)

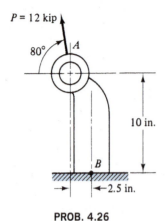

P = 12 kip

80°   A

10 in.

B

2.5 in.

PROB. 4.26

4.27 For the beam column shown, replace the forces by a resultant force at M and a couple. (*Hint:* Replace each force by a force at M and a couple. Add the forces and couples to find a single force at M and a single couple.)

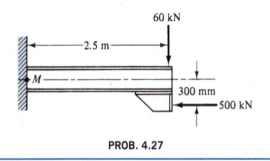

60 kN

2.5 m

M

300 mm

500 kN

PROB. 4.27

4.9 RESULTANT OF DISTRIBUTED LOADING

In many practical problems, the surface or volume of a body may be subjected to distributed loading. The loading may be the result of wind, water, the weight of the material supported by the body, or the weight of the body itself.

Consider a beam supporting a distributed load. In Fig. 4.17(a), we show the beam with the load curve plotted over the length of the beam. Each unit length of the beam supports a load w, which may be expressed in N/m or lb/ft. The area of a typical narrow strip of width Δx under the load curve is $w\Delta x$ and is equal to ΔF, which has units of force. We can consider each narrow strip under the load curve as representing a series of parallel loads. The resultant and the location of the resultant may be found by the methods of integral calculus and Sec. 4.5. However, we see that the resultant concentrated load is equal to the total area under the load curve. In addition, we can say, without proof, that the line of action of the resultant passes through the geometric center or centroid of that area [Fig. 4.17(b)]. Centroids are discussed in Chapter 9. The resultant of a distributed load can be used to find external reactions for the beam but cannot be used to find internal reactions or deflections.

We will consider two kinds of distributed loads: the *uniform load* and the *triangular load.*

Resultant of a Uniform Load

A uniform load may be represented by a rectangle. The height of the rectangle represents the load per unit of length. It may be expressed in lb/ft or N/m. The length of the rectangle is the distance over which the load acts and may be expressed in feet or meters. The resultant of a uniformly distributed load is equal in magnitude to the area of that rectangle, that is, $R = wL$, and it acts at the geometric center in the middle of the rectangle a distance from either end of $L/2$. See Fig. 4.18(a).

Resultant of a Triangular Load

A triangular load varies uniformly from zero on one end to a maximum on the other end. The height represents the load per unit of length, and the length is the distance over which the load acts. The resultant of a triangular load is equal in magnitude to the area of the triangle, $R = wL/2$, and it acts at the geometric center of the triangle a distance $2L/3$ from the pointed end of the triangle and a distance of $L/3$ from the flat end of the triangle, as shown in Fig. 4.18(b).

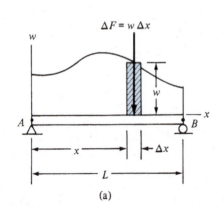

$\Delta F = w\,\Delta x$

w

w

A

B

x

x

Δx

L

(a)

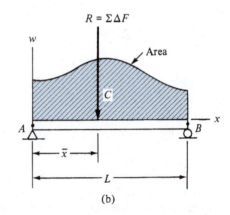

$R = \Sigma \Delta F$

w

Area

C

A

B

x

$\bar{x}$

L

(b)

FIGURE 4.17

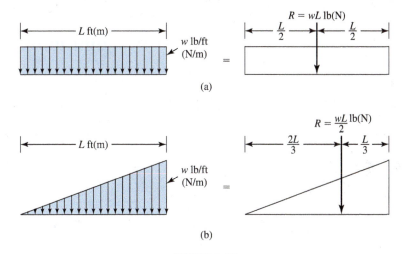

FIGURE 4.18

Example 4.8   Determine the resultant of the distributed load shown in Fig. 4.19(a).

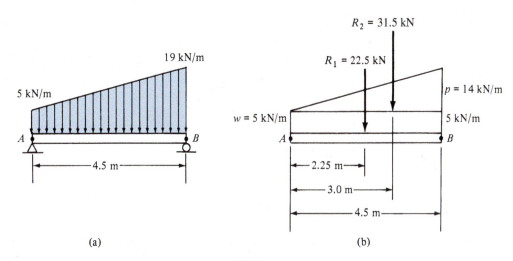

(a)   (b)

FIGURE 4.19

Solution: The distributed load is divided in Fig. 4.19(b) into a uniform and a triangular load. The uniform load has an area $wL = (5\,\text{kN/m})(4.5\,\text{m}) = 22.5\,\text{kN}$ and acts at the middle of the rectangle a distance $L/2 = (4.5\,\text{m})/2 = 2.25\,\text{m}$ from A. The triangular load has an area $wL/2 = (14\,\text{kN/m})(4.5\,\text{m})/2 = 31.5\,\text{kN}$ and acts a distance $2L/3 = 2(4.5\,\text{m})/3 = 3.0\,\text{m}$ from A. The resultant of the two parallel forces is found by the method of Sec. 4.5. Therefore, the resultant force

$$R = \Sigma F = 22.5 + 31.5 = 54.0\,\text{kN}$$

and $$\Sigma M_A = 22.5(2.25) + 31.5(3.0) = 145.1\,\text{kN·m} = R\bar{r}$$

or $$\bar{r} = \frac{145.1\,\text{kN·m}}{54.0\,\text{kN}} = 2.69\,\text{m}$$

The resultant of 54.0 kN acts downward 2.69 m to the right of A.   **Answer**

PROBLEMS

4.28 through 4.33 (a) Determine the resultant of the force system shown. Locate the resultant with respect to point A. (b) Replace the resultant force by a force at A and a couple.

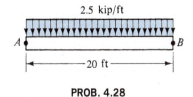

PROB. 4.28

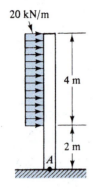

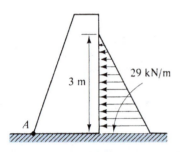

PROB. 4.29

PROB. 4.30

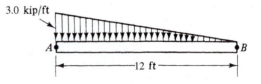

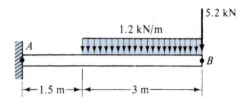

PROB. 4.31

PROB. 4.32

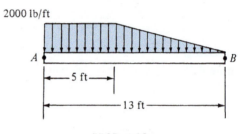

PROB. 4.33

EQUILIBRIUM OF A RIGID BODY

CHAPTER OBJECTIVES

In this chapter, we examine the conditions required for
equilibrium of a rigid body. Before proceeding to
Chapter 6, you should be able to

- Draw complete and accurate free-body diagrams for
  rigid bodies that are constrained by various types of
  supports.

- Write and solve the equations of equilibrium for a
  rigid body that is acted on by one or more external
  loads.

- Describe the special conditions of equilibrium for
  two- and three-force bodies.

- Identify the conditions under which rigid bodies are
  classified as constrained or partially constrained,
  statically determinate or statically indeterminate.

5.1 INTRODUCTION

A particle is in *equilibrium* if the resultant force acting on the
particle is equal to zero. In the case of a rigid body, equilibrium requires that both the resultant force and the resultant
moment acting on the body be equal to zero. The rigid body,
such as a structure or machine, may consist of a single member *AB* [Fig. 5.1(a)] or several members such as *CD* and *DE*
joined together to form a single system [Fig. 5.1(b)].

    Forces and moments may be classified as *external* or
internal. Forces that hold a member or members together are

internal and may occur either at a hinge such as *D* or between any two parts of a single member. External forces can
be either *applied* or *reactive*. Applied forces act directly on
the member, such as F_1, F_2, and F_3, and the weight of the
members, W_1, W_2, and W_3. Reactive forces are those produced by the members' supports, R_A, R_B, R_C, and R_E. Applied
forces or loads are usually known quantities, whereas reactive forces or reactions are unknown quantities.

5.2 SUPPORT CONDITIONS FOR BODIES IN A PLANE

The supports develop reactions in response to the weight
of the body and to loads (external forces or moments) that
are applied to the body. They prevent the body from moving. That is, the body is in equilibrium under the action of
the loads and reactions. There are several types of supports
for bodies loaded by forces *acting in a plane*. They may be
classified by the kind of resistance they offer to the forces,
as follows.

One Reactive Force

These supports can prevent translational motion along a
specific line of action. They include smooth surfaces, rollers,
cables, and links. The smooth surface and cable provide resistance in one direction only, whereas the roller and link
provide resistance in either direction along the same line of
action.

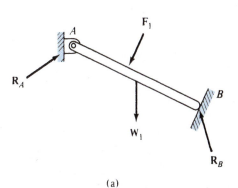

(a)

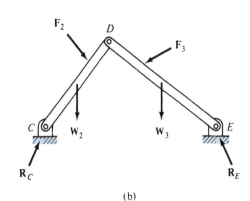

(b)

FIGURE 5.1

Two Reactive Forces

These supports prevent translational motion in a plane, that is, motion in two perpendicular directions. They include smooth pins or hinges and rough surfaces. The smooth pin provides resistance in any direction in a plane, whereas the rough surface provides resistance normal to the surface and tangent to the surface.

Three Reactive Forces

These supports prevent translational motion in two directions, as well as rotational motion. Such a support is called fixed, built in, or clamped. The actual support is produced by building the body into a wall, casting the body as part of a larger body, or welding or mechanically attaching the body to a larger body. These supports are shown in Fig. 5.2, together with the number and kind of unknown reactions.

5.3 CONSTRUCTION OF FREE-BODY DIAGRAMS

Here, we consider various physical problems as described by space diagrams. In each problem, one or more of the supports shown in Fig. 5.2 will be used. A free body will be isolated, and all the loads and reactions acting on the body will be shown.

FIGURE 5.2

Type of Support	Reaction Force	Number and Kind of Unknown Reaction
 Smooth surface	 $\uparrow$ F	One unknown—magnitude of force **F**. Force normal to surface and toward body.
 Roller	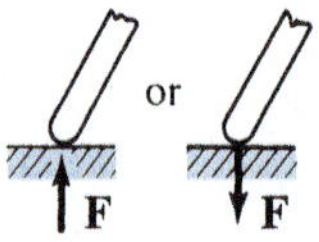 $\uparrow$ F or $\downarrow$ F	One unknown—magnitude of force **F**. Force normal to surface and directed toward body or away from body.
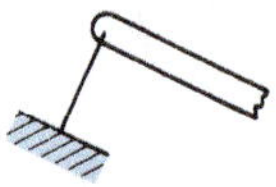 Cable	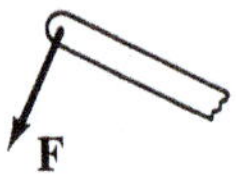 F	One unknown—magnitude of force **F**. Force acts in the direction of the cable away from the body.
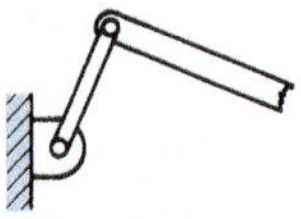 Weightless link	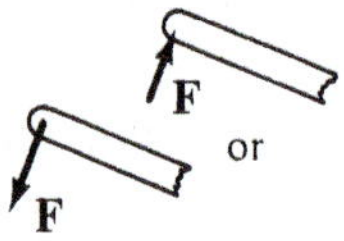 F or F	One unknown—magnitude of force **F**. Force acts in the direction of the link directed either toward or away from body.
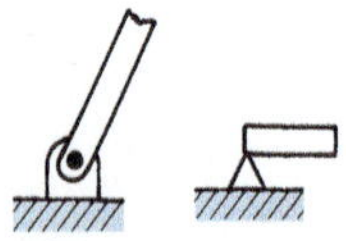 Smooth pin or hinge	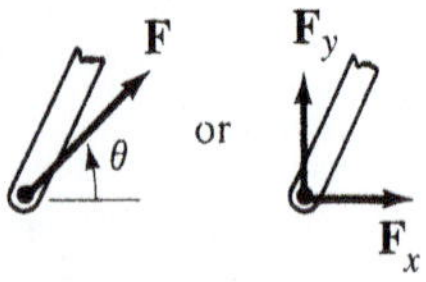 F θ or F_y F_x	Two unknowns—magnitude of the force **F** and direction θ or components of the force F_x and F_y.
 Rough surface	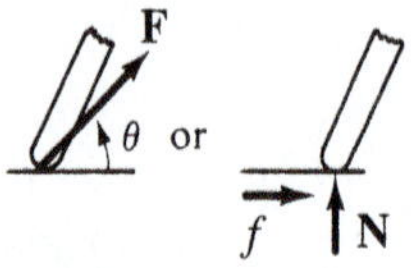 F θ or f $\uparrow$ N	Two unknowns—magnitude and direction of resultant force or the force **N** normal to the surface toward the body and the friction force **f** tangent to surface.
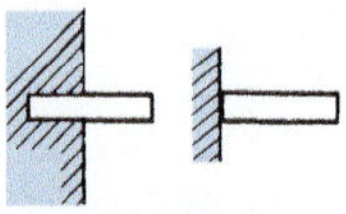 Fixed or clamped	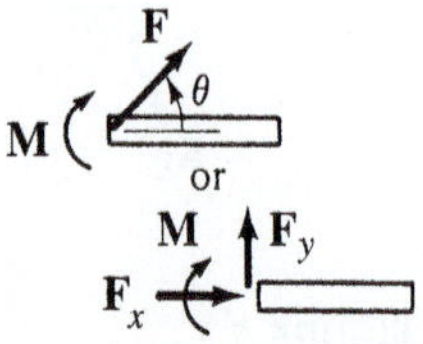 M $\overbrace{}$ F θ or M $\uparrow F_y$ $F_x \leftarrow$	Three unknowns—magnitude of the couple **M** and the direction and magnitude of the resultant force **F** or the magnitude of the couple and the components of the force F_x and F_y.

Example 5.1   Loads **P₁** and **P₂** are applied to a bent that may be supported in four different ways, as shown in Fig. 5.3(a)–(d). Draw the free-body diagram and determine the total number of unknown forces and moments. Neglect the weight of the bent.

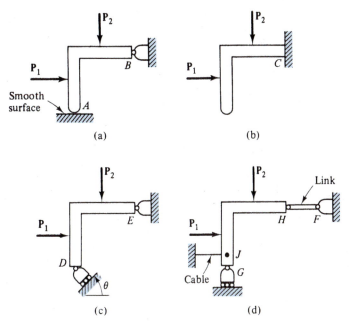

FIGURE 5.3

Solution: The free-body diagrams for the various parts of Fig. 5.3(a)–(d) correspond with the same part of Fig. 5.4(a)–(d). Because the support at *A* is a *smooth surface*, the reaction is perpendicular to the surface and directed toward the bent. The *rollers* at *D* and *G* provide reactions that are perpendicular to the surface on which

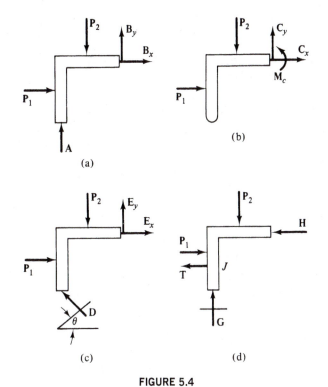

FIGURE 5.4

they roll and are directed toward the bent or away from the bent, as required by equilibrium. The reaction of the *link* at *H* is directed along the link and can be directed toward the bent or away from the bent as required by equilibrium. The *cable* at *J* is in tension, and the reaction is in the direction of the cable and away from the bent. For the *pins* at *B* and *E*, a reaction of unknown magnitude and direction occurs. For convenience, the *x* and *y* components of the reaction are shown. The direction of the components was assumed. Their actual direction will depend on the requirements of equilibrium. The support at *C* is *fixed;* therefore, three reactions occur. For convenience, the components of the reaction and the moment of the couple were assumed in a positive direction. Their actual direction will be found from equilibrium. In each bent, there are three unknown reactions, consisting of either three forces or two forces and a moment couple.

Example 5.2   Draw the free-body diagram for the *uniform* beam shown in Fig. 5.5(a). The total weight of the beam is 100 lb.

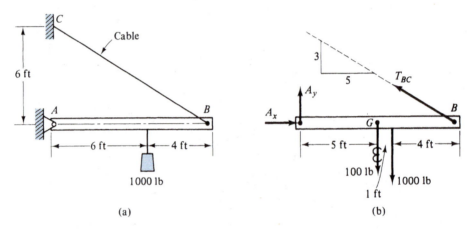

(a)   (b)

FIGURE 5.5

Solution: In Fig. 5.5(b), we draw the free-body diagram of the beam. The 100-lb force represents the weight of the beam. The beam is uniform; therefore, the geometric center and center of gravity *G* are at the same point. The 100-lb weight is in effect a *distributed load* and is indicated in Fig. 5.5 by a superimposed script capital E. The 1000-lb force represents the tension in the cable supporting the 1000-lb weight. The unknown force T_{BC} represents the tension in the cable *BC*, and the two unknown forces A_x and A_y represent the horizontal and vertical components of the pin reaction at *A*. There are three unknown reactions, T_{BC}, A_x, and A_y.

Example 5.3   The beam shown in Fig. 5.6(a) carries a distributed load as shown. Neglect the weight of the beam. Draw the free-body diagram for the beam.

Solution: The free-body diagram for the beam is shown in Fig. 5.6(b). In the diagram, the uniform load was replaced by its resultant

$$F_1 = 300\frac{\text{lb}}{\text{ft}} \times 8\,\text{ft} = 2400\,\text{lb}$$

at the middle of the load 4 ft from *A*. The triangular load was replaced by its resultant

$$F_2 = \frac{1}{2}(300)\frac{\text{lb}}{\text{ft}} \times 12\,\text{ft} = 1800\,\text{lb}$$

at two-thirds of the distance from *C* to *B* or 8 ft. The resultants F_1 and F_2 of the distributed loads are indicated in Fig. 5.6(b) and (c) by superimposed script capital Es. The unknown force F_C at the roller *C* is normal to the surface on which the rollers move and is assumed to be directed toward the beam. The two unknown forces A_x and A_y represent the horizontal and vertical components of the pin reaction at *A*. There are three unknown reactions, F_C, A_x, and A_y, as shown in Fig. 5.6(c). It should be noted that a distributed load can be replaced by its resultant to find the reactions. However, the distributed load still acts over the length of the beam as given.

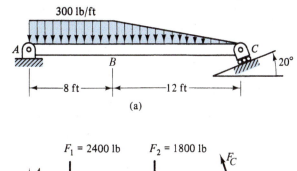

300 lb/ft

A

B

C

20°

8 ft

12 ft

(a)

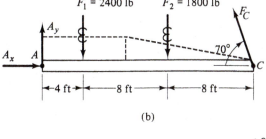

$F_1 = 2400$ lb

$F_2 = 1800$ lb

F_C

A_y

A_x

A

70°

C

4 ft

8 ft

8 ft

(b)

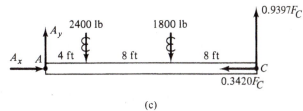

$0.9397F_C$

2400 lb

1800 lb

A_y

A_x

A

4 ft

8 ft

8 ft

C

$0.3420F_C$

(c)

FIGURE 5.6

5.4 EQUATIONS FOR EQUILIBRIUM OF A RIGID BODY

Equilibrium of a rigid body requires that the resultant force and resultant moment acting on the body vanish. After resolving forces into their x and y components, the conditions for equilibrium in a plane can be expressed in terms of three equations:

$$\Sigma F_x = 0, \quad \Sigma F_y = 0, \quad \text{and} \quad \Sigma M_A = 0 \quad (5.1)$$

where A is any point in the plane of the body. These three equations are independent of each other. That is, no two of the equations can be combined to form the third equation. In fact, for equilibrium in two directions, we can write only three independent equations. However, an alternative set of equilibrium equations may be useful. One of the force equations has been replaced by a moment equation:

$$\Sigma F_x = 0, \quad \Sigma M_A = 0, \quad \text{and} \quad \Sigma M_B = 0 \quad (5.2)$$

where A and B are any two points in the plane of the body not on a line perpendicular to the x axis.

In the third possible set of equations, both force equations have been replaced by a moment equation. That is,

$$\Sigma M_A = 0, \quad \Sigma M_B = 0, \quad \text{and} \quad \Sigma M_C = 0 \quad (5.3)$$

where A, B, and C are any three points in the plane of the body not on a straight line.

Any one of the three sets of equations may be used to solve an equilibrium problem. The problem can be simplified, however, if we select equations of equilibrium that result in only one unknown in each equation. Moment equations can frequently be used to eliminate unknowns by summing moments about the point of intersection of the lines of action of two unknown forces. In the problem shown in Fig. 5.6, unknowns A_x and A_y can be eliminated if we sum moments with respect to A. Unknowns A_x and F_C can be eliminated if we sum moments with respect to C. In fact, we can write any number of force and moment equations. However, *only three of the equations can be independent.*

Regardless of the specific combination of equations used, remember that *each equation should be written to agree with your free-body diagram.* If the solution to these equations yields negative values for one or more reaction forces, it simply means that these forces were drawn in the wrong directions on the free-body diagram. This does not affect the validity of your numerical results.

Also, when analyzing a problem in static equilibrium, you should try to develop a sense of the physical effects taking place on an object under load. Most of us have considerable experience with forces and moments in our everyday lives: lifting boxes, moving furniture, changing a flat tire, or helping a friend whose automobile is stuck in a snowbank. Use this knowledge to look beyond the algebraic considerations of trigonometry, vectors, and simultaneous equations; learn to visualize and anticipate the effects of forces on real objects.

Example 5.4   A simple balance beam rests on the smooth support at *B*, as shown in Fig. 5.7(a). A weight is suspended at each end of the beam, but the beam itself is weightless.

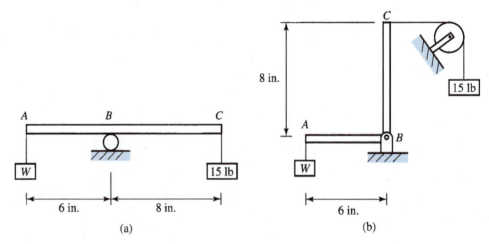

FIGURE 5.7

a.   Draw a free-body diagram of this beam.
b.   Write the necessary equations of equilibrium.
c.   Solve your equations for the unknown weight, *W*, and the reaction provided by the support.
d.   Repeat (a) through (c) if the beam is bent as shown in Fig. 5.7(b) and the cord attached to the 15-lb weight is supported by a frictionless pulley. Note that the support at *B* is now a hinge or pin connection.

Solution: a.   This beam, similar to a playground seesaw or the type of weighing scale found in most laboratories, acts as a *simple lever*. Because *B* is a smooth support, the beam can tip in either direction, slide from side to side, or be lifted off the support completely. The only possible constraining effect that this support can produce on the beam, then, is an upward force, F_B. The complete free-body diagram is shown in Fig. 5.8(a).

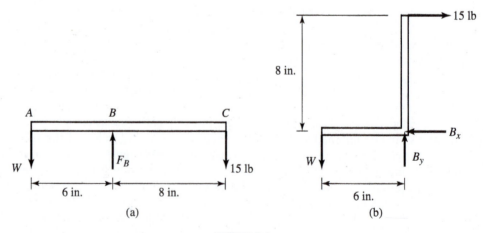

FIGURE 5.8

b.   Whatever the value of *W*, the sum of both weights is ultimately carried by the support. For equilibrium in the vertical direction,

$$\Sigma F_y = F_B - W - 15\,\text{lb} = 0$$

But each weight creates a moment about point *B*, and these clockwise and counterclockwise "tipping" effects must also balance:

$$\Sigma M_B = (W \times 6\,\text{in.}) - (15\,\text{lb} \times 8\,\text{in.}) = 0$$

Note that our moment equation could have been written about any point. By selecting a point (*A* or *B*) through which one of the unknown forces passes, we are able to eliminate that unknown from the moment equation.

c. Solving the moment equation yields

$$W = 20\,\text{lb}$$ **Answer**

Substituting this value into the vertical force equation,

$$F_B = 35\,\text{lb}$$ **Answer**

d. In this configuration, the 15-lb weight exerts a horizontal force on the beam, so a pin connection is required at *B*. Although the beam can still pivot about *B*, it is now constrained from moving either horizontally or vertically. The completed free-body diagram is shown in Fig. 5.8(b). The three equations of equilibrium for this beam are

$$\Sigma F_x = -B_x + 15\,\text{lb} = 0$$
$$\Sigma F_y = +B_y - W = 0$$
$$\Sigma M_B = (W \times 6\,\text{in.}) - (15\,\text{lb} \times 8\,\text{in.}) = 0$$

The moment equation again yields

$$W = 20\,\text{lb}$$ **Answer**

Substituting into the vertical force equation,

$$B_y = 20\,\text{lb}$$ **Answer**

From the horizontal force equation,

$$B_x = 15\,\text{lb}$$ **Answer**

Many statics problems involve nothing more than simple levers. Some are bent into curious shapes, others are disguised in the forms of real objects such as automobiles or house trusses, but a surprising number resemble the two examples analyzed in Example 5.4.

Example 5.5   For the beam in Example 5.3, find the reactions at *A* and *B*.

Solution: 1. Draw the free-body diagram shown in Fig. 5.6(b).

2. We replace the unknown reaction F_C by its *x* and *y* components [Fig. 5.6(c)].

$$F_C \cos 70° = 0.3420 F_C \quad \text{and} \quad F_C \sin 70° = 0.9397 F_C$$

3. Summing moments about *A*, we have

$$\Sigma M_A = -2400(4) - 1800(12) + 0.9397 F_C (20) = 0$$
$$= -9600 - 21{,}600 + 18.80\,F_C = 0$$

or $$18.80 F_C = 31{,}200; \quad F_C = 1660\,\text{lb}$$ **Answer**

4. Summing moments about *C*, we have

$$\Sigma M_C = -A_y(20) + 2400(16) + 1800(8) = 0$$
$$= -20A_y + 38{,}400 + 14{,}400 = 0$$
$$20A_y = 52{,}800; \quad A_y = 2640\,\text{lb}$$ **Answer**

5. Summing forces in the *x* direction, we have

$$\Sigma F_x = A_x - 0.3420 F_C = 0$$

Substituting for F_C yields

$$A_x = 0.3420 F_C = 0.3420(1660) = 568\,\text{lb}$$ **Answer**

6. We check our answer by using one additional equilibrium equation. Summing forces in the y direction, we have

$$\Sigma F_y = A_y - 2400 - 1800 + 0.9397 F_C$$

or $2640 - 2400 - 1800 + 1560 = 0$ OK

Example 5.6   A force of 6500 N is applied to the beam, as shown in Fig. 5.9(a). The uniform beam weighs 600 N. What are the reactions at A and B?

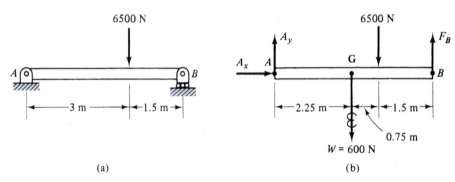

(a)   (b)

FIGURE 5.9

Solution: The solution will be outlined step-by-step.

1. Draw the free-body diagram shown in Fig. 5.9(b). The beam is uniform; therefore, the weight of 600 N acts at its geometric center. The unknown reactions of the pin at A are A_x and A_y, and the unknown force F_B is perpendicular to the surface on which the roller moves.

2. Summing moments about A, we have

$$\Sigma M_A = -600(2.25) - 6500(3) + F_B(4.5) = 0$$
$$= -1350 - 19,500 + 4.5 F_B = 0$$

or $4.5 F_B = 20,850;$ $F_B = 4630\,\text{N}$ **Answer**

By summing moments about A we have eliminated all but one unknown, F_B.

3. Summing moments about B, we have

$$\Sigma M_B = -A_y(4.5) + 600(2.25) + 6500(1.5) = 0$$
$$= -4.5 A_y + 1350 + 9750 = 0$$

or $4.5 A_y = 11,100;$ $A_y = 2470\,\text{N}$ **Answer**

Similarly, by summing moments about B we have eliminated all but one unknown, A_y. Thus, F_B and A_y are independent of each other.

4. Summing forces in the x direction, we have

$$\Sigma F_x = A_x = 0; \quad A_x = 0$$ **Answer**

5. We check our answers by using one additional equilibrium equation. Summing forces in the y direction, we have

$$\Sigma F_y = A_y - 600 - 6500 + F_B = 0$$

or $2470 - 600 - 6500 + 4630 = 0$ OK

Example 5.7   Determine the reaction at the fixed support A for the loaded bent [Fig. 5.10(a)].

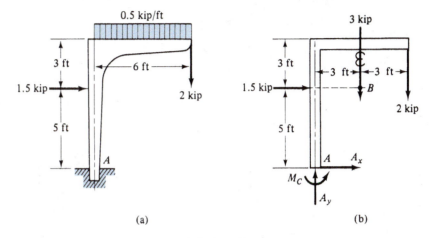

(a)   (b)

FIGURE 5.10

Solution: 1.   The free-body diagram is shown in Fig. 5.10(b). The uniformly distributed load was replaced by its resultant. At the fixed support A we have three unknowns, the two forces A_x and A_y and the moment couple M_C.

2.   Summing moments about A, we have

$$\Sigma M_A = M_C - 1.5(5) - 3(3) - 2(6) = 0$$
$$= M_C - 7.5 - 9 - 12 = 0$$
$$M_C = 28.5 \text{ kip-ft} \circlearrowright \qquad \qquad \text{\bf Answer}$$

3.   Summing the forces in the x direction,

$$\Sigma F_x = 1.5 + A_x = 0$$
$$A_x = -1.5 \text{ kip} = 1.5 \text{ kip} \leftarrow \qquad \qquad \text{\bf Answer}$$

4.   Summing forces in the y direction,

$$\Sigma F_y = A_y - 3 - 2 = 0$$
$$A_y = 5 \text{ kip} = 5 \text{ kip} \uparrow \qquad \qquad \text{\bf Answer}$$

5.   We *check* our answer by using one additional equilibrium equation. Recall that the moment of a couple has the same value about any point in the plane of the couple. Summing moments about B,

$$\Sigma M_B = A_x(5) + M_C - 2(3) - A_y(3) = 0$$
$$= -1.5(5) + 28.5 - 6 - 5(3)$$
$$= -7.5 + 28.5 - 6 - 15 = 0 \quad \text{OK}$$

PROBLEMS

For each of the following problems, draw an appropriate free-body diagram and use the equations of static equilibrium to find the unknown quantities specified. Neglect the weights of all members and objects unless given as part of the problem statement.

5.1   A race car is driven onto a set of weighing scales as shown. If the front wheels together carry 6250 N of weight, and the rear wheels together carry 5350 N of weight, find the horizontal distance, x, that locates the vehicle's center of gravity (c.g.).

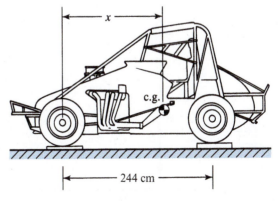

PROB. 5.1

5.2 A 12-ft beam of uniform cross section weighing 108 lb sits atop smooth supports as shown.
   a. If $F = 240$ lb, find the reactions at B and C.
   b. Find the *minimum* value of F for which the beam remains in equilibrium.
   c. Find the *maximum* value of F for which the beam remains in equilibrium.

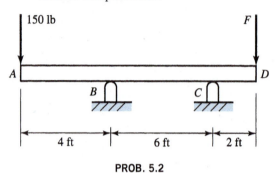

PROB. 5.2

5.3 The portable engine hoist shown in the figure weighs 275 lb. This weight acts as if concentrated at the machine's center of gravity (c.g.). A chain at E is used to support the 640-lb load.
   a. Find the support forces provided at wheels A and F.
   b. How far can boom member CDE be extended from its present position before the hoist tips over?

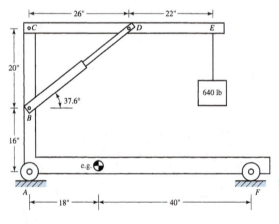

PROB. 5.3

5.4 Find the support reactions developed at A and C for the rigid truss loaded as shown.

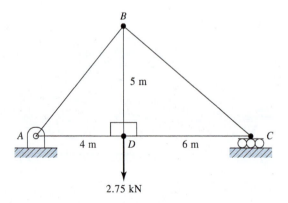

PROB. 5.4

5.5 Same as Prob. 5.4.

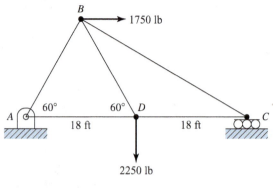

PROB. 5.5

5.6 The metal plate shown is held in position by a pin connection at B and a roller at C. Find the reactions at B and C for the equilibrium of the plate.

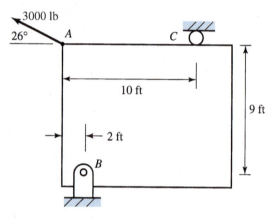

PROB. 5.6

5.7 Determine the reactions at A and B for the simply supported beam shown.

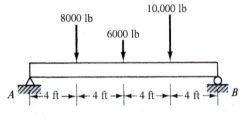

PROB. 5.7

5.8 The cantilever beam supports two loads as shown. Determine the reactions at A.

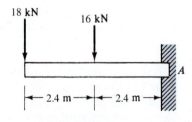

PROB. 5.8

5.9 The total weight for the uniform beam *ABC* is 200 lb. Determine the reactions at *A* and the tension in the cable *BD*.

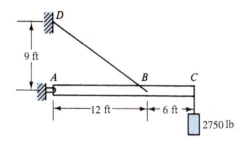

PROB. 5.9

5.10 Determine the reactions at *A* and the tension or compression in member *IJ* for the pin-connected tower shown.

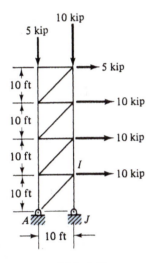

PROB. 5.10

5.11 The T-shaped member is supported as shown. Determine the reactions at *A* and the tension or compression in link *BC*.

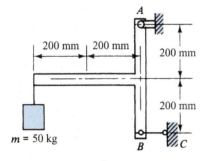

PROB. 5.11

5.12 Determine the reactions at *A* and *B* for the beam with the overhang as shown.

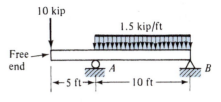

PROB. 5.12

5.13 Determine the reactions at *A* and *B* for the beam with the overhang as shown.

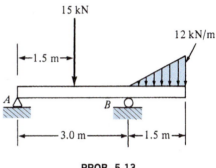

PROB. 5.13

5.14 The cantilever beam supports a uniform load as shown. Find the reactions at *B*.

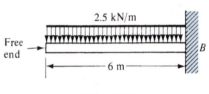

PROB. 5.14

5.15 For the cantilever beam with loads as shown, find the reactions at *A*.

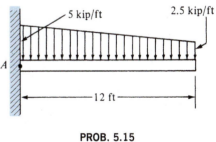

PROB. 5.15

5.16 The beam is supported by a pin and cable as shown on the next page. Find the reactions at *A* and the tension in the cable *BC*.

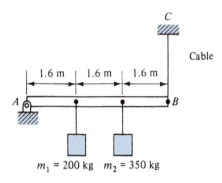

PROB. 5.16

5.17 The pin-connected tower supports a horizontal load as shown. Determine the reactions at A and the tension or compression in member HJ if P = 20 kip.

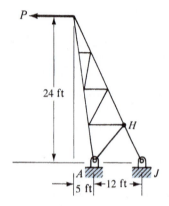

PROB. 5.17

5.18 The boom is supported by a pin and cable as shown. Find the reactions at A and the tension in the cable BC.

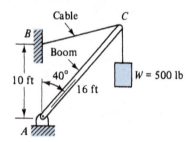

PROB. 5.18

5.19 The frame is acted on by two forces and a couple as shown. Determine the reactions at A and B.

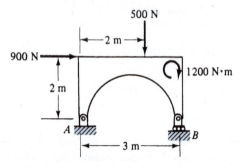

PROB. 5.19

5.20 The loaded bracket shown is supported by a pin at A and a frictionless roller at B. Find the reactions at A and B.

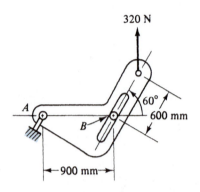

PROB. 5.20

5.21 The triangular truss supports loads as shown. Determine the reactions at E and G.

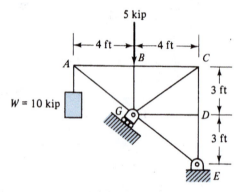

PROB. 5.21

5.22 The four-panel truss supports loads as shown. Find the reactions at A and E.

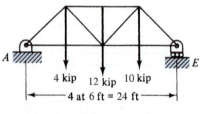

PROB. 5.22

5.23 The boom OA supports a load of 10 kN as shown. The uniform boom weighs 2 kN. Find the reactions at O and the tension T in the lifting cable.

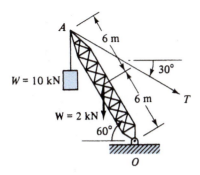

PROB. 5.23

5.24 Member *ABC* has a uniform cross section and weighs W_2. Determine the reactions at *C* and *B* if $W_1 = 2400$ lb and $W_2 = 300$ lb.

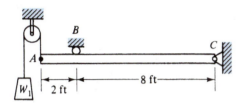

PROB. 5.24

5.25 Beam *ABC* has a uniform cross section and weighs 80 N. Determine (a) the reactions at *A* and the tension in the cable and (b) the normal reaction between the block and the beam. (*Hint:* Draw a free-body diagram of the beam and the block together. Solve for the tension in the cable and the reactions at *A*. With the tension known, draw a free-body diagram of the block and solve for the normal reaction.)

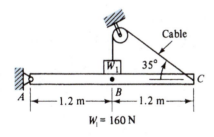

PROB. 5.25

5.26 The hook-shaped bar is acted on by a force as shown. Find the reactions at *B* and *C*.

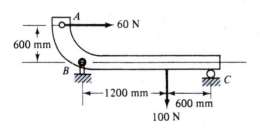

PROB. 5.26

5.27 The boom is supported by a cable and pin as shown. Neglect the size of the pulley at *D*. Find the reactions at *C* and the tension in the cable if $W_1 = 35$ kN and $W_2 = 50$ kN.

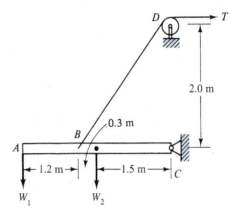

PROB. 5.27

5.28 The truss supports a sign board as shown in the figure. Calculate the reactions at *A* and the force in member *FG* produced by the horizontal wind load of 60 lb per vertical foot of sign.

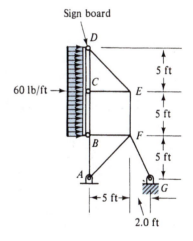

PROB. 5.28

5.29 The lift force on the wing of an aircraft is approximated by the distributed load shown. Determine the reactions at *A*.

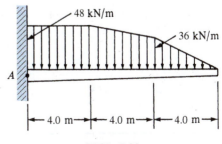

PROB. 5.29

5.30 A uniform member *ABC* weighing 500 lb is being lifted by a cable as shown. When $\theta = 55°$, find the tension in the lifting cable and the tension in the anchor cable. Neglect the size of the pulley at *D*.

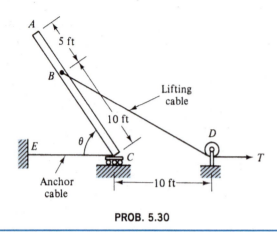

PROB. 5.30

5.5 EQUILIBRIUM OF A TWO-FORCE BODY

A body acted on by two forces is called a *two-force body*. For the equilibrium of a two-force body the force acting at one point must be *equal in magnitude, opposite in direction, and have the same line of action* as the force acting at the other point. For proof, consider the two-force body acted on by forces **P** and **Q**, which are directed at angles θ_A and θ_B as shown in Fig. 5.11(a). We replace **P** and **Q** by components parallel and perpendicular to the line joining *A* and *B* [Fig. 5.11(b)]. Because the body is in equilibrium, moments about any point and the resultant in any direction must vanish. Summing moments about *A* and then about *B*, $Q_y = 0$ and $P_y = 0$. Summing forces in the direction of the line *AB*, $P_x = Q_x$. Therefore, only two possibilities exist for a two-force body, as shown in Fig. 5.12(a)

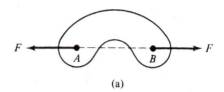

(a)

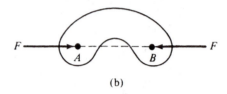

(b)

FIGURE 5.12

and (b). The forces must be equal, be opposite, and have the same line of action.

Several common examples of two-force bodies are shown in Fig. 5.13: (a) The members (called *links*) of a truss are connected to each other only at their endpoints, and (b) hydraulic and pneumatic cylinders are generally attached to supports by pins at each end of the cylinder. In both cases, the weight of the member is neglected. This is a valid assumption if the load carried by the member is very large compared with the weight.

Note that, in theory, two-force members experience no bending effects. In practice, however, real truss links that are long and slender and carry compressive loads are subject to the buckling effects described in Chapter 17. Also, as Fig. 5.14 illustrates, temporary accidental side loads can be applied to two-force members with disastrous results.

5.6 EQUILIBRIUM OF A THREE-FORCE BODY

A body acted on by three forces is called a *three-force body*. For equilibrium, the line of action of the three forces must be either *concurrent* or *parallel*.

Consider a body in equilibrium acted on by three non-parallel forces **P, Q**, and **S** [Fig. 5.15(a)]. The line of action of two of the forces must intersect at some point *B*. Because the three forces are in equilibrium, the sum of their moments about *B* must be equal to zero. Therefore, the line of action of **P** must also act through *B* as in Fig. 5.15(b). Thus, the line of action of all three forces **P, Q**, and **S** act through *B*; they are therefore concurrent.

This type of three-force body problem can be solved by the triangular method. The three forces are concurrent and thus form a closed triangle. The problem can also be solved by the equilibrium equations of Sec. 5.4. In the case of a three-force body loaded by parallel forces, use of the equations of equilibrium is always required. (See Examples 5.4 and 5.5.)

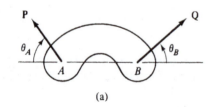

(a)

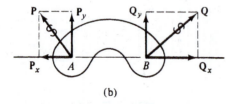

(b)

FIGURE 5.11

FIGURE 5.13. (a) Each steel link of this bridge truss is loaded only through the two endpoints at which it is attached to other members of the truss. (b) Many frames and machines contain hydraulic or pneumatic cylinders that are pinned to supports at each end. This cylinder has an extended length of approximately 36 in. and operates the front blade on a piece of construction equipment.

(a)

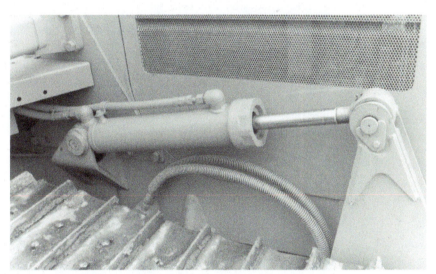

(b)

FIGURE 5.14. Although this hydraulic cylinder can exert many tons of axial force, its thin-walled tube is extremely susceptible to side loads.

TWO-FORCE MEMBERS

SB.4(a)

SB.4(b)

London's Millennium Footbridge, pictured in Figure SB.4(a), opened on June 10, 2000. The first new bridge to be built across the Thames River in more than a century, this pedestrian walkway immediately exhibited severe lateral movement, making it virtually impossible for pedestrians to walk across the structure. Two days later, on June 12, 2000, the bridge was closed until this problem of excessive sway could be reduced or eliminated. The eventual solution involved installation of 37 *two-force members* at various points on the bridge. Ranging in length from 0.7 m to 8.3 m, these *viscous damping devices*, like the approximately 8-m long pier damper shown in Figure SB.4(b), effectively act like the shock absorbers on an automobile to damp out unwanted motion. The bridge itself was reopened on February 22, 2002, and is used by more than four million pedestrians each year.

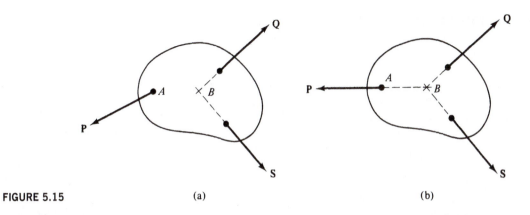

FIGURE 5.15   (a)   (b)

Example 5.8   A truss supports a sign board as shown in Fig. 5.16(a). Calculate the reactions at A and the force in member FG produced by the horizontal wind load of 0.8 kN per vertical meter of sign.

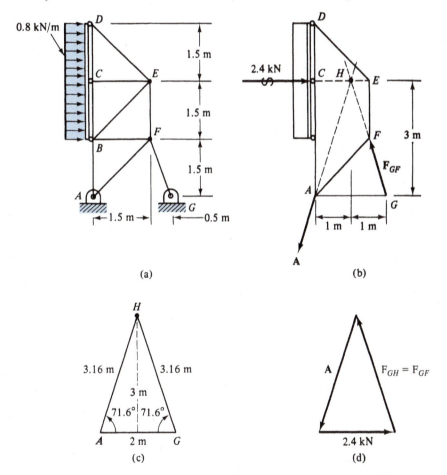

FIGURE 5.16

Solution: The truss $ADEF$ is a three-force body acted on by the resultant of the wind forces, the force along the two-force member FG, and the pin reaction at A. The resultant of the wind force $R = (0.8\,\text{kN/m})$ $(3\,\text{m}) = 2.4\,\text{kN}$ acts at the center C of the sign board. The forces are concurrent and intersect at H as shown [Fig. 5.16(b)]. Dimensions of the triangle AHG are shown in Fig. 5.16(c). The force triangle is constructed around AHG in Fig. 5.16(d). The triangles are similar; therefore, $A = F_{GH}$ and

$$\frac{A}{2.4} = \frac{AH}{AG} = \frac{3.16}{2} \quad \text{or} \quad A = 3.79\,\text{kN}$$

Thus,

$$F_{GH} = 3.79\,\text{kN}; \qquad \mathbf{A} = 3.79\,\text{kN} \; \nearrow 71.6° \qquad \textbf{Answer}$$

5.7 STATICAL DETERMINACY AND CONSTRAINT OF A RIGID BODY

For equilibrium, a rigid body must satisfy the equations of equilibrium (Sec. 5.4). In addition, the body must also be properly constrained or supported. In this section, we examine various bodies to determine whether the supports provide complete constraints against motion and whether the reactions at the supports can be determined from the equations of equilibrium. The analysis is simplified if we replace the supports by a set of links or hinged bars that have actions that are *equivalent* to those of the supports. The replacement is made as follows.

A *single-reaction support* such as a smooth surface, roller, or cable is replaced by a single link. The link must have the same direction as the reaction [Fig. 5.17(a)].

A *two-reaction support,* the smooth pin or hinge, is replaced by two links at right angles to each other as in Fig. 5.17(b).

A *three-reaction support,* the fixed or clamped end, is replaced by three links as shown in Fig. 5.17(c). The horizontal force F_x and couple M are first replaced by two horizontal forces, F_1 and F_2. The three forces are then replaced by two horizontal links and a vertical link as shown in the figure. We emphasize again that *the links have actions that are equivalent to those of the supports they replace.*

With the aid of linkage diagrams, we consider the bodies shown in Figs. 5.18 and 5.19. We want to determine if the supports provide complete constraints against motion and if the reactions at the supports can be found from the equations of equilibrium.

With two links at point A [Fig. 5.18(a)], the only possibility of motion is rotation of the body about A. Any arrangement of the link BF that prevents motion perpendicular to line AB will constrain the body. With the link as shown, the body is *constrained.* The three equations of equilibrium can be used to solve for the three unknown reactions. Therefore, the body is statically *determinate.* If link BF is rotated so that it lies along the line through A and B, the body can rotate and would only be *partially constrained.* The three reactions are

concurrent, and we have only two equations of equilibrium. Thus, the body is statically *indeterminate.*

In Fig. 5.18(b), we add one additional link GC. Support of the body exceeds that required for complete constraint, and one of the supports is *redundant.* With three equilibrium equations and four unknown reactions, the body is statically *indeterminate* to the first degree.

With the three parallel links as shown in Fig. 5.19(a), the body is free to move to the right. Therefore, the body is *partially constrained.* Although there are only three reactions, the body is statically *indeterminate.* This is true because one of the equilibrium equations, $\sum F_x = 0$, does not contain any of the unknown reactions. We are left with two equilibrium equations and three unknown reactions. If link BF is removed, the body is supported by two parallel links, and the body is still *partially constrained.* However, the two reactions can be determined from the three equations of equilibrium. Thus, the body is statically *determinate.*

In Fig. 5.19(b), the body is supported by four links. One of the links is *redundant,* and the body is statically *indeterminate.* By removing any one of the links, the body remains constrained and becomes statically *determinate.*

It follows from the preceding examples that complete constraint is provided if a body is supported by three links whose axes are neither parallel nor intersect at a common point. In such cases, the three reactions can be found by the equilibrium equations, and the body is statically determinate.

PROBLEMS

In Probs. 5.31 through 5.36, use the triangular method and the property of three-force bodies in equilibrium.

5.31 Solve Prob. 5.6.

5.32 Solve Prob. 5.11.

5.33 Solve Prob. 5.17.

5.34 Solve Prob. 5.20.

5.35 Solve Prob. 5.25.

5.36 Solve Prob. 5.26.

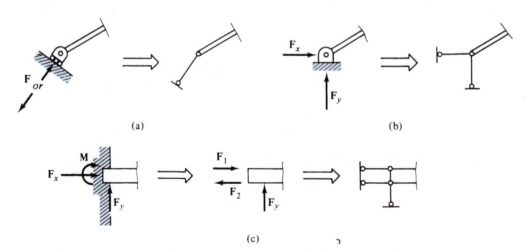

FIGURE 5.17

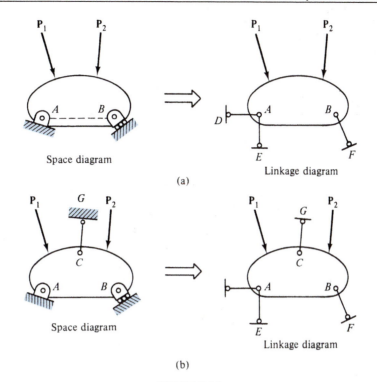

FIGURE 5.18

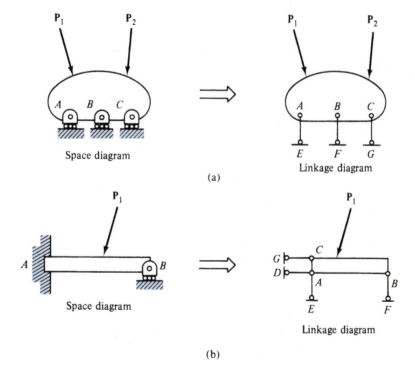

FIGURE 5.19

5.37 A triangular-shaped member has dimensions and loads as shown in (a) of the figure. The member is supported in various ways as shown in figures (b) through (i). In each case, determine whether the member is completely or partially restrained and whether the reactions are statically determinate or indeterminate. If determinate, find the reactions if $P = 100\,\text{N}$ and $Q = 50\,\text{N}$.

5.38 The truss (a) of the figure has loads and dimensions as shown. The truss is supported in various ways, as in figures (b) through (i). In each case, determine whether the truss is completely or partially restrained and whether the reactions are statically determinate or indeterminate. If determinate, find the reactions if $P = 5\,\text{kip}$ and $W = 10\,\text{kip}$.

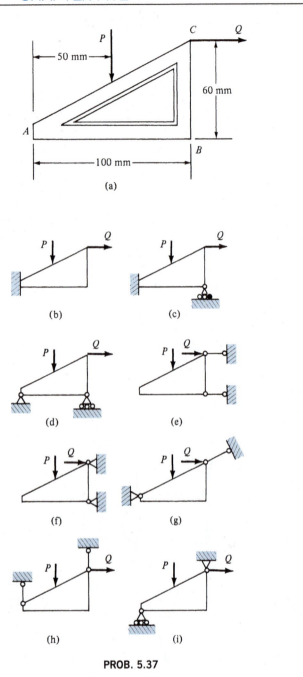

(a)

(b)

(c)

(d)

(e)

(f)

(g)

(h)

(i)

PROB. 5.37

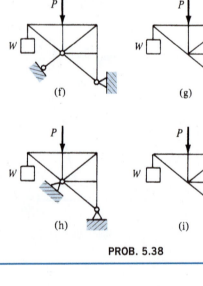

(a)

(b)

(c)

(d)

(e)

(f)

(g)

(h)

(i)

PROB. 5.38

FORCE ANALYSIS OF STRUCTURES AND MACHINES

6.1 INTRODUCTION

In Chapter 5, we were concerned mainly with a single structural member or machine element. Here, we study problems involving the force analysis of structures and machines.

A *structure* consists of a series of connected structural members or rigid bodies that are designed to *support* loads or forces. The loads may be either stationary or moving, but the structure is usually at rest. If the structure is moved, the motion is executed slowly. *Machines* are made up of several machine elements connected together. They are designed to *transmit* and *transform* loads or forces rather than support them. The machine generally has moving parts. With structures and machines, we must determine not only the external forces acting on them, but also the forces that hold the parts together.

In this chapter, we consider simple trusses, pin-connected frames, and machines or mechanisms that can be analyzed in a plane. Examples of each are shown in Fig. 6.1(a)–(d), respectively.

6.2 SIMPLE PLANE TRUSSES

A *truss* is a structure usually consisting of straight uniform bars or members arranged in a series of adjoining triangles and fastened together at their ends to form *joints*. The joints

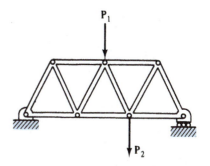

(a) Truss

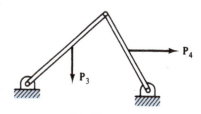

(b) Frame

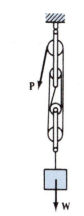

(c) Pulley system

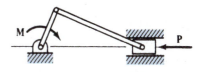

(d) Crank mechanism

FIGURE 6.1

FIGURE 6.2. Wooden roof trusses are engineered and prefabricated in a wide variety of configurations for use on residential and light commercial structures. (Note that the man standing at the left of each truss is 6 ft tall.)

may be pinned, welded, riveted, bolted, or nailed, depending on the type of truss. They may be used for such things as bridges [see Fig. 5.13(a)], towers, and buildings (see Fig. 6.2). In most cases, a truss has two general purposes: (a) to *distribute any applied loads* among the many members of the truss, and (b) to *span greater distances* than that allowed by the longest single member in the truss. From the late 1800s to the early 1900s, a great deal of work was done on the design and analysis of trusses. Those configurations that were particularly effective or aesthetic were often named after the engineers and architects who developed them. These include the Fink, Howe, Pratt, and Warren trusses. To simplify our calculations, we make three assumptions about the *simple plane truss:*

1. The bars are connected at their ends to form joints that behave like frictionless pins.

2. All forces acting on the truss are applied at the joints.

3. The members, joints, and loads all lie in a plane.

The first assumption represents a substantial departure from the real truss. In fact, the ends of the members are attached to each other so that bending of the member must occur. However, the forces in the members calculated on the basis of pin construction do not differ substantially from their true value. The second assumption is justified in most practical cases. Except for the weight of a member that acts at its center of gravity, the loads are usually trans-

mitted by other structures or members to the joints of the truss. The weight of a member is small in comparison with other loads on the truss and can be either neglected or divided equally between the joints at its ends. With careful design and fabrication of the truss, the third assumption is also justified.

The first and second assumption ensures that all bars are two-force bodies or members. Thus, no bending can occur in the members. Each member must be either in tension or compression. (See Sec. 5.5 and Fig. 5.14.) The third assumption ensures that we have a problem in plane statics.

A *simple truss* can be constructed by the following method. As shown in Fig. 6.3, we start with the simplest rigid structure, consisting of three bars and three joints, triangle *ABC*. By adding pairs of bars *AD* and *CD*, the structure remains rigid. To the rigid structure *ABCD*, we add the

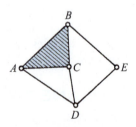

FIGURE 6.3

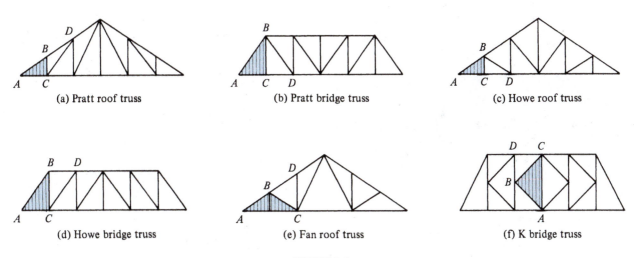

(a) Pratt roof truss (b) Pratt bridge truss (c) Howe roof truss

(d) Howe bridge truss (e) Fan roof truss (f) K bridge truss

FIGURE 6.4

pair of bars BE and DE, and the structure remains rigid. The process is continued until the truss is complete. The term *rigid* here means that the truss functions as a whole like a rigid body.

The trusses shown in Fig. 6.4(a)–(f) were all constructed in this manner. The shaded triangle ABC has been taken as the starting point for each truss. Adding pairs of members CD and BD, we establish joint D. Through the addition of each new pair of members, we establish the remaining joints. The trusses shown are all simple trusses. Whatever the complexity of a truss, remember that the structure's ability to distribute loads among its members depends on the shape of the truss (angles between adjacent links) and the points at which loads are applied. The size of the truss, link material, and cross-sectional geometry of the links do not affect the load carried by any individual link.

Now consider the Warren truss shown in Fig. 6.5(a). The pin joints are labeled A, B, C, D, and E. The member joining pins A and B is referred to as AB, and the force in the member is F_{AB}. Using these symbols, the free-body diagram for the entire truss is shown in Fig. 6.5(b), and the free-body diagram for each joint and member is shown in Fig. 6.5(c). Here, all members of the truss are shown in tension. If, in fact, the member is in tension, the equilibrium equations give a positive value for the force in the member. If in compression, the equilibrium equations give a negative value. Notice that between each joint and member the forces are in opposite directions. For example, the force of member AB on joint A is equal in magnitude and opposite in direction to the force of joint A on member AB. This is in agreement with Newton's third law, which states that action equals reaction.

6.3 MEMBERS UNDER SPECIAL LOADING

Consider three members of a truss joined together at a joint with two of the members lying along a straight line [Fig. 6.6(a)]. Summing forces perpendicular to the straight

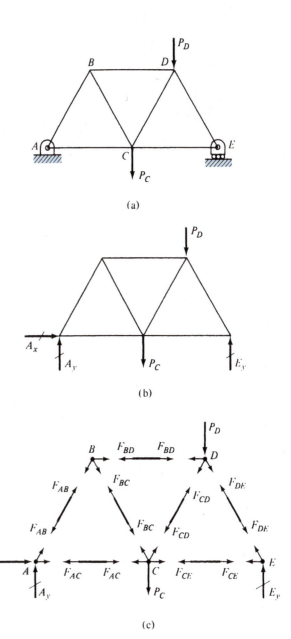

FIGURE 6.5

TRUSS FABRICATION

SB.5(a)

SB.5(b)

Modern *truss fabrication* techniques involve varying degrees of automation. Truss design is accomplished by using computer software that specifies lumber size and grade for each member, as well as the size and placement of connector plates (rectangular steel plates with protruding teeth) that are used at each joint. Truss assembly generally takes place on large segmented metal tables that are often 100 to 200 ft in length. In many plants, one or more laser heads located on the ceiling of the facility are controlled by computer and project full-size outlines of the truss onto this table. Workers use these images as guides while they fasten metal pieces to the table, forming a jig or fixture in which the trusses will be assembled. The projected images [SB.5(a)] trace the required locations for connector plates and truss members, thus eliminating the need for squaring and measuring during fabrication. (It is estimated that laser projection heads can reduce setup and changeover time between different truss designs by as much as 70 percent and can increase overall productivity by 25 percent compared with conventional assembly techniques.) Once the truss members and connector plates are placed in the fixtures, the joints are pinched between heavy rollers or hydraulic presses to squeeze each plate into the wooden members. A lifting device in the steel table then raises the completed trusses and moves them to a stacking area, where they are bundled with other trusses. In SB.5(b), the table fixtures, stacks of connector plates, and bundled trusses are all visible. *(Courtesy of the Wood Truss Council of America, Madison, Wisconsin, www.woodtruss.com)*

line, we see that a component of $\mathbf{F}_{AB}$ is equal to zero. Therefore, F_{AC} must be zero, provided there are no external forces applied to the joint. Because $F_{AC} = 0$, $F_{AB} = F_{AD}$, member AC is a *zero-force member*.

When two members are joined together, as shown in Fig. 6.6(b), the force in each member is zero. If the members are colinear, as shown in Fig. 6.6(c), the forces are equal.

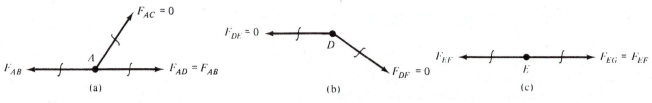

FIGURE 6.6

6.4 METHOD OF JOINTS

In the method of joints a free-body diagram of each joint is drawn. Because the forces at each joint are concurrent, equilibrium requires that $\Sigma F_x = 0$ and $\Sigma F_y = 0$. Thus, only two unknowns can be determined at each joint. Therefore, we begin our analysis by drawing a free-body diagram of a joint with no more than two unknown forces, and then proceeding from joint to joint so that each new free-body diagram contains no more than two unknown forces. The method is self-checking because the last joint will involve only known forces, and they must be in equilibrium. Each joint or pin connection is represented as a point, so either the equations of equilibrium or the triangular method can be used to obtain a solution.

The method of joints will be illustrated by several examples.

Example 6.1   A simply supported plane truss is loaded as shown in Fig. 6.7. For θ values of 30°, 40°, 45°, 50°, and 60°, find the force carried by each link and indicate whether the link is in tension (T) or compression (C).

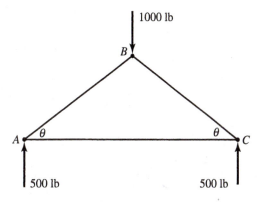

FIGURE 6.7

Solution: A free-body diagram of joint A is shown in Fig. 6.8(a), while Fig. 6.8(b) shows the same point with force F_{AB} reduced to its rectangular components. From the equations of equilibrium for a point,

$$\Sigma F_y = 0 \qquad F_{AB}\sin\theta + 500\,\text{lb} = 0$$

or

$$F_{AB} = \frac{-500\,\text{lb}}{\sin\theta}$$

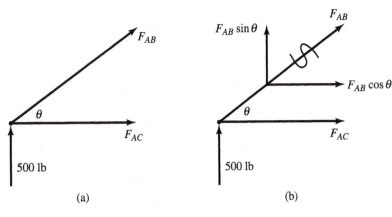

(a)   (b)

FIGURE 6.8

Because $\sin\theta > 0$ for the range of values given, the negative value of F_{AB} tells us that this force was drawn in the wrong direction on our free-body diagram. Link AB, then, must be in compression. From the remaining equation of equilibrium,

$$\Sigma F_x = 0 \qquad F_{AB}\cos\theta + F_{AC} = 0$$

or

$$F_{AC} = -F_{AB}\cos\theta = -\frac{(-500\,\text{lb})}{\sin\theta} \times \cos\theta$$

$$F_{AC} = \frac{500\,\text{lb}}{\tan\theta}$$

Because the value of F_{AC} must be positive, we know that link AC is in tension. We could also have achieved the same results by using the triangular method. An incomplete free-body diagram (no arrowheads to indicate directions of forces) is shown in Fig. 6.9(a). The resulting vector addition, drawn using the methods of Sec. 3.5, is shown in Fig. 6.9(b). From this triangle, we see that the direction of force F_{AC} must be to the right, whereas that of force F_{AB} must be downward to the left. Writing the trigonometric equations for this right triangle, we obtain

$$\frac{500 \text{ lb}}{F_{AB}} = \sin \theta$$

$$\frac{500 \text{ lb}}{F_{AC}} = \tan \theta$$

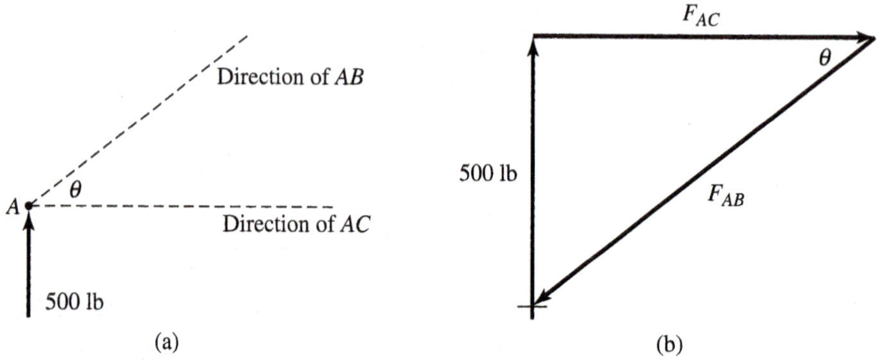

(a)   (b)

FIGURE 6.9

Using either set of equations, the values of F_{AB} and F_{AC} shown in Table (A) may be computed. Because of symmetry, the force in link AB must equal that in link BC. Note that the forces in all members of the truss decrease as angle θ increases and that the analysis was not dependent on truss size or link material.

TABLE (A) Example 6.1		
θ	$F_{AB} = F_{BC}$ (lb)	F_{AC} (lb)
30°	1000 (C)	866 (T)
40°	778 (C)	596 (T)
45°	707 (C)	500 (T)
50°	653 (C)	420 (T)
60°	577 (C)	289 (T)

Because of the forces in members AB and BC, we see that the pins at A and C want to move apart, thus placing link AC in tension. In practice, several methods can be used to provide the horizontal forces required at these joints. One of the oldest techniques involves the use of heavy masses of material (called *buttresses*) placed outside the truss, as shown in Fig. 6.10(a). In modern residential construction, wooden *ceiling joists* function as the horizontal link between joints, with *collar ties* used to increase rigidity of the structure [Fig. 6.10(b)]. Finally, because link AC is in tension, a *flexible member* (such as steel cable) can be used to tie the ends of a truss together; this may be accomplished by fastening the cable ends to bearing plates placed on the outside surfaces at the top of each supporting wall [Fig. 6.10(c)]. On multistoried buildings such as cathedrals, which have vaulted ceilings and large wall openings for stained-glass windows, common practice was to brace the upper floors and walls using *flying buttresses*, as shown in Fig. 6.11. These structural members, either straight or arched, transmitted the building's lateral thrust to external piers or buttresses.

Not all truss links are in tension or compression. For instance, if an additional link, BD, is added to our original truss as shown in Fig. 6.12(a), we might expect this member to carry a substantial portion of the 1000-lb applied load. However, the free-body diagram of pin D given in Fig. 6.12(b) shows that for equilibrium in the vertical direction the force in BD must be zero.

In contrast, if our 1000-lb load is applied at D instead of B as shown in Fig. 6.12(c), the new free-body diagram of pin D [Fig. 6.12(d)] shows that link BD now carries a 1000-lb tensile force. Obviously, then, the location of an applied load can have a significant effect on the forces carried by each link in a truss.

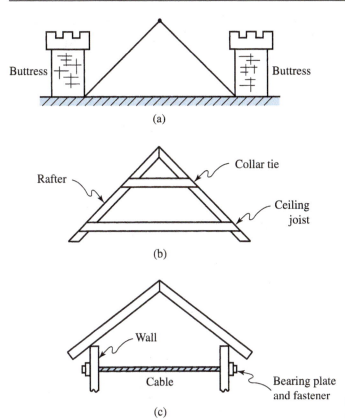

FIGURE 6.11. The *flying buttress* is a straight or arched member used to brace the upper floors and walls of a structure, transmitting lateral thrust from the main building to external piers or buttresses. Introduced during the *Gothic period* (1200–1300 A.D.), these members were used primarily on castles and churches, such as the historic British cathedral shown here.

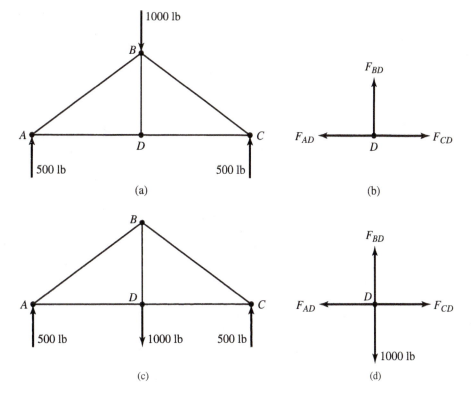

FIGURE 6.12

Example 6.2   Shown in Fig. 6.13(a) is a simple plane truss. Using the method of joints, find the force in each member of the truss.

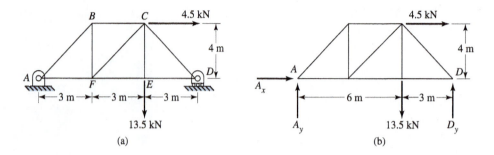

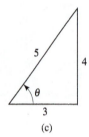

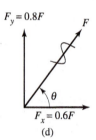

(c)

(d)

FIGURE 6.13

Solution: The free-body diagram of the entire truss is shown in Fig. 6.13(b). The x and y components of the reaction at pin A and the reaction at roller D are shown. For equilibrium of the entire truss,

$$\Sigma M_A = D_y 9 - 4.5(4) - 13.5(6) = 0$$
$$D_y = 11.0 \,\text{kN} \uparrow$$
$$\Sigma M_D = 13.5(3) - A_y(9) - 4.5(4) = 0$$
$$A_y = 2.5 \,\text{kN} \uparrow$$
$$\Sigma F_x = A_x + 4.5 = 0$$
$$A_x = -4.5 \,\text{kN} = 4.5 \,\text{kN} \leftarrow$$

Check:

$$\Sigma F_y = A_y + D_y - 13.5 = 2.5 + 11.0 - 13.5 = 0 \quad \text{OK}$$

The slope of all diagonal members is 4 to 3. In Fig. 6.13(c), the sine and cosine of the slope angle are indicated, and in Fig. 6.13(d), the x and y components of a force with the same slope are shown. In Fig. 6.14(a)–(f), the free-body diagrams for the joints are drawn so that each new free-body diagram contains no more than two unknown forces. A diagonal line superimposed on a force means that the force is one of the external reactions acting on the truss. The internal forces have all been assumed to be tensile. Starting with joint A, we write the equilibrium equations for each new joint in the sequence.

Joint A [Fig. 6.14(a)]:

$$\Sigma F_y = 2.5 + 0.8F_{AB} = 0$$
$$F_{AB} = -3.125 \,\text{kN} \quad \text{(compression)} \qquad \text{Answer}$$
$$\Sigma F_x = -4.5 + F_{AF} + 0.6F_{AB} = -4.5 + F_{AF} + 0.6(-3.125) = 0$$
$$F_{AF} = 6.375 \,\text{kN} \quad \text{(tension)} \qquad \text{Answer}$$

Joint B [Fig. 6.14(b)]:

$$\Sigma F_x = F_{BC} - 0.6F_{AB} = F_{BC} - 0.6(-3.125) = 0$$
$$F_{BC} = -1.875 \,\text{kN} \quad \text{(compression)} \qquad \text{Answer}$$

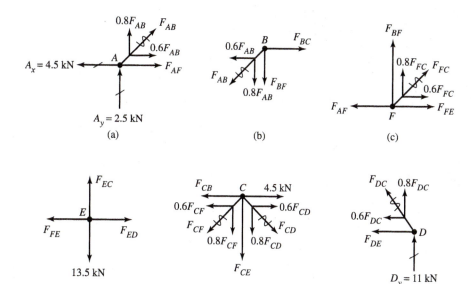

FIGURE 6.14

$$\Sigma F_y = -F_{BF} - 0.8F_{AB} = -F_{BF} - 0.8(-3.125) = 0$$
$$F_{BF} = 2.5\,\text{kN} \quad \text{(tension)}$$ **Answer**

Joint F [Fig. 6.14(c)]:

$$\Sigma F_y = F_{BF} + 0.8F_{FC} = 2.5 + 0.8F_{FC} = 0$$
$$F_{FC} = -3.125\,\text{kN} \quad \text{(compression)}$$ **Answer**
$$\Sigma F_x = 0.6F_{FC} + F_{FE} - F_{AF} = 0.6(-3.125) + F_{FE} - 6.375 = 0$$
$$F_{FE} = 8.25\,\text{kN} \quad \text{(tension)}$$ **Answer**

Joint E [Fig. 6.14(d)]:

$$\Sigma F_x = F_{ED} - F_{FE} = F_{ED} - 8.25 = 0$$
$$F_{ED} = 8.25\,\text{kN} \quad \text{(tension)}$$ **Answer**
$$\Sigma F_y = F_{EC} - 13.5 = 0$$
$$F_{EC} = 13.5\,\text{kN} \quad \text{(tension)}$$ **Answer**

Joint C [Fig. 6.14(e)]:

$$\Sigma F_x = 4.5 + 0.6F_{CD} - F_{CB} - 0.6F_{CF}$$
$$= 4.5 + 0.6F_{CD} - (-1.875) - 0.6(-3.125) = 0$$
$$F_{CD} = -13.75\,\text{kN} \quad \text{(compression)}$$ **Answer**

First check:

$$\Sigma F_y = -0.8F_{CF} - F_{CE} - 0.8F_{CD}$$
$$= -0.8(-3.125) - 13.5 - 0.8(-13.75) = 0 \quad \text{OK}$$

Joint D [Fig. 6.14(f)]:
Second check:

$$\Sigma F_x = -0.6F_{DC} - F_{DE} = -0.6(-13.75) - 8.25 = 0 \quad \text{OK}$$

Third check:

$$\Sigma F_y = 0.8F_{DC} + 11 = 0.8(-13.75) + 11 = 0 \quad \text{OK}$$

The results are shown on the truss in Fig. 6.15. Positive values indicate tension and negative values compression.

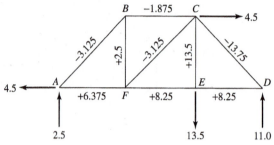

(All forces are expressed in kilonewtons.)

FIGURE 6.15

Example 6.3   Using the method of joints, find the force in each member of the truss shown in Fig. 6.16(a).

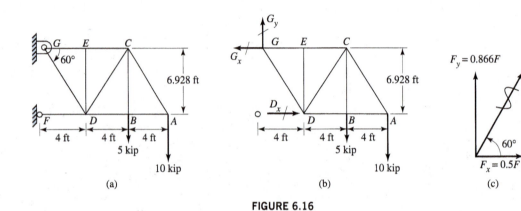

(a)   (b)   (c)

FIGURE 6.16

Solution: The member DF is a two-force member, and the reaction at D is directed along the member. Members EG and EC lie along a straight line; therefore, the force in member ED is zero, and the forces in EG and EC are equal. The free-body diagram of the entire truss is shown in Fig. 6.16(b). For equilibrium of the *entire truss*, we have

$$\Sigma M_G = D_x(6.928) - 5(8) - 10(12) = 0$$
$$D_x = 23.09 \text{ kip} \rightarrow$$
$$\Sigma F_x = -G_x + D_x = 0$$
$$G_x = 23.09 \text{ kip} \leftarrow$$
$$\Sigma F_y = G_y - 5 - 10 = 0$$
$$G_y = 15 \text{ kip} \uparrow$$

Check:

$$\Sigma M_D = G_x(6.928) - G_y(4) - 10(8) - 5(4)$$
$$= 160 - 60 - 80 - 20 = 0 \quad \text{OK}$$

In Fig. 6.16(c), the x and y components of a force with a slope of 60° are indicated, and in Fig. 6.17(a)–(e), the free-body diagrams of the joints are drawn so that each new free-body diagram contains no more than two unknown forces. The internal forces are all assumed to be tensile. Starting with joint A, we write the equilibrium equations for each new joint in the sequence.

Joint A [Fig. 6.17(a)]:

$$\Sigma F_y = 0.866F_{AC} - 10 = 0$$
$$F_{AC} = 11.55 \text{ kip} \quad \text{(tension)} \qquad \qquad \textbf{Answer}$$
$$\Sigma F_x = -0.5F_{AC} - F_{AB} = 0$$
$$F_{AB} = -5.77 \text{ kip} \quad \text{(compression)} \qquad \qquad \textbf{Answer}$$

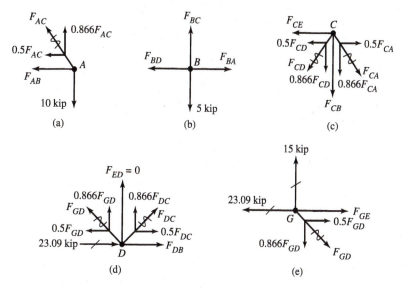

FIGURE 6.17

Joint B [Fig. 6.17(b)]:

$$\Sigma F_x = F_{BA} - F_{BD} = 5.77 - F_{BD} = 0$$

$$F_{BD} = 5.77 \text{ kip} \quad \text{(tension)} \qquad \qquad \textbf{Answer}$$

$$\Sigma F_y = F_{BC} - 5 = 0$$

$$F_{BC} = 5 \text{ kip} \quad \text{(tension)} \qquad \qquad \textbf{Answer}$$

Joint C [Fig. 6.17(c)]:

$$\Sigma F_y = -0.866 F_{CA} - F_{CB} - 0.866 F_{CD}$$

$$= -0.866(11.55) - 5 - 0.866 F_{CD} = 0$$

$$F_{CD} = -17.32 \text{ kip} \quad \text{(compression)} \qquad \qquad \textbf{Answer}$$

$$\Sigma F_x = 0.5 F_{CA} - 0.5 F_{CD} - F_{CE}$$

$$= 0.5(11.55) - 0.5(-17.32) - F_{CE} = 0$$

$$F_{CE} = 14.44 \text{ kip} \quad \text{(tension)} \qquad \qquad \textbf{Answer}$$

Joint E: Member *ED* is a zero-force member, and *CE* and *EG* are collinear; therefore,

$$F_{ED} = 0 \qquad \qquad \textbf{Answer}$$

and

$$F_{CE} = F_{EG} = 14.44 \text{ kip} \quad \text{(tension)} \qquad \qquad \textbf{Answer}$$

Joint D [Fig. 6.17(d)]:

$$\Sigma F_y = 0.866 F_{DC} + 0.866 F_{GD} = 0$$

$$F_{GD} = -F_{DC} = -(-17.32) = 17.32 \text{ kip} \quad \text{(tension)} \qquad \qquad \textbf{Answer}$$

First check:

$$\Sigma F_x = F_{DB} + 0.5 F_{DC} + 23.09 - 0.5 F_{GD}$$

$$= -5.78 + 0.5(-17.32) + 23.09 - 0.5(17.32) = 0 \quad \text{OK}$$

Joint G [Fig. 6.17(e)]:
Second check:

$$\Sigma F_y = 15 - 0.866 F_{GD} = 15 - 0.866(17.32) = 0 \quad \text{OK}$$

Third check:

$$\Sigma F_x = F_{GE} + 0.5 F_{GD} - 23.09$$

$$= 14.44 + 0.5(17.32) - 23.09 = 0 \quad \text{OK}$$

The results are shown on the truss in Fig. 6.18. Positive values indicate tension and negative values compression.

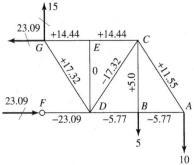

(All forces are expressed in kip.)

FIGURE 6.18

PROBLEMS

6.1 through 6.16   Using the method of joints, determine the force in each member of the truss shown. Indicate whether the member is in tension or compression.

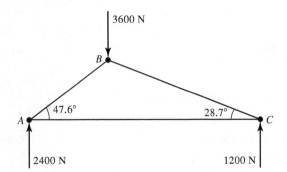

PROB. 6.1

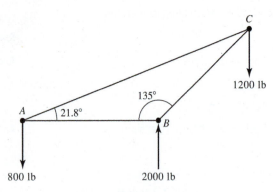

PROB. 6.2

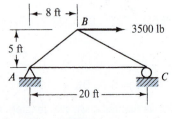

PROB. 6.3

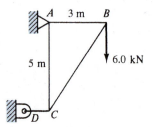

PROB. 6.4

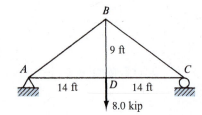

PROB. 6.5

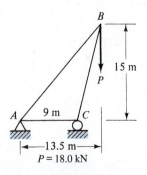

PROB. 6.6

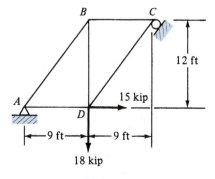

PROB. 6.7

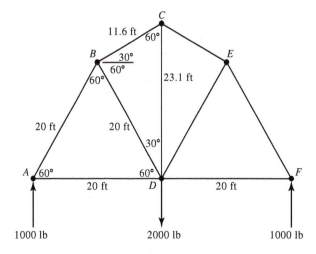

PROB. 6.11

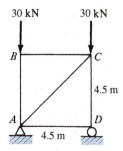

PROB. 6.8

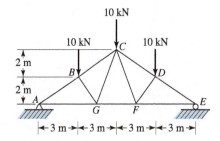

PROB. 6.12

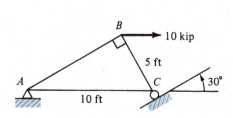

PROB. 6.9

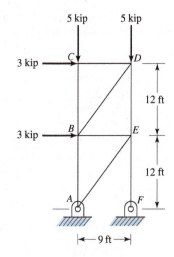

PROB. 6.13

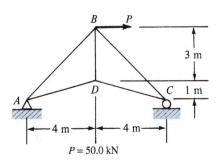

PROB. 6.10

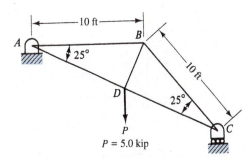

PROB. 6.14

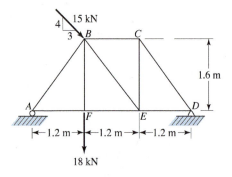

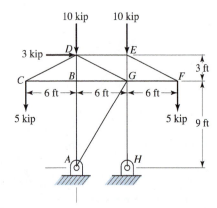

15 kN

18 kN

PROB. 6.15

PROB. 6.16

6.17 through 6.20   By inspection, determine the zero-force members in the trusses loaded as shown.

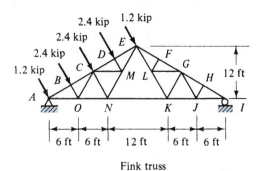

Fink truss

PROB. 6.17

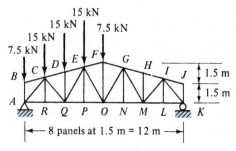

Warren roof truss

PROB. 6.18

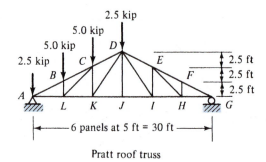

Pratt roof truss

PROB. 6.19

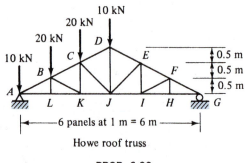

Howe roof truss

PROB. 6.20

6.5 GRAPHICAL METHOD OF JOINTS

A graphical truss analysis may also be performed by applying the *triangular method* or *polygon method* of Chapter 2 to each pin connection in the truss. Because the forces at these points are concurrent and in equilibrium, they form a closed figure when added together. For analysis of an entire truss, the procedure may be streamlined by using a special labeling system, known as *Bow's notation,* and by combining the force polygons for all joints into a single construction, known as a *Maxwell diagram.* Generally attributed to the Scottish physicist James Clerk Maxwell (1831–79), this technique suffers from the same inaccuracies as all graphical vector additions (see Sec. 2.3) and is seldom used in place of the algebraic methods described in Sec. 6.4.

6.6 METHOD OF SECTIONS

A second method for finding the forces in the members of a truss consists of cutting a section through the truss and drawing a free-body diagram of either part of the truss. The forces in the members cut become external forces. The forces are nonconcurrent. Therefore, the three independent equations of equilibrium can be used to solve for no more than three unknown member forces. In Fig. 6.19(a), we show a section cutting through the truss, and in Fig. 6.19(b), we show free-body diagrams of both parts of the truss. Forces in the members are shown in tension. The method will be used to solve several examples.

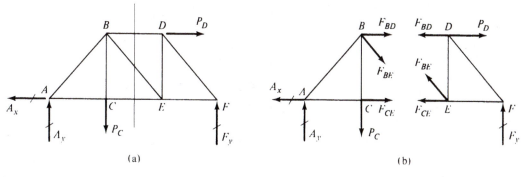

FIGURE 6.19

Example 6.4   Use the method of sections to find the forces in members *BC*, *BG*, *HG*, and *DG* for the truss shown in Fig. 6.20(a).

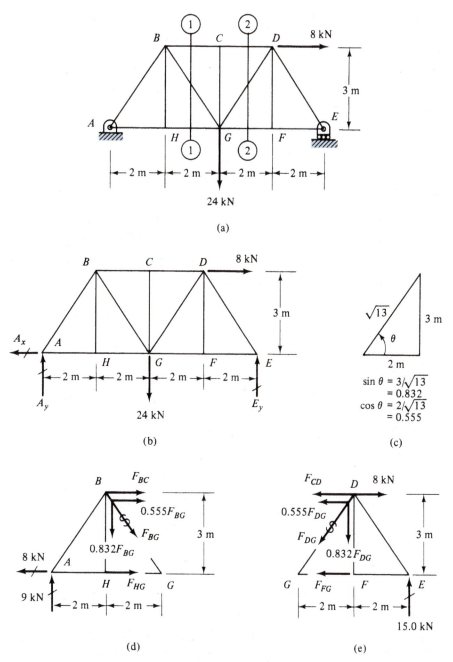

FIGURE 6.20

Solution: The slope of the diagonal members of the truss is calculated in Fig. 6.20(c). The free-body diagram of the entire truss is shown in Fig. 6.20(b). For equilibrium of the *entire truss,*

$$\Sigma M_A = E_y(8) - 24(4) - 8(3) = 0$$
$$E_y = 15.0\,\text{kN}\uparrow$$
$$\Sigma M_E = -A_y(8) - 8(3) + 24(4) = 0$$
$$A_y = 9.0\,\text{kN}\uparrow$$
$$\Sigma F_x = -A_x + 8 = 0$$
$$A_x = 8.0\,\text{kN}\leftarrow$$

Check:

$$\Sigma F_y = A_y + E_y - 24 = 9 + 15 - 24 = 0 \quad \text{OK}$$

The free-body diagram for part of the truss to the left of section 1-1 is shown in Fig. 6.20(d). The *x* and *y* components of the force $\mathbf{F_{BG}}$ are shown on the diagram. For equilibrium of the part of the truss to the left of section 1-1,

$$\Sigma M_B = F_{HG}(3) - 8(3) - 9(2) = 0$$
$$F_{HG} = 14.0\,\text{kN} \quad \text{(tension)} \qquad \textbf{Answer}$$
$$\Sigma M_G = -F_{BC}(3) - 9(4) = 0$$
$$F_{BC} = -12.0\,\text{kN} \quad \text{(compression)} \qquad \textbf{Answer}$$
$$\Sigma F_y = 9.0 - 0.832F_{BG} = 0$$
$$F_{BG} = 10.82\,\text{kN} \quad \text{(tension)} \qquad \textbf{Answer}$$

Check:

$$\Sigma F_x = F_{HG} + 0.555F_{BG} + F_{BC} - 8$$
$$= 14.0 + 0.555(10.82) + (-12) - 8 = 0 \quad \text{OK}$$

The free-body diagram for part of the truss to the right of section 2-2 is shown in Fig. 6.20(e). The *x* and *y* components of the force $\mathbf{F_{DG}}$ are shown on the diagram. For equilibrium of the part of the truss to the right of section 2-2,

$$\Sigma F_y = 15 - 0.832F_{DG} = 0$$
$$F_{DG} = 18.03\,\text{kN} \quad \text{(tension)} \qquad \textbf{Answer}$$

Example 6.5   Use the method of sections to find the force in members *BC, BG,* and *HG* of the truss shown in Figure 6.21(a).

Solution: The free-body diagram of the entire truss is shown in Fig. 6.21(b). The slope of the diagonal members is calculated in Fig. 6.21(c). For equilibrium of the *entire truss,*

$$M_A = E_y(16) - 2(4) - 4(8) - 4(12) = 0$$
$$E_y = 5.5\,\text{kip}\uparrow$$
$$\Sigma M_E = -A_y(16) + 2(12) + 4(8) + 4(4) = 0$$
$$A_y = 4.5\,\text{kip}\uparrow$$

Check:

$$\Sigma F_y = A_y + E_y - 2 - 4 - 4 = 5.5 + 4.5 - 2 - 4 - 4 = 0 \quad \text{OK}$$

The truss to the left of section 1-1 is shown in Fig. 6.21(d). For equilibrium of the part of the truss to the left of section 1-1,

$$\Sigma M_A = -0.658F_{BG}(4.0) - 0.753F_{BG}(3.5) - 2(4) = 0$$
$$F_{BG} = -1.52\,\text{kip} \quad \text{(compression)} \qquad \textbf{Answer}$$

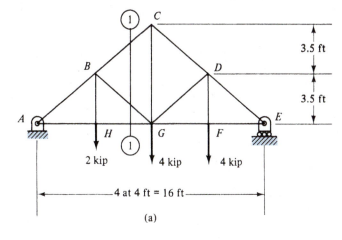

(a)

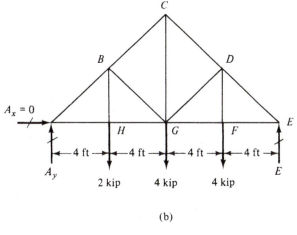

(b)

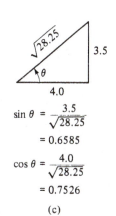

(c)

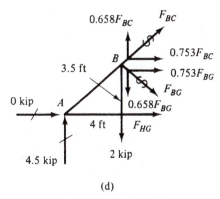

(d)

FIGURE 6.21

$$\Sigma M_B = -4.5(4) + F_{HG}(3.5) = 0$$

$$F_{HG} = 5.14 \text{ kip} \quad \text{(tension)} \qquad \textbf{Answer}$$

$$\Sigma M_G = -4.5(8) - 0.658 F_{BC}(4) - 0.753 F_{BC}(3.5) + 2(4) = 0$$

$$F_{BC} = -5.32 \text{ kip} \quad \text{(compression)} \qquad \textbf{Answer}$$

Check:

$$\Sigma F_x = F_{HG} + 0.753 F_{BC} + 0.753 F_{BG}$$

$$= 5.14 + 0.753(-5.32) + 0.753(-1.52) = 0 \quad \text{OK}$$

PROBLEMS

6.21 For the truss shown, determine by the method of sections the force in members (a) *CD*, *CF*, and *GF*, and (b) *BC*, *BG*, and *AG*.

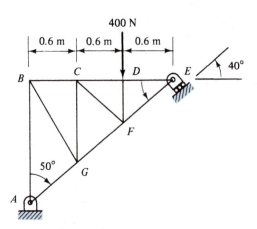

PROB. 6.21

6.22 Determine the force in members *BC*, *BE*, and *FE* of the truss shown. Use the method of sections.

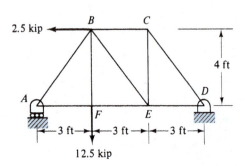

PROB. 6.22

6.23 Using the method of sections, determine the force in members (a) *AB*, *AF*, and *GF*, and (b) *BC*, *BE*, and *FE* of the truss shown.

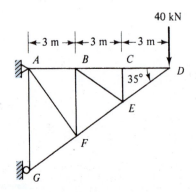

PROB. 6.23

6.24 For the truss shown, determine by the method of sections the force in members *BC*, *BF*, and *AF*.

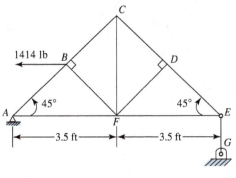

PROB. 6.24

6.25 Determine the force in members *BC*, *BG*, and *HG* of the truss shown. Use the method of sections.

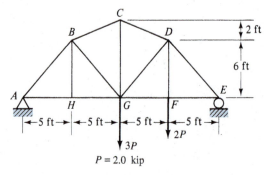

PROB. 6.25

6.26 For the truss shown, determine by the method of sections the force in members (a) *BC*, *GC*, and *GF*, and (b) *CD*, *FD*, and *FE*.

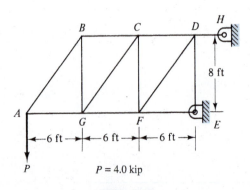

PROB. 6.26

6.27 Determine the force in members *BC*, *BE*, and *EF* of the truss shown. Use the method of sections.

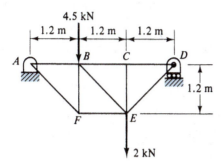

4.5 kN

PROB. 6.27

6.28 Determine the force in members *BC*, *BE*, and *EF* of the truss shown. Use the method of sections.

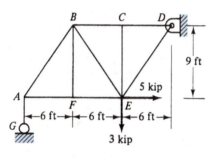

PROB. 6.28

6.7 FRAMES AND MACHINES

Like a truss, the *pin-connected frame* is a rigid structure that is designed to support external forces. However, while trusses consist only of two-force members connected at their endpoints, frames have at least one member that is acted on by three or more forces.

A *machine,* which also contains members that carry more than two forces, is designed to transmit forces from the input to the output of the device, often changing the magnitude and direction of these forces as well. To accomplish this, machines have at least one movable member and, unlike rigid pin-connected frames, are themselves nonrigid structures. Machines may vary in size and complexity, from the small, precise robot arm of Fig. 6.22 to the extremely large and durable mining truck shown in Fig. 6.23.

The same methods are used for the solution of both frames and machines. Consider the frame of Fig. 6.24(a), which supports load *W*. The frame consists of three members joined together by frictionless pins. It is supported by a pin at *A* and a roller at *D*.

A free-body diagram of the entire frame is shown in Fig. 6.24(b). The external loads and reactions appear on the frame. They consist of the weight *W* at *F*, A_x, and A_y, the

FIGURE 6.22. This electro-pneumatic robot arm is a precision machine that is capable of the rapid, repetitive, and often delicate *pick-and-place* movements required in many types of high-volume manufacturing operations.

horizontal and vertical components of the pin reaction at *A*, and the reaction of the roller, D_x. The internal reactions holding the frame together are not shown on the free-body diagram. To determine the forces that hold the frame together, we separate the members of the frame and draw free-body diagrams for each member. The forces at the hinges are now external and must be shown on the free-body diagrams.

The free-body diagrams for the three parts of the frame are shown in Fig. 6.24(c)–(e). In accordance with Newton's third law, the forces at *C* on member *AD* are in opposite directions from the forces on member *CF*. Similarly, the force at *B* on member *AD* is opposite the force on member *BE*, and the force at *E* on member *CF* is opposite the force on member *BE*. Member *BE* is a short link—a special case of a two-force member—and the force in the member acts along the member. If the three free-body diagrams for the parts of the frame are combined, the forces at *B*, *C*, and *E* cancel, and we have a free-body diagram for the entire frame.

This method of disassembling frames and machines can also be used for problems that are otherwise statically indeterminate. Consider the frame of Fig. 6.25(a), which is identical to that of Fig. 6.24(a), except that the roller at *D* has been replaced by a pin connection. From the free-body diagram of Fig. 6.25(b), notice that there are now *four* unknown external forces acting on the assembled structure, but only *three* equations are available to describe the conditions for its equilibrium. Mathematically, this problem is not solvable because the requirement for *complete* solution is that the number of unknowns must be equal to the number of equations relating these unknowns.

To obtain a solution (and to determine the forces that hold the frame together), we again disassemble the frame and draw free-body diagrams for each member, as shown in

FIGURE 6.23. This "extreme machine" is the largest mining truck manufactured anywhere in the world and is designed to carry up to 360 tons of coal, copper ore, or gold ore. When fully loaded, the vehicle has a total weight of 1.23 *million* pounds (40 percent heavier than a fully loaded Boeing 747 airplane) and can reach speeds of 40 mph on tires that are 13 ft tall and 5 ft wide. The truck is 47 ft long, 30 ft wide, and 23 ft tall and is powered by a 3400 (gross)-hp engine that is fed by a 1000-gal fuel tank. (*Reprinted courtesy of Caterpillar Inc.*)

Fig. 6.25(c)–(e). This process has increased the number of unknown forces to eight: A_x, A_y, B, C_x, C_y, D_x, D_y, and E. However, each free-body diagram now has its own set of equilibrium equations: *three* for Fig. 6.25(c), *three* for Fig. 6.25(d), and *two* for Fig. 6.25(e) (no moment equation). Because the total number of equations available equals the number of unknowns, the problem is mathematically solvable.

As you can see, disassembling a frame or machine into several component parts increases the number of unknowns by exposing the internal forces that exist at each point of connection. However, because each internal force appears on *two* free-body diagrams, the total number of equilibrium

equations available increases faster than the number of unknowns. For statically determinate problems, this will allow for a complete solution.

Also, although six, seven, or eight equilibrium equations involving six, seven, or eight unknown forces might seem like a formidable problem in algebra, usually these equations need not be solved simultaneously. By working from one free-body diagram to another as necessary, it is generally possible to solve one linear equation at a time.

The following example problems illustrate the step-by-step procedure required to solve problems involving frames and machines.

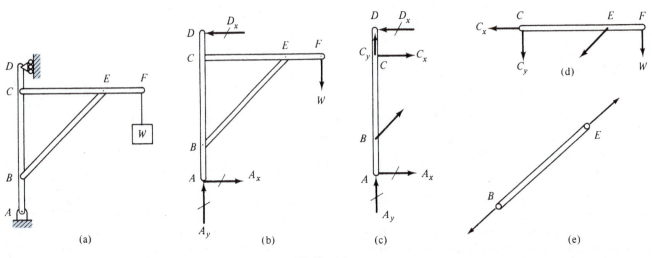

FIGURE 6.24

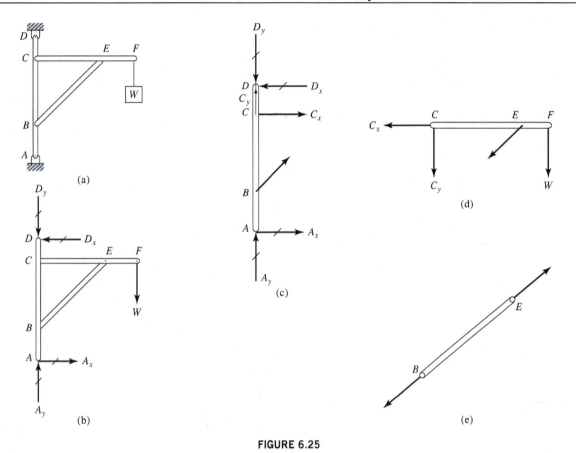

FIGURE 6.25

Example 6.6   The frame in Fig. 6.26(a) is supported by a pin at A and a link DE. Neglect the weight of the members. Determine the reaction at pin A and the tensile or compressive force in members DE, BD, and CD.

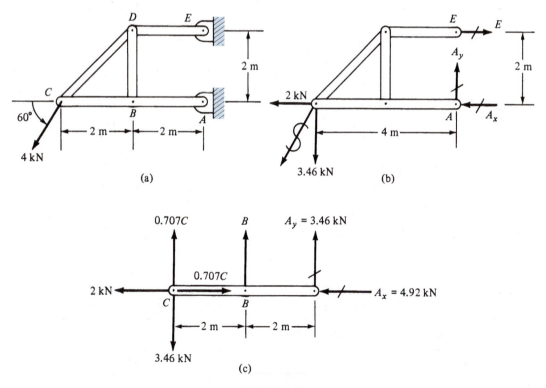

FIGURE 6.26

Solution: **Entire structure:** A free-body diagram of the entire structure is drawn in Fig. 6.26(b). The x and y components of the 4-kN force are shown in the diagram. Member DE is a two-force member. From equilibrium,

$$\Sigma M_A = 3.46(4) - E(2) = 0 \qquad E = 6.92\,\text{kN} \qquad \textbf{Answer}$$

$$\text{Member } DE \text{ is in tension}$$

$$\Sigma F_x = E - 2.0 - A_x = 0 \qquad A_x = 4.92\,\text{kN} \leftarrow \qquad \textbf{Answer}$$

$$\Sigma F_y = A_y - 3.46 = 0 \qquad A_y = 3.46\,\text{kN} \uparrow \qquad \textbf{Answer}$$

Check:

$$\Sigma M_E = 3.46(4) - 2.0(2) - A_x 2$$
$$= 13.84 - 4.0 - 9.84 = 0 \qquad \text{OK}$$

Member AC: Members BD and CD are two-force members; therefore, the directions of the forces in the members are known, and we draw the free-body diagram of member AC in Fig. 6.26(c). Components of C are shown in the diagram. From equilibrium,

$$\Sigma M_C = B(2) + 3.46(4) = 0 \qquad B = -6.92\,\text{kN} \qquad \textbf{Answer}$$

$$\text{Member } BD \text{ is in compression}$$

$$\Sigma M_B = -0.707C(2) + 3.46(2) + 0.346(2) = 0$$
$$C = 9.79\,\text{kN} \qquad \textbf{Answer}$$

$$\text{Member } CD \text{ is in tension}$$

Check:

$$\Sigma F_x = -2.0 + 0.707C - 4.92$$
$$= -2.0 + 6.92 - 4.92 = 0 \qquad \text{OK}$$

Example 6.7   The frame shown in Fig. 6.27(a) is supported by a pin at A and rollers at E. Find the components of all forces acting on each member of the frame.

Solution: We draw free-body diagrams for the entire structure and for each member of the structure. They are shown in Fig. 6.27(b)–(e).

Entire Structure
For equilibrium,

$$\Sigma M_A = -1000(9) - 600(3) + E_y(9) = 0$$
$$E_y = 1200\,\text{lb} \qquad \textbf{Answer}$$
$$\Sigma F_x = 1000 - A_x = 0$$
$$A_x = 1000\,\text{lb} \qquad \textbf{Answer}$$
$$\Sigma F_y = A_y - 600 + E_y = A_y - 600 + 1200 = 0$$
$$A_y = -600\,\text{lb} \qquad \textbf{Answer}$$

Member AC:

$$\Sigma M_B = -1000(4) - C_x(8) - A_x(5) = 0$$

Substituting the value of $A_x = 1000\,\text{lb}$ yields

$$C_x = -1125\,\text{lb} \qquad \textbf{Answer}$$
$$\Sigma M_C = 1000(4) + B_x(8) - A_x(13) = 0$$
$$B_x = 1125\,\text{lb} \qquad \textbf{Answer}$$
$$\Sigma F_y = A_y - B_y - C_y = -600 - B_y - C_y = 0$$
$$B_y + C_y = -600 \qquad \textbf{(a)}$$

Check:

$$\Sigma F_x = C_x + 1000 + B_x - A_x$$
$$= -1125 + 1000 + 1125 - 1000 = 0 \qquad \text{OK}$$

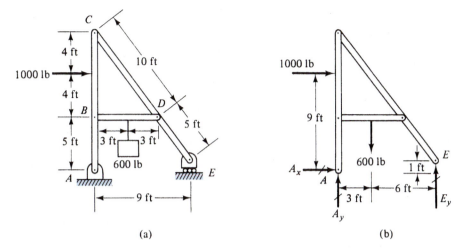

(a)   (b)

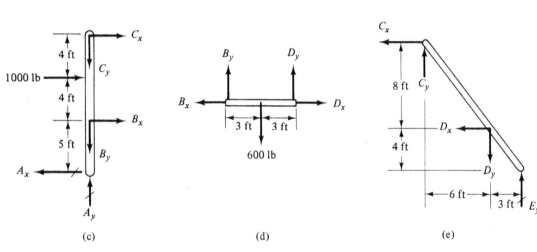

(c)   (d)   (e)

FIGURE 6.27

Member BD:

$$\Sigma M_D = -B_y(6) + 600(3) = 0$$
$$B_y = 300 \text{ lb} \qquad\qquad \textbf{Answer}$$

Substituting into Eq. (a), we obtain

$$C_y = -900 \text{ lb} \qquad\qquad \textbf{Answer}$$
$$\Sigma F_y = B_y + D_y - 600 = 300 + D_y - 600 = 0$$
$$D_y = 300 \text{ lb} \qquad\qquad \textbf{Answer}$$
$$\Sigma F_x = -B_x + D_x = -1125 + D_x = 0$$
$$D_x = 1125 \text{ lb} \qquad\qquad \textbf{Answer}$$

Member CE:
Check:

$$\Sigma F_x = -C_x - D_x = -(-1125) - 1125 = 0 \quad \text{OK}$$

and

$$\Sigma F_y = E_y - D_y + C_y = 1200 - 300 + (-900) = 0 \quad \text{OK}$$

The pin reactions on member AC are

$$A_x = 1000 \text{ lb} \leftarrow \qquad B_y = 300 \text{ lb} \downarrow$$
$$A_y = 600 \text{ lb} \downarrow \qquad C_x = 1125 \text{ lb} \rightarrow$$
$$B_x = 1125 \text{ lb} \rightarrow \qquad C_y = 900 \text{ lb} \uparrow$$

The pin reactions on member *BD* are

$$B_x = 1125\,lb \leftarrow \qquad D_x = 1125\,lb \rightarrow$$
$$B_y = 300\,lb \uparrow \qquad D_y = 300\,lb \uparrow$$

The pin reactions on member *CE* are

$$C_x = 1125\,lb \leftarrow \qquad D_y = 300\,lb \downarrow$$
$$C_y = 900\,lb \downarrow \qquad E_y = 1200\,lb \uparrow$$
$$D_x = 1125\,lb \leftarrow$$

Example 6.8 The three-hinged structure in Fig. 6.28(a) is loaded as shown. Find the components of all forces acting on each member of the frame.

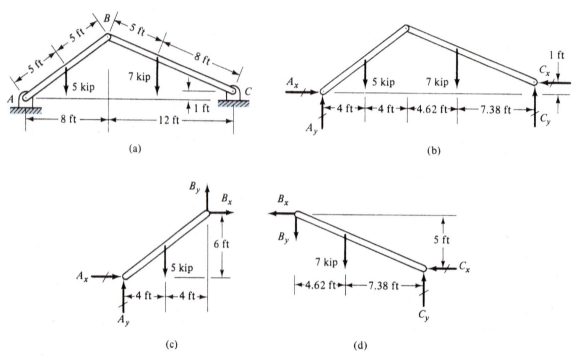

(a)   (b)   (c)   (d)

FIGURE 6.28

Solution: The free-body diagram of the entire structure and members *AB* and *BC* are shown in Fig. 6.28(b)–(d).

Entire Structure:
For equilibrium,

$$\Sigma M_A = -5(4) - 7(12.615) + C_x + C_y(20) = 0$$
$$C_x + 20C_y = 108.3 \tag{a}$$

$$\Sigma M_C = -A_y(20) + A_x(1) + 5(16) + 7(7.38) = 0$$
$$A_x - 20A_y = -131.7 \tag{b}$$

$$\Sigma F_x = A_x - C_x = 0$$
$$A_x = C_x \tag{c}$$

Member *AB*:

$$\Sigma M_B = A_x(6) + 5(4) - A_y(8) = 0$$
$$6A_x - 8A_y = -20 \tag{d}$$

From Eq. (b),

$$A_x = 20A_y - 131.7 \tag{e}$$

Substituting for A_x in Eq. (d), we have

$$6(20A_y - 131.7) - 8A_y = -20$$
$$120A_y - 790 - 8A_y = -20$$
$$A_y = 6.88 \text{ kip} \qquad \textbf{Answer}$$

From Eq. (e),

$$A_x = 20(+6.88) - 131.7$$
$$= +5.84 \text{ kip} \qquad \textbf{Answer}$$

From Eq. (c),

$$C_x = +5.84 \text{ kip} \qquad \textbf{Answer}$$

From Eq. (a),

$$20C_y = 108.3 - C_x$$
$$C_y = 5.12 \text{ kip} \qquad \textbf{Answer}$$
$$\Sigma F_x = A_x + B_x = 5.84 + B_x = 0$$
$$B_x = -5.84 \text{ kip} \qquad \textbf{Answer}$$
$$\Sigma F_y = A_y + B_y - 5 = 6.88 + B_y - 5 = 0$$
$$B_y = -1.88 \text{ kip} \qquad \textbf{Answer}$$

Member BC:
Check:

$$\Sigma F_x = -B_x - C_x = -(-5.84) - (5.84) = 0 \quad \text{OK}$$
$$\Sigma F_y = -B_y - 7 + C_y = -(-1.88) - 7 + 5.12 = 0 \quad \text{OK}$$
$$\Sigma M_B = -7(4.62) - C_x(5) + C_y(12)$$
$$= -7(4.62) - 5.84(5) + 5.12(12) = 0 \quad \text{OK}$$

The pin reactions on member AB are

$$\mathbf{A_x} = 5.84 \text{ kip} \rightarrow \qquad \mathbf{B_x} = 5.84 \text{ kip} \leftarrow$$
$$\mathbf{A_y} = 6.88 \text{ kip} \uparrow \qquad \mathbf{B_y} = 1.88 \text{ kip} \downarrow$$

The pin reactions on member BC are

$$\mathbf{B_x} = 5.84 \text{ kip} \rightarrow \qquad \mathbf{C_x} = 5.84 \text{ kip} \leftarrow$$
$$\mathbf{B_y} = 1.88 \text{ kip} \uparrow \qquad \mathbf{C_y} = 5.12 \text{ kip} \uparrow$$

Example 6.9 Determine the force required to maintain equilibrium of the frictionless pulley system shown in Fig. 6.29(a) and the reaction at the ceiling.

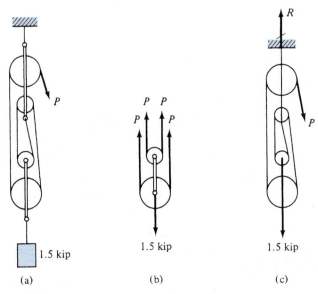

FIGURE 6.29 (a)   (b)   (c)

Solution: The cable is continuous, and the pulleys are frictionless; therefore, the cable has a constant tension. The free-body diagram for the lower two pulleys is shown in Fig. 6.29(b). Equilibrium requires that

$$\Sigma F_y = 4P - 1.5 = 0$$
$$P = 0.375 \text{ kip}$$ **Answer**

The free-body diagram for the entire system is shown in Fig. 6.29(c). Equilibrium requires that

$$\Sigma F_y = R - P - 1.5 = R - 0.375 - 1.5 = 0$$
$$R = 1.875 \text{ kip}$$ **Answer**

Example 6.10   A force is applied to the handle of the pliers shown in Fig. 6.30(a). Find the force applied to the bolt and the horizontal and vertical reactions at the hinge A.

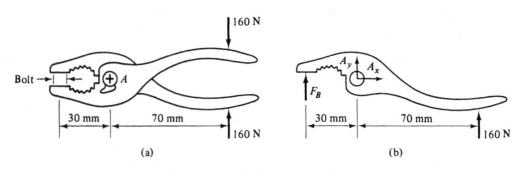

FIGURE 6.30

Solution: A free-body diagram for one member of the pliers is shown in Fig. 6.30(b). Equilibrium requires that

$$\Sigma M_A = 160(70) - F_B(30) = 0$$
$$F_B = 373 \text{ N}$$ **Answer**
$$\Sigma F_x = A_x = 0$$ **Answer**
$$\Sigma F_y = F_B + A_y + 160 = 373 + A_y + 160 = 0$$
$$A_y = -533 \text{ N}$$ **Answer**

PROBLEMS

6.29 A building contractor uses his utility trailer to bring tools and materials to various job sites. The trailer's loaded weight, L, acts as if concentrated at its center of gravity, E. The pickup truck used to tow this trailer has a total weight, W, and a center of gravity at B. Ball hitch D, between trailer and truck, acts like a pin connection. If the truck and trailer are at rest and have the given dimensions and weights:

a. Determine the force (called the *tongue weight*) that the trailer exerts on ball hitch D.
b. Find the force exerted by the ground on each of the two tires at A, C, and F.

   $a = 1.17 \text{ m}$ $b = 2.24 \text{ m}$ $c = 1.42 \text{ m}$
   $d = 2.59 \text{ m}$ $e = 0.61 \text{ m}$ $L = 13.5 \text{ kN}$
   $W = 28.5 \text{ kN}$

6.30 Same as Prob. 6.29 but with these dimensions and weights:

   $a = 46 \text{ in.}$ $b = 88 \text{ in.}$ $c = 56 \text{ in.}$
   $d = 102 \text{ in.}$ $e = 24 \text{ in.}$ $L = 3000 \text{ lb}$
   $W = 6400 \text{ lb}$

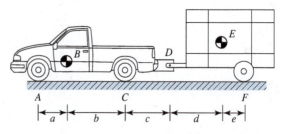

PROB. 6.29 and PROB. 6.30

6.31 A decorative mobile is to be constructed using three geometric shapes suspended from weightless rods

BCD and EFG by light cords as shown. The rectangle, circle, and triangle are cut from sheets of different colored glass, each having the same uniform thickness, so that the weight of each shape is directly proportional to its area. Find dimensions d_1 and d_2 required for the mobile to balance.

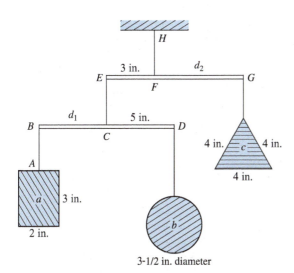

PROB. 6.31

6.32 through 6.38   Determine the components of the forces acting on each member of the pin-connected frame shown.

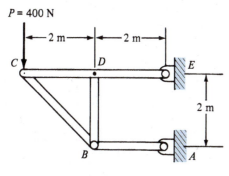

PROB. 6.32

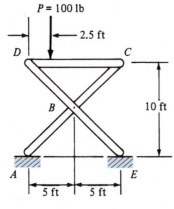

PROB. 6.33

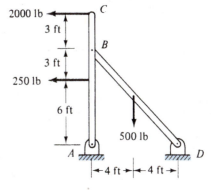

PROB. 6.34

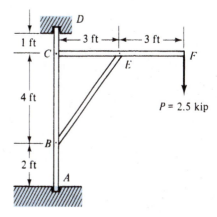

PROB. 6.35

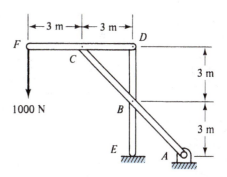

PROB. 6.36

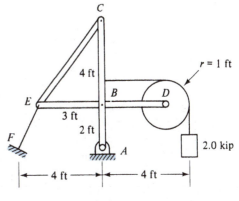

PROB. 6.37

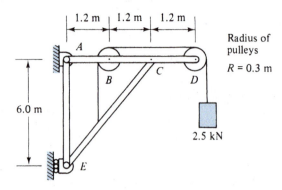

PROB. 6.38

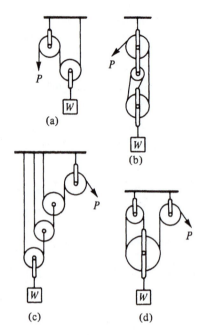

PROB. 6.42

6.39 A crate of weight W is supported by boom member ABC, which can be raised or lowered as hydraulic cylinder DB extends or retracts. For the position shown, with $W = 18.5$ kN, $a = 1.75$ m, $b = 2.50$ m, $c = 0.75$ m, and $d = 1.40$ m:
  a. Find the force exerted on ABC by DB.
  b. Determine the forces exerted by pin A on ABC.

6.40 Same as Prob. 6.39, but $W = 425$ lb, $a = 26.0$ in., $b = 38.5$ in., $c = 14.0$ in., and $d = 20.0$ in.

6.43 A force $P = 46$ N is applied to the claw hammer shown. What force Q is exerted on the nail?

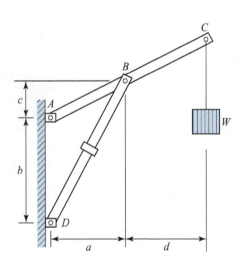

PROB. 6.39 and PROB. 6.40

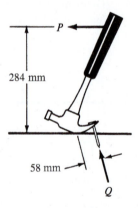

PROB. 6.43

6.44 The foot lever shown is used to transmit an applied force P at A to the rod at C. Determine the applied force and the hinge reaction at B if (a) $a = 230$ mm, $b = 115$ mm, and the required force $Q = 920$ N and (b) $a = 9$ in., $b = 4.5$ in., and the required force $Q = 200$ lb.

6.41 For the portable engine hoist shown in Prob. 5.3 of Chapter 5:
  a. Find the force exerted by hydraulic cylinder BD on member CDE.
  b. Determine the pin reaction at C.

6.42 If $W = 1.25$ kip, determine the force P required to maintain equilibrium of the frictionless pulley systems shown and the reactions at the ceiling.

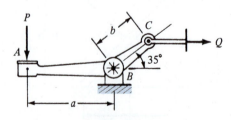

PROB. 6.44

6.45 For the crank mechanism shown, a force $P = 1800$ lb is applied to the piston at C. Determine the required couple M at A for equilibrium and the corresponding force on the pins at A, B, and C. Neglect the weight of the members.

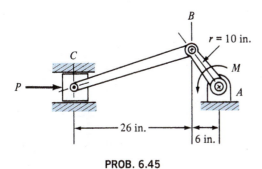

PROB. 6.45

6.46 In the device shown, called a toggle joint, a small force P can be applied at B to produce a much larger force Q at A. If $P = 2.2$ kN, determine the force Q and the force on the pins at A, B, and C. Neglect the weight of the members.

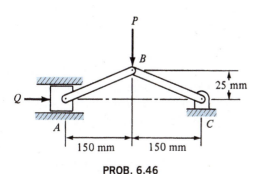

PROB. 6.46

6.47 Each handle of the bolt cutters shown acts like a *simple lever*. For the dimensions/input force given, find:
  a. The force exerted by pin B on handle ABD.
  b. The force exerted by handle ABD on the bolt at A.

    $a = 1\text{-}7/8$ in. $b = 28$ in. $P_i = 40$ lb

Compare this result with that of Prob. 6.49(c).

6.48 Same as Prob. 6.47 but with the following dimensions/input force.

    $a = 42$ mm $b = 668$ mm $P_i = 225$ N

Compare your answer for part (b) with that of Prob. 6.50(c).

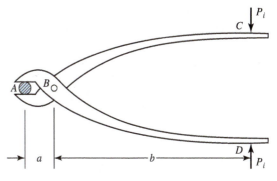

PROB. 6.47 and PROB. 6.48

6.49 To increase their mechanical advantage, bolt cutters are often designed as shown in (i) to use the action of a *compound lever*. For such a machine, simplified in (ii), the output of one lever (handle DFG) becomes the input of a second lever (jaw ABD). For the dimensions/input force given, find
  a. The force exerted by pin F on handle DFG.
  b. The force exerted by pin B on jaw ABD.
  c. The force exerted on the bolt at A by jaw ABD.
  $a = 1\text{-}5/8$ in. $b = 4$ in. $c = 3/4$ in.
  $d = 21\text{-}1/2$ in. $P_i = 40$ lb

Compare this result with that of Prob. 6.47(b).

6.50 Same as Prob. 6.49 but with these dimensions/input force:
  $a = 42$ mm $b = 102$ mm $c = 20$ mm
  $d = 546$ mm $P_i = 225$ N

Compare your results for part (c) with those of Prob. 6.48(b).

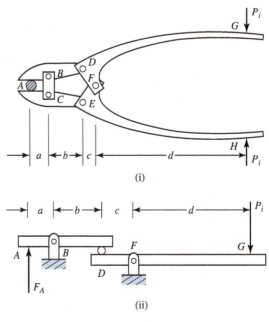

PROB. 6.49 and PROB. 6.50

6.51 The gripper shown was designed for use by individuals such as wheelchair patients and invalids whose mobility or range of motion may be restricted. Such devices generally have an overall length of 24 to 36 in. (610 to 915 mm) and allow users to pick up or retrieve articles that would otherwise remain out of reach. If a squeezing force, *P*, is exerted at pistol grip *A* and trigger *B*, member *BCD* pivots about pin *C* and creates a tension in cord *DEFH*. This cord passes over smooth pins at *E* and *F* and is attached to the movable jaw *GHJ* at pin *H*. Tension in the cord causes *GHJ* to pivot about pin *G* and grip the desired article (shaded). Find the forces exerted on the cylindrical object at *J* and *K* for this dimensional data and squeezing force:

a = 2 in.	*b* = 1/4 in.
c = 2 in.	*d* = 1-1/4 in.
e = 1-1/4 in.	*f* = 1-1/2 in.
g = 2-1/2 in.	*P* = 5 lb

6.52. Same as Prob. 6.51 but with the following data:

a = 51 mm	*b* = 6.5 mm
c = 51 mm	*d* = 32 mm
e = 32 mm	*f* = 38 mm
g = 64 mm	*P* = 22 N

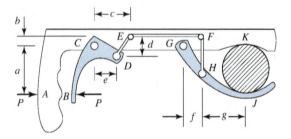

PROB. 6.51 and PROB. 6.52

6.53 A box of weight *W* = 2000 N is lifted by tongs that cross without touching as shown. Determine the force on link *AE*. Neglect the weight of the members.

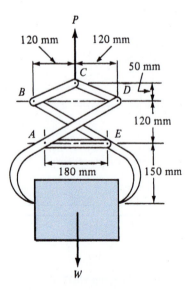

PROB. 6.53

6.54 The hydraulically controlled backhoe exerts a horizontal back force *P* = 15 kN. With the members in the positions shown, determine the forces on the pins at *B* and *E*. Neglect the weight of the members.

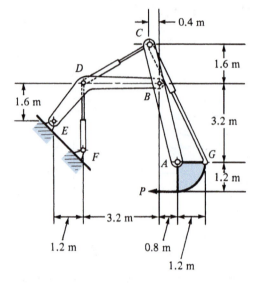

PROB. 6.54

*6.8 CABLES

Flexible members such as cables, chains, and wire ropes represent an important class of structural elements and are found in a wide variety of applications. These include electrical transmission lines (see Application Sidebar 6); cable-stayed and suspension bridges; tower guy wires (see Application Sidebar 8); and some roof suspension systems.

In this section, we will discuss three common types of cable loads. The first of these involves single or multiple *concentrated loads applied randomly along the cable's length.* As shown in Fig. 6.31, such loads pull the cable into a series

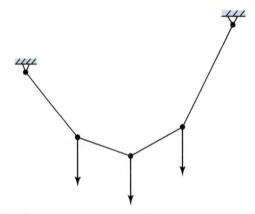

FIGURE 6.31. Concentrated loads applied at *random* points along a cable create straight-line segments that form part of a *polygon.*

---

*Sections denoted by an asterisk indicate material that can be omitted without loss of continuity.

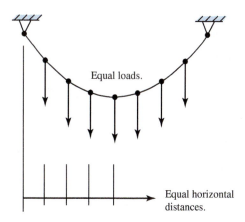

FIGURE 6.32. *Equal* concentrated loads applied at *equal* horizontal distances across the span of a cable yield the characteristic shape of a *parabola*.

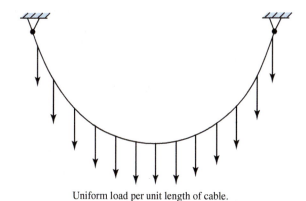

FIGURE 6.33. Loads that are *uniformly distributed* along a cable's length produce the smooth shape known as a *catenary curve*.

of straight-line segments so that the cable's shape resembles some portion of a *polygon*.

The second type of load consists of *equal concentrated loads applied at equal horizontal distances across the cable span*. These loads produce a cable shape whose smooth curve (Fig. 6.32) is defined mathematically as a *parabola*. The analyses of both types of concentrated loads described here generally assume that the weight of the cable is negligible compared to the applied loads.

The third type of load may be described as *equal loads applied at equal distances along the cable itself*. This description includes cables having a uniform weight per length and external loads such as snow and ice, which may also be considered as being uniform along the cable. These types of loads pull the cable into a smooth curve called a *catenary curve* (see Application Sidebar 7), as shown in Fig. 6.33.

Let us now investigate the behavior of cables under these three different types of loads.

Concentrated Loads at Random Locations

For loads similar to those shown in Fig. 6.31, our analysis makes the following assumptions:

1. All loads are in the plane of the cable (i.e., the system is *coplanar*).

2. The cable is *flexible* and produces no resistance to bending.

3. *Weight* of the cable is *negligible* compared to the applied loads.

4. From assumptions (2) and (3), the section of cable between any two adjacent loads behaves like a *two-force member*.

Because the reactions at each end of a cable generally have both vertical and horizontal components, a free-body diagram of the entire cable has four unknowns, and the system is statically indeterminate. As is the case with frames and machines, this often requires the use of multiple free-body diagrams, here involving sections of the cable. In addition, sufficient geometric information regarding cable placement and location of loads is usually necessary for a complete analysis of most cable systems. The general method of solution is shown by the following example.

Example 6.11   Cable *ABCD* supports two concentrated loads and spans a distance of 28 ft, as shown in Fig. 6.34. The distance along the cable from support *A* to the 150-lb load is 10 ft. For this cable, determine the following:

a.   Tensions in portions *AB*, *BC*, and *CD* of the cable.

b.   Maximum cable sag "*h.*"

c.   Total length of the cable.

Solution: a.   The free-body diagram for this entire cable is shown in Fig. 6.35, and our three equations of equilibrium are

$$\Sigma F_x = 0 \qquad\qquad A_x = D_x$$
$$\Sigma F_y = 0 \qquad A_y + D_y = 150 + 250$$
$$\Sigma M_A = 0 \qquad D_y \times (28\,\text{ft}) = 150 \times (6\,\text{ft}) + 250 \times (18\,\text{ft})$$

These equations yield: $D_y = 192.9$ lb   and: $A_y = 207.1$ lb

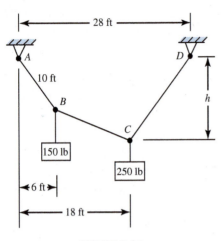

FIGURE 6.34

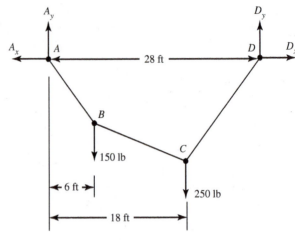

FIGURE 6.35

but no values for A_x and D_x. Using the same technique as in our analysis of structures and machines, then, we take a free-body diagram of portion AB of the cable, as shown in Fig. 6.36. The equilibrium equations become:

$$\Sigma M_B = 0 \qquad 207.1 \times (6\,\text{ft}) = A_x \times (8\,\text{ft}) \quad \rightarrow \quad A_x = 155.3\,\text{lb}$$
$$\Sigma F_x = 0 \qquad A_x = 155.3 = T_{BC}\cos\theta$$
$$\Sigma F_y = 0 \qquad 207.1 = 150 + T_{BC}\sin\theta$$

Simplifying the horizontal and vertical force equations:

$$T_{BC}\sin\theta = 57.1 \qquad \text{and:} \qquad T_{BC}\cos\theta = 155.3$$

Dividing one equation by the other yields

$$\frac{T_{BC}\sin\theta}{T_{BC}\cos\theta} = \frac{57.1}{155.3} = \tan\theta = 0.36768$$

Then $\theta = 20.19°$, and:

$$T_{BC} = \frac{57.1}{\sin 20.19°} \quad \rightarrow \quad T_{BC} = 165.4\,\text{lb} \qquad \textbf{Answer}$$

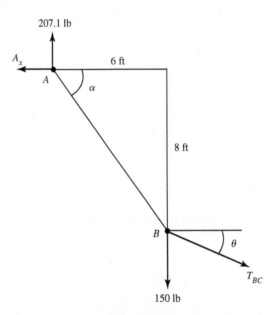

FIGURE 6.36

From the free-body diagram of point A shown in Fig. 6.37:

$$R_A = T_{AB} = \sqrt{A_x^2 + A_y^2} = \sqrt{(155.3)^2 + (207.1)^2}$$

or:
$$T_{AB} = 258.9 \text{ lb}$$ **Answer**

Similarly, Fig. 6.35 tells us that $A_x = D_x = 155.3$ lb, so the free-body diagram of point D shown in Fig. 6.38 yields

$$R_D = T_{CD} = \sqrt{D_x^2 + D_y^2} = \sqrt{(155.3)^2 + (192.9)^2}$$

or:
$$T_{CD} = 247.6 \text{ lb}$$ **Answer**

Our free-body diagrams also yield geometric information that may be useful in parts (b) and (c) of this problem. From Fig. 6.37:

$$\tan\alpha = A_y/A_x = (207.1)/(155.3) = 1.3335 \quad \rightarrow \quad \alpha = 53.13°$$

Likewise, from Fig. 6.38:

$$\tan\beta = D_y/D_x = (192.9)/(155.3) = 1.2421 \quad \rightarrow \quad \beta = 51.16°$$

Notice that although the vertical force components are different for each portion of the cable, *the horizontal force components are the same at all points along the cable*. In other words:

$$A_x = T_{AB}\cos\alpha = T_{BC}\cos\theta = T_{CD}\cos\beta = D_x$$

From this we see that for a cable carrying vertical loads, the *tension is greatest* in that portion of the cable, which is at the *largest angle from a horizontal reference line*. As this angle increases, the value of the cosine decreases, whereas overall tension must be large enough to maintain the necessary horizontal force component. This condition, shown graphically in Fig. 6.39, generally produces a maximum tension at one of the supports.

b. The geometry of cable segment CD is shown in Fig. 6.40, where angle β is 51.16°. Trigonometry yields

$$CD\cos\beta = 10 \quad \rightarrow \quad CD = 15.95 \text{ ft}$$

and:
$$CD\sin\beta = h \quad \rightarrow \quad h = 12.42 \text{ ft}$$ **Answer**

c. Similarly, the geometry of cable segment BC is shown in Fig. 6.41, where angle θ is 20.19°. Then:

$$BC\cos\theta = 12 \quad \rightarrow \quad BC = 12.79 \text{ ft}$$

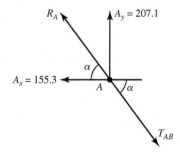

FIGURE 6.37

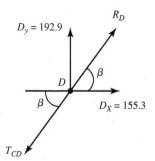

FIGURE 6.38

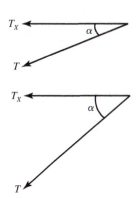

FIGURE 6.39. For a cable at angle α and tension T, as shown, the horizontal component of tension is $T_x = T \cos \alpha$. If *angle α increases*, the value of cos α decreases, and *tension T must also increase* to maintain a constant value of T_x. For a given cable, then, we expect the highest tension to occur in that cable segment that is at the steepest angle relative to a horizontal reference line, generally at one of the supports.

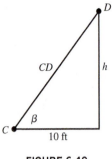

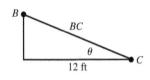

FIGURE 6.40   **FIGURE 6.41**

and total cable length, L_C, becomes:

$$L_c = AB + BC + CD = 10\,\text{ft} + 12.79\,\text{ft} + 15.95\,\text{ft}$$

$$L_c = 38.74\,\text{ft}$$ **Answer**

Equal Loads at Equal Horizontal Distances

The most common example for this type of loading is the basic *suspension bridge*, as shown in Fig. 6.42. Here the weights of bridge deck, roadway, and vehicular traffic are transmitted to the main cables by equispaced vertical cables (called *suspenders*) of varying lengths. As mentioned previously, all the cable weights are generally considered negligible compared to those of the deck and roadway and are therefore neglected. Also, it is common practice to describe the weights

of deck and roadway as a *uniform load per unit length* (in lb/ft or N/m), rather than to designate the concentrated load applied to the main cables by each vertical suspender cable. This uniform load is determined by dividing the total weight of deck and roadway by the span of the bridge.

To help analyze the parabolic curve formed by the main cables, we may locate the *origin* of a rectangular coordinate system at the lowest point (called the *vertex*) on each cable, as shown in Fig. 6.42. For the free-body diagram of a cable

CABLE PRACTICES

SB.6(a)

SB.6(b)

Many cables contain wires of different materials. The 450,000-volt electrical transmission line shown in SB.6 (a) is made primarily of aluminum wires for conductivity, but most of its strength comes from a seven-wire steel core. Overall diameter of this cable is approximately 2.25 in.; each steel wire is about the same diameter as the 8d common nail shown in the photo.

The main cables on suspension bridges are often inspected for corrosion as shown in SB.6 (b). First, an individual cable is unwrapped, then wooden wedges are used to separate the thousands of parallel wires that make up the cable. (The main cables on California's Golden Gate Bridge, for example, are over 36 in. in diameter, and each contains more than 27,000 wires!) The interior of a cable is then inspected visually, and a linseed-oil solution is applied as necessary to control corrosion. The 30-in. cable shown here is located on the Benjamin Franklin Bridge, which crosses the Delaware River between Philadelphia, PA and Camden, NJ. *(Courtesy of Piasecki Steel Construction Corp., Castleton, New York, www.piaseckisteel.com)*

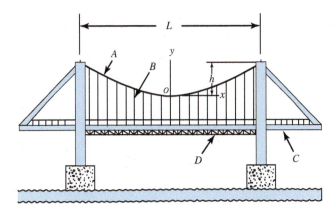

FIGURE 6.42. For a typical suspension bridge, roadway (*C*) rests on a steel deck (*D*), which is often trussed for stiffness. Weights of deck and roadway are transmitted to the main cables (*A*) by suspenders (*B*) of varying lengths. Each main cable generally consists of numerous bundles of parallel wires (see Application Sidebar 6). The suspenders are formed from groups of wires that have been twisted into *strands,* which in turn have been twisted about a metal or natural-fiber core to form *wire rope.*

section of any horizontal span, *x,* and having one end at the origin (see Fig. 6.43(a)):

1. Cable tension at the origin is horizontal and is shown as T_0.

2. Total applied load, *W,* is the product of the uniform load per unit length, *w,* described earlier, and the length, *x,* of the cable section. This load $W = wx$ acts as if concentrated at the midpoint of our cable section (i.e., at a distance $x/2$ from the origin).

3. Cable tension at distance *x* is designated *T* and acts in a direction *tangent* to the curve at that point. The direction is specified by horizontal angle θ.

From the free-body diagram of Fig. 6.43(a), the horizontal and vertical equilibrium equations become:

$$T\cos\theta = T_0 \quad \text{and} \quad T\sin\theta = W$$

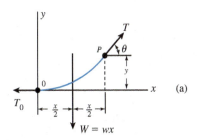

(a)

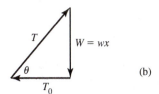

(b)

FIGURE 6.43. (a) Free-body diagram of any general cable section having one end at the vertex and the other at some horizontal span, *x.* (b) Equilibrium force triangle for the cable section of (a).

The force triangle of Fig. 6.43(b) yields these useful relationships:

$$\tan\theta = \frac{wx}{T_0} \tag{6.1}$$

and:

$$T = \sqrt{(T_0)^2 + (wx)^2} \tag{6.2}$$

We may apply our third equation of equilibrium by taking moments about the general point *P* located at the right end of the cable section shown in Fig. 6.43(a). If the coordinates of that point are (*x, y*), then the moment balance yields:

$$T_0(y) = W\left(\frac{x}{2}\right) = wx\left(\frac{x}{2}\right) = \left(\frac{w}{2}\right)x^2$$

Solving for *y* results in:

$$y = \left(\frac{w}{2T_0}\right)x^2 \tag{6.3}$$

For any given cable system, both *w* and T_0 are constants, so Eq. (6.3) actually has the form $y = ax^2$, which is the formula for a parabola whose vertex is at the origin. Note that if *w* has units of lb/ft or N/m, and T_0 is in lb or N, then the constant factor in Eq. (6.3) that is equivalent to "a" has units of ft^{-1} or m^{-1} respectively.

From Eq. (6.2), we can see that the minimum cable tension exists at the vertex, while maximum tension occurs at a support. In the case of symmetrical cables (supports at same elevation), we may substitute the coordinates of a support ($x = L/2, y = h$) in Eq. (6.3) to obtain this relationship for T_{min}:

$$T_0 = T_{min} = \frac{wL^2}{8h} \tag{6.4}$$

Substituting Eq. (6.4) into Eq. (6.2) yields an equation for T_{max} in symmetrical cables:

$$T_{max} = T_{L/2} = \sqrt{\left(\frac{wL^2}{8h}\right)^2 + w^2\left(\frac{L}{2}\right)^2}$$

or:

$$T_{max} = \frac{wL}{2}\sqrt{1 + \left(\frac{L}{4h}\right)^2} \tag{6.5}$$

Determining cable lengths for parabolic shapes can be quite a lengthy process and generally involves the use of calculus. For these reasons, most analyses use standard formulas to *approximate* values for length of cable and amount of sag. These formulas are known as *infinite series* and are subject to certain conditions. For a symmetrical cable having span, *L,* and sag, *h,* cable length, L_C, may be approximated using the first three terms of the series:

$$L_C = L\left[1 + \frac{8}{3}\left(\frac{h}{L}\right)^2 - \frac{32}{5}\left(\frac{h}{L}\right)^4 + \cdots\cdots\right] \tag{6.6}$$

Quantity (h/L) is known as the *sag ratio* and generally has a value between 0.05 and 0.20. Eq. (6.6) is valid for small sag ratios where the slope of the cable is 45° or less.

For nonsymmetrical cable systems as shown in Fig. 6.44, the approximate length of each cable section may be computed separately using the first three terms of the series:

$$L_A = a\left[1 + \frac{2}{3}\left(\frac{h_A}{a}\right)^2 - \frac{2}{5}\left(\frac{h_A}{a}\right)^4 + \cdots \right] \quad (6.7)$$

$$L_B = b\left[1 + \frac{2}{3}\left(\frac{h_B}{b}\right)^2 - \frac{2}{5}\left(\frac{h_B}{b}\right)^4 + \cdots \right]$$

Eq. (6.7) may also be used for symmetrical cable systems, but in all applications is valid only if (h_A/a) and (h_B/b) have values that are less than 0.5.

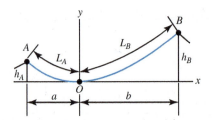

FIGURE 6.44. For nonsymmetrical cable systems, each portion, L_A and L_B, of the total cable length is computed separately.

Example 6.12   The symmetrical cable of Fig. 6.45 has a span of $L = 30$ m, has a sag of $h = 2$ m, and carries a uniform external load of $w = 3500$ N/m.

    a.   Find T_{min} and T_{max} for this cable.
    b.   Determine the required cable length, L_C.
    c.   Repeat (a) and (b) if $h = 4$ m.

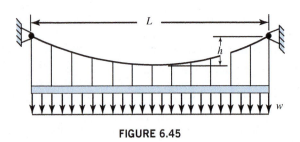

FIGURE 6.45

Solution: a.   Because the cable system is symmetrical, Eqs. (6.4) and (6.5) apply.

$$T_{min} = \frac{wL^2}{8h} = \frac{(3500 \text{ N/m})(30 \text{ m})^2}{8(2 \text{ m})}$$

$$T_{min} = 196.9 \text{ kN} \qquad \qquad \textbf{Answer}$$

$$T_{max} = \frac{wL}{2}\sqrt{1 + \left(\frac{L}{4h}\right)^2}$$

$$T_{max} = \frac{(3500 \text{ N/m})(30 \text{ m})}{2}\sqrt{1 + \left(\frac{30 \text{ m}}{4(2 \text{ m})}\right)^2}$$

$$T_{max} = 203.8 \text{ kN} \qquad \qquad \textbf{Answer}$$

    b.   Eq. (6.6) can be used to compute L_C, but first we must check the restriction for a maximum cable slope of 45° or less.

Letting $x = x_{max} = 15$ m in Eq. (6.1):

$$\tan\theta = \frac{wx}{T_0} = \frac{(3500 \text{ N/m})(15 \text{ m})}{196,900 \text{ N}} = 0.2666$$

$$\theta = 14.9° < 45° \quad \text{OK}$$

Then for the sag ratio $(h/L) = (2 \text{ m}/30 \text{ m}) = 0.066667$:

$$L_C = (30\text{m})\left[1 + \frac{8}{3}(0.066667)^2 - \frac{32}{5}(0.066667)^4\right]$$

$$L_C = 30.4 \text{ m} \qquad \qquad \textbf{Answer}$$

An alternate solution involves use of Eq. (6.7) to compute half the cable length, L_A, with $a = 15$ m and $h_A = 2$ m. Then $(h_A/a) = (2\text{ m}/15\text{ m}) = (0.1333)$, and

$$L_A = (15\,\text{m})\left[1 + \frac{2}{3}(0.1333)^2 - \frac{2}{5}(0.1333)^4\right]$$

$$L_A = 15.18\,\text{m}$$

Then: $L_C = 2L_A = 2(15.18\,\text{m}) = 30.4\,\text{m}$.

Note that Eq. (6.7) is valid here because its use restriction is satisfied, i.e.:

$$(h_A/a) = (2\,\text{m}/15\,\text{m}) = 0.1333 < 0.5$$

c.   Following the procedures of (a) and (b) above with $h = 4$ m and $(h/L) = (4\text{ m}/30\text{ m}) = 0.1333$:

$$T_{min} = 98.4\,\text{kN} \qquad \textbf{Answer}$$
$$T_{max} = 111.6\,\text{kN} \qquad \textbf{Answer}$$
$$L_C = 31.4\,\text{m} \qquad \textbf{Answer}$$

Notice the effect that a doubling of the sag ratio has on cable tensions and length. Do the results agree with what you would expect on an intuitive basis?

Despite the many formulas available for parabolic cables, some systems defy exact analysis, primarily for reasons of geometry. In the case of nonsymmetrical cables, for example, the exact location of the vertex is often unknown and must be determined from other data. Even the relatively simple problem described in Example 6.13 that follows is not easily solved.

Example 6.13   The cable shown in Fig. 6.45 is 275 ft long and spans a distance of 250 ft. Find the maximum sag in this cable. Assume that the load pulls this cable into a parabolic shape, and that the conditions for use of Eq. (6.6) are satisfied.

Solution: Without using the calculus, only an approximate value of sag, h, may be obtained. A first approximation involves simplifying Eq. (6.6) by dropping the third term, leaving this result:

$$L_C = L\left[1 + \frac{8}{3}\left(\frac{h}{L}\right)^2\right]$$

(Note that this simplification is often used in industry, even when computing values for L_C.) Solving the equation for h yields:

$$h = L\sqrt{\frac{3}{8}\left[\frac{L_C}{L} - 1\right]}$$

For $L = 250$ ft, and $L_C = 275$ ft:

$$h = (250\,\text{ft})\sqrt{\frac{3}{8}\left[\frac{275\,\text{ft}}{250\,\text{ft}} - 1\right]}$$

$$h = 48.4\,\text{ft} \qquad \textbf{Answer}$$

A better approximation is obtained by using all three terms in Eq. (6.6), but setting the square of the sag ratio equal to some arbitrary quantity such as "R." In other words, let:

$$R = \left(\frac{h}{L}\right)^2$$

Eq. (6.6) now becomes

$$L_C = L\left[1 + \frac{8}{3}R - \frac{32}{5}R^2\right]$$

Rearranging yields:

$$\frac{32}{5}R^2 - \frac{8}{3}R + \left[\frac{L_C}{L} - 1\right] = 0$$

This equation is a standard quadratic equation of the form:

$$ax^2 + bx + c = 0$$

where $a = 32/5$, $b = -8/3$, and $c = [(275 \text{ ft}/250 \text{ ft}) - 1]$.
Substituting into the quadratic formula:

$$R = x = \frac{-b \pm \sqrt{b^2 - 4ac}}{2a}$$

yields the two values $R = 0.375$ and $R = 0.041667$.
Because $R = (h/L)^2$, then

$$h = L\sqrt{R}$$

The four possible values of h become

$$h = \pm 153.1 \text{ ft} \quad \text{and} \quad h = \pm 51.0 \text{ ft}$$

Negative values do not apply here, and we can see from Fig. 6.46 that the 153.1 ft solution is incorrect. (Even if both halves of the cable are straight-line sections, the Pythagorean theorem yields a total cable length of 2 $d = 395$ ft.) Our approximate sag, then, is

$$h = 51.0 \text{ ft}$$ **Answer**

Finally, a good trial-and-error solution may be obtained using spreadsheets and the iterative method outlined in Section 15.7. (As an alternative, the calculations described here could be done manually.) First, select a range of assumed values for h. Next, compute the sag ratio (h/L). Third, use Eq. (6.6) to compute corresponding values of L_C, examining your results for the value of h that yields a cable length closest to the given value of 275 ft. A typical set of values is given in Table 1. The top portion of the table uses sag increments of 10 ft and shows that the desired value for cable length is between h-values of 50 ft and 60 ft (arrows). In the lower portion of the table, sag increments have been reduced to 0.5 ft, and the results indicate a most likely sag of 51.0 ft.

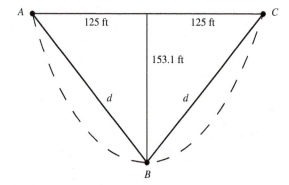

FIGURE 6.46. The shortest possible cable that will span 250 ft and pass through points A, B, and C has a length equal to 2d. From the Pythagorean theorem, this cable length is 395 ft, well in excess of the 275-ft length specified in Example 6.13.

TABLE 1

h (ft)	h/L	L_C (ft)	
10	0.04	251.1	
20	0.08	254.2	
30	0.12	259.3	
40	0.16	266.0	
50	0.20	274.1	←
60	0.24	283.1	←
70	0.28	292.4	
50.5	0.202	274.5	
51.0	0.204	275.0	← Answer
51.5	0.206	275.4	

Uniformly Distributed Load Along Cable

In the analysis of cables loaded by any type of concentrated loads, the weight of the cable itself is generally neglected. However, cables loaded solely by their own weight, or by other uniformly distributed loads such as ice or snow, require a different type of analysis, one that is more complicated mathematically than what we have seen so far. Cables hanging in a *parabolic* shape may be described by the equation:

$$y = ax^2$$

where, as derived earlier, the constant $a = (w/2 \ T_0)$. But loads that are distributed uniformly along the cable itself produce a shape known as a *catenary* curve, from the Latin word *catena*, or *chain*, a reference to the shape assumed by such a member when it is supported at both ends and hangs under its own weight. The simplest form of equation used to describe a catenary curve is

$$y = \frac{e^x + e^{-x}}{2}$$

The right-hand side of this equation is called the *hyperbolic cosine*, or *cosh*, and is often written in shorthand form as:

$$y = \cosh x$$

In the long form of this equation, "e", like π (pi), is a constant that occurs so frequently in science and mathematics that it is assigned its own symbol and has a numerical value that is known to an incredibly large number of decimal places. Although e is not always given its own key on hand calculators, its value may be obtained by installing "1" (one) on the display, and then asking the calculator for the inverse natural logarithm (INV LN, or LN^{-1}). On a 10-digit display, $e = 2.718281828$.

Parabolic and catenary curves have very similar shapes, as shown graphically in Fig. 6.47. (In fact, for taut catenary cables with low sag ratios, parabolic analysis is often used with good results. Large sag ratios, however, generally yield unacceptable errors.) Notice that when $x = 0$, $\cosh x = 1$, so that the curves would normally be displaced from each other along the vertical axis. To superimpose the two curves for comparison purposes, it is necessary to shift the catenary curve by representing it as:

$$y = (\cosh x) - 1$$

Shifting also allows us to place the origin of a coordinate system at the vertex of a catenary curve, and for this reason, the more general form of equation used in catenary cable analysis is

$$y = a\left(\cosh\frac{x}{a} - 1\right)$$

where a is a constant related to both the load per unit length of cable and the minimum tension in the cable itself.

Consider the free-body diagram of Fig. 6.48 showing a section of catenary cable with its left-hand end at the origin. If the load per unit length of cable is w_C and the length of this cable section is s, then the total load is $W = w_C s$. The equilibrium force triangle for this cable section would be similar to Fig. 6.43(b), but with load wx replaced by $w_C s$, resulting in these two equations as counterparts for Eqs. (6.1) and (6.2):

$$\tan\theta = \frac{w_C s}{T_0} \tag{6.8}$$

$$T = \sqrt{(T_0)^2 + (w_C s)^2} \tag{6.9}$$

Notice that in Fig. 6.48, distance d to the total load W is no longer $(x/2)$, and that the dimension is, in fact, unknown. This prevents us from writing a moment equation about point 0 and developing equations for y, T, and cable length L, as we did in the parabolic case. Instead, derivation of such relationships requires use of the calculus, and so the resulting equations are presented here without proof.

$$y = \frac{T_0}{w_C}\left(\cosh\frac{w_C x}{T_0} - 1\right) \tag{6.10}$$

Note that Eq. (6.10) fits the general (shifted) form of the catenary equation given earlier, with $a = (T_0/w_C)$.

Tension in the cable may be determined at any x-location as:

$$T = T_0 \cosh\frac{w_C x}{T_0} \tag{6.11}$$

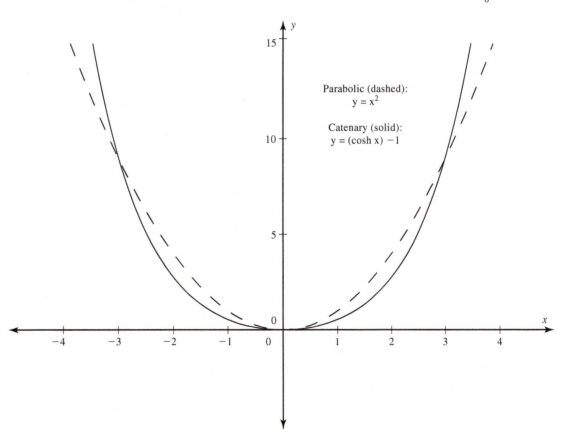

Parabolic (dashed):
$y = x^2$

Catenary (solid):
$y = (\cosh x) - 1$

FIGURE 6.47. At small values of x, parabolic and catenary curves differ only slightly. For taut cables, or those having sag ratios of 1:8 or less, parabolic analysis provides a good approximation to catenary behavior.

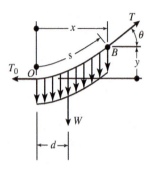

FIGURE 6.48. In the free-body diagram for a section of catenary cable, distance d to the resultant load, W, is unknown. For a parabolic cable, this dimension is $(x/2)$.

but if Eqs. (6.11) and (6.10) are combined:

$$T = T_0 + w_C y \qquad (6.12)$$

Finally, cable length, s, from the origin to point B in Fig. 6.48 (also referred to as L_B for the parabolic case) is computed as:

$$s = L_B = \frac{T_0}{w_C} \sinh \frac{w_C x}{T_0} \qquad (6.13)$$

where *sinh* is called the *hyperbolic sine* and is defined as:

$$\sinh = \frac{e^x - e^{-x}}{2}$$

The following example demonstrates use of these equations and, particularly, the computational steps involved in their evaluation.

Example 6.14   A cable weighing 30 N/m is suspended symmetrically from supports at the same elevation. This cable is 150 m long, spans a distance of 130 m, and is loaded solely by its own weight.

   a.   Find the minimum and maximum tensions in the cable.
   b.   Determine the sag of this cable.
   c.   Compare your results for (a) and (b) with those obtained by parabolic approximation.

Solution: a.   Minimum tension is found by the numerical methods shown in Example 6.13. Rearranging catenary Eq. (6.13):

$$w_C s = T_0 \sinh \frac{w_C x}{T_0}$$

Now $w_C = 30$ N/m, and from the vertex to a support $x = 65$ m and $s = 75$ m. Then:

$$w_C s = 30\,\text{N/m}\,(75\,\text{m}) = 2250\,\text{N}$$
$$w_C x = 30\,\text{N/m}\,(65\,\text{m}) = 1950\,\text{N}$$

and our equation becomes:

$$2250 = T_0 \sinh \frac{1950}{T_0}$$

For a trial-and-error solution, we now select values of T_0 and evaluate the right-hand side of this equation, looking for the tension that yields a result that is close to 2250. Selecting an arbitrary starting value of $T_0 = 1000$ N, the computations proceed as follows, using $e = 2.718$:

$$\sinh \frac{1950}{1000} = \frac{2.718^{(1950/1000)} - 2.718^{-(1950/1000)}}{2}$$
$$= \frac{7.027 - 0.1423}{2} = 3.442$$

Then:

$$T_0 \sinh \frac{1950}{T_0} = (1000\,\text{N})(3.442) = 3442\,\text{N}$$

Using this same method yields the values in Table 2. The 200-N increments (top of table) indicate that a minimum tension exists between 2000 N and 2200 N (arrows), whereas the smaller 25-N increments used at the bottom of this table converge on the value $T_0 = T_{min} = 2075$ N.

Maximum cable tension occurs at a support, and Eq. (6.9) yields:

$$T = T_{max} = \sqrt{(2075\,\text{N})^2 + [(30\,\text{N/m})(75\,\text{m})]^2}$$
$$T_{max} = 3060\,\text{N} \qquad \qquad \textbf{Answer}$$

TABLE 2

T_0 (N)	sinh	T_0 sinh (N)	
1000	3.442	3442	
1200	2.440	2928	
1400	1.889	2644	
1600	1.543	2470	
1800	1.308	2354	
2000	1.137	2274	←
2200	1.007	2215	←
2025	1.119	2265	
2050	1.101	2257	
2075	1.084	2250	← Answer

b. From Eq. (6.12), $T = T_{max}$ when $y = y_{max} = h$ (sag).
    Then:

$$y_{max} = h = \frac{T - T_0}{w_C} = \frac{(3060 - 2075)\,\text{N}}{30\,\text{N/m}}$$

$$h = 32.8\,\text{m} \qquad\qquad\text{**Answer**}$$

c. We may approximate sag h using parabolic Eq. (6.6).
    Discarding the third term and rearranging:

$$\left(\frac{h}{L}\right)^2 = \frac{3}{8}\left[\frac{L_C}{L} - 1\right]$$

With $L = 130$ m and $L_C = 150$ m:

$$h = 31.2\,\text{m} \qquad\qquad\text{**Answer**}$$

To compute maximum and minimum tensions, a value for load w is needed. With taut cables, it might be possible to obtain satisfactory results by using $w = w_C$ directly. However, in this case, we have a cable whose sag ratio is approximately 1:4, so a better choice for w is to divide the total cable weight by its span. In other words:

$$w = \frac{w_C L_C}{L} = \frac{(30\,\text{N/m})(150\,\text{m})}{130\,\text{m}} = 34.6\,\text{N/m}$$

Then from Eq. (6.4):

$$T_{min} = T_0 = \frac{wL^2}{8h} = \frac{(34.6\,\text{N/m})(130\,\text{m})^2}{8(31.2\,\text{m})}$$

$$T_{min} = 2343\,\text{N} \qquad\qquad\text{**Answer**}$$

Finally, from Eq. (6.5):

$$T_{max} = \frac{wL}{2}\sqrt{1 + \left(\frac{L}{4h}\right)^2}$$

$$T_{max} = \frac{(34.6\,\text{N/m})(130\,\text{m})}{2}\sqrt{1 + \left(\frac{130\,\text{m}}{4(31.2\,\text{m})}\right)^2}$$

$$T_{max} = 3248\,\text{N} \qquad\qquad\text{**Answer**}$$

The errors in our parabolic approximations compared to the catenary results are h: 4.9% T_{min}: 13% T_{max}: 6.1%

PARABOLIC AND CATENARY CURVES

SB.7(a)

SB.7(b)

As we saw in Fig. 6.47, there is little visual difference between *parabolic* and *catenary* curves, yet each has its own areas of application. The parabola, for example, has a singular point, called the *focus*, along its axis of symmetry. Under certain conditions, any incoming energy waves (whether sound, light, or electronic signals) are directed to this point. Because of this ability to "capture" energy, parabolic reflectors are often used as TV satellite dishes, microphones, solar collectors, and radio telescopes such as the 84-ft-diameter array shown in SB.7(a). Likewise, a source of sound, light, or radio waves placed at the focus will reflect from a parabolic shape to produce a relatively uniform energy field at the open end of the parabola. This makes them well suited for use in loudspeakers and light sources (floodlights, automobile headlights), as well as radar and radio transmitters.

Uses of the catenary curve are a bit more esoteric. For instance, shapes of the fronts and backs of arch-top or f-hole musical instruments such as violins, violas, and guitars are often derived using catenary curves. Similarly, a U.S. patent was issued in 2004 for a golf ball having surface "dimples" whose profiles were those of a catenary curve. This particular dimple shape was thought to vary the flight performance of a golf ball, based on the ball's spin characteristics and the speed of the golf club itself. Finally, catenary curves are sometimes used as structural elements, the most famous of which is the Gateway Arch shown in SB.7(b). Designed by architect Eero Saarinen and completed in 1965, this inverted catenary curve rises 630 ft above the ground and is a focal point along the Missippi River in St. Louis, Missouri.

PROBLEMS

6.55 A 75-lb horizontal force holds the cable system in the position shown. Find:
  a. The tensions in cable sections *AB* and *BC*.
  b. Total sag, *h*.
  c. Span, *L*.

6.56 For the cable system shown, determine:
  a. The tensions in cable sections *AB*, *BC*, and *CD*.
  b. Vertical distance, *h*.
  c. Total cable length, L_C.

6.57 A symmetrical cable is used to support a pedestrian walkway between two office buildings as shown. The structure has a span of $L = 94$ ft; a sag of $h = 10$ ft; and the walkway itself exerts a uniform external load of $w = 150$ lb/ft.

  Assuming that this load pulls the cable into a parabolic shape, find
  a. T_{min} and T_{max} for the cable.
  b. Required cable length, L_C.

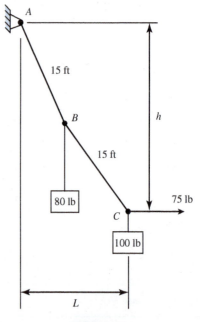

PROB. 6.55

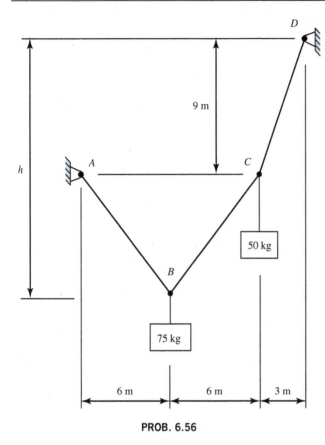

PROB. 6.56

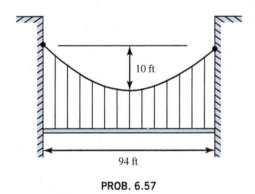

PROB. 6.57

6.58 The nonsymmetrical cable system shown carries elevated chemical distribution pipes across an on-site railroad track at a manufacturing facility. This structure has the following geometry: $a = 15$ m; $h_A = 7.2$ m; $b = 25$ m; $h_B = 20$ m. If the cable assumes a parabolic shape under a uniform external load of $w = 175$ N/m, find

  a. Minimum tension in the cable.
  b. Maximum tension in cable section AC.
  c. Maximum tension in cable section BC.
  d. Total cable length, L_C.

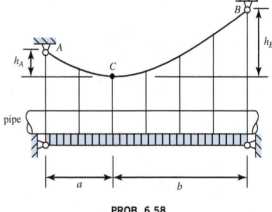

PROB. 6.58

6.59 A 450-ft long electrical transmission line is suspended symmetrically between two towers and spans a distance of $L = 300$ ft. If the cable has a uniform weight of $w_C = 35$ lb/ft and arranges itself in the shape of a catenary curve, find

  a. The maximum and minimum tension in the cable;
  b. Maximum cable sag, h.

6.60 A symmetrical cable similar to that of Prob. 6.59 spans a distance of $L = 200$ m and has a uniform weight of $w_C = 50$ N/m. Determine

  a. Cable length L_C, which produces a minimum cable tension of 9500 N.
  b. For your value of L_C from part (a), compute the maximum tension developed in this cable and the magnitude of cable sag, h.

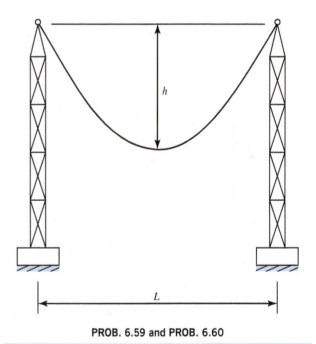

PROB. 6.59 and PROB. 6.60

FORCES IN SPACE

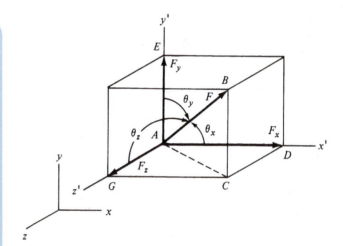

FIGURE 7.1

7.1 INTRODUCTION

In previous chapters, we considered coplanar force systems, that is, force systems in which the forces act in a single plane. In this chapter, we consider force systems in space. Although the solution of complex problems is best accomplished through the use of techniques involving vector multiplication, many equilibrium conditions can be analyzed using the trigonometric methods presented here.

7.2 COMPONENTS OF A FORCE IN SPACE

A force in space is a vector that has three rectangular components. To visualize the force, let us construct a rectangular transparent box with the force $\mathbf{F}$ as the diagonal AB of the box as shown in Fig. 7.1. The back edge of the box lies along the x', y', and z' axes, and the projection of $\mathbf{F}$ on these axes forms rectangular components F_x, F_y, and F_z.

Although the perspective of the box distorts angles, clearly a line from B to D is perpendicular to the x' axis, a line from B to E is perpendicular to the y' axis, and a line

from B to G is perpendicular to the z' axis. The force $\mathbf{F}$ forms the *hypotenuse* of three right triangles, where F_x, F_y, and F_z are the *adjacent sides*. Therefore, the angles between the force and one of its components can be expressed in terms of the cosine of *direction angles* θ_x, θ_y, and θ_z. That is,

$$\cos\theta_x = \frac{F_x}{F}; \quad \cos\theta_y = \frac{F_y}{F}; \quad \cos\theta_z = \frac{F_z}{F} \quad (7.1)$$

$$F_x = F\cos\theta_x; \quad F_y = F\cos\theta_y; \quad F_z = F\cos\theta_z \quad (7.2)$$

To obtain a relationship between the force and its component, we use the *Pythagorean theorem* to add the two forces F_x and F_z to find their resultant (AC) in the $x'z'$ plane and then add AC to the force F_y to find the resultant F as follows:

$$(AC)^2 = F_x^2 + F_z^2$$
$$F^2 = (AC)^2 + F_y^2 = F_x^2 + F_y^2 + F_z^2 \quad (7.3)$$

$$F = \sqrt{F_x^2 + F_y^2 + F_z^2} \quad (7.4)$$

A relationship between the direction cosines is found by substituting Eq. (7.2) into Eq. (7.3):

$$F^2 = F_x^2 + F_y^2 + F_z^2 = F^2(\cos^2\theta_x + \cos^2\theta_y + \cos^2\theta_z)$$
$$\cos^2\theta_x + \cos^2\theta_y + \cos^2\theta_z = 1 \quad (7.5)$$

The direction cosines can also be found from the coordinates of points A and B. From Fig. 7.2, the rectangular coordinates for A and B are x_A, y_A, z_A and x_B, y_B, z_B, respectively. The length of the line from A to B is equal to d and its projection on the x', y', and z' axes are d_x, d_y, and d_z. The relationship between d and its projections is given by an equation of the same form as Eq. (7.4).

$$d = \sqrt{d_x^2 + d_y^2 + d_z^2} \qquad (7.6)$$

where

$$d_x = x_B - x_A, \quad d_y = y_B - y_A, \quad d_z = z_B - z_A \quad (7.7)$$

The line $AB = d$ is the *hypotenuse* of three right triangles with *adjacent sides* d_x, d_y, and d_z. Therefore,

$$\cos\theta_x = \frac{d_x}{d} \qquad \cos\theta_y = \frac{d_y}{d} \qquad \cos\theta_z = \frac{d_z}{d} \qquad (7.8)$$

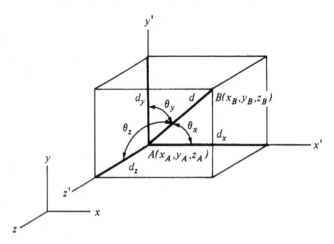

FIGURE 7.2

Example 7.1   Determine the components of the forces shown in Fig. 7.3.

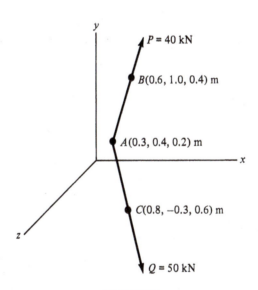

$P = 40$ kN

$B(0.6, 1.0, 0.4)$ m

$A(0.3, 0.4, 0.2)$ m

$C(0.8, -0.3, 0.6)$ m

$Q = 50$ kN

FIGURE 7.3

Solution: Force $P = 40$ kN: From Eqs. (7.6) and (7.7),

$$d_x = x_B - x_A = 0.6 - 0.3 = 0.3\,\text{m}$$
$$d_y = y_B - y_A = 1.0 - 0.4 = 0.6\,\text{m}$$
$$d_z = z_B - z_A = 0.4 - 0.2 = 0.2\,\text{m}$$
$$d = \sqrt{d_x^2 + d_y^2 + d_z^2} = \sqrt{(0.3)^2 + (0.6)^2 + (0.2)^2} = 0.7\,\text{m}$$

The components of the force P are given by Eqs. (7.2) and (7.8):

$$P_x = P\cos\theta_x = P\frac{d_x}{d} = \frac{40(0.3)}{0.7} = 17.1\,\text{kN} \qquad \textbf{Answer}$$

$$P_y = P\cos\theta_y = P\frac{d_y}{d} = \frac{40(0.6)}{0.7} = 34.3\,\text{kN} \qquad \textbf{Answer}$$

$$P_z = P\cos\theta_z = P\frac{d_z}{d} = \frac{40(0.2)}{0.7} = 11.4\,\text{kN} \qquad \textbf{Answer}$$

Check: Eq. (7.4)

$$P = \sqrt{P_x^2 + P_y^2 + P_z^2}$$
$$= \sqrt{(17.1)^2 + (34.3)^2 + (11.4)^2} = 40.0 \, \text{kN} \qquad \text{OK}$$

Force $Q = 50 \, \text{kN}$: From Eqs. (7.6) and (7.7)

$$d_x = x_C - x_A = 0.8 - 0.3 = 0.5 \, \text{m}$$
$$d_y = y_C - y_A = -0.3 - 0.4 = -0.7 \, \text{m}$$
$$d_z = z_C - z_A = 0.6 - 0.2 = 0.4 \, \text{m}$$
$$d = \sqrt{d_x^2 + d_y^2 + d_z^2} = \sqrt{(0.5)^2 + (-0.7)^2 + (0.4)^2} = \sqrt{0.9} \, \text{m}$$

The components of the force Q are given by Eqs. (7.2) and (7.8):

$$Q_x = Q\cos\theta_x = Q\frac{d_x}{d} = \frac{50(0.5)}{\sqrt{0.9}} = 26.4 \, \text{kN} \qquad \textbf{Answer}$$

$$Q_y = Q\cos\theta_y = Q\frac{d_y}{d} = \frac{50(-0.7)}{\sqrt{0.9}} = -36.9 \, \text{kN} \qquad \textbf{Answer}$$

$$Q_z = Q\cos\theta_z = Q\frac{d_z}{d} = \frac{50(0.4)}{\sqrt{0.9}} = 21.1 \, \text{kN} \qquad \textbf{Answer}$$

Check: Eq. (7.4)

$$Q = \sqrt{Q_x^2 + Q_y^2 + Q_z^2}$$
$$= \sqrt{(26.4)^2 + (-36.9)^2 + (21.1)^2} = 50.0 \, \text{kN} \qquad \text{OK}$$

7.3 RESULTANT OF CONCURRENT FORCES IN SPACE

The resultant of two or more concurrent forces in space is found by summing the components of the forces in the x, y, and z directions. That is,

$$R_x = \Sigma F_x \qquad R_y = \Sigma F_y \qquad R_z = \Sigma F_z \qquad (7.9)$$

By arguments similar to those in Sec. 7.2, the magnitude and direction of the resultant are then found by the following equations:

$$R = \sqrt{R_x^2 + R_y^2 + R_z^2} \qquad (7.10)$$

$$\cos\theta_x = \frac{R_x}{R} \qquad \cos\theta_y = \frac{R_y}{R} \qquad \cos\theta_z = \frac{R_z}{R} \qquad (7.11)$$

Example 7.2   Determine the resultant of the two forces in Example 7.1. (Force $P = 40 \, \text{kN}$ had components $P_x = 17.1 \, \text{kN}$, $P_y = 34.3 \, \text{kN}$, $P_z = 11.4 \, \text{kN}$; force $Q = 50 \, \text{kN}$ had components $Q_x = 26.4 \, \text{kN}$, $Q_y = -36.9 \, \text{kN}$, $Q_z = 21.1 \, \text{kN}$. Both forces act through point A.)

Solution: Summing forces in the x, y, and z directions, we have

$$R_x = \Sigma F_x = P_x + Q_x = 17.1 + 26.4 = 43.5 \, \text{kN}$$
$$R_y = \Sigma F_y = P_y + Q_y = 34.3 - 36.9 = -2.6 \, \text{kN}$$
$$R_z = \Sigma F_z = P_z + Q_z = 11.4 + 21.1 = 32.5 \, \text{kN}$$

and the magnitude and direction of the resultant are given by

$$R = \sqrt{R_x^2 + R_y^2 + R_z^2}$$
$$= \sqrt{(43.5)^2 + (-2.6)^2 + (32.5)^2} = 54.4 \, \text{kN} \qquad \textbf{Answer}$$

and

$$\cos\theta_x = \frac{R_x}{R} = \frac{43.5}{54.4} = 0.7996 \qquad\qquad \theta_x = 36.9° \qquad \textbf{Answer}$$

$$\cos\theta_y = \frac{R_y}{R} = \frac{-2.6}{54.4} = -0.0478 \qquad\qquad \theta_y = 92.7° \qquad \textbf{Answer}$$

$$\cos\theta_z = \frac{R_z}{R} = \frac{32.5}{54.4} = 0.5974 \qquad\qquad \theta_z = 53.3° \qquad \textbf{Answer}$$

Check:

$$\cos^2\theta_x + \cos^2\theta_y + \cos^2\theta_z = (0.7996)^2 + (-0.0478)^2 + (0.5974)^2 = 1.000 \quad \text{OK}$$

Example 7.3   Determine the resultant of the two forces $S = 1.5$ kip and $T = 2.0$ kip shown in Fig. 7.4.

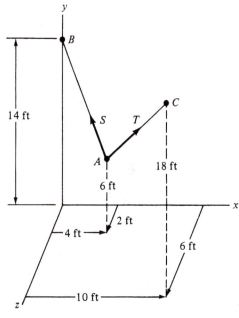

FIGURE 7.4

Solution: Force $S = 1.5$ kip acts from A to B. The coordinates of B and A are $B(0, 14, 0)$ ft and $A(4, 6, 2)$ ft. Subtracting the coordinates of A from the coordinates of B, we have

$$d_x = x_B - x_A = \quad 0 - 4 = -4 \text{ ft}$$
$$d_y = y_B - y_A = 14 - 6 = \quad 8 \text{ ft}$$
$$d_z = z_B - z_A = \quad 0 - 2 = -2 \text{ ft}$$

and
$$d = \sqrt{d_x^2 + d_y^2 + d_z^2} = \sqrt{(-4)^2 + (8)^2 + (-2)^2} = \sqrt{84} \text{ ft}$$

Therefore, the components of S are

$$S_x = \frac{Sd_x}{d} = \frac{1.5(-4)}{\sqrt{84}} = -0.655 \text{ kip}$$

$$S_y = \frac{Sd_y}{d} = \frac{1.5(8)}{\sqrt{84}} = \quad 1.309 \text{ kip}$$

$$S_z = \frac{Sd_z}{d} = \frac{1.5(-2)}{\sqrt{84}} = -0.327 \text{ kip}$$

Force $T = 2.0$ kip acts from A to C. The coordinates of C and A are $C(10, 18, 6)$ ft and $A(4, 6, 2)$ ft. Subtracting the coordinates of A from the coordinates of C, we have

$$d_x = x_C - x_A = 10 - 4 = \quad 6 \text{ ft}$$
$$d_y = y_C - y_A = 18 - 6 = 12 \text{ ft}$$
$$d_z = z_C - z_A = \quad 6 - 2 = \quad 4 \text{ ft}$$

and
$$d = \sqrt{d_x^2 + d_y^2 + d_z^2} = \sqrt{(6)^2 + (12)^2 + (4)^2} = 14 \text{ ft}$$

Therefore, the components of **T** are

$$T_x = \frac{Td_x}{d} = \frac{2.0(6)}{14} = 0.857 \text{ kip}$$

$$T_y = \frac{Td_y}{d} = \frac{2.0(12)}{14} = 1.714 \text{ kip}$$

$$T_z = \frac{Td_z}{d} = \frac{2.0(4)}{14} = 0.571 \text{ kip}$$

Summing forces in the x, y, and z directions,

$$R_x = \Sigma F_x = S_x + T_x = -0.655 + 0.857 = 0.202 \text{ kip}$$
$$R_y = \Sigma F_y = S_y + T_y = 1.309 + 1.714 = 3.023 \text{ kip}$$
$$R_z = \Sigma F_z = S_z + T_z = -0.327 + 0.571 = 0.244 \text{ kip}$$

The magnitude and direction of the resultant are given by

$$R = \sqrt{R_x^2 + R_y^2 + R_z^2}$$
$$= \sqrt{(0.202)^2 + (3.023)^2 + (0.244)^2} = 3.04 \text{ kip} \qquad \textbf{Answer}$$

and
$$\cos\theta_x = \frac{R_x}{R} = \frac{0.202}{3.04} = 0.0665 \qquad \theta_x = 86.2° \qquad \textbf{Answer}$$

$$\cos\theta_y = \frac{R_y}{R} = \frac{3.023}{3.04} = 0.9944 \qquad \theta_y = 6.06° \qquad \textbf{Answer}$$

$$\cos\theta_z = \frac{R_z}{R} = \frac{0.244}{3.04} = 0.0803 \qquad \theta_z = 85.4° \qquad \textbf{Answer}$$

Check:

$$\cos^2\theta_x + \cos^2\theta_y + \cos^2\theta_z = (0.0665)^2 + (0.9944)^2 + (0.0803)^2 = 1.000 \qquad \text{OK}$$

7.4 EQUILIBRIUM OF A CONCURRENT FORCE SYSTEM IN SPACE

A three-dimensional concurrent force system is in equilibrium if the sum of the external forces is equal to zero. Therefore, from Eq. (7.9), we see that equilibrium requires that the following three conditions be satisfied:

$$\Sigma F_x = 0 \qquad \Sigma F_y = 0 \qquad \Sigma F_z = 0 \qquad (7.12)$$

Example 7.4   The mast AO is supported by a ball and socket at A and by two cables OB and OC, as shown in Fig. 7.5(a). A horizontal load of 15 kip acts at O. Determine the tension in each cable and the compression in the mast. Neglect the weight of the mast.

Solution: Mast AO is a two-force member. Therefore, the force in the member acts in the direction of the member, and we have a system of concurrent force acting through O.

Because mast OA lies in the xy coordinate plane and has a slope of 4 to 3, the components of the force at A are $A_x = 0.6A$, $A_y = 0.8A$ and $A_z = 0$. Cable OB lies in the yz coordinate plane and has a slope of 12 to 5, and the components of the force at B are $B_x = 0$, $B_y = 12/13B = 0.923B$, and $B_z = 5/13B = 0.385B$. Cable OC lies in the yz plane and has a slope of 4 to 3, and the components of the force at C are $C_x = 0$, $C_y = 0.8C$, and $C_z = 0.6C$. The mast, cable, and various force components are shown on the free-body diagrams in the xy and yz coordinate planes of Fig. 7.5(b) and (c). Notice that force components that are perpendicular to a coordinate plane do not appear in that plane.

The force components are all shown in tension. Therefore, a positive answer indicates tension and a negative answer compression.

From Fig. 7.5(b) and equilibrium,

$$\Sigma F_x = 15 + 0.6 A = 0 \quad A = -25 \text{ kip} \qquad \textbf{Answer}$$

and from Fig. 7.5(c),

$$\Sigma F_y = 0.923B + 0.8C + 0.8A = 0 \qquad \text{(a)}$$

$$\Sigma F_z = 0.385B - 0.6C = 0 \qquad \text{(b)}$$

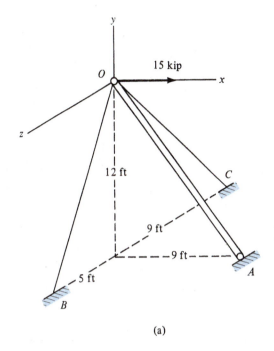

(a)

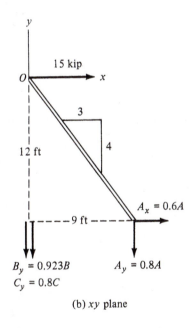

(b) xy plane

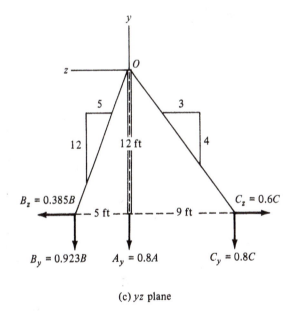

(c) yz plane

FIGURE 7.5

Substituting $A = -25$ kip into Eq. (a) and solving for B and C, we have

$$B = 13.93 \text{ kip} \qquad C = 8.93 \text{ kip}$$ **Answer**

Check:

$$\Sigma F_x = 0.6A + 15 = 0.6(-25) + 15 = 0 \quad \text{OK}$$

$$\Sigma F_y = 0.923B + 0.8A + 0.8C$$
$$= 0.923(13.93) + 0.8(-25) + 0.8(8.93) = 0.001 \quad \text{OK}$$

$$\Sigma F_z = 0.385B - 0.6C$$
$$= 0.385(13.93) - 0.6(8.93) = 0.005 \quad \text{OK}$$

The tension in cables OB and OC is 13.93 kip and 8.93 kip, respectively, and compression in the mast is 25 kip.

Example 7.5 The boom *AD* shown in Fig. 7.6(a) is supported by a ball and socket at *D* and two cables *AC* and *AB*. The boom supports a vertical load $P_y = 1.0\,\text{kN}$ at *A*. Determine the tension in the cables and compression in the boom. Neglect the weight of the boom.

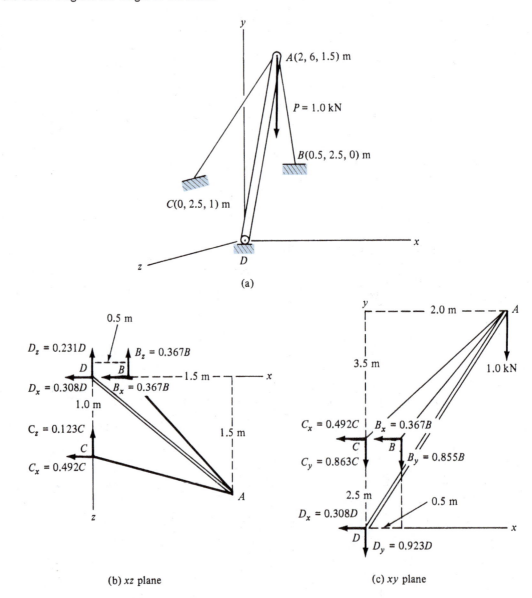

FIGURE 7.6

Solution: Member *AD* is a two-force member. Therefore, the force in the member acts in the direction of the member, and we have a system of concurrent forces acting through *A*.

   The force in the cables and boom are all three-dimensional; therefore, we determine the components of the forces in each of them by the methods of Sec. 7.2 as follows.

Cable *AB*:

$$d_x = 0.5 - 2 = -1.5 \qquad d_y = 2.5 - 6 = -3.5 \qquad d_z = 0 - 1.5 = -1.5$$
$$d = \sqrt{(-1.5)^2 + (-3.5)^2 + (-1.5)^2} = \sqrt{16.75}$$

Therefore, the components of *B* are

$$B_x = \frac{d_x}{d}B = \frac{-1.5}{\sqrt{16.75}}B = -0.367B \qquad B_y = \frac{d_y}{d}B = \frac{-3.5}{\sqrt{16.75}}B = -0.855B$$

$$B_z = \frac{d_z}{d}B = \frac{-1.5}{\sqrt{16.75}}B = -0.367B$$

Cable AC:

$$d_x = 0 - 2 = -2 \qquad d_y = 2.5 - 6 = -3.5 \qquad d_z = 1 - 1.5 = -0.5$$
$$d = \sqrt{(-2)^2 + (-3.5)^2 + (-0.5)^2} = \sqrt{16.5}$$

Therefore, the components of C are

$$C_x = \frac{d_x}{d}C = \frac{-2}{\sqrt{16.5}}C = -0.492C \qquad C_y = \frac{d_y}{d}C = \frac{-3.5}{\sqrt{16.5}}C = -0.862C$$

$$C_z = \frac{d_z}{d}C = \frac{-0.5}{\sqrt{16.5}}C = -0.367C$$

Boom AD:

$$d_x = -2 \qquad d_y = -6 \qquad d_z = -1.5$$
$$d = \sqrt{(-2)^2 + (-6)^2 + (-1.5)^2} = 6.5$$

Therefore, the components of D are

$$D_x = \frac{d_x}{d}D = \frac{-2}{6.5}D = -0.308D \qquad D_y = \frac{d_y}{d}D = \frac{-6}{6.5}D = -0.923D$$

$$D_z = \frac{d_z}{d}D = \frac{-1.5}{6.5}D = -0.231D$$

The algebraic sign of the component reactions at B, C, and D were used to draw the free-body diagrams of Fig. 7.6(b) and (c). Therefore, the equilibrium equations already include their sign. From Fig. 7.6(b) and (c), we have

$$\Sigma F_x = 0.367B + 0.492C + 0.308D = 0 \tag{a}$$

$$\Sigma F_y = 0.855B + 0.862C + 0.923D + 1.0 = 0 \tag{b}$$

$$\Sigma F_z = 0.367B + 0.123C + 0.231D = 0 \tag{c}$$

These three equations in three unknowns can be solved manually by elimination or by determinants. They can also be solved by computer or an advanced calculator. (See Sec. 1.8 in Chapter 1.) Solving by elimination, we have

$$B = 2.11\,\text{kN}, \qquad C = 0.786\,\text{kN}, \quad \text{and} \quad D = -3.77\,\text{kN} \qquad \textbf{Answers}$$

PROBLEMS

7.1 Determine the components of the force shown.

7.2 through 7.5 For the force system shown, determine
a. the components of the forces, and
b. the magnitude and direction angles of the resultant.

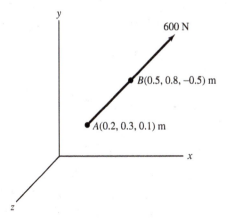

PROB. 7.1

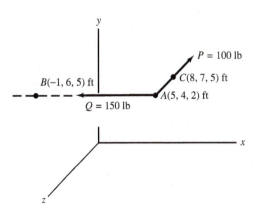

PROB. 7.2

CABLE-STAYED TOWERS

SB.8(a)

SB.8(b)

The television tower shown in Fig. 3.22 provides a good example of the equilibrium of forces in space. As shown in SB.8(a), each of the four guy wires connected to concrete pier *A* is separately tensioned by a cable arrangement identical to *B–C–D*. Point *C* is a roller-and-pulley assembly that rides on the guy wire and carries the tensioning cable from a fixed anchor point at *B* to a self-adjusting anchor point at *D*. Each of the four boxes at *D* contain an automotive shock absorber assembly that not only acts like a spring to maintain tension in the guy wire, but also supplies a damping force to reduce vibrations in the guy wire caused by the wind. Although SB.8(b) shows that the guy wire and tensioning cable forces acting on the roller and pulley do not pass *exactly* through a single point, the assembly is usually modeled as a point in space for which the resultant force must equal zero. *(Courtesy of WAND-TV, Decatur, Illinois, www.wandtv.com)*

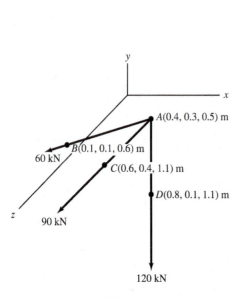

PROB. 7.3

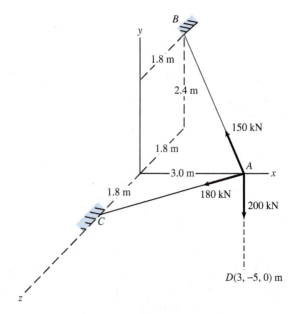

PROB. 7.4

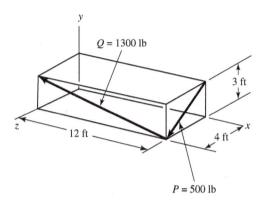

PROB. 7.5

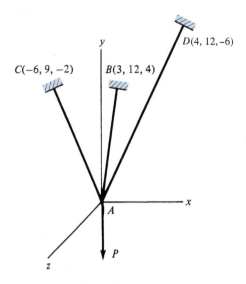

PROB. 7.7 and PROB. 7.8

7.6 A weight is supported by a tripod as shown. Neglect the weight of the tripod. Determine the force in each leg of the tripod for the following conditions.

a. $r = 16$ in., $h = 62$ in., and $W = 75$ lb

b. $r = 400$ mm, $h = 1570$ mm, and $W = 350$ N

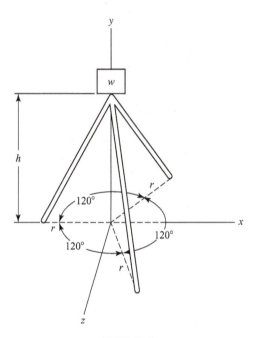

PROB. 7.6

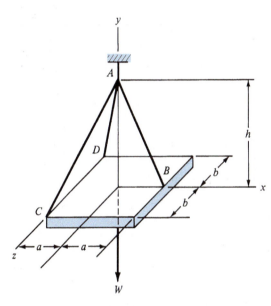

PROB. 7.9

7.7 A vertical force $P = 1000$ lb is supported by the three cables AB, AC, and AD as shown. Determine the tension in the cables.

7.8 A vertical force $P = 1200$ N is supported by the three cables AB, AC, and AD. Determine the tension in the cables.

7.9 A uniform plate is lifted by three cables that are joined at A directly over the center of the plate as shown. Determine the tension in the cables for the following conditions.

a. $W = 2$ kip, $a = 3$ ft, $b = 4$ ft, and $h = 12$ ft

b. $W = 10$ kN, $a = 0.8$ m, $b = 1.2$ m, and $h = 2.4$ m

7.10 The three bars shown have balls and sockets at B, C, and D and are joined at A by a ball and socket. Determine the force in the three bars. Neglect the weight of the bars.

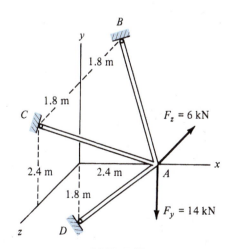

PROB. 7.10

7.11 Part of the landing gear for a lunar spacecraft is simulated by the three struts (bars) shown. The three struts have balls and sockets at B, C, and D and are joined at A by a ball and socket. When the spacecraft touches down, a force $F = 2.8\,\text{kN}$ acts upward at A. Determine the force in the three struts. Neglect the weight of the struts.

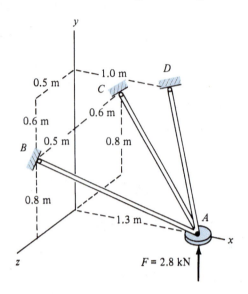

PROB. 7.11 and PROB. 7.12

7.12 Repeat Prob. 7.11 if the coordinates of D are $x = 0.6\,\text{m}$, $y = 0$, $z = 0$ and $F = 2.1\,\text{kN}$.

7.5 MOMENT OF A FORCE ABOUT AN AXIS

The tendency of a force to produce rotation about an axis is called the moment of the force about that axis. Consider the forces F_x, F_y, and F_z acting at a point in space with coordinates x, y, and z [Fig. 7.7(a)]. Because the force F_x is parallel with the x axis, it cannot produce rotation about the x axis. The same is true for the force F_y and the y axis and the force F_z and the z axis. However, forces that can produce rotation about an axis are F_y and F_z about the x axis, F_x and F_z about the y axis, and F_x and F_y about the z axis. Consistent with

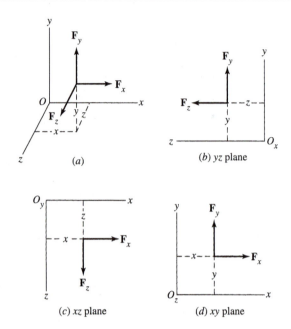

these observations, *the moment of a force about an axis is defined as the product of the force and the perpendicular distance from the line of action of the force to the axis.*

To determine the perpendicular distances, we project the forces onto the coordinate planes—the planes formed by the coordinate axes [Fig. 7.7(b)–(d)]. If a force in the projected view tends to produce counterclockwise rotation about the axis, it is defined as positive; if it tends to produce clockwise rotation, it is defined as negative. For example, in Fig. 7.7(b), the moment of F_z about the x axis is counterclockwise (positive), and the moment of F_y is clockwise (negative). Therefore,

$$M_x = F_z y - F_y z$$

Similarly, the moments about the y axis and z axis from Fig. 7.7(c) and (d) are as follows:

$$M_y = F_x z - F_z x$$

and

$$M_z = F_y x - F_x y$$

FIGURE 7.7. Forces in space and projections on coordinate planes.

Example 7.6   Find the moments of the three forces $F_x = 5\,\text{kN}$, $F_y = 15\,\text{kN}$, and $F_z = 10\,\text{kN}$ located at $x = 3\,\text{m}$, $y = 5\,\text{m}$, and $z = 2\,\text{m}$ about the x, y, and z axes as shown in Fig. 7.8(a).

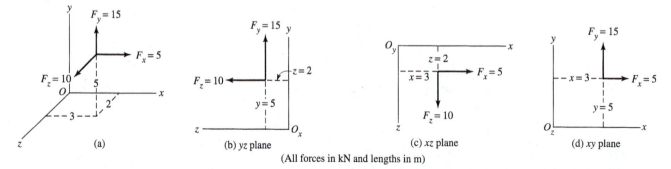

(All forces in kN and lengths in m)

FIGURE 7.8

Solution: From the projection of the forces on the three coordinate planes in Fig. 7.8(b)–(d), we have

$$M_x = 10(5) - 15(2) = \quad 20\,\text{kN·m}$$ **Answer**

$$M_y = \quad 5(2) - 10(3) = -20\,\text{kN·m}$$ **Answer**

$$M_z = 15(3) - \quad 5(5) = \quad 20\,\text{kN·m}$$ **Answer**

7.6 RESULTANT OF PARALLEL FORCES IN SPACE

The resultant of a parallel system of forces must have the same or equivalent static effect as the parallel system. That is, *the resultant force must be equal to the vector sum of the parallel forces, and the moment of the resultant force about any axis must be equal to the sum of the moments of each parallel force about the same axis.*

Not all parallel force systems have a single force as a resultant. If the resultant of the positive forces is equal to the resultant of the negative forces but has a different line of action (noncollinear), the system represents a resultant couple. If they have the same line of action (collinear), the system is in equilibrium.

Example 7.7 A system of forces parallel to the *y* axis acts on a plate as shown in Fig. 7.9(a). The coordinates (*x*, *z*) locate each force. Find the magnitude and location of the resultant.

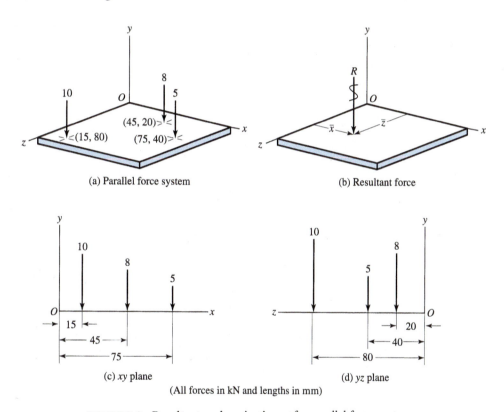

(a) Parallel force system

(b) Resultant force

(c) *xy* plane

(d) *yz* plane

(All forces in kN and lengths in mm)

FIGURE 7.9. Resultant and projections of a parallel force system.

Solution: The forces on the plate are projected onto the *xy* and *yz* planes in Fig. 7.9(c) and (d). To determine the magnitude of the resultant, we sum forces in the *y* direction.

$$\Sigma F = R = 10 + 8 + 5 = 23.0\,\text{kN}$$ **Answer**

To find the location of the resultant ($\bar{x}$, $\bar{z}$) shown in Fig. 7.9(b), we sum moments about the *x* and *z* axes. From Fig. 7.9(c) and (b),

$$\Sigma M_z = 10(15) + 8(45) + 5(75) = 885\,\text{kN·mm}$$

$$\bar{x} = \frac{\Sigma M_z}{R} = \frac{885}{23.0} = 38.5\,\text{mm}$$ **Answer**

and from Fig. 7.9(d) and (b),

$$\Sigma M_x = 8(20) + 5(40) + 10(80) = 1160\ \text{kN·mm}$$

$$\bar{z} = \frac{\Sigma M_x}{R} = \frac{1160}{23.0} = 50.4\ \text{mm} \qquad \textbf{Answer}$$

7.7 SUPPORT CONDITIONS FOR BODIES IN SPACE

The supports react to the weight of the body and to loads (external forces and moments) that act on the body. The supports prevent translation and rotation. Translation can occur in the x, y, and z directions and rotation about the x, y, and z axes. Various supports are pictured and described in Fig. 7.10(a)–(i). They are also described in the following, beginning with Fig. 7.10(a):

a.   The ball, smooth surface, and roller on a smooth surface prevent translation along a single line of action

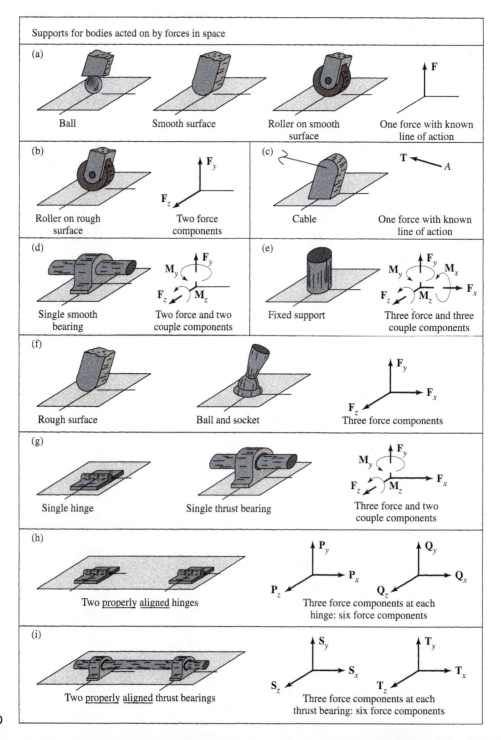

FIGURE 7.10

perpendicular to the surface. They develop one force with a known line of action.

b. The roller on a rough surface prevents translation along two lines of action—one perpendicular to the surface and one parallel to the axis of the roller. It develops two forces with known lines of action.

c. The cable prevents translation along the direction of the cable. It develops one force along the cable away from the body on which it acts.

d. The single smooth bearing prevents translation along the orthogonal axes perpendicular to the axis of the shaft and provides weak resistance to rotation about the same axes. It does not prevent rotation or translation along the axis of the shaft. It develops two forces and two moments along axes perpendicular to the axis of the shaft.

e. The fixed support prevents translation along and rotation about the three coordinate axes. It develops reactions that have three component forces and three component moments (couples).

f. The rough surface and the ball and socket prevent translation along the three coordinate axes. They develop reactions that have three component forces.

g. The single hinge and single thrust bearing prevent translation along the three coordinate axes and rotation about the two coordinate axes that are perpendicular to the axis of the shaft. The hinge and thrust bearing develop reactions that have three component forces and two component moments about axes that are perpendicular to the axis of the shaft.

h. and (i) The *properly aligned* hinges and thrust bearings prevent translation and rotation about the three coordinate axes. Each hinge or thrust bearing develops three component forces. Rotation is prevented by the forces, and so no reactive moments develop.

7.8 EQUILIBRIUM OF A RIGID BODY IN SPACE

A body under the action of a general three-dimensional force system is in equilibrium if the sum of the external forces and the sum of the moments of the external forces and couples are equal to zero. Therefore, equilibrium requires that we satisfy the following six conditions:

$$\Sigma F_x = 0 \qquad \Sigma F_y = 0 \qquad \Sigma F_z = 0 \qquad (7.13)$$

$$\Sigma M_x = 0 \qquad \Sigma M_y = 0 \qquad \Sigma M_z = 0 \qquad (7.14)$$

Example 7.8 The uniform 90-lb plate in Fig. 7.11(a) is supported by three cables. We place a small 90-lb cylinder on the plate and move it until the tension in the three cables is equal. What is the tension in the cables and to what location was the cylinder moved?

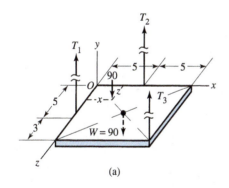 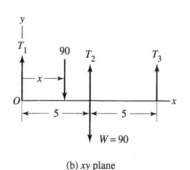 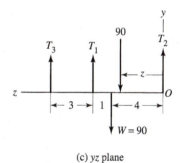

(a) (b) *xy* plane (c) *yz* plane

(All forces in lb and length in ft)

FIGURE 7.11

Solution: The forces are projected onto the *xy* and *yz* planes in Fig. 7.11(b) and (c). To determine the magnitude of the tension in the cables, we sum forces in the *y* direction. Because the tension in the cables is equal,

$$\Sigma F = T + T + T - 90 - 90 = 0$$

$$3T = 180 \quad \text{or} \quad T = 60 \,\text{lb} \qquad \textbf{Answer}$$

To determine the location of the 90-lb cylinder, we sum moments about the *x* and *z* axes. From Fig. 7.11(c) and (b) and with $T = 60 \,\text{lb}$,

$$\Sigma M_x = -60(8) - 60(5) + 90(4) + 90z = 0$$

$$z = \frac{240}{90} = 2.67 \,\text{ft} \qquad \textbf{Answer}$$

$$\Sigma M_z = -90x - 90(5) + 60(5) + 60(10) = 0$$

$$x = \frac{450}{90} = 5.0 \, \text{ft}$$ **Answer**

Example 7.9   The uniform sign in Fig. 7.12(a) is supported by a ball and socket at E and two cables DA and CB. The sign weighs 800 N. Determine the reaction at E and the tension in the cables.

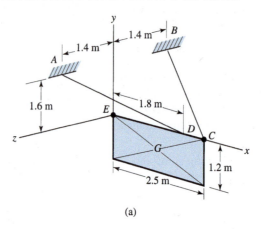

(a)

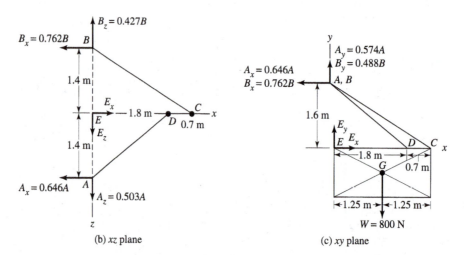

(b) xz plane

(c) xy plane

FIGURE 7.12

Solution: The ball and socket support provides reactions in the x, y, and z directions as in Fig. 7.10(f). The tension in DA acts in the direction of the cable from D toward A [Fig. 7.10(c)]. The coordinates of D and A are $D(1.8, 0, 0)$ m and $A(0, 1.6, 1.4)$ m. Subtracting the coordinates of D from A, we have

$$d_x = 0 - 1.8 = -1.8 \, \text{m} \quad D_y = 1.6 - 0 = 1.6 \, \text{m} \quad d_z = 1.4 - 0 = 1.4 \, \text{m}$$

and $$d = \sqrt{(-1.8)^2 + (1.6)^2 + (1.4)^2} = 2.786 \, \text{m}$$

Therefore, the components of the tension in DA are

$$A_x = \frac{d_x}{d} A = \frac{-1.8}{2.786} A = -0.646A \quad A_y = \frac{d_y}{d} A = \frac{1.6}{2.786} A = 0.574A$$

$$A_z = \frac{d_z}{d} A = \frac{1.4}{2.786} A = 0.503A$$

The force in CB acts in the direction of the cable from C toward B [Fig. 7.10(c)]. The coordinates of C and B are $C(2.5, 0, 0)$ m and $B(0, 1.6, -1.4)$ m. Subtracting the coordinates of C from B, we have

$$d_x = 0 - 2.5 = -2.5 \, \text{m} \quad d_y = 1.6 - 0 = 1.6 \, \text{m} \quad d_z = -1.4 - 0 = -1.4 \, \text{m}$$

and $$d = \sqrt{(-2.5)^2 + (1.6)^2 + (-1.4)^2} = 3.28 \, \text{m}$$

Therefore, the components of the tension in *CB* are

$$B_x = \frac{d_x}{d} B = \frac{-2.5}{3.28} B = -0.762B \qquad B_y = \frac{d_y}{d} B = \frac{1.6}{3.28} B = 0.488B$$

$$B_z = \frac{d_z}{d} B = \frac{-1.4}{3.28} B = -0.427B$$

With the reaction components at *A* and *B* known, we draw the free-body diagrams in Fig. 7.12(b) and (c). Notice that the algebraic sign (direction) of the component reactions was used to draw the free-body diagrams. Thus, the equilibrium equations already include their directions. From the free-body diagram of Fig. 7.12(b), we sum moments about point *E*:

$$\Sigma M_E = B_x(1.4) - A_x(1.4) = 0 \qquad A_x = B_x \tag{a}$$

From Fig. 7.12(c), we write a moment equation about point *E*:

$$\Sigma M_E = (A_x + B_x)1.6 - 800(1.25) = 0 \tag{b}$$

From equations (a) and (b),

$$A_x = 312.5 \text{ N} \qquad B_x = 312.5 \text{ N}$$

Because $A_x = 0.646A$ and $B_x = 0.762B$, we have the tension in cables *DA* and *CB*.

$$A = \frac{A_x}{0.646} = \frac{312.5}{0.646} = 484 \text{ N} \qquad B = \frac{B_x}{0.762} = \frac{312.5}{0.762} = 410 \text{ N} \qquad \textbf{Answers}$$

With *A* and *B* known, we solve for the *y* and *z* components of *A* and *B*, as follows:

$$A_y = 0.574A = 0.574(484) = 278 \text{ N}$$
$$A_z = 0.503A = 0.503(484) = 243 \text{ N}$$
$$B_y = 0.488B = 0.488(410) = 200 \text{ N}$$
$$B_z = 0.427B = 0.427(410) = 175 \text{ N}$$

From the free-body diagram of Fig. 7.12(c), we sum moments about point *A* and sum forces in the *y* direction.

$$\Sigma M_A = E_x(1.6) - 800(1.25) = 0$$
$$E_x = 625 \text{ N} \qquad \textbf{Answer}$$

$$\Sigma F_y = A_y + B_y + E_y - 800 = 0$$
$$E_y = 800 - A_y - B_y$$
$$E_y = 800 - 278 - 200 = 322 \text{ N} \qquad \textbf{Answer}$$

Summing forces in the *z* direction [Fig. 7.12(b)], we have

$$\Sigma F_z = A_z + E_z - B_z = 0$$
$$E_z = B_z - A_z = 175 - 243$$
$$= -68 \text{ N} \qquad \textbf{Answer}$$

Example 7.10   The schematic of a crank and pulley is shown in Fig. 7.13(a). The crank and pulley are supported by a properly aligned smooth bearing at *A* and a thrust bearing at *B*. Determine the tension *Q* required to maintain equilibrium and the bearing reactions at *A* and *B*.

Solution: Free-body diagrams of the crank and pulley are shown in the projections of Fig. 7.13(b)–(d). Because the bearings at *A* and *B* are properly aligned, they can support forces only. Bearing *A* can support reactions in the *xy* plane only. Thrust bearing *B* can support reactions in the *xy* plane and in the *z* direction for stability. From Fig. 7.13(b), we write moment equations about point *C*:

$$\Sigma M_C = 2.1(125) - Q(75) = 0, \qquad Q = 3.5 \text{ kN} \qquad \textbf{Answer}$$

Summing moments about points *B* and *A* for the free-body diagram of Fig. 7.13(c), we have

$$\Sigma M_B = A_x(200) - 2.1(100) = 0, \qquad A_x = 1.05 \text{ kN} \qquad \textbf{Answer}$$
$$\Sigma M_A = 2.1(100) - B_x(200) = 0, \qquad B_x = 1.05 \text{ kN} \qquad \textbf{Answer}$$

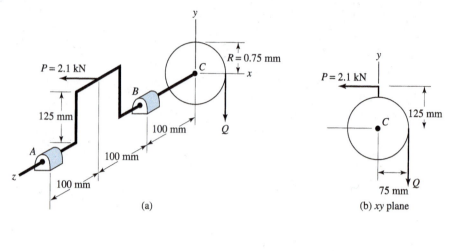

(a)

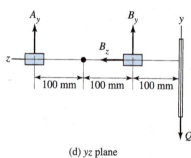

(b) xy plane

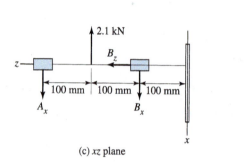

(c) xz plane

(d) yz plane

FIGURE 7.13

For the free-body diagram of Fig. 7.13(d), we sum moments about points B and A as follows:

$$\Sigma M_B = -A_y(200) - Q(100) = 0$$
$$\Sigma M_A = B_y(200) - Q(300) = 0$$

Because $Q = 3.5\,kN$, we have

$$A_y = -1.75\,kN \qquad B_y = 5.25\,kN \qquad \text{**Answers**}$$

To conclude our solution, we sum forces in the z direction:

$$\Sigma F_z = B_z = 0 \qquad \text{**Answer**}$$

Check:

$$\Sigma F_x = A_x + B_x - P = 1.05 + 1.05 - 2.1 = 0 \quad OK$$
$$\Sigma F_y = A_y + B_y - Q = -1.75 + 5.25 - 3.5 = 0 \quad OK$$

PROBLEMS

7.13 Determine the moment of the force shown in Prob. 7.1 about the x, y, and z axes.

7.14 Determine the moment of the forces shown in Prob. 7.5 about the x, y, and z axes.

7.15 and 7.16 Determine the moment of the forces shown about the x, y, and z axes.

7.17 and 7.18 Determine the resultant of the parallel force system shown.

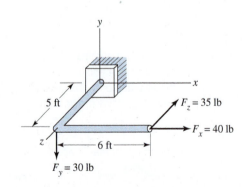

PROB. 7.15

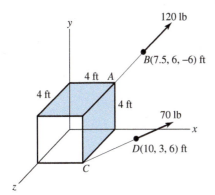

PROB. 7.16

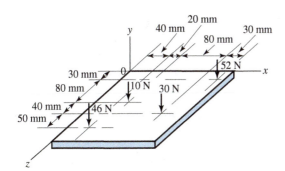

PROB. 7.17

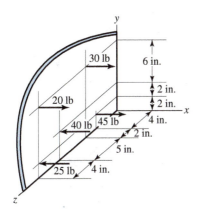

PROB. 7.18

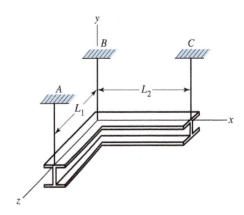

PROB. 7.19 and PROB. 7.20

7.21 The uniform aluminum plate supported by three cables has a weight of 705 lb. Determine the tension in the cables.

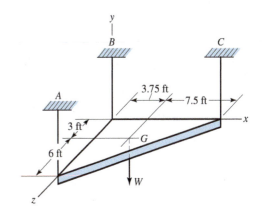

PROB. 7.21

7.22 A boom *DE* is supported by a ball and socket at *D* and cables *AC* and *AB*. The boom has a length of 7 m. Determine the tensions in the cables and the socket reactions at *D*. Neglect the weight of the boom.

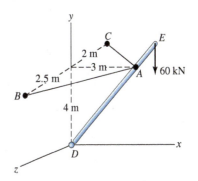

PROB. 7.22

7.19 Two W10 × 100 steel beams (see Sec. 9.6 in Chap. 9) are welded together to form an L-shaped member. If $L_1 = 6$ ft and $L_2 = 8$ ft, determine the tension in the cables. (The beam weighs 100 lb/ft.)

7.20 Two W270 × 149 steel beams (see Sec. 9.6 in Chap. 9) are welded together to form an L-shaped member. If $L_1 = 2$ m and $L_2 = 3$ m, determine the tension in the cables. (The beam has a mass of 149 kg/m.)

7.23 A shaft with pulleys is supported by a properly aligned smooth bearing at *A* and a thrust bearing at *B*. Determine the magnitude of the couple *M* for equilibrium and the bearing reactions at *A* and *B*.

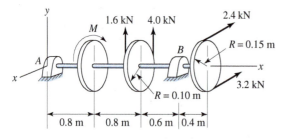

PROB. 7.23

7.24 The schematic of a 15,000-lb rotor is shown in the diagram. The rotor supports consist of a properly aligned smooth bearing at A and a thrust bearing at B. When the spinning rotor is in the position shown, it produces unbalanced forces $P_x = 5000$ lb and $Q_y = 5000$ lb at the ends of the rotor. Determine the bearing reactions at A and B. [The unbalanced forces are dynamic (centrifugal) forces. Their positions change as the rotor turns.]

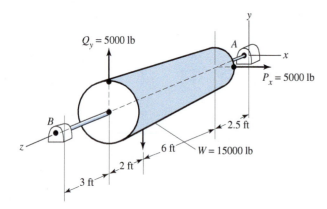

PROB. 7.24

7.25 The schematic of a crank shaft, connecting rod, piston, and pulley is shown in the diagram. The crankshaft is supported by a properly aligned smooth bearing at B and a thrust bearing at A. Determine the force P on the pulley for equilibrium and the bearing reactions at A and B.

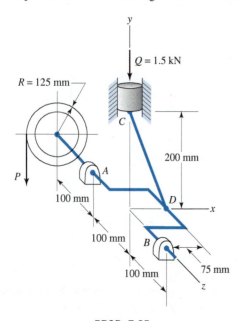

PROB. 7.25

7.26 A uniform trap door weighing 350 N is tied open by a rope as shown. The hinges are properly aligned. Assume that the hinge at A does not support an axial reaction (z direction). Determine the tension in the rope and the reactions at the hinges.

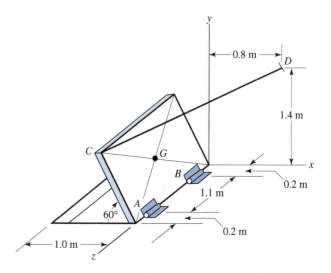

PROB. 7.26

7.27 The power pole AC is supported by the ground (xz plane) at A and by guy wire CDE at E. Two cables that are *parallel* to the ground exert forces on the pole of 270 N each as shown. If the tension in the guy wire is 350 N, determine the reactions at A.

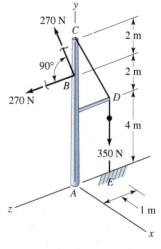

PROB. 7.27

FRICTION

CHAPTER OBJECTIVES

In this chapter, we investigate the frictional forces that can develop between two surfaces in contact with each other. After completing this material, you should be able to

- Define the coefficients of static friction and kinetic friction, and give an approximate range of numerical values for each.

- Describe the three stages of dry friction as a stationary object is set in motion by external forces.

- Explain the relationship between angle of repose and coefficient of static friction.

- Solve problems of static equilibrium for which frictional effects must be considered.

- Analyze common mechanical elements such as wedges, bearings, screws, and belts whose operations are affected by friction.

- Define the coefficient of rolling resistance and apply it to simple problems involving uniform or impending motion.

8.1 INTRODUCTION

In Chapters 5 and 7, we considered various support conditions. Among these are the smooth and rough surfaces. For the *smooth surface,* the reactive force is normal to the surface at the point of contact. The body and surface on which the body rests are free to move with respect to each other. With the *rough surface,* the reactive force is oblique to the surface at the point of contact. For convenience, the reactive force can be resolved into components normal and tangential to

the surface. The tangential component acts to prevent the free motion of the body and surface with respect to each other. The direction of the friction force acting on the body is opposed to the motion of the body over the surface.

Friction can be both a liability and an asset. With such things as brakes, belt drives, and walking, friction is indispensable. However, the loss of power and wear of surfaces in contact for gears, bearings, and other machine elements, due to friction, is undesirable.

In this chapter, we consider dry friction, that is, friction that occurs between a body and surface that are in contact along dry surfaces. Dry friction should not be confused with friction between surfaces that are completely or partially separated by a fluid.

8.2 DRY OR COULOMB FRICTION

The nature of *dry friction* is complex and not well understood. However, it does result from the interactions between the surface layers in contact. The interactions are made up of a number of processes, including the effects of surface roughness and the molecular attraction between surfaces. Although dry friction is complex, empirical laws are available that can be used to determine values of the dry friction force. The laws are usually attributed to Coulomb, who published the results of numerous friction experiments in 1781.

A simple example involving the sliding of a block along a rough surface will serve to introduce Coulomb's law of dry friction. In Fig. 8.1(a), we show a block of weight W resting on a rough surface. A gradually increasing force P is applied to the block, and the resulting friction force is measured.

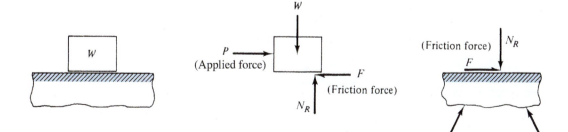

(a)	(b)	(c)

FIGURE 8.1

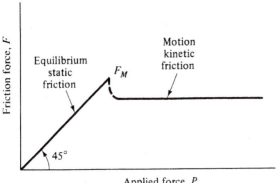

FIGURE 8.2

TABLE 8.2   Coefficient of Static Friction

Substance	f_s
Metal on metal	0.15–0.75
Metal on wood	0.25–0.65
Metal on stone	0.25–0.70
Metal on ice	0.02–0.03
Wood on wood	0.40–0.70
Rubber on concrete	0.60–0.90

Note: Values of the coefficient of kinetic friction are approximately 75 percent of those for the corresponding coefficient of static friction.

Free-body diagrams for the block and rough surface are shown in Fig. 8.1(b) and (c). As we increase P, the frictional force increases to maintain equilibrium, $F = P$. When the friction reaches a maximum value F_m, the block is about to move. Motion of the block is *impending* at that point $P = F_m$. For values of $P > F_m$, the block moves, and the friction force drops rapidly to a kinetic value F_k. After motion occurs, the friction force remains essentially constant for increasing values of P. The relationship between the applied force and friction force as described is shown in Fig. 8.2.

It has been found from experiments that for a given pair of surfaces, the maximum value of friction F_m is proportional to the normal reaction N_R. That is,

$$F_m = f_s N_R \qquad (8.1)$$

where f_s is the *coefficient of static friction.* Also, from experiments, the kinetic value of friction F_k is proportional to the normal reaction N_R. Thus,

$$F_k = f_k N_R \qquad (8.2)$$

where f_k is the *coefficient of kinetic friction.*

When a force is applied to a block on a dry (nonlubricated) surface, three results are possible:

1. The block is in equilibrium and does not move. The friction force is equal to the applied force P. ($F = P \le F_m = f_s N_R$)

2. The block is about to move—motion is impending (imminent). The friction force is equal to the product of the coefficient of *static* friction f_s and the normal force N_R. ($F = f_s N_R$)

3. The block is moving, and the friction force decreases. The force F is equal to the product of the coefficient of *kinetic* friction f_k and the normal force N_R. ($F = f_k N_R$)

The three stages for dry friction are summarized in Table 8.1.

TABLE 8.1   Stages of Friction for a Block on a Dry Surface

1. Block does not move.	$F = P \le F_m = f_s N_R$
2. Block is about to move.	$F = F_m = f_s N_R$
3. Block is moving.	$F = F_k = f_k N_R$

Both coefficients of friction are almost independent of the area of contact between the block and the surface. Thus, if the block tipped and only an edge was in contact with the surface, we would have nearly the same coefficients of static friction for the same dry surfaces. Some typical values of the coefficient of static friction are listed in Table 8.2. The values shown are representative only. Experimentation would be required to determine values for an actual engineering system.

8.3 ANGLE OF FRICTION

In Fig. 8.3, we show that the friction force F and the normal reaction N_R are components of the total reaction R of the surface on the block. The angle between R and N_R represented by the Greek letter φ (phi) increases with increasing values of F. When F reaches the maximum value F_m, motion impends. The angle $\varphi = \varphi_s$ between F_m and N_R is given by

$$\tan \varphi_s = \frac{F_m}{N_R} \qquad (a)$$

Substituting for F_m from Eq. (8.1) in Eq. (a), we have

$$\tan \varphi_s = \frac{f_s N_R}{N_R} = f_s \qquad (8.3)$$

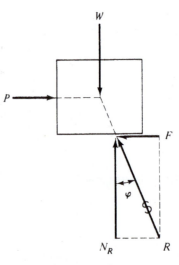

FIGURE 8.3

SURFACE ROUGHNESS

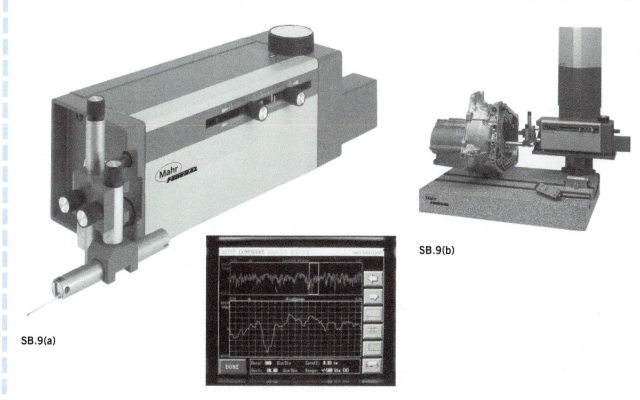

SB.9(a)

SB.9(b)

SB.9(c)

Frictional effects, as well as a type of material behavior known as *fatigue strength* (see Sec. 11.11), are each influenced by an object's surface quality. To evaluate the condition of a surface, an instrument called a *profilometer* is often used. The device shown in SB.9(a) uses a 0.0002-in.-diameter diamond stylus that is drawn across the surface to be tested. This stylus can trace irregular profiles, such as the automatic transmission housing in SB.9(b), and can store the data or generate visual images in the form of a graph or computer display SB.9(c). The process is similar to that of a phonograph needle as it follows the groove on a record and reproduces sound information contained in the record's surface variations. *(Courtesy of Mahr Federal, Inc., Providence, Rhode Island, www.mahrfederal.com)*

The angle φ_s is called the *angle of friction.* The tangent of the angle of friction is equal to the coefficient of friction. If the force P changes direction but remains horizontal, the reaction R sweeps out a cone that is called the *cone of friction.*

Example 8.1   A block rests on a rough inclined plane. Determine the angle of inclination of the plane if the block is on the verge of sliding down the plane.

Solution: The free-body diagram for the block is shown in Fig. 8.4(a). The block is in equilibrium under the action of forces N_R, W, and the maximum static friction F_m. In Fig. 8.4(b), the forces have been moved along their line of action until they are directed away from a common point. From equilibrium,

$$\Sigma F_x = 0 \qquad -F_m \cos\theta + N_R \sin\theta = 0$$

$$\frac{\sin\theta}{\cos\theta} = \frac{F_m}{N_R} \qquad \text{or} \qquad \tan\theta = \frac{F_m}{N_R}$$

Substituting for F_m from Eq. (8.1), we write

$$\tan\theta = \frac{f_s N_R}{N_R} = f_s$$

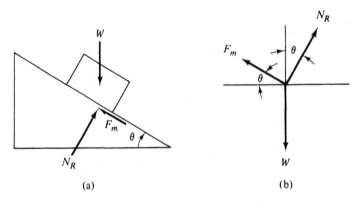

(a)   (b)

FIGURE 8.4

Comparing this result with Eq. (8.3), we see that

$$\tan\theta = \tan\varphi_s \quad \text{or} \quad \theta = \varphi_s \qquad \qquad \textbf{Answer}$$

Therefore, if the plane is inclined at the angle of friction φ_s, the body will be on the verge of motion. For this reason, the angle of friction is also called the *angle of repose*.

Example 8.2 A ladder with a mass of 8.66 kg and a length of 2.5 m rests in a vertical plane with its lower end on a horizontal floor ($f_s = 0.3$) and its upper end against a vertical wall ($f_s = 0.2$). Find the smallest angle θ that the ladder can make with the floor before slipping begins. The center of gravity of the ladder is two-fifths of the length of the ladder from the bottom.

Solution: The free-body diagram for the ladder is shown in Fig. 8.5. The mass of the ladder is 8.66 kg. From Eq. (1.1), the weight of the ladder is

$$W = mg = 8.66(9.81) = 85.0 \text{ N}$$

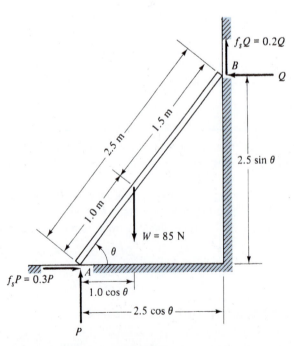

FIGURE 8.5

Because the ladder is about to move, the friction forces will have reached their maximum values and will be directed to oppose impending motion. Apply the equations of equilibrium:

$$\Sigma F_x = 0 \qquad 0.3P - Q = 0 \tag{a}$$

$$\Sigma F_y = 0 \qquad P + 0.2Q - 85 = 0 \tag{b}$$

$$\Sigma M_A = 0$$

$$Q(2.5\sin\theta) + 0.2Q(2.5\cos\theta) - 85(1.0\cos\theta) = 0 \tag{c}$$

Eliminate P between Eqs. (a) and (b):

$$Q = 24.1 \text{ N}$$

Substitute Q in Eq. (c):

$$\frac{\sin\theta}{\cos\theta} = \tan\theta = 1.2108$$

$$\theta = 50.5° \qquad\qquad \textbf{Answer}$$

Example 8.3 A packing crate weighing 150 lb and resting on a floor is pushed by a horizontal force P, as shown in Fig. 8.6(a). The coefficient of friction between the crate and floor is 0.35. How high above the floor can the force P act so that the crate moves without tipping?

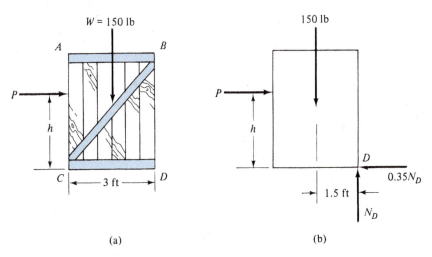

FIGURE 8.6

Solution: The free-body diagram for conditions that give the limiting position of P is shown in Fig. 8.6(b). Impending tipping requires that the friction and normal reaction be concentrated at D—the right edge of the crate. Impending slipping requires that the friction force $F_m = f_s N_R = 0.35N_D$. From the equations of equilibrium, we write

$$\Sigma F_y = 0 \qquad N_D - 150 = 0$$
$$N_D = 150 \text{ lb}$$

$$\Sigma F_x = 0 \qquad P - 0.35N_D = 0$$
$$P = 0.35(150) = 52.5 \text{ lb}$$

$$\Sigma M_D = 0 \qquad -Ph + 150(1.5) = 0$$
$$h = \frac{150(1.5)}{52.5} = 4.29 \text{ ft}$$

The height must be less than 4.29 ft. **Answer**

PROBLEMS

8.1 A 600-lb block rests on a horizontal surface as shown. Determine the friction force if the applied load P is equal to (a) 100 lb, (b) 200 lb, and (c) 300 lb. ($f_s = 0.4$ and $f_k = 0.3$.)

8.2 The block resting on a horizontal surface as shown has a mass of 100 kg. Determine the friction force if the applied load P is equal to (a) 150 N, (b) 390 N, and (c) 600 N. ($f_s = 0.4$ and $f_k = 0.3$.)

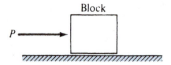

PROB. 8.1 and PROB. 8.2

8.3 A 100-lb block rests on the inclined plane shown. Determine the maximum and minimum value of P for which the block is in equilibrium. ($f_s = 0.25$ and $\theta = 35°$.)

8.4 The block resting on the inclined plane shown has a mass of 40 kg. Determine the maximum and minimum value of P for which the block is in equilibrium. ($f_s = 0.35$ and $\theta = 25°$.)

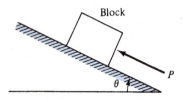

PROB. 8.3 and PROB. 8.4

8.5 The block resting on the inclined plane shown has a mass of 60 kg. Determine the maximum and minimum value of P for which the block is in equilibrium. ($f_s = 0.35$.)

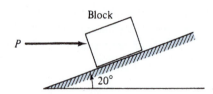

PROB. 8.5 and PROB. 8.6

8.6 A 200-lb block rests on the inclined plane shown. Determine the maximum and minimum value of P for which the block is in equilibrium. ($f_s = 0.25$.)

8.7 The 400-kg trunk rests on the floor as shown. If $b = 1$ m, $H = 2.5$ m, and $P = 1400$ N, determine the maximum height h the force can be moved above the floor before the trunk starts to tip over. ($f_s = 0.3$.)

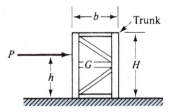

PROB. 8.7 and PROB. 8.8

8.8 The 1000-lb trunk rests on the floor as shown. If $b = 36$ in., $H = 75$ in., and $P = 280$ lb, determine the maximum height h the force can be moved above the floor before the trunk starts to tip over. ($f_s = 0.3$.)

8.9 The drawer shown has a weight of 25 lb. If the coefficient of friction between the drawer and surface A is 0.4, what minimum horizontal force P is required to close the drawer?

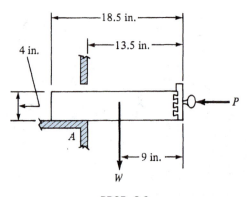

PROB. 8.9

8.10 A 10-kg uniform ladder 2.5 m long rests against a smooth wall as shown. An 80-kg man climbs up the ladder to a point A, a distance of $L = 2.0$ m from the bottom of the ladder, before the ladder slips. What is the coefficient of friction between the ladder and the ground if the angle between the ladder and ground is 60°?

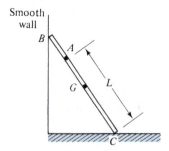

PROB. 8.10 and PROB. 8.11

8.11 A 25-lb uniform ladder 8 ft long rests against a smooth wall as shown. A 175-lb man climbs up the ladder to a point A, a distance L from the bottom of the ladder, before the ladder slips. If the coefficient of friction between the ladder and the ground is $f_s = 0.4$, what is the distance L if the angle between the ladder and ground is 60°?

8.12 A movable bracket can be raised or lowered on the vertical 1-in.-diameter rod shown. If $P = 80$ lb, $H = 3$ in., and $L = 10$ in., what minimum coefficient of friction is required to prevent the bracket from slipping downward? Assume that contact is made between the bracket and rod at points M and N.

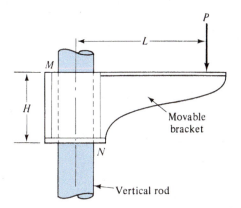

PROB. 8.12 and PROB. 8.13

8.13 A movable bracket can be raised or lowered on the 50-mm-diameter rod as shown. If $H = 75$ mm, $L = 275$ mm, and $P = 350$ N, what is the largest coefficient of friction for which the bracket will slip down? Assume that contact is made between the bracket and the rod at points M and N.

8.14 A gate is supported by two rings at A and B that fit over a vertical rod as shown. If the coefficient of friction between the rings and the rod is 0.2, find the maximum distance H between the rings for which the gate is in equilibrium. The reaction at the upper ring is directed away from the gate and at the lower ring toward the gate.

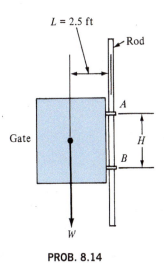

PROB. 8.14

8.4 WEDGES

The *wedge* is a simple machine. It may be used to transform horizontal input forces into vertical output forces. The angle of friction may be used to simplify the solution, as will be seen in the following example.

Example 8.4 A wedge B of negligible weight is to be used to move block A upward [Fig. 8.7(a)]. The coefficient of friction between all surfaces of contact is 0.3. Determine the input force P required to produce an output force $Q = 2150$ lb. (Q includes the weight of block A.)

Solution: The free-body diagrams together with their force diagrams for impending motion of block A and wedge B have been shown in Fig. 8.7(b) and (c). The maximum static friction forces and normal reactions have been replaced by their resultants acting at the angle of friction $\varphi_s = \arctan 0.3 = 16.7°$ with the normal. For equilibrium of body A,

$$\Sigma F_x = 0 \quad R_2 \cos 65.3° - R_1 \cos 16.7° = 0$$
$$R_1 = 0.4363 R_2 \quad \text{(a)}$$
$$\Sigma F_y = 0 \quad R_2 \sin 65.3° - R_1 \sin 16.7° = 2150 \quad \text{(b)}$$

Substituting Eq. (a) for R_1 in Eq. (b), we have

$$R_2 = 2745 \text{ lb}$$

For equilibrium of body B,

$$\Sigma F_y = 0 \quad R_3 \sin 73.3° - R_2 \sin 65.3° = 0$$

Substituting for R_2, we have

$$R_3 = 2604 \text{ lb}$$
$$\Sigma F_x = 0 \quad P - R_3 \cos 73.3° - R_2 \cos 65.3° = 0$$

Substituting for R_2 and R_3 yields

$$P = 1895 \text{ lb} \qquad \text{**Answer**}$$

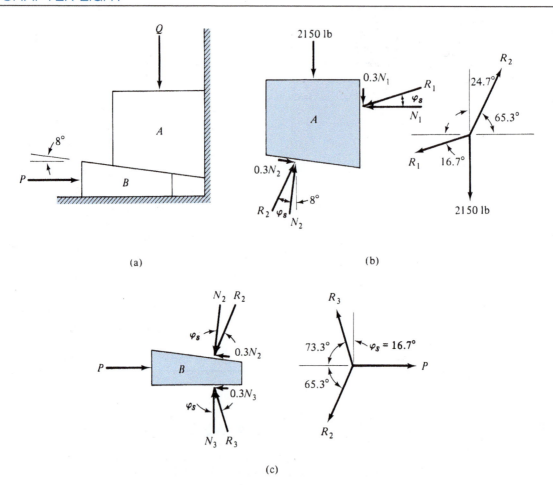

(a)

(b)

(c)

FIGURE 8.7

8.5 SQUARE-THREADED SCREWS: SCREW JACKS

The *screw jack* has a threaded stem that turns in a fixed base [Fig. 8.8(a)]. As a force P is applied to the handle, the stem turns and moves up (or down) lifting (or lowering) the load W. The screw can be analyzed by considering the screw thread as an inclined plane rolled on a cylinder [Fig. 8.8(b)]. The inclination of the plane is given by

$$\tan \theta = \frac{p}{\pi d} \qquad (8.4)$$

where p is the pitch of the thread and d the mean diameter of the threads.

We isolate a small element of the screw and draw a free-body diagram for raising the load as shown in Fig. 8.8(b). The corresponding force diagram is shown in Fig. 8.8(c). The friction force F_m and normal force N_R have been replaced by their resulting R, which acts at an angle φ_s (angle of static friction) from the normal N_R. From the equations of equilibrium,

$$\Sigma F_x = 0 \qquad F_x - R \sin (\theta + \varphi_s) = 0 \qquad \textbf{(a)}$$

$$\Sigma F_y = 0 \qquad R \cos (\theta + \varphi_s) - W = 0 \qquad \textbf{(b)}$$

Eliminating R between Eqs. (a) and (b), we obtain

$$F_x = \frac{W \sin (\theta + \varphi_s)}{\cos (\theta + \varphi_s)}$$

$$F_x = W \tan (\theta + \varphi_s) \qquad \textbf{(c)}$$

To produce motion, the moment of the force P on the handle of the screw jack must be equal to the moment of the force F_x at the mean radius $d/2$ of the screw—both about the axis of the screw. Thus,

$$M = PL = F_x \frac{d}{2}$$

Substituting for F_x from Eq. (c), we obtain

$$M = PL = W \frac{d}{2} \tan (\theta + \varphi_s) \qquad \textbf{(8.5)}$$

where P is the force applied to the handle, L the length of the handle, W the load to be *lifted* by the jack screw, θ the inclination of the plane of the screw given by Eq. (8.4), and φ_s the angle of friction. To lower the load, the friction force is reversed, and Eq. (8.5) must be modified as follows:

$$M = PL = W \frac{d}{2} \tan (\varphi_s - \theta) \qquad \textbf{(8.6)}$$

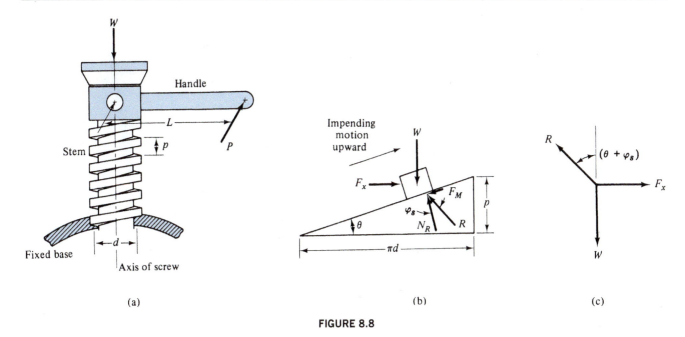

FIGURE 8.8

If $\varphi_s \geq \theta$, then M is a positive number, and the screw is said to be *self-locking*. This means that the weight will remain in place, even if M is removed. If $\varphi_s < \theta$, then M is a negative number, and the weight will lower by itself unless M continues to be applied.

Example 8.5   A screw jack having two threads per 10 mm and a mean diameter of 25 mm is used to lift a weight. If a force of 2.5 kN is applied at the end of a handle 300 mm long, what weight can be lifted by the screw jack? ($f_s = 0.35$.)

Solution: Because the screw moves through two turns and travels 10 mm, the pitch $p = 10/2 = 5$ mm. From Eq. (8.4),

$$\tan\theta = \frac{p}{\pi d} = \frac{5}{\pi(25)}$$

$$\theta = 3.64°$$

The angle of friction

$$\varphi_s = \arctan f_s = \arctan 0.35$$
$$= 19.29°$$

From Eq. (8.5),

$$W = \frac{2PL}{d\tan(\theta + \varphi_s)} = \frac{2(2.5)(300)}{25\tan(3.64° + 19.29°)}$$
$$= 141.8 \text{ kN} \qquad \text{Answer}$$

PROBLEMS

In Probs. 8.15 through 8.24, assume that all screw threads are square.

8.15 Determine the force P required to move the 500-lb block shown if the coefficient of friction between all contact surfaces is 0.3.

8.16 The coefficient of friction between all contact surfaces is 0.4, except as noted. Determine the force P required to move the 2000-N block.

8.17 A 5° wedge is used to raise the end of a beam as shown. If the coefficient of friction between all surfaces is $f_s = 0.25$, determine the force P required to move the wedge. Assume that the beam is horizontal and does not move.

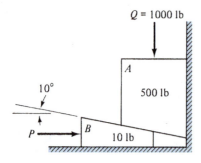

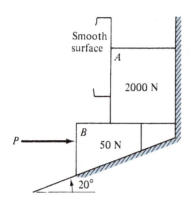

PROB. 8.15

PROB. 8.16

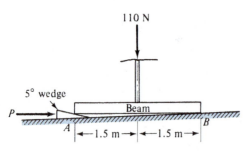

PROB. 8.17

8.18 The base of a torsion machine is to be lifted by a 7°
wedge as shown. If the coefficient of friction between
all surfaces is $f_s = 0.3$, determine the force P required
to move the wedge. Assume that the machine base is
horizontal and does not move.

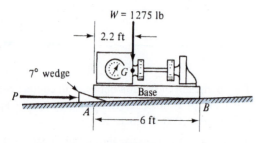

PROB. 8.18

8.19 Determine the vertical force Q required to move block
B as shown in the figure. ($f_s = 0.2$ for all surfaces.)

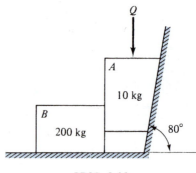

PROB. 8.19

8.20 A steel wedge is used to split a log as shown. Wedges
are designed to be self-locking, that is, so the wedge
will not slip out of the log. Determine the largest angle
θ for which the wedge is self-locking if the coefficient
of friction between the wedge and the log is 0.3.

PROB. 8.20

8.21 A screw jack having four threads per inch and a mean
diameter of 1.2 in. is used to raise or lower a weight of
4500 lb as shown. What force must be applied at the
end of a 12-in.-long handle to (a) raise the weight and
(b) lower the weight? ($f_s = 0.4$.)

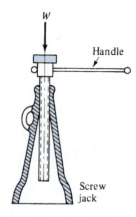

PROB. 8.21 and PROB. 8.22

8.22 A screw jack with two threads per 15 mm, a mean radius
of 30 mm, and a handle length of 0.3 m is used to raise
or lower a weight as shown. (a) If a force of 0.3 kN is
applied to the handle to raise a weight of 20 kN, what
is the coefficient of friction? (b) If a force of 0.25 kN is
applied to the handle to lower a weight of 20 kN, what
is the coefficient of friction?

8.23 The C-clamp is used to clamp two blocks of plastic together as shown with a force of 700 lb. The clamp has 10 threads per inch and a mean diameter of 0.5 in. If the coefficient of friction is 0.25, what moment is required?

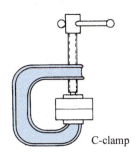

PROB. 8.23 and PROB. 8.24

8.24 The C-clamp shown is used to hold two blocks together. The clamp has a pitch of 0.3 mm and a mean diameter of 20 mm. The coefficient of friction is 0.2. What force will be applied to the blocks if a moment of 6.0 N·m is applied to the handle?

8.6 AXLE FRICTION: JOURNAL BEARINGS

Journal bearings are commonly used to provide lateral support to rotating machines such as shafts and axles. If the bearing is partially lubricated or not lubricated at all, the methods of this chapter may be applied.

The journal bearing shown in Fig. 8.9(a) supports a shaft that is rotating at a constant speed. To maintain the rotation, a torque or moment couple M must be applied. The load W and moment couple M cause the shaft to touch the bearing at A. This can be explained by the fact that the shaft climbs up on the bearing until the forces and couple are in equilibrium. The shaft touching the bearing at A gives rise to

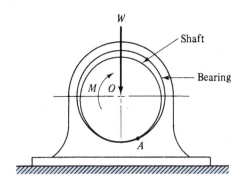

(a)

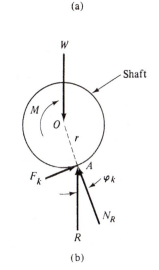

(b)

FIGURE 8.9

reaction N_R and friction force F_k as shown in the free-body diagram of Fig. 8.9(b). For equilibrium,

$$\Sigma F_y = 0 \qquad W = R$$
$$\Sigma M_O = 0 \qquad M = Rr\sin\varphi_k \qquad (8.7)$$

For small angles of friction, $\sin\varphi_k \approx \tan\varphi_k = f_k$. Thus,

$$M \approx Rrf_k \qquad (8.8)$$

Example 8.6 A drum of 300-mm radius on a shaft with a diameter of 75 mm and supported by two bearings as shown in Fig. 8.10(a) is used to hoist a 250-kg load. The drum and shaft have a mass of 50 kg, and the coefficient of friction for the bearing is 0.25. Determine the moment that must be applied to the drum shaft to raise the load at uniform speed.

Solution: The masses are changed to weights by Eq. (1.1):

$$W_1 = m_1g = 250(9.81) = 2450\,\text{N}$$
$$W_2 = m_2g = 50(9.81) = 490\,\text{N}$$

The free-body diagram of the drum and shaft is shown in Fig. 8.10(b). From equilibrium,

$$\Sigma F_y = 0 \qquad R - W_1 - W_2 = 0$$
$$R - 2450 - 490 = 0$$
$$R = 2940\,\text{N}$$

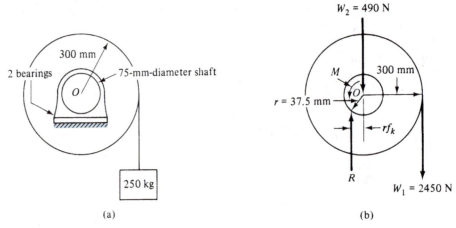

FIGURE 8.10

$$\Sigma M_O = 0 \qquad M - Rrf_k - W_1L = 0$$

$$M - (2940)(37.5)0.25 - 2450(300) = 0$$

$$M = 763 \text{ N·m} \qquad \textbf{Answer}$$

WOOD BEARINGS

SB.10

Although most bearings are made of metal or plastic, a surprising number of industrial applications use *wood bearings*. The basic material, typically a dense, close-grained hardwood such as maple, is impregnated with a blend of oil lubricants and is then machined into the desired shapes, usually to tolerances between 0.002 and 0.005 in. The finished products (SB.10) are self-lubricating; perform well in gritty or abrasive environments; and are not harmed by water, mild acids, alkalis, and most caustic chemicals. For these reasons, wood bearings are well suited to equipment that operates in harsh environments, such as those found in agriculture, food processing, paper production, wire and cable making, sewage treatment, and marine operations. They are also found on a variety of individual products such as turnstiles, conveyors, exercise equipment, and vending machines. *(Courtesy of POBCO, Inc., Worcester, Massachusetts, www.pobcoplastics.com)*

8.7 SPECIAL APPLICATIONS

We will now consider the application of dry friction laws to several special devices. The derivations of the governing equations depend on calculus and will not be considered here.

Disk Friction, Thrust Bearings

Consider a hollow rotating shaft whose end is in contact with a fixed surface, as shown in Fig. 8.11. We assume that the force between the rotating shaft and fixed surface is distributed uniformly. The external torque M required to cause slipping to occur is given by

$$M = f_k P \frac{2(R_o^3 - R_i^3)}{3(R_o^2 - R_i^2)} \quad (8.9)$$

where P is the thrust force, R_o the outside radius, and R_i the inside radius of the hollow shaft. For a solid shaft of radius R,

$$M = f_k P (2/3) R \quad (8.10)$$

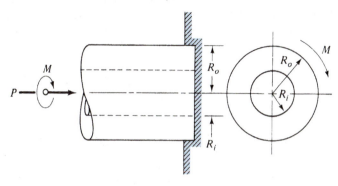

FIGURE 8.11

The largest torque, without slipping, and transmitted by a disk clutch can be found from Eqs. (8.9) and (8.10), where f_k has been replaced by f_s.

Example 8.7   Determine the torque capacity of the clutch shown in Fig. 8.12 if the coefficient of friction $f_k = 0.25$ and the axial force is 1580 lb. Assume uniform force distribution on the contact surfaces.

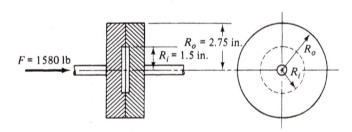

$R_o = 2.75$ in.
$R_i = 1.5$ in.
$F = 1580$ lb

FIGURE 8.12

Solution: From Eq. (8.9),

$$M = f_k P \frac{2(R_o^3 - R_i^3)}{3(R_o^2 - R_i^2)} = 0.25(1580) \frac{2[(2.75)^3 - (1.5)^3]}{3[(2.75)^2 - (1.5)^2]}$$

$$= 864 \text{ lb-in.} \qquad \text{Answer}$$

Belt Friction

Consider a flat belt passing over a fixed rough cylinder as shown in Fig. 8.13. The belt has impending motion as shown. The relationship between the tensions T_L (larger tension) and T_S (smaller tension) at the ends of the belt is given by

$$\frac{T_L}{T_S} = e^{\pi f \beta / 180°} \quad (8.11)$$

where e is the base of the natural logarithm ($e = 2.718$) and the Greek letter β (beta) is the angle of contact between the belt and cylinder in degrees.

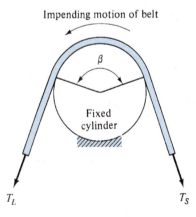

Impending motion of belt
β
Fixed cylinder
T_L T_S

FIGURE 8.13

Example 8.8   Three rough round pins are attached to a 150-kg block B, which can move vertically between smooth guides. A rope is placed over the pins as shown in Fig. 8.14(a). Determine the mass of the block A if block B is on the point of impending motion. ($f_s = 0.35$.)

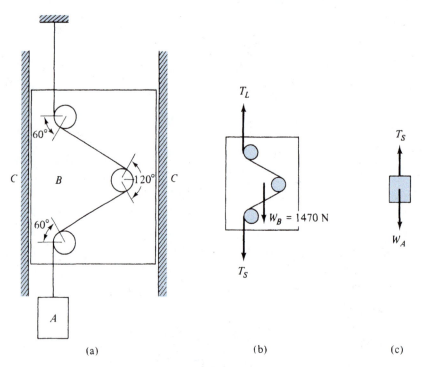

(a)   (b)   (c)

FIGURE 8.14

Solution: Free-body diagrams of blocks B and A are shown in Fig. 8.14(b) and (c). The weight of block B is found from Eq. (1.1).

$$W_B = m_B g = 150(9.81) = 1470\,\text{N}$$

The total angle in contact between the rope and the three pins is

$$\beta = 60° + 120° + 60° = 240°$$

For equilibrium of block B,

$$\Sigma F_y = 0 \qquad T_L - 1470 - T_S = 0 \tag{a}$$

From Eq. (8.11),

$$\frac{T_L}{T_S} = e^{\pi f \beta/180°} = e^{\pi(0.35)240°/180°}$$

$$= e^{1.466} = 4.332 \tag{b}$$

Eliminating T_L between Eqs. (a) and (b), we obtain

$$4.332 T_S - T_S = 1470$$

$$T_S = 441\,\text{N}$$

For equilibrium of block A, $T_S = W$. Therefore, $W_A = 441\,\text{N}$. From Eq. (1.1),

$$W_A = m_A g \qquad \text{or} \qquad m_A = \frac{W_A}{g} = \frac{441}{9.81} = 45\,\text{kg} \qquad \textbf{Answer}$$

FLAT BELTS

SB.11(a)

SB.11(b)

Most conveyor systems, many pieces of power transmission equipment, and some machine tools found on production lines today operate using *flat belts.* Modern belts are generally constructed in layers: an inner surface of leather (for friction), a central core of nylon (for strength), and an outer layer of polyester or rubber (to protect the nylon core). These belts are capable of transmitting 150 hp/in. of belt width at linear speeds of up to 20,000 ft/min (about 225 mph!). Because of the various layer materials and belt thicknesses and widths available, belt tensions are not specified in pounds but rather as *strains,* expressed in percent (see Sec. 11.2). Typically, a belt is cut to its required length and strung loosely along the pulley system. Two lines 25 in. apart are then marked off at any convenient location along the belt; these serve as a gauge length. Next, wooden blocks are clamped to the free ends of the belt, and the blocks are pulled together using threaded rods, as shown in SB.11(a). With this apparatus, belt tension is increased until the desired strain is reached; every 1/4-in. elongation in the 25-in. gauge length represents a 1 percent strain. (Light-duty belts are generally installed at 1–2 percent strain, heavy duty belts at 3–4 percent strain.) Finally, adhesive is applied to the belt ends, which are then clamped at a specified pressure in an electric heater [SB.11(b)] and "cooked" for 10 to 60 min. The 2-in.-wide power transmission belt shown in these photos connects 24-in.-diameter and 34-in.-diameter cone pulleys used on the production line of a paper mill. Its narrow width allows this belt to be shifted several inches in either direction along the pulley axes; this creates different diameter ratios for the driver and the driven pulleys, thus allowing some degree of speed variation in the production line itself. *(Courtesy of Marceau's Industrial Belt Services, Ashburnham, Massachusetts)*

8.8 ROLLING RESISTANCE

If a wheel is moved without slipping over a horizontal surface while supporting a load, a force is required to maintain uniform motion. Thus, some kind of rolling resistance must be present. The usual method for describing rolling resistance is based on the deformation of the surface as shown in Fig. 8.15. Let the resultant reaction between the surface and rollers act at point A. From equilibrium,

$$\Sigma M_A = 0 \qquad Wa - Ph = 0$$

Because h is nearly equal to r, we write

$$Wa - Pr = 0$$

$$P = \frac{Wa}{r} \qquad (8.12)$$

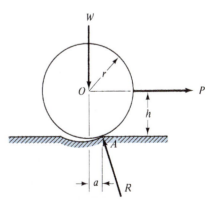

FIGURE 8.15

The distance *a* is called the *coefficient of rolling resistance.* The values of *a* are given in millimeters or inches. They vary substantially for the same material with different values of *W* and *r*. Therefore, caution must be used with this method for finding rolling resistance.

Example 8.9   What horizontal force must be applied to the center of a wheel with a mass of 250 kg and a diameter of 0.7 m to just cause motion of the wheel? The coefficient of rolling resistance is 0.25 mm.

Solution: From Eq. (1.1), the weight of the wheel

$$W = mg = 250(9.81) = 2450 \text{ N}$$

From Eq. (8.12),

$$P = \frac{Wa}{r} = \frac{2450(0.25)}{350}$$

$$= 1.750 \text{ N} \qquad \qquad \textbf{Answer}$$

PROBLEMS

8.25   A 20-lb rotor is attached to a 2-in.-diameter shaft that is supported by two bearings as shown. If the moment required to start the rotor turning is 1.5 lb-ft, determine the coefficient of friction between the shaft and bearings.

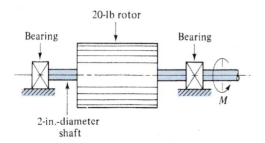

PROB. 8.25 and PROB. 8.26

8.26   If the coefficient of friction for Prob. 8.25 is 0.3, what is the required starting moment?

8.27   A drum of 300-mm radius on a shaft with a diameter of 80 mm is supported by two bearings as shown. It is used to hoist a 500-kg load. The shaft and drum have a mass of 75 kg. The coefficient of friction for the bearings is 0.3. Determine the moment that must be applied to the drum shaft to raise the load at uniform speed.

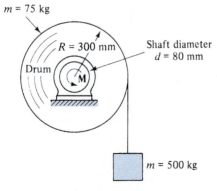

PROB. 8.27 and PROB. 8.28

8.28   What moment must be applied to the drum shaft in Prob. 8.27 to lower the load at uniform speed?

8.29   A thrust of 1000 lb is supported by a collar bearing as shown. Determine the moment required to maintain a constant speed of rotation if the coefficient of friction is 0.15.

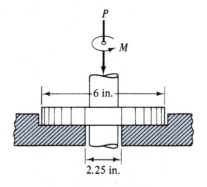

PROB. 8.29 and PROB. 8.30

8.30   If the moment for Prob. 8.29 is 300 lb-in., what is the coefficient of friction?

8.31   A floor polisher with a mass of 25 kg is operated on a surface with a coefficient of friction of 0.175. (a) Determine the moment required to keep the polisher from turning. (b) If the handles are separated by 0.45 m, what force on each handle is required to keep the handles from turning?

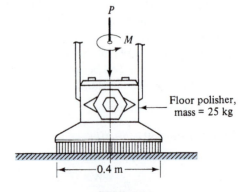

PROB. 8.31

8.32 A disk sander with a diameter $D = 6$ in. is used to refinish a teak boat deck. The downward force P of the operator together with the weight of the sander is 33 lb. Determine the moment M (torque) required of the sander motor to maintain a constant speed if the coefficient of kinetic friction between the teak and the sander is 0.4.

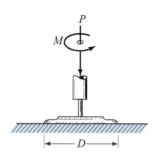

PROB. 8.32

8.33 A rope is wrapped around a capstan to secure a ship. The force on the free end of the rope is 800 lb and on the ship end 20 kip. Determine (a) the coefficient of friction between the rope and capstan if the rope makes three full turns and (b) the number of full turns required if the coefficient of friction is $f_s = 0.3$.

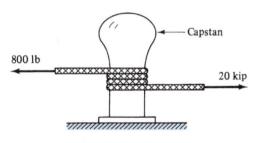

PROB. 8.33

8.34 A flat leather belt is placed around two cast iron pulleys as shown. Determine the largest moment M (torque) that can be transmitted from A to B if the radius of the pulleys is $R = 6$ in., the distance between the pulleys is $H = 24$ in., and the coefficient of friction between the belt and the pulleys is 0.6. The allowable belt tension is 700 lb.

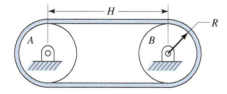

PROB. 8.34

8.35 A nylon rope is looped around two pulleys in a figure eight as shown. The pulleys are made of oak and have a steel rim. Determine the maximum moment M (torque) that can be transmitted from A to B if the radius of the pulleys is $R = 150$ mm and the distance

between the pulleys is $H = 0.6$ m. The coefficient of friction between the nylon rope and steel rim is 0.35 and the allowable rope tension is 3.2 kN. Assume no interference occurs between parts of the rope.

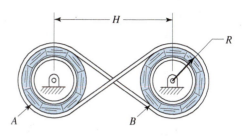

PROB. 8.35

8.36 What horizontal force must be applied at the center of a wheel with a diameter of 1.2 m and a mass of 750 kg to just cause it to roll if the coefficient of rolling resistance is 0.40 mm?

8.37 If a 5-in.-diameter cylinder rolls at a constant speed down an inclined plane that has a slope of 1 to 50, what is the coefficient of rolling resistance between the cylinder and inclined plane?

8.38 The lawn roller shown weighs 250 lb and has a radius of $R = 8$ in. A force of $F = 30$ lb along the handle at an angle $\theta = 25°$ with the horizontal is required to push the roller at a constant speed. What is the coefficient of rolling resistance for the roller? Neglect the axle friction.

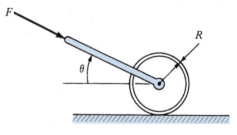

PROB. 8.38

8.39 A plate weighing 25 kN rests on two rollers with diameters of 0.625 m and a weight of 0.9 kN each. If the coefficient of rolling resistance between the plate and rollers is 3.0 mm and between the rollers and rail 3.8 mm, find the force F required to just cause motion.

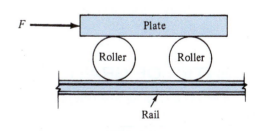

PROB. 8.39

CENTER OF GRAVITY, CENTROIDS, AND MOMENTS OF INERTIA OF AREAS

CHAPTER OBJECTIVES

In this chapter, we define several mathematical quantities needed for the analysis of various structural and mechanical elements. At the conclusion of these sections, you should be able to

- Locate the centroid (geometric center) for plane areas consisting of one or more simple geometric shapes.

- Use tabulated data to compute the centroid location for various combinations of standard structural shapes.

- Calculate the rectangular moment of inertia for any simple plane area about its centroidal axes.

- Use the parallel-axis theorem to determine rectangular moment of inertia for a simple or composite area about any axis contained in the plane of that area.

- Compute the radius of gyration for simple and composite plane areas.

- Calculate the polar moment of inertia for a simple or composite area about any axis perpendicular to the plane of that area.

- (Optional) Use integral calculus to locate the centroid and compute the moment of inertia for any general plane area.

cases the forces may be considered to be parallel. As discussed in Sec. 3.4, the resultant of this parallel force system acts through a point called its *center of gravity*. Accordingly, the center of gravity is the point where the total weight may be assumed to act.

For bodies of uniform thickness the center of gravity can be located by a simple experiment. In the experiment the body is suspended by two successive points as shown in Fig. 9.1. The body is first suspended from point A [Fig. 9.1(a)]. For equilibrium, the tension in the cable must be equal to the weight W, and the weight must be collinear with the tension in the cable; that is, it must lie along a line through A and D as shown. Then, in Fig. 9.1(b), the body is suspended from point B. In the new position, the tension in the cable must again be equal to the weight, and the weight must lie along a line through B and D. The intersection of the two lines [Fig. 9.1(a) and (b)] defines the location of the center of gravity G. Notice here that the center of gravity is located outside the body.

To determine the center of gravity of a thin homogeneous plate by analysis, we divide the plate into n small particles as in Fig. 9.2(a). The particles have weights $\Delta w_1, \Delta w_2, \ldots, \Delta w_n$, and the coordinates of the weights are $x_1, y_1; x_2, y_2; \ldots; x_n, y_n$. Figure 9.2(b) shows the resultant W, and the coordinates of the resultant $\bar{x}$ and $\bar{y}$ (x bar and y bar).

The resultant weight must be equal to the sum of the particle weights. Therefore,

$$W = \Delta w_1 + \Delta w_2 + \cdots + \Delta w_n = \Sigma \Delta w_i \quad (9.1)$$

9.1 INTRODUCTION

In previous chapters, we defined the center of gravity of a body to be the point at which the total weight of the body acts. The center of gravity is considered in more detail in this chapter, as are centroids and moments of inertia. Later, when we analyze and design various structural and machine members, it may be necessary to determine the moment of inertia and the centroid of a member's cross section.

9.2 CENTER OF GRAVITY

The force of gravity acts on each particle in a body and the weight of the body consists of a system of small forces acting on each particle of the body. The forces are all directed toward the center of the earth; however, for most practical

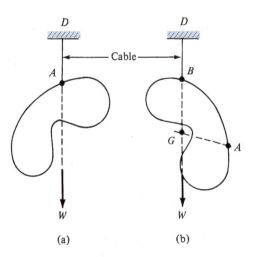

FIGURE 9.1

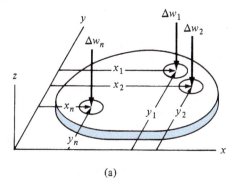

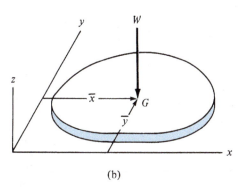

(a)

(b)

FIGURE 9.2

SB.12

Like two-dimensional shapes, a *three-dimensional* object will balance about any point along an imaginary vertical line drawn through the object's center of gravity. For homogeneous solids having uniform cross sections, such as a right circular cylinder, this line is simply the object's long axis passing through the centroid of its cross section. However, for irregular, nonhomogeneous solids, such as the welded steel sculpture shown in SB.12, locating this axis of balance is largely a matter of trial and error. Commissioned by a large New England company for installation in its corporate plaza, the sculpture, titled *Leading Edge*, is 12-1/2 ft tall and weighs more than 3000 lb. Working from a scale model, fabricators made small adjustments to the original design to ensure that the finished piece would stand alone safely on its base before being attached to the pedestal and underlying footing in the plaza. *(Courtesy of the artist, Frances G. Pratt, Cambridge, Massachusetts)*

and the moment of the resultant weight about an axis must be equal to the sum of the moments of each particle weight about the same axis. For the y axis,

$$W\bar{x} = x_1 \Delta w_1 + x_2 \Delta w_2 + \cdots + x_n \Delta w_n = \Sigma x_i \Delta w_i$$
$$(9.2)$$

and for the x axis,

$$W\bar{y} = y_1 \Delta w_1 + y_2 \Delta w_2 + \cdots + y_n \Delta w_n = \Sigma y_i \Delta w_i$$
$$(9.3)$$

Accordingly, the coordinates for the center of gravity G are given by

$$\bar{x} = \frac{\Sigma x_i \Delta w_i}{W} \quad \text{and} \quad \bar{y} = \frac{\Sigma y_i \Delta w_i}{W}$$

where
$$W = \Sigma \Delta w_i \qquad (9.4)$$

If we let the number of particles increase without limit, we obtain the exact center of gravity, a topic considered in integral calculus.

9.3 CENTROID OF A PLANE AREA

In Fig. 9.3(a) and (b), a plane figure and a thin homogeneous plate of uniform thickness in the same shape as the plane figure are shown. The *center of gravity* of the thin plate is on its middle plane. The *centroid* is at the same location in the plane of the thin homogeneous plate. The terms *centroid* and *center of gravity* are often used to indicate the same point. They are at the same location only if the plate is homogeneous.

To determine the centroid of a plane area by analysis, we replace the plate of Fig. 9.2 with the middle area of the plate as

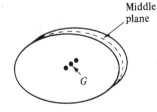

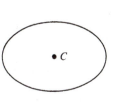

(a) Plane figure   (b) Thin plate

FIGURE 9.3

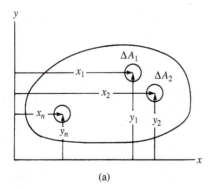

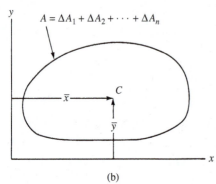

$$A = \Delta A_1 + \Delta A_2 + \cdots + \Delta A_n$$

(b)

FIGURE 9.4

shown in Fig. 9.4(a). To effect the replacement, we let t be the thickness of the plate and γ (the Greek lowercase letter gamma) be the specific weight or weight per unit volume. Then, if the surface area of the plate and the particle are A and ΔA_i, the total weight $W = \gamma t A$ and the particle weight $\Delta w_i = \gamma t \Delta A_i$. Substituting for W and Δw_i in Eq. (9.1), we have

$$\gamma t A = \gamma t \Delta A_1 + \gamma t \Delta A_2 + \cdots + \gamma t \Delta A_n = \Sigma \gamma t \Delta A_i$$

or dividing by γt it follows that the total area A is equal to the sum of the areas ΔA_i, that is,

$$A = \Delta A_1 + \Delta A_2 + \cdots + \Delta A_n = \Sigma \Delta A_i$$

In a similar way, we replace Δw_i by $\gamma t \Delta A_i$ in Eqs. (9.2) and (9.3) and divide by γt. It follows that the moment of the total area A about an axis must be equal to the sum of the moments of the areas ΔA_i about the same axis. For the y axis,

$$A\bar{x} = x_1 \Delta A_1 + x_2 \Delta A_2 + \cdots + x_n \Delta A_n = \Sigma x_i \Delta A_i$$

and for the x axis,

$$A\bar{y} = y_1 \Delta A_1 + y_2 \Delta A_2 + \cdots + y_n \Delta A_n = \Sigma y_i \Delta A_i$$

Accordingly, the coordinates of the centroid [see Fig. 9.4(b)] are given by

$$\bar{x} = \frac{\Sigma x_i \Delta A_i}{A} \quad \text{and} \quad \bar{y} = \frac{\Sigma y_i \Delta A_i}{A}$$

where $\quad A = \Sigma \Delta A_i$ **(9.5)**

If we let the number of areas increase without limit, we obtain the exact centroid of the area, a topic considered in Sec. 9.12 on the determination of centroids by integration. The section is marked with an asterisk and can be omitted without a loss of continuity in the text.

9.4 CENTROIDS BY INSPECTION

For a plane area with *one axis of symmetry* [Fig. 9.5(a)], every element of area ΔA_i has a corresponding element with a coordinate that is equal in magnitude and opposite in sign.

Because of this, the sum $\Sigma x_i \Delta A_i$ is equal to zero and $\bar{x} = 0$ from Eq. (9.5). Hence, *the centroid lies on the axis of symmetry.*

With *two axes of symmetry*, the centroid must lie on both axes; accordingly, *the centroid lies on the intersection of the two axes* [Fig. 9.5(b)].

With a *center of symmetry*, each element of area has a corresponding element with coordinates that are equal in

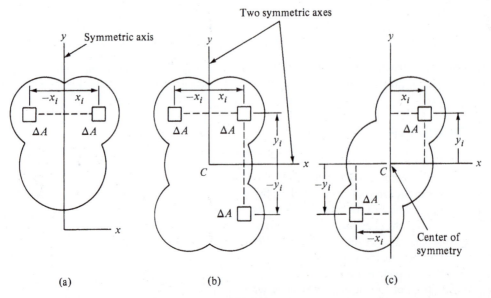

(a)   **(b)**   **(c)**

FIGURE 9.5

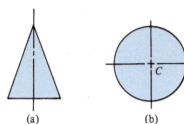
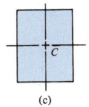

(a)   (b)   (c)

Symmetric axes

FIGURE 9.6

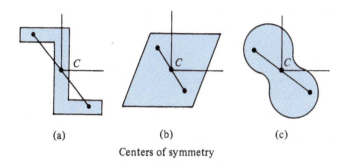

(a)   (b)   (c)

Centers of symmetry

FIGURE 9.7

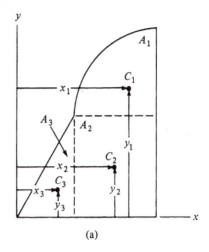

(a)

$$A = A_1 + A_2 + A_3$$

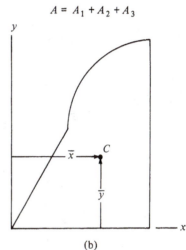

(b)

FIGURE 9.8

magnitude and opposite in sign [Fig. 9.5(c)]. Because of this, the two sums $\Sigma x_i \, \Delta A_i$ and $\Sigma y_i \, \Delta A_i$ are both equal to zero. Therefore, *the centroid and center of symmetry are at the same location.*

Figures 9.6(a)–(c) and 9.7(a)–(c) show examples of plane areas with one and two axes of symmetry and with centers of symmetry at point C.

Formulas for the areas and centroids of the rectangle, triangle, circle, semicircle, and quarter circle are given in Table A.1 of the Appendix.

9.5 CENTROIDS OF COMPOSITE AREAS

When a composite area can be broken up into simple areas whose centroids are known, the centroid of the composite area can be determined. The method is similar to the method of finding the centroid of a plane area. We replace each element of area ΔA_i by a simple area A_i and the coordinates of the element by the centroid of that simple area.

In Fig. 9.8(a), we consider a composite area that can be broken into three simple areas A_1, A_2, and A_3 whose centroids C_1, C_2, and C_3 are known or can be calculated. We also consider in Fig. 9.8(b) the composite system made up of the total area with an unknown centroid. By construction, the

sum of the simple areas must be equal to the composite or total area.

$$A = \Sigma A = A_1 + A_2 + A_3$$

The sum of the moments of the simple areas about an axis must be equal to the moment of the composite area about the same axis. For the y axis,

$$M_y = A_1 x_1 + A_2 x_2 + A_3 x_3 = (A_1 + A_2 + A_3)\bar{x}$$

or $$\bar{x} = \frac{A_1 x_1 + A_2 x_2 + A_3 x_3}{A_1 + A_2 + A_3} = \frac{\Sigma Ax}{\Sigma A} \qquad (9.6)$$

For the x axis,

$$M_x = A_1 y_1 + A_2 y_2 + A_3 y_3 = (A_1 + A_2 + A_3)\bar{y}$$

or $$\bar{y} = \frac{A_1 y_1 + A_2 y_2 + A_3 y_3}{A_1 + A_2 + A_3} = \frac{\Sigma Ay}{\Sigma A} \qquad (9.7)$$

Example 9.1   Find the centroid for the composite area shown in Fig. 9.9(a).

Solution: Because AB is an axis of symmetry, the centroid lies on AB and by inspection $\bar{y} = 3$ in. The composite area is made up of three simple shapes: a rectangle, a circular hole, and a triangle. Areas and centroidal distances are determined from formulas given in Table A.1 of the Appendix. Note that the area of the circle is negative

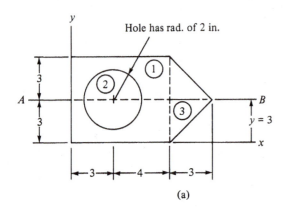

(a)

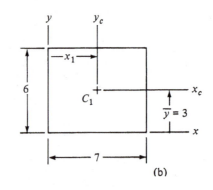

$A_1 = bh = 7(6) = 42$ in.^{2}

$x_1 = \dfrac{b}{2} = \dfrac{7}{2} = 3.5$ in.

(b)

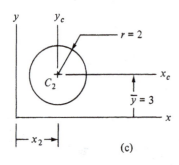

$A_2 = -\pi r^2 - \pi(2)^2 = -12.56$ in.^{2}

(Area removed from rectangle, therefore negative)

$x_2 = 3$ in.

(c)

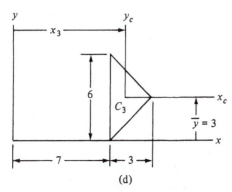

$A_3 = \dfrac{bh}{2} = \dfrac{6(3)}{2} = 9$ in.^{2}

$x_3 = 7 + \dfrac{h}{3} = 7 + \dfrac{3}{3} = 8$ in.

(d)

(All dimensions are in inches.)

FIGURE 9.9

because it is removed from the rectangle. The areas and centroidal distances are calculated in Fig. 9.9(b), (c), and (d), and the areas, centroidal distances, and required calculations are listed in Table (A).

TABLE (A)	Example 9.1			
	Shape	A (in.^{2})	x (in.)	Ax (in.^{3})
①	Rectangle	42.0	3.5	147
②	Circular hole	−12.56	3.0	−37.7
③	Triangle	9.0	8.0	72
	Σ	38.43 in.^{2}	—	181.3 in.^{3}

From Eq. (9.6), the sums in the table and symmetry

$$\bar{x} = \frac{\Sigma Ax}{\Sigma A} = \frac{181.3}{38.43} = 4.72 \text{ in.} \qquad \bar{y} = 3 \text{ in.} \qquad \textbf{Answer}$$

Example 9.2   A cross section of the model of an arched dam is shown in Fig. 9.10(a). Determine the centroid of the cross-sectional area.

Solution: The composite cross-sectional area of the model is made up of two rectangles and a triangle. Areas and centroidal distances are determined from the figures and formulas given in Table A.1 of the Appendix. The areas and centroids are calculated in Fig. 9.10(b)–(d) and listed in Table (A), together with the required calculations. From Eqs. (9.6) and (9.7) and the sums in the table,

$$\bar{x} = \frac{\Sigma Ax}{\Sigma A} = \frac{1.498 \times 10^6}{48.6 \times 10^3} = 30.8 \text{ mm} \qquad \textbf{Answer}$$

$$\bar{y} = \frac{\Sigma A_y}{\Sigma A} = \frac{9.490 \times 10^6}{48.6 \times 10^3} = 195.3 \text{ mm} \qquad \textbf{Answer}$$

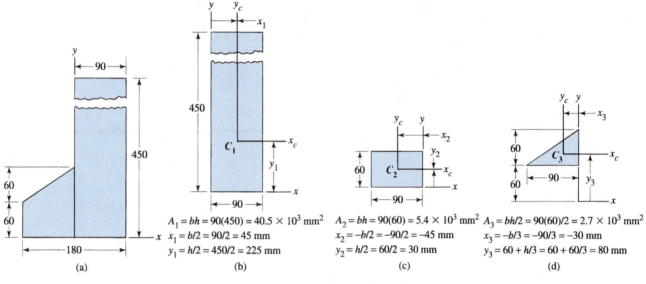

(All dimensions are in mm.)

FIGURE 9.10

TABLE (A)   Example 9.2

	Shape	A (mm^2)	x (mm)	Ax (mm^3)	y (mm)	Ay (mm^3)
①	Rectangle	40.5×10^3	45	1.822×10^6	225	9.112×10^6
②	Rectangle	5.4×10^3	−45	-0.243×10^6	30	0.162×10^6
③	Triangle	2.7×10^3	−30	-0.081×10^6	80	0.216×10^6
	Σ	$48.6 \times 10^3 \text{ mm}^2$	—	$1.498 \times 10^6 \text{ mm}^3$	—	$9.490 \times 10^6 \text{ mm}^3$

9.6 CENTROIDS OF STRUCTURAL CROSS SECTIONS

Structural beams and columns are fabricated in various configurations of steel or wood. Some of the more common steel shapes are shown in Fig. 9.11, and a brief discussion of their designations is given in the next section.

Typical Shapes for Structural Steel

Steel is rolled into a variety of shapes and sizes. Steel sections are designated by the shape and size of their cross section.

The properties of selected steel sections are given in Tables A.3 through A.8 of the Appendix. Some of the typical shapes may be described as follows:

1. *Wide-flange beam.* This section has a shape similar to an I. The two parallel horizontal parts of the cross section are called *flanges,* and the vertical part of the cross section is called the *web.* This beam is designated by the symbol *W,* the nominal depth in inches or mm, and weight in lb per ft or mass in kg per m. *Example:* W16 × 100; (SI units) W410 × 149.

2. *American Standard beam.* This section is commonly called an *I-beam.* The flanges are narrower than those

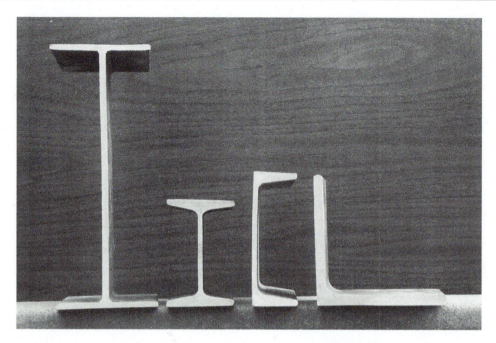

FIGURE 9.11. Some common structural steel shapes *(l-to-r)*: wide-flange beam (W-member), American Standard beam (S-member), American Standard channel (C-member), and an equal-leg angle (L-member).

of the wide-flange beam, and the inner flange surface has a slope of approximately 2 to 12 in. It is designated by the symbol S, the nominal depth in inches or mm, and the weight in lb per ft or mass in kg per m. *Example:* S18 × 70; (SI units) S310 × 74.

3. *American Standard channel.* This section is similar to the American Standard beam with the flanges removed from one side. It is designated by the symbol C, the nominal depth in inches or mm, and the weight in lb per ft or mass in kg per m. *Example:* C15 × 40; (SI units) C380 × 60.

4. *Angles.* This section has a shape like an *L*. The horizontal and vertical parts are called *legs*. It is designated by the letter L, the length of the long or equal leg, the length of the short or equal leg, and the thickness of the legs. The lengths and thickness are given in inches or mm. *Examples:* L8 × 8 × 1 and L5 × 3 × 1/2; (SI units) L203 × 203 × 25.4 and L127 × 76 × 12.7.

5. *Plates.* A plate has a rectangular cross section, usually 6 in. (152 mm) or more in width. It is designated by the symbol PL, the thickness of the plate, and the width of the plate. The thickness and width are given in inches or mm. *Example:* PL1/2 × 12; (SI units) PL12.7 × 305.

6. *Pipes.* These steel members, often filled with concrete, are most commonly used as columns. They are designated by their nominal diameters and are available in several weights. Dimensions and properties for standard weight and extra strong pipe are given in Table A.8 of the Appendix. *Example:* 4-in. standard-weight pipe has an outside diameter of 4.500 in., an inside diameter of 4.026 in., and weighs 10.79 lb/ft.

The methods of Sec. 9.5, together with the properties of structural sections, are used in the following examples to determine the centroid of composite structural cross sections.

Example 9.3 Find the centroid of the composite area made up of structural shapes as shown in Fig. 9.12(a).

Solution: Properties of the American Standard beam S10 × 25.4 and American Standard channel C8 × 11.5 from Tables A.4 and A.5 of the Appendix are shown in Fig. 9.12(b) and (c). Note that the channel in the figure is rotated through 90° from the channel shown in the table, and the centroid $\bar{x}$ of the channel is measured from the back of the channel. From Fig. 9.12(c), we see that when measured from the x axis the centroid of the channel $y_2 = d_S/2 + t_{web} - \bar{x} = 10/2 + 0.220 - 0.571 = 4.649$ in. The areas, centroidal distances, and required calculations are listed in Table (A). From symmetry, Eq. (9.7), and the sums in the table,

$$\bar{x} = 0 \qquad \bar{y} = \frac{\Sigma Ay}{\Sigma A} = \frac{15.71}{10.84} = 1.45 \text{ in.} \qquad \textbf{Answer}$$

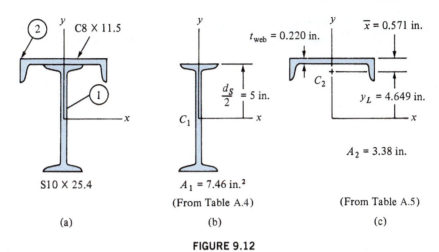

FIGURE 9.12

	Shape	A (in.²)	y (in.)	Ay (in.³)
①	S10 × 25.4	7.46	0	0
②	C8 × 11.5	3.38	4.649	15.71
	Σ	10.84 in.²	—	15.71 in.³

TABLE (A)   Example 9.3

Example 9.4   Find the centroid of the composite area made up of structural steel shapes as shown in Fig. 9.13(a).

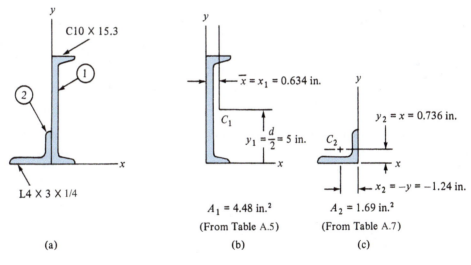

FIGURE 9.13

Solution: Properties of the American Standard channel C10 × 15.3 and angle L4 × 3 × 1/4 from Tables A.5 and A.7 of the Appendix are shown in Fig. 9.13(b) and (c). Note that the angle in the figure is rotated through 90° from the angle shown in the table. Accordingly, the x and y axes are interchanged. Also note that x_2 is negative. The areas, centroidal distances, and the required calculations are listed in Table (A). From Eqs. (9.6) and (9.7) and the sums in the table,

$$\bar{x} = \frac{\Sigma Ax}{\Sigma A} = \frac{0.744}{6.17} = 0.121 \text{ in.} \qquad \bar{y} = \frac{\Sigma Ay}{\Sigma A} = \frac{23.64}{6.17} = 3.83 \text{ in.} \qquad \textbf{Answer}$$

	Shape	A ($in.^2$)	x ($in.$)	Ax ($in.^3$)	y ($in.$)	Ay ($in.^3$)
TABLE (A)	**Example 9.4**					
①	C10 × 15.3	4.48	0.634	2.840	5.0	22.4
②	L4 × 3 × 1/4	1.69	−1.24	−2.096	0.736	1.244
	Σ	6.17 in.²	—	0.744 in.³	—	23.64 in.³

Wooden Structural Members

Unlike steel, wood is an organic material that can have wide variations in its physical properties. However, in recent years, the trend has been toward standardizing structural lumber and manufacturing *engineered wood* products that possess more uniform characteristics than natural wood. A few of these practices and products are discussed as follows:

1. *Visually graded and machine-graded lumber.* For solid lumber, individual boards are generally designated by their nominal size, occasionally, but not always, followed by S4S to indicate that the board has been dressed (*surfaced* on all *four sides*) to the standard dimensions given in Table A.9 of the Appendix. *Example:* A 2 × 6 S4S board has a 1.5-in. × 5.5-in. cross section.

   Lumber has traditionally been graded by *visual inspection,* with designations such as "select structural," No. 1, No. 2, and No. 3 assigned according to visible characteristics including grain, and number and size of knots. Today, lumber can also be *machine graded,* a process in which each individual piece is nondestructively tested for stiffness, strength in bending, and strength in tension (see Application Sidebar SB.13). The Southern Forest Products Association, for instance, uses two different machine-grading designations on Southern Pine boards: MSR for *machine stress-rated* lumber, and MEL for *machine-evaluated lumber.* MSR boards are primarily rated according to two physical properties known as bending strength and modulus of elasticity. *Example:* MSR boards designated as 1950f—1.7E have a minimum bending strength of 1950 psi and a modulus of elasticity of 1,700,000 psi. MEL lumber is rated in grades ranging from M–5 through M–28, with individual grades having minimum values for each of several structural properties; an M–23 board has a bending strength of 2400 psi, a tensile strength (parallel to the grain) of 1900 psi, and a modulus of elasticity of 1,800,000 psi. (These properties will be discussed fully in Chapters 10, 11, and 14.) Typical grading marks as stamped on each board are shown in Fig. 9.14, and structural properties are listed in Table A.11 of the Appendix.

   Machine-graded lumber is particularly advantageous in engineered products such as trusses (see Fig. 6.2), and fabricated members.

2. *Laminated veneer lumber.* This engineered product, known as LVL, consists of thin wood-veneer laminations (generally less than 1/4 in. thick) that are placed vertically, run the length of each board, and are glued together using exterior (weather-resistant) adhesives. Cross sections for solid wood and LVL boards are shown in Fig. 9.15(a) and (b), respectively. Although the standard width of LVL boards is 1-3/4 in., they are available from specific manufacturers in widths ranging from 3/4 in. to 7 in. Depths between 5-1/2 in. and 20 in. are the most common, but larger sizes may be special ordered. Because of this wide range of sizes for LVL boards, structural properties for only the most common dimensions are included in Table A.12 of the Appendix.

3. *Glue-laminated lumber.* Called *glulam,* this manufactured product consists of several layers of solid structural

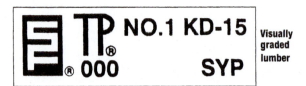

FIGURE 9.14. Typical grading marks for visually graded lumber (*top*), machine stress-rated (MSR) lumber (*middle*), and machine-evaluated lumber (MEL) (*bottom*). (*Used with permission of Southern Forest Products Association, Kenner, Louisiana*)

MACHINE-GRADED LUMBER

SB.13(a)

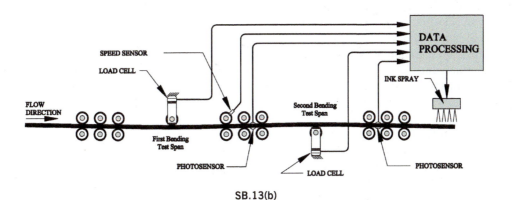

SB.13(b)

The two most common types of machine-graded structural lumber produced in North America today are machine stress-rated (MSR) and machine-evaluated lumber (MEL). Each board is graded and marked in an automated process using a machine similar to the one shown in SB.13(a). Boards enter one end of the machine and pass through a series of rollers at linear speeds of 2200 ft/min or more. As indicated in the schematic diagram of SB.13(b), the machine also contains two load cells that push on the board in both directions perpendicular to the flat surfaces, measuring the force required to produce a given deflection. Force measurements are made at intervals of approximately 1/2 in. along the length of each board, and the data are fed into a computer for processing. Photosensors detect the leading and trailing edges of boards to ensure that all data supplied are for a specific piece of lumber. These measured values of force are used to compute a *stiffness,* or *modulus of elasticity,* for the board, and design values for tensile strength, compressive strengths parallel and perpendicular to the grain, shear strength, and fiber strength in bending are then assigned along with the grade. As they leave the machine, boards are color coded by an ink sprayer and must pass a visual inspection for knots and other surface defects before the structural properties are stamped onto the piece. In another process, an X-ray source is placed on one side of the board, and sensors behind the board are used to measure how much of the emitted energy passes through the piece. These data are directly related to the *density* (weight or mass per unit volume) of the material, which in turn may be used to predict structural values. Machines of the type shown here are capable of producing *both* MSR and MEL lumber as needed. *(Courtesy of Metriguard, Inc., Pullman, Washington, www.metriguard.com)*

lumber stacked horizontally and laminated together, as shown in Fig. 9.15(c). Available in widths from 3-1/8 in. to 6-3/4 in. and depths up to 50 in., glulam members allow great flexibility to both architects and builders.

4. *Wooden I-beams.* Because these members are used primarily as floor joists, they are commonly referred to as *I-joists* rather than I-beams. Although handmade versions were used as early as the 1920s, mass-produced members did not appear until 1969. I-joists

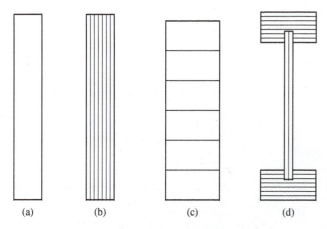

FIGURE 9.15. Cross sections of structural lumber: (a) solid lumber, (b) laminated veneer lumber (LVL), (c) glue-laminated lumber (glulam), (d) I-beam or I-joist.

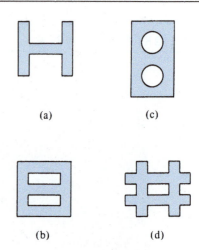

PROB. 9.2

are fabricated using a vertical web of either plywood or oriented-strand board (OSB) and flanges at top and bottom of either LVL or solid structural lumber, as shown in Fig. 9.15(d). Unlike steel I-beams, there is, at this time, little standardization of sizes for these engineered wood products. Although depths up to 16 in. are available (the most common being 9-1/2, 11-7/8, 14, and 16 in.), other dimensions may vary considerably between manufacturers. There are, however, some general guidelines to which the majority of commercial products adhere: a minimum web thickness of 3/8 in., a minimum flange width of 1-1/2 in., and a minimum flange thickness of 1-5/16 in. Structural properties for some of the more popular I-joist sizes are given in Table A.13 of the Appendix.

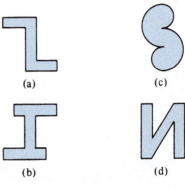

PROB. 9.3

PROBLEMS

9.1 through 9.4.   Locate the centroid of the plane areas shown, by inspection.

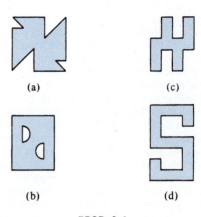

PROB. 9.4

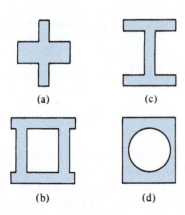

PROB. 9.1

9.5 *Norman architecture* is characterized by rounded arches over doors and windows. Locate the centroid on a Norman window that consists of a semicircle atop a rectangle as shown.

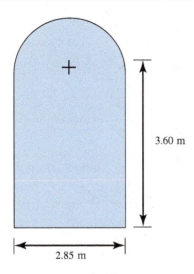

PROB. 9.5

9.6 through 9.19   Locate the centroid of the plane composite area shown.

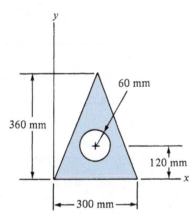

PROB. 9.6

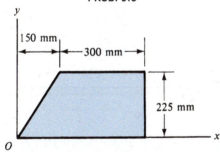

PROB. 9.7

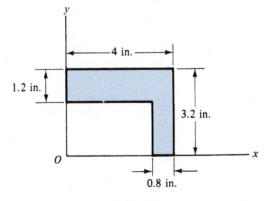

PROB. 9.8

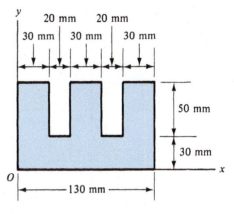

PROB. 9.9

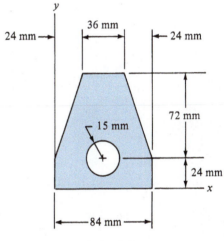

PROB. 9.10

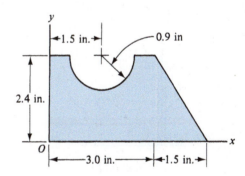

PROB. 9.11

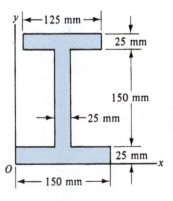

PROB. 9.12

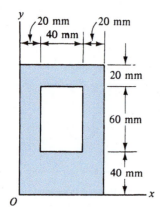

PROB. 9.13

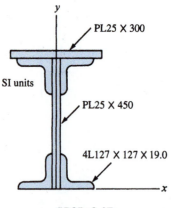

PROB. 9.17

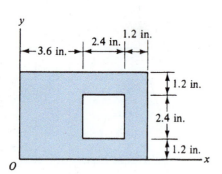

PROB. 9.14

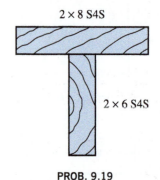

PROB. 9.18

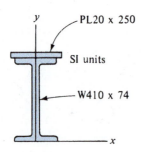

PROB. 9.15

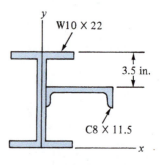

PROB. 9.19

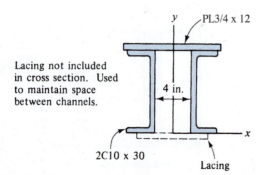

PROB. 9.16

9.7 MOMENT OF INERTIA OF A PLANE AREA

In the study of the strength of beams and columns and statics of fluids, the moment of inertia or second moment of a plane area is required. The numerical value of this quantity is used to indicate how the area is *distributed* around a specified axis. If the axis lies within the plane of an area, that area's second moment about the axis is called the *rectangular* moment of inertia and is represented by the letter *I*. For any axis that is perpendicular to the plane of an area, the area's second moment is known as the *polar* moment of inertia and is indicated as *J* (see Sec. 9.10). A low value for *I* or *J* is used to describe an area whose elements are closely grouped about an axis, while a high value of *I* or *J* indicates that much of the area is located at some distance from the selected axis.

Consider the plane area shown in Fig. 9.16. The area is divided into n thin strips parallel to the x axis of areas $\Delta A_1, \Delta A_2, \ldots, \Delta A_n$ with centroidal distances $y_1, y_2, \ldots, y_n$ from the x axis to the centroid of the area.

The rectangular moment of inertia of the area ΔA about the x axis is equal to the product of the square of the centroidal distance y and the area ΔA. With n areas, we add up the moments of inertia for each area to find the total approximate moment of inertia. Thus, the approximate moment of inertia about the x axis I_x is given by

$$I_x = y_1^2\,\Delta A_1 + y_2^2\,\Delta A_2 + \cdots + y_n^2\,\Delta A_n = \Sigma y_i^2\,\Delta A_i \quad (9.8)$$

The approximate moment of inertia about the y axis I_y is given by

$$I_y = x_1^2\,\Delta A_1 + x_2^2\,\Delta A_2 + \cdots + x_n^2\,\Delta A_n = \Sigma x_i^2\,\Delta A_i \quad (9.9)$$

where the thin strips of area ΔA_i are parallel to the y axis and x_i is the distance from the y axis to the centroid of the area ΔA_i.

If we let the number of areas increase without limit, we obtain the exact moment of inertia. This requires integral calculus and is considered in Sec. 9.13 on the determination

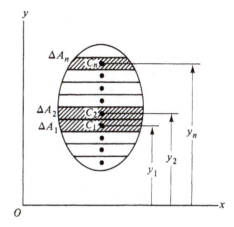

FIGURE 9.16

of moments of inertia by integration. Marked with an asterisk, the section can be omitted without a loss of continuity in the text. Formulas for the moment of inertia of simple areas are given in Table A.2 of the Appendix.

Example 9.5   For the area shown in Fig. 9.17(a), determine the approximate value and exact value for I_x.

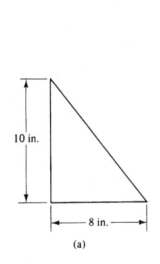

(a)

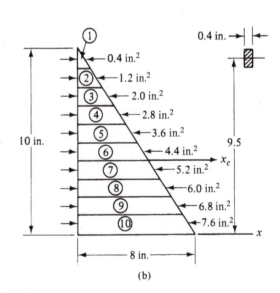

(b)

FIGURE 9.17

Solution: As shown in Fig. 9.17(b), divide the triangle into 10 horizontal strips 1 in. wide. Replace each strip with a small rectangle whose length is the average length of the strip and whose centroidal distance is the mean distance of the rectangle from the x axis.

The area of strip ①, $\Delta A_1 = 0.4$ in.^{2} and the centroidal distance $y_1 = 9.5$ in. from the x axis. The moment of inertia, I_{x_1} is

$$I_{x_1} = y_1^2\,\Delta A_1 = (9.5)^2(0.4) = 36.1 \text{ in.}^4$$

The remaining moments of inertia are calculated in the same way. The calculations are tabulated in Table (A). From the sum shown in the table,

$$I_x = 670 \text{ in.}^4 \qquad\qquad \textbf{Answer}$$

The exact value is calculated from the formula given in Table A.2 of the Appendix.

$$I_x = \frac{bh^3}{12} = \frac{8(10)^3}{12} = 666.7 \text{ in.}^4$$ **Answer**

The error in the approximation is only 0.5 percent. If we divide the area into more than 10 strips, the error will decrease and be less than 0.5 percent.

This method for finding the approximate moment of inertia can be useful when the areas are irregular and formulas for finding their moments of inertia are not available.

TABLE (A) Example 9.5			
	y_i (in.)	$\Delta A_i (in.^2)$	$I_{x_i} = y_i^2 \Delta A_i (in.^4)$
①	9.5	0.4	36.1
②	8.5	1.2	86.7
③	7.5	2.0	112.5
④	6.5	2.8	118.3
⑤	5.5	3.6	108.9
⑥	4.5	4.4	89.1
⑦	3.5	5.2	63.7
⑧	2.5	6.0	37.5
⑨	1.5	6.8	15.3
⑩	0.5	7.6	1.9
			$\Sigma I_{x_i} = 670 \text{ in.}^4$

PROBLEMS

9.20 through 9.23   For the area shown, determine (a) the exact value of the moment of inertia I_x by formula, (b) the approximate value of the moment of inertia I_x if the area is divided into 10 horizontal strips of equal width, and (c) the percent error in the approximation. Percent error is equal to 100 (approximate I_x – exact I_x)/exact I_x.

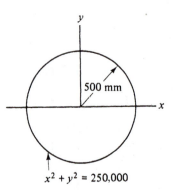

$x^2 + y^2 = 250{,}000$

PROB. 9.22

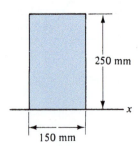

PROB. 9.20

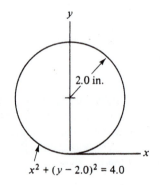

$x^2 + (y - 2.0)^2 = 4.0$

PROB. 9.23

9.24 through 9.27   For the areas shown, determine (a) the centroidal moment of inertia I_{x_c} and I_{y_c} and (b) the moment of inertia I_x and I_y.

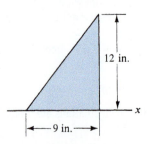

PROB. 9.21

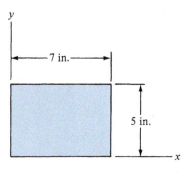

PROB. 9.24

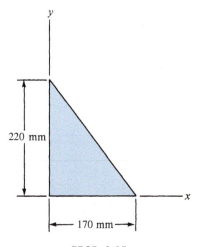

PROB. 9.25

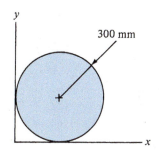

PROB. 9.26

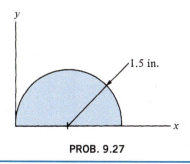

PROB. 9.27

9.8 PARALLEL-AXIS THEOREM

The *parallel-axis theorem* or transfer formula gives a relationship between the moment of inertia with respect to any axis and the moment of inertia with respect to a parallel axis through the centroid. Consider the area shown in Fig. 9.18. The *bb'* axis is a centroidal axis, that is, through the centroid. The *aa'* axis is any axis parallel to the centroidal axis. The moment of inertia of the element of area ΔA_i for the *aa'* axis is given by

$$I_i = (y_i + d)^2 \Delta A_i$$

The moment of inertia for the *aa'* axis of the total area A is found by summing I_i for all the elements of area ΔA_i.

$$I = \Sigma I_i = \Sigma (y_i + d)^2 \Delta A_i$$

Expanding and distributing terms, we have

$$I = \Sigma(y_i^2 + 2dy_i + d^2)\Delta A_i$$
$$= \Sigma y_i^2 \, \Delta A_i + \Sigma 2dy_i \, \Delta A_i + \Sigma d^2 \, \Delta A_i$$

The first term on the right is the moment of inertia, I_c, for the centroidal axis *bb'*. The second term is equal to $2d\,\Sigma y_i \, \Delta A_i$. Because the moment of area for the centroidal axis, $\Sigma y_i \, \Delta A_i$, is equal to zero, the second term is equal to zero. The last term is equal to $d^2 \Sigma \Delta A_i$, and $\Sigma \Delta A_i$ is the total area A. Therefore,

$$I = I_c + Ad^2 \qquad (9.10)$$

where I_c is the moment of inertia for the centroidal axis, A the total area, and d the distance between the two parallel axes *aa'* and the centroidal axis *bb'*. Equation (9.10) is known as the *parallel-axis theorem*.

We notice that the term Ad^2 is *added* when the transfer is made *from* the centroidal axis but *subtracted* when the transfer is made *to* the centroidal axis. Therefore, the moment of inertia for the centroidal axis is always smaller than for any other parallel axis.

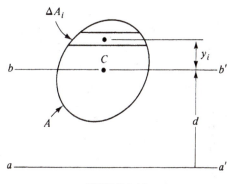

FIGURE 9.18

Example 9.6   Find the moment of inertia of the right triangle shown in Fig. 9.19(a) for the x and x_c axes.

Solution: From Table A.1 of the Appendix the centroid of the triangle is $h/3 = 10/3 = 3.33$ in. above the base of the triangle [Fig. 9.19(b)]. The distance between the parallel axes is $d = 4.0 + 3.33 = 7.33$ in. The area of the triangle is $A = bh/2 = 6(10)/2 = 30$ in.^{2}, and from Table A.2 of the Appendix the moment

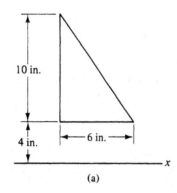

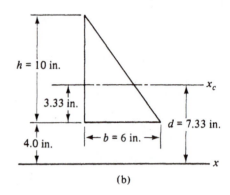

(a)   (b)

FIGURE 9.19

of inertia about the centroidal axis is $I_c = bh^3/36 = 6(10)^3/36 = 166.7$ in.^{4} From the parallel-axis theorem, Eq. (9.10),

$$I_x = I_c + Ad^2 = 166.7 + 30(7.33)^2$$
$$= 1780 \text{ in.}^4$$ **Answer**

9.9 MOMENT OF INERTIA OF COMPOSITE AREAS

When a composite area can be broken up into simple areas whose centroid and moment of inertia with respect to the centroid are known or can be calculated, the moment of in-

ertia of the composite area can be determined. The following example will illustrate the method.

Example 9.7   Find the moment of inertia about the y axis for the composite area shown in Fig. 9.20(a).

Solution: The composite area is divided into three simple shapes (a triangle, rectangle, and quarter circle), as shown in Fig. 9.20(b)–(d). The centroidal distance, area, and centroidal moment of inertia are calculated for each shape from the formulas given in Table A.2 of the Appendix and shown in Fig. 9.20. To find the moment of inertia about the y axis for the three areas, the parallel-axis theorem is used as follows:

$$(I_y)_1 = (I_c)_1 + (Ad^2)_1 = 0.6 \times 10^6 + 3 \times 10^3 (40)^2$$
$$= 5.4 \times 10^6 \text{ mm}^4$$
$$(I_y)_2 = (I_c)_2 + (Ad^2)_2 = 14.4 \times 10^6 + 12 \times 10^3 (120)^2$$
$$= 187.2 \times 10^6 \text{ mm}^4$$
$$(I_y)_3 = (I_c)_3 + (Ad^2)_3 = 5.5 \times 10^6 + 7.85 \times 10^3 (222.4)^2$$
$$= 394.1 \times 10^6 \text{ mm}^4$$

The total moment of inertia about the y axis for the composite area is found by adding the moment of inertia for the three areas.

$$I_y = (I_y)_1 + (I_y)_2 + (I_y)_3$$
$$= 5.4 \times 10^6 + 187.2 \times 10^6 + 394.1 \times 10^6$$
$$= 586.7 \times 10^6 \text{ mm}^4$$ **Answer**

For convenience, the calculations have also been listed in Table (A). The column for $I_y = I_c + Ad^2$ could have been omitted from the table and the moment of inertia obtained by adding the sums ΣI_c and $\Sigma (Ad^2)$. That is,

$$I_y = \Sigma I_c + \Sigma (Ad^2) = 20.5 \times 10^6 + 566.2 \times 10^6$$
$$= 586.7 \times 10^6 \text{ mm}^4$$

which is the same result as before.

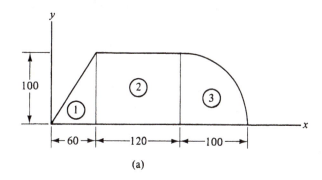

(a)

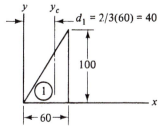

$A_1 = bh/2 = 60(100)/2 = 3 \times 10^3$ mm^2

$I_{c_1} = bh^3/36 = 100(60)^3/36 = 0.6 \times 10^6$ mm^4

(b)

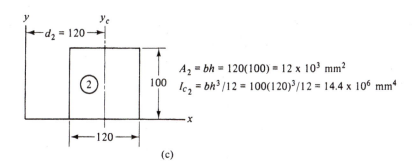

$A_2 = bh = 120(100) = 12 \times 10^3$ mm^2

$I_{c_2} = bh^3/12 = 100(120)^3/12 = 14.4 \times 10^6$ mm^4

(c)

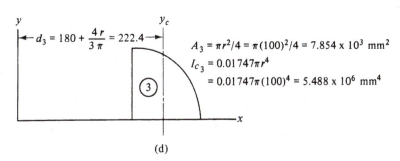

$A_3 = \pi r^2/4 = \pi(100)^2/4 = 7.854 \times 10^3$ mm^2

$I_{c_3} = 0.01747\pi r^4$

$= 0.01747\pi(100)^4 = 5.488 \times 10^6$ mm^4

(d)

(All dimensions are in mm.)

FIGURE 9.20

TABLE (A)   Example 9.7

	Shape	A (mm^2)	d (mm)	I_c (mm^4)	Ad^2 (mm^4)	$I_y = I_c + Ad^2$ (mm^4)
①	Triangle	3×10^3	40	0.6×10^6	4.8×10^6	5.4×10^6
②	Rectangle	12×10^3	120	14.4×10^6	172.8×10^6	187.2×10^6
③	Quarter-circle	7.85×10^3	222.4	5.5×10^6	388.6×10^6	394.1×10^6
				$\Sigma I_c = 20.5 \times 10^6$ (mm^4)	$\Sigma(Ad^2) = 566.2 \times 10^6$ (mm^4)	$\Sigma I_y = 586.7 \times 10^6$ (mm^4)

Example 9.8   Find the moments of inertia of the composite area shown in Fig. 9.21(a) for the x and y centroidal axes.

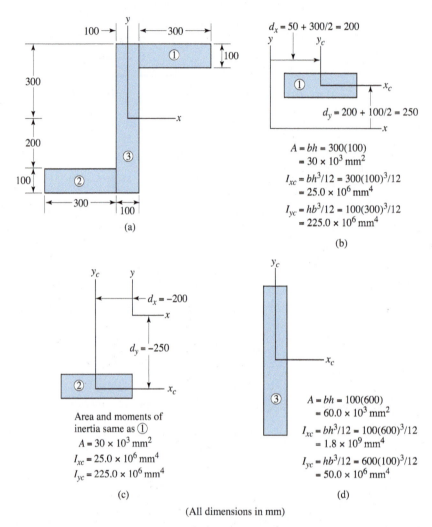

(a)

$d_x = 50 + 300/2 = 200$

$A = bh = 300(100)$
$\quad = 30 \times 10^3 \text{ mm}^2$

$I_{xc} = bh^3/12 = 300(100)^3/12$
$\quad = 25.0 \times 10^6 \text{ mm}^4$

$I_{yc} = hb^3/12 = 100(300)^3/12$
$\quad = 225.0 \times 10^6 \text{ mm}^4$

$d_y = 200 + 100/2 = 250$

(b)

$d_x = -200$

$d_y = -250$

Area and moments of
inertia same as ①
$A = 30 \times 10^3 \text{ mm}^2$
$I_{xc} = 25.0 \times 10^6 \text{ mm}^4$
$I_{yc} = 225.0 \times 10^6 \text{ mm}^4$

(c)

$A = bh = 100(600)$
$\quad = 60.0 \times 10^3 \text{ mm}^2$

$I_{xc} = bh^3/12 = 100(600)^3/12$
$\quad = 1.8 \times 10^9 \text{ mm}^4$

$I_{yc} = hb^3/12 = 600(100)^3/12$
$\quad = 50.0 \times 10^6 \text{ mm}^4$

(d)

(All dimensions in mm)

FIGURE 9.21

Solution: The composite area is made up of three rectangles. The areas, centroids, and centroidal moments of inertia are calculated in Fig. 9.21(b)–(d) from formulas given in Tables A.1 and A.2 of the Appendix. They are listed in Table (A) together with the required calculations. From the sums in the table and the parallel-axis theorem, we have

$$I_x = I_{x_c} + Ad^2$$
$$= 1.85 \times 10^9 + 3.75 \times 10^9 = 5.60 \times 10^9 \text{ mm}^4 \qquad \textbf{Answer}$$

$$I_y = I_{y_c} + Ad^2$$
$$= 0.5 \times 10^9 + 2.4 \times 10^9 = 2.90 \times 10^9 \text{ mm}^4 \qquad \textbf{Answer}$$

TABLE (A)   Example 9.8

	Shape	A (mm^2)	d_y (mm)	I_{x_c} (mm^4)	Ad_y^2 (mm^4)	d_x (mm)	I_{y_c} (mm^4)	Ad_x^2 (mm^4)
①	Rectangle	30×10^3	250	25.0×10^6	1.875×10^9	200	225×10^6	1.2×10^9
②	Rectangle	30×10^3	−250	25.0×10^6	1.875×10^9	−200	225×10^6	1.2×10^9
③	Rectangle	60×10^3	0	1.8×10^9	0	0	50×10^6	0
			Σ	$1.85 \times 10^9 \text{ mm}^4$	$3.75 \times 10^9 \text{ mm}^4$	—	$0.5 \times 10^9 \text{ mm}^4$	$2.4 \times 10^9 \text{ mm}^4$

Example 9.9   A structural member made up of two American Standard channels and a plate are fastened together as shown in Fig. 9.22(a). Determine the moment of inertia of the composite section about a centroidal axis parallel with the plate.

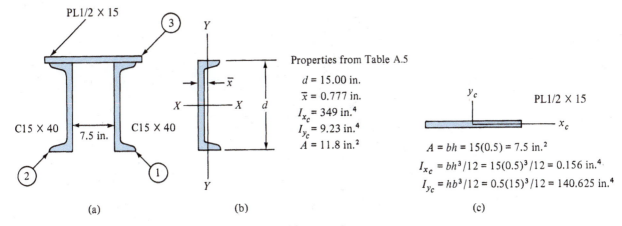

Properties from Table A.5

$d = 15.00$ in.
$\bar{x} = 0.777$ in.
$I_{x_c} = 349$ in.^{4}
$I_{y_c} = 9.23$ in.^{4}
$A = 11.8$ in.^{2}

$A = bh = 15(0.5) = 7.5$ in.^{2}
$I_{x_c} = bh^3/12 = 15(0.5)^3/12 = 0.156$ in.^{4}
$I_{y_c} = hb^3/12 = 0.5(15)^3/12 = 140.625$ in.^{4}

(a)   (b)   (c)

FIGURE 9.22

Solution: Properties of the C15 × 40 (Table A.5 of the Appendix) and calculated properties for the plate (rectangle) are shown in Fig. 9.22(b) and (c). The origin of the coordinate axes for the cross section is located with the centroid of the two channels on the x axis and the y axis, an axis of symmetry (Fig. 9.23). Because of the symmetry, $\bar{x}$ is equal to zero.

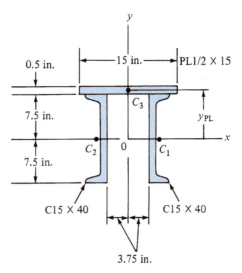

FIGURE 9.23

Locating the Centroid: To find $\bar{y}$, we use the method of Sec. 9.5. The centroid of the two channels is on the x axis; therefore, the y centroid for both channels is zero. The centroid for plate $y_{PL} = (d_C + t_{PL})/2 = (15 + 0.5)/2 = 7.75$ in. (See Fig. 9.23.) Areas of the channel and plate are shown in Fig. 9.22(b) and (c), and the areas, centroidal distances, and calculations are listed in Table (A). From Eq. (9.7) and the sums in the table

$$\bar{y} = \frac{\Sigma Ay}{\Sigma A} = \frac{58.12}{31.1} = 1.869 \text{ in.}$$

TABLE (A)	Example 9.9			
	Shape	A (in.^{2})	y (in.)	Ay (in.^{3})
①	C15 × 40(R)	11.8	0	0
②	C15 × 40(L)	11.8	0	0
③	PL 1/2 × 15	7.5	7.75	58.12
	Σ	31.1 in.^{2}	—	58.12 in.^{3}

Finding the Centroidal Moment of Inertia: Method 1. To find the moment of inertia for the centroidal axis x_c, we first find the moment of inertia for the x axis for the composite section. We then use the parallel-axis theorem to transfer the moment of inertia to the centroidal axis.

Properties of the channel and plate are displayed in Fig. 9.22(b) and (c). The transfer distances d are shown in Fig. 9.23 and listed in Table (B) as y. In this example, y and d have the same value. From the sums in Table (B) and the parallel-axis theorem,

$$I_x = \Sigma I_{x_c} + \Sigma (Ad^2) = 698.2 + 450.5 = 1148.7 \text{ in.}^4$$

TABLE (B)   Example 9.9

	Shape	A (in.²)	y (in.)†	I_{x_c}(in.⁴)	Ad² (in.⁴)
①	C15 × 40(R)	11.8	0	349.0	0
②	C15 × 40(L)	11.8	0	349.0	0
③	PL 1/2 × 15	7.5	7.75	0.156	450.5
	Σ	31.1 in.²	—	698.2 in.⁴	450.5 in.⁴

†In this example, y and d have the same value.

Now, we use the parallel-axis theorem to find the composite moment of inertia, I_C, for the centroidal axis x_c.

$$I_C = I_x - A\bar{y}^2$$
$$= 1148.7 - 31.1(1.869)^2 = 1040 \text{ in.}^4 \qquad \textbf{Answer}$$

Notice that we use the negative sign with the transfer term because the transfer is made to the centroidal axis.

Method 2. We apply the parallel-axis theorem to transfer the moment of inertia of each channel and the plate directly to the composite centroidal axis x_c. The transfer distances for each area in the cross section are $d = y - \bar{y}$. For the channels $d_1 = d_2 = y_1 - \bar{y} = 0 - 1.869 = -1.869$ in., and for the plate $d_3 = y_3 - \bar{y} = 7.75 - 1.869 = 5.881$ in. (See Fig. 9.24.)

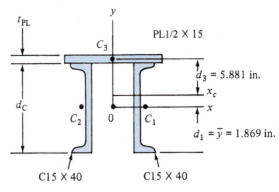

FIGURE 9.24

Properties of the channel and plate are shown in Fig. 9.22(b) and (c) and areas, moments of inertia, transfer distances, and calculations are listed in Table (C). From the parallel-axis theorem and the sums shown in the table, we have the moment of inertia with respect to the centroidal axis.

$$I_C = \Sigma I_{x_c} + \Sigma (Ad^2) = 698.2 + 341.8 = 1040 \text{ in.}^4 \qquad \textbf{Answer}$$

TABLE (C)   Example 9.9

	Shape	A (in.²)	d (in.)	I_{x_c}(in.⁴)	Ad² (in.⁴)
①	C15 × 40(R)	11.8	−1.869	349	41.22
②	C15 × 40(L)	11.8	−1.869	349	41.22
③	PL 1/2 × 15	7.5	5.881	0.156	259.40
	Σ	31.1 in.²	—	698.2 in.⁴	341.8 in.⁴

Example 9.10   A structural member made up of a channel and a wide-flange section are fastened together as shown in Fig. 9.25(a). Determine the moment of inertia of the composite section about the centroidal axis parallel with the web of the channel.

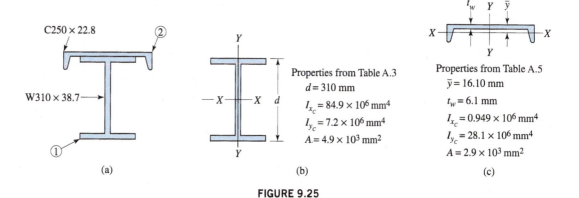

FIGURE 9.25

Solution: Properties of the W310 × 38.7 and C250 × 22.8 (Tables A.3 and A.5 of the Appendix) are shown in Fig. 9.25(b) and (c). (Notice that the channel has been turned 90° from the one shown in the table.) The origin of the coordinate axes of the composite section is located at the centroid of the W shape, and the y axis is an axis of symmetry. Therefore, we see that $\bar{x} = 0$.

Locating the Centroid: To find $\bar{y}$, we use the method of Sec. 9.5. The x and y centroids for the W shape are equal to zero. The centroid for the channel is $y_c = d/2 + t_w - \bar{y} = 155 + 6.1 - 16.10 = 145$ mm. (See Fig. 9.26.) The areas, centroidal distances, and calculations for Ay are listed in Table (A). From Eq. (9.7) and the sums in the table,

$$\bar{y} = \frac{\Sigma Ay}{\Sigma A} = \frac{420.5 \times 10^3}{7.84 \times 10^3} = 53.6 \text{ mm}$$

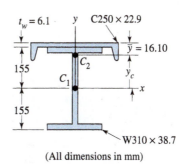

(All dimensions in mm)

FIGURE 9.26

	Shape	A (mm^2)	y (mm)[†]	Ay (mm^3)	I_{x_c}(mm^4)	Ad^2 (mm^4)
①	W310 × 38.7	4.94 × 10³	0	0	84.9 × 10⁶	0
②	C250 × 22.8	2.90 × 10³	145	420.5 × 10³	0.949 × 10⁶	61.0 × 10⁶
	Σ	7.84 × 10³	—	420.5 × 10³	85.8 × 10⁶	61.0 × 10⁶

TABLE (A)   Example 9.10

[†]In this example, y and d have the same value.

Finding the Centroidal Moment of Inertia: Method 1. To find the moment of inertia about the centroidal axis, we first find the moment of inertia about the x axis and then use the parallel-axis theorem to transfer the moment of inertia to the centroidal axis. Properties of the W shape and channel are shown in Fig. 9.25(b) and (c). The transfer distances d are the same as y in the table. Areas, moments of inertia, transfer distances, and calculations for Ad^2 are shown in Table (A). From the sums in the table,

$$I_x = (85.8 + 61.0) \times 10^6 = 146.8 \times 10^6 \, mm^4$$

Transferring the moment of inertia from the x axis to the centroidal axis with the parallel-axis theorem, we have

$$I_C = I_x - A\bar{y}^2 = 146.8 \times 10^6 - 7.84 \times 10^3(53.6)^2$$
$$= 124.3 \times 10^6 \, mm^4 \qquad \textbf{Answer}$$

Method 2. We apply the parallel-axis theorem to transfer the moment of inertia of the channel and W shape directly to the composite centroidal axis x_c. The transfer distances are $d = y - \bar{y}$ for each area in the cross section: For the W shape $d_1 = y_1 - \bar{y} = 0 - 53.6 = -53.6$ mm, and for the channel $d_2 = y_2 - \bar{y} = 145 - 53.6 = 91.4$ mm, where $y_2 = y_c$. (See Fig. 9.27.)

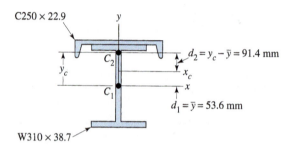

C250 × 22.9

$d_2 = y_c - \bar{y} = 91.4$ mm

$d_1 = \bar{y} = 53.6$ mm

W310 × 38.7

FIGURE 9.27

Properties of the W shape and channel are shown in Fig. 9.25(b) and (c), and areas, moments of inertia, transfer distances, and calculations are listed in Table (B). From the parallel-axis theorem and the sums in the table, we have the moment of inertia, I_C, with respect to the centroidal axis x_c.

$$I_C = \Sigma I_{x_c} + \Sigma(Ad^2) = (85.85 + 38.42) \times 10^6$$
$$= 124.3 \times 10^6 \, mm^4 \qquad \textbf{Answer}$$

TABLE (B)   Example 9.10

	Shape	A (mm^2)	d (mm)	$I_{x_c}(mm^4)$	Ad^2 (mm^4)
①	W310 × 38.7	4.94×10^3	−53.6	84.9×10^6	14.19×10^6
②	C250 × 22.8	2.90×10^3	91.4	0.949×10^6	24.23×10^6
			Σ	85.85×10^6	38.42×10^6

PROBLEMS

9.28 Strips of expanded plastic foam having 200-mm-by-50-mm rectangular cross sections are used in the packaging of fragile electronic equipment. To increase the rigidity of these strips, they have been redesigned to include four stiffening ribs, as shown. Compute the rectangular moment of inertia about a horizontal axis through the centroid for both the original and redesigned strips.

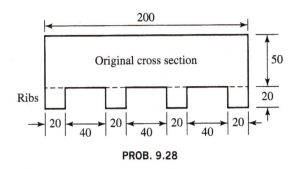

PROB. 9.28

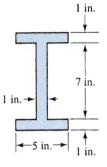

PROB. 9.31

9.29 through 9.44   For the composite area shown, determine the moment of inertia with respect to the centroidal axes (a) I_{x_C} and (b) I_{y_C}.

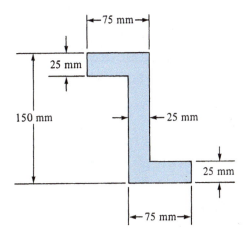

PROB. 9.32

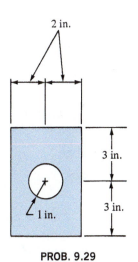

PROB. 9.29

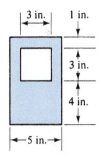

PROB. 9.33

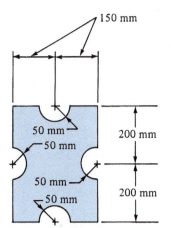

PROB. 9.30

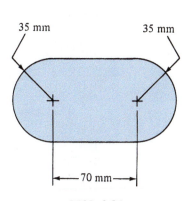

PROB. 9.34

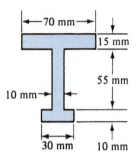

PROB. 9.35

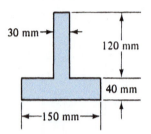

PROB. 9.36

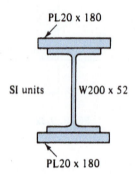

PROB. 9.37

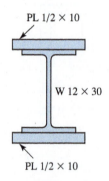

PROB. 9.38

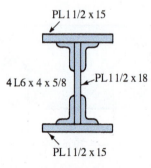

PROB. 9.39

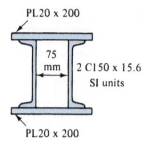

PROB. 9.40

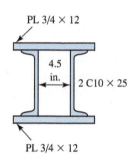

PROB. 9.41

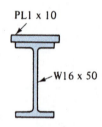

PROB. 9.42

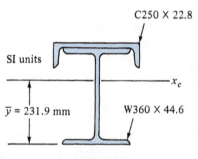

PROB. 9.43

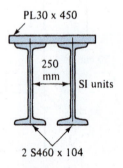

PROB. 9.44

9.45 A particular manufacturer offers wooden I-joists with the following dimensions: W = 2-5/16 in., d = 16 in., t = 3/8 in., f = 1-1/2 in. Find the moment of inertia about the centroidal x-axis for this cross section.

9.46 Same as Prob. 9.45 but with the following dimensions: W = 58.7 mm, d = 406 mm, t = 9.5 mm, f = 38 mm.

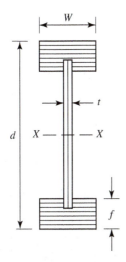

PROB. 9.45 and PROB. 9.46

9.47 A hollow box beam is fabricated from two nominal 2 × 6 boards and two 14-in.-wide strips of 5/8-in.-thick plywood (*shaded*). Compute the moment of inertia about the centroidal x-axis for this section.

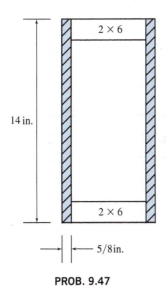

PROB. 9.47

9.48 To form a wooden beam, three 2 × 8 S4S boards may be nailed together in any of the three configurations shown. For each cross-sectional area, compute the rectangular moment of inertia about a horizontal axis through the centroid.

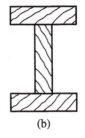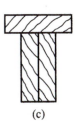

(a)   (b)   (c)

PROB. 9.48

9.10 POLAR MOMENT OF INERTIA

In the analysis and design of circular bars in torsion, the polar moment of inertia of the cross-sectional area of the bar is required.

Consider the cross section shown in Fig. 9.28. The product of the area ΔA_i and the square of the distance r_i from the origin O (or z axis) to the centroid of the area is called the *polar moment of inertia* for the area ΔA_i. Adding the polar moments of inertia for each area, we find the total approximate polar moment of inertia J, that is,

$$J = \Sigma r_i^2 \Delta A_i \tag{9.11}$$

From the Pythagorean theorem, $r_i^2 = x_i^2 + y_i^2$. Therefore,

$$J = \Sigma(x_i^2 + y_i^2)\Delta A_i = \Sigma x_i^2 \Delta A_i + \Sigma y_i^2 \Delta A_i$$

From Eqs. (9.8) and (9.9),

$$J = I_x + I_y \tag{9.12}$$

A relationship similar to the parallel-axis theorem for rectangular moments of inertia exists for polar moments of inertia. That is,

$$J_o = J_c + Ad^2 \tag{9.13}$$

where J_o is the polar moment of inertia for an area A about the origin O (or z axis), J_c is the polar moment of inertia for the area about the centroid C (or z_c axis), and d is the distance between the origin and the centroid.

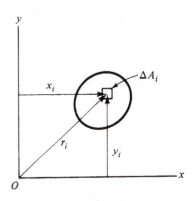

FIGURE 9.28

Example 9.11   For the wide-flange-beam cross section shown in Fig. 9.29, find the polar moment of inertia.

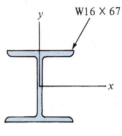

FIGURE 9.29

Solution: From Table A.3 of the Appendix, the W16 × 67 has moments of inertia $I_x = 954$ in.^{4} and $I_y = 119$ in.^{4} From Eq. (9.12),

$$J = I_x + I_y = 954 + 119 = 1073 \text{ in.}^4$$ **Answer**

9.11 RADIUS OF GYRATION

In the analysis and design of columns, the radius of gyration of the cross-sectional area of the column is required. The radius of gyration r of an area with respect to a given axis is defined by the relationship

$$r = \sqrt{\frac{I}{A}} \quad \text{or} \quad I = Ar^2 \qquad (9.14)$$

where I is the moment of inertia with respect to the given axis, and A is the cross-sectional area. The radius of gyration

for standard structural shapes is tabulated in Tables A.3 through A.8 of the Appendix.

We also have a parallel-axis theorem for the radius of gyration. That is,

$$r_o^2 = r_c^2 + d^2 \qquad (9.15)$$

where r_o is the radius of gyration for an area about the origin O (or z axis), r_c is the radius of gyration for the area about the centroid C (or z_c axis), and d is the distance between the origin and the centroid.

Example 9.12   For the cross section of Example 9.11, determine the radius of gyration with respect to the x and y axes.

Solution: From Table A.3 of the Appendix, the W16 × 67 has a cross-sectional area of $A = 19.7$ in.^{2} From Eq. (9.14),

$$r_x = \sqrt{\frac{I_x}{A}} = \sqrt{\frac{954}{19.7}} = 6.96 \text{ in.} \qquad r_y = \sqrt{\frac{I_y}{A}} = \sqrt{\frac{119}{19.7}} = 2.46 \text{ in.} \qquad \textbf{Answer}$$

These values for the radius of gyration are also given in Table A.3 of the Appendix.

PROBLEMS

In Probs. 9.49 through 9.52, determine the polar centroidal moment of inertia J_c *and the polar moment of inertia* J *for the area shown.*

9.49 Use the figure for Prob. 9.24.

9.50 Use the figure for Prob. 9.25.

9.51 Use the figure for Prob. 9.26.

9.52 Use the figure for Prob. 9.27.

In Probs. 9.53 through 9.56, determine the radius of gyration for the centroidal axes of the area shown.

9.53. Use the figure for Prob. 9.31.

9.54. Use the figure for Prob. 9.47.

9.55. Use the figure for Prob. 9.37.

9.56. Use the figure for Prob. 9.39.

*9.12 DETERMINATION OF CENTROIDS BY INTEGRATION

The centroid of a plane area was discussed in Sec. 9.3. If we let the number of areas increase without limit in Eq. (9.5), we have the following equations for the centroids—in integral form.[†]

$$\bar{x} = \frac{\int x \, dA}{\int dA} \qquad \bar{y} = \frac{\int y \, dA}{\int dA} \qquad (9.16)$$

*Sections marked with an asterisk can be omitted without a loss of continuity.

[†]The integral sign is a long S used by early writers to denote the first letter of the word *sum*.

or $$\bar{x} = \frac{Q_y}{A} \qquad \bar{y} = \frac{Q_x}{A} \qquad\qquad (9.17)$$

where $$A = \int dA \quad Q_x = \int y\, dA \quad Q_y = \int x\, dA$$

In most cases, the centroid of an area is found by dividing the area into elemental strips or rectangles of width dx or d_y and area dA. The strips are parallel with the x or y axis. The centroid x_c and y_c for a typical strip is then determined. The total area A is found by integrating the area of the strips over the complete area. The moment of the elemental area about the x axis dQ_x is found by integrating the product $y_c\, dA$ over the complete area. Similarly, the moment of the area about the y axis dQ_y is found by integrating the product $x_c\, dA$ over the complete area.

As an illustration, we determine the centroid of a rectangle [Fig. 9.30(a)] by integration. The rectangle is divided into horizontal strips parallel with the x axis. The typical strip shown in Fig. 9.30(b) has a length b and width dy. The element of area $dA = b\, dy$ and the centroids $x_c = b/2$ and $y_c = y$. The limits of integration for y are 0 and h. Therefore,

$$dA = b\, dy$$

$$A = \int_0^h b\, dy = b[y]_0^h = bh$$

$$dQ_x = y_c\, dA = yb\, dy$$

$$Q_x = \int_0^h by\, dy = b\left[\frac{y^2}{2}\right]_0^h = \frac{bh^2}{2}$$

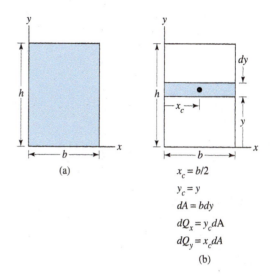

FIGURE 9.30

$x_c = b/2$
$y_c = y$
$dA = b\, dy$
$dQ_x = y_c\, dA$
$dQ_y = x_c\, dA$

(b)

$$dQ_y = x_c\, dA = \frac{b}{2} b\, dy$$

$$Q_y = \int_0^h \frac{b^2}{2}\, dy = \frac{b^2}{2}[y]_0^h = \frac{b^2 h}{2}$$

From Eq. (9.17),

$$\bar{x} = \frac{Q_y}{A} = \frac{b^2 h/2}{bh} = \frac{b}{2} \qquad \textbf{Answer}$$

$$\bar{y} = \frac{Q_x}{A} = \frac{bh^2/2}{bh} = \frac{h}{2} \qquad \textbf{Answer}$$

Example 9.13 Determine the centroids of the triangle shown in Fig. 9.31(a) by integration. From the equation for the line that forms the top of the triangle,

$$y = \frac{h}{b}x \quad \text{or} \quad x = \frac{b}{h}y \qquad\qquad (a)$$

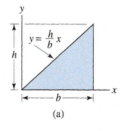

(a)

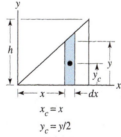

$x_c = x$
$y_c = y/2$
$dA = y\, dx$
$dQ_x = y_c\, dA$
$dQ_y = x_c\, dA$

(b)

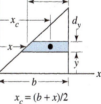

$x_c = (b+x)/2$
$y_c = y$
$dA = (b-x)\, dy$
$dQ_x = y_c\, dA$
$dQ_y = x_c\, dA$

(c)

FIGURE 9.31

Solution 1: The triangle is divided into vertical strips. The typical strip shown in Fig. 9.31(b) has an area $dA = y\, dx$, and the centroids are $x_c = x$ and $y_c = y/2$. From Eq. (a), $y = hx/b$. The limits of integration for x are 0 and b. Therefore,

$$dA = y\, dx = \frac{h}{b}x\, dx$$

$$A = \frac{h}{b}\int_0^b x\, dx = \frac{h}{2b}[x^2]_0^b = \frac{bh}{2}$$

$$dQ_x = y_c\, dA = \frac{y}{2}y\, dx = \frac{1}{2}\left(\frac{h^2}{b^2}x^2\right)dx$$

$$Q_x = \frac{1}{2}\int_0^b \left(\frac{h^2}{b^2}x^2\right)dx = \frac{1}{2}\left[\frac{h^2}{b^2}\frac{x^3}{3}\right]_0^b = \frac{bh^2}{6}$$

$$dQ_y = x_c\, dA = xy\, dx = \frac{h}{b}x^2\, dx$$

$$Q_y = \int_0^b \frac{h}{b}x^2\, dx = \frac{h}{b}\left[\frac{x^3}{3}\right]_0^b = \frac{b^2 h}{3}$$

From Eq. (9.17),

$$\bar{x} = \frac{Q_y}{A} = \frac{b^2 h/3}{bh/2} = \frac{2}{3}b \qquad\qquad \textbf{Answer}$$

$$\bar{y} = \frac{Q_x}{A} = \frac{bh^2/6}{bh/2} = \frac{1}{3}h \qquad\qquad \textbf{Answer}$$

Solution 2: The triangle is divided into horizontal strips. The typical strip shown in Fig. 9.31(c) has a length $(b - x)$ and a width dy. The element of area $dA = (b - x)\, dy$, and the centroids $x_c = x + (b - x)/2 = (b + x)/2$ and $y_c = y$. From Eq. (a), $x = by/h$. The limits of integration for y are 0 and h. Therefore,

$$dA = (b - x)dy = \left(b - \frac{b}{h}y\right)dy$$

$$A = \int_0^h \left(b - \frac{b}{h}y\right)dy = \left[by - \frac{by^2}{2h}\right]_0^h = \frac{bh}{2}$$

$$dQ_x = y_c\, dA = y\left(b - \frac{b}{h}y\right)dy = b\left(y - \frac{y^2}{h}\right)dy$$

$$Q_x = b\int_0^h \left(y - \frac{y^2}{h}\right)dy = b\left[\frac{y^2}{2} - \frac{y^3}{3h}\right]_0^h = \frac{bh^2}{6}$$

$$dQ_y = x_c\, dA = \frac{1}{2}(b + x)(b - x)dy = \frac{1}{2}(b^2 - x^2)dy$$

From Eq. (a), $x^2 = b^2 y^2/h^2$. Therefore,

$$Q_y = \frac{1}{2}\int_0^h \left(b^2 - \frac{b^2}{h^2}y^2\right)dy = \frac{1}{2}\left[b^2 y - \frac{b^2 y^3}{h^2 3}\right]_0^h = \frac{b^2 h}{3}$$

From Eq. (9.17),

$$\bar{x} = \frac{Q_y}{A} = \frac{b^2 h/3}{bh/2} = \frac{2}{3}b \qquad\qquad \textbf{Answer}$$

$$\bar{y} = \frac{Q_x}{A} = \frac{bh^2/6}{bh/2} = \frac{1}{3}h \qquad\qquad \textbf{Answer}$$

Example 9.14 Determine the centroids of the parabolic spandrel shown in Fig. 9.32(a) by integration.

Solution: The parabolic spandrel is divided into vertical strips. A typical strip is shown in Fig. 9.32(b). The area of strip $dA = y\, dx$, and the centroids are $x_c = x$ and $y_c = y/2$. The equation for the parabolic top of the spandrel is $y = hx^2/b^2$, and the limits of integration for x are 0 and b. Therefore,

$$dA = y\, dx = \left(\frac{hx^2}{b^2}\right)dx$$

$$A = \frac{h}{b^2}\int_0^b x^2\, dx = \frac{h}{b^2}\left[\frac{x^3}{3}\right]_0^b = \frac{bh}{3}$$

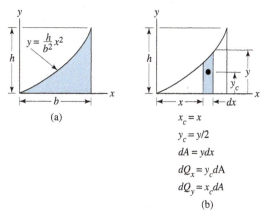

FIGURE 9.32

$$dQ_x = y_c\, dA = \frac{y^2}{2}dx = \left(\frac{h^2}{2b^4}x^4\right)dx$$

$$Q_x = \frac{h^2}{2b^4}\int_0^b x^4\, dx = \frac{h^2}{2b^4}\left[\frac{x^5}{5}\right]_0^b = \frac{bh^2}{10}$$

$$dQ_y = x_c\, dA = xy\, dx = \left(\frac{h}{b^2}x^3\right)dx$$

$$Q_y = \frac{h}{b^2}\int_0^b x^3\, dx = \frac{h}{b^2}\left[\frac{x^4}{4}\right]_0^b = \frac{b^2h}{4}$$

From Eq. (9.17),

$$\bar{x} = \frac{Q_y}{A} = \frac{b^2h/4}{bh/3} = \frac{3}{4}b \qquad \textbf{Answer}$$

$$\bar{y} = \frac{Q_x}{A} = \frac{bh^2/10}{bh/3} = \frac{3}{10}h \qquad \textbf{Answer}$$

*9.13 DETERMINATION OF MOMENTS OF INERTIA BY INTEGRATION

The moment of inertia of a plane area was discussed in Sec. 9.7. If we let the number of areas increase without limit in Eqs. (9.8) and (9.9), we have the following equations for the moments of inertia—in integral form.

$$I_x = \int y^2\, dA \qquad \textbf{(9.18)}$$

$$I_y = \int x^2\, dA \qquad \textbf{(9.19)}$$

The moment of inertia of an area is found by dividing the area into elemental strips or rectangles of width dx or dy and area dA. When the strip is parallel with the x axis, a centroidal distance y_c is determined, and the moment of inertia of the strip about the x axis is $dI_x = y_c^2\, dA$. The moment of inertia of the same strip about the y axis can be found from the formula for the moment of inertia of a rectangle about its base. When the strip is parallel with the y axis, the moments

of inertia are found by a method similar to the one previously described, with an interchange of x and y.

Moment of Inertia of a Rectangular Area

As an illustration, we determine the moment of inertia of a rectangle about its base [Fig. 9.33(a)] by integration. The rectangle is divided into horizontal strips parallel with the x axis. The typical strip shown in Fig. 9.33(b) has an element of area $dA = b\, dy$, and the centroidal distance from the x axis $y_c = y$. The limits of integration for y are 0 and h. For the moment of inertia of the strip about the x axis,

$$dI_x = y^2\, dA = y^2 b\, dy$$

$$I_x = b\int_0^h y^2\, dy = b\left[\frac{y^3}{3}\right]_0^h = \frac{bh^3}{3} \qquad \textbf{Answer}$$

The distance d between the x axis and x_c axis is $h/2$, and the area A of the rectangle is bh. The transfer is made *to* the centroidal axis; therefore, we *subtract* the term Ad^2 when applying the parallel-axis theorem.

$$I_{x_c} = I_x - Ad^2$$

$$I_{x_c} = \frac{bh^3}{3} - bh\left(\frac{h}{2}\right)^2 = \frac{bh^3}{12} \qquad \textbf{Answer}$$

*Sections marked with an asterisk can be omitted without a loss of continuity.

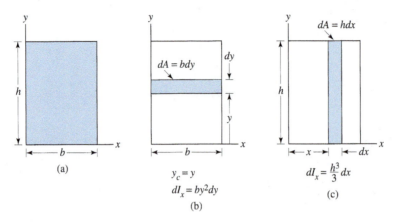

FIGURE 9.33

In Fig. 9.33(c), we divide the rectangle into vertical strips parallel with the y axis. The formula for the moment of inertia of a rectangle about its base $I_x = bh^3/3$ can be used to determine the moment of inertia of the typical vertical strip about the x axis. The base of the strip is dx and the height h; therefore,

$$dI_x = \frac{h^3}{3} dx$$

Integrating x from 0 to b, we have

$$I_x = \int_0^b \frac{h^3}{3} dx = \frac{bh^3}{3}$$ **Answer**

Example 9.15   Determine the moment of inertia about the x axis for the triangle shown in Fig. 9.34(a).

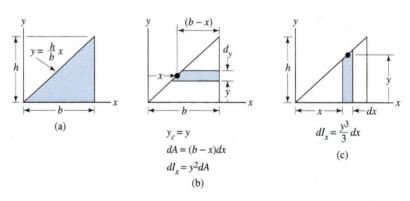

FIGURE 9.34

Solution 1:   The triangle is divided into horizontal strips parallel with the x axis. A typical strip shown in Fig. 9.34(b) has an element of area $dA = (b - x)\, dy$. The centroidal distance is $y_c = y$. The moment of inertia of the strip about the x axis is given by

$$dI_x = y^2 dA = y^2(b - x)dy$$

From the equation for the top of the triangle $x = by/h$, we have

$$dI_x = y^2 \left(b - \frac{b}{h} y \right) dy$$

Integrating y from 0 to h, the moment of inertia we get is

$$I_x = b \int_0^h y^2 \, dy - \frac{b}{h} \int_0^h y^3 \, dy$$

$$I_x = \left[\frac{by^3}{3} - \frac{by^4}{4h} \right]_0^h = \frac{bh^3}{3} - \frac{bh^3}{4}$$

$$I_x = \frac{bh^3}{12}$$ **Answer**

Solution 2:   The triangle is divided into vertical strips parallel with the y axis, as shown in Fig. 9.34(c). From the formula for the moment of inertia of a rectangle about its own base and the equation for the top of the triangle $y = hx/b$, we have the moment of inertia of the strip

$$dI_x = \frac{y^3}{3}\,dx = \frac{h^3}{3b^3}x^3\,dx$$

Integrating x from 0 to b,

$$I_x = \frac{h^3}{3b^3}\int_0^b x^3\,dx = \frac{h^3}{3b^3}\left[\frac{x^4}{4}\right]_0^b = \frac{bh^3}{12} \qquad \textbf{Answer}$$

Example 9.16   Determine the moment of inertia about the x axis for the parabolic area shown in Fig. 9.35(a).

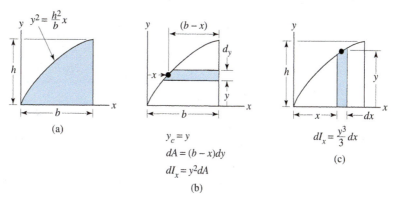

$$y_c = y$$
$$dA = (b - x)dy$$
$$dI_x = y^2 dA$$
(b)

$$dI_x = \frac{y^3}{3}dx$$
(c)

FIGURE 9.35

Solution 1:   The parabolic area is divided into horizontal strips parallel with the x axis. A typical strip shown in Fig. 9.35(b) has an element of area $dA = (b - x)\,dy$, and the centroidal distance is $y_c = y$. The moment of inertia of the strip about the x axis is as follows:

$$dI_x = y^2\,dA = y^2(b - x)\,dy$$

From the equation of the parabola, $x = by^2/h^2$, we have

$$dI_x = y^2\left(b - \frac{b}{h^2}y^2\right)dy$$

Integrating y from 0 to h, we find the moment of inertia:

$$I_x = \int_0^h by^2\,dy - \int_0^h \frac{b}{h^2}y^4\,dy$$
$$= \left[\frac{by^3}{3} - \frac{by^5}{5h^2}\right]_0^h = \frac{bh^3}{3} - \frac{bh^3}{5}$$
$$I_x = \frac{2}{15}bh^3 \qquad \textbf{Answer}$$

Solution 2:   The parabolic area is divided into vertical strips as shown in Fig. 9.35(c). From the formula for the moment of inertia of a rectangle about its own base and the equation of the parabola $y = hx^{1/2}/b^{1/2}$, we have the moment of inertia of the strip.

$$dI_x = \frac{y^3}{3}dx = \frac{h^3}{3b^{3/2}}x^{3/2}\,dx$$

Integrating x from 0 to b,

$$I_x = \frac{h^3}{3b^{3/2}}\int_0^b x^{3/2}\,dx = \frac{h^3}{3b^{3/2}}\left[\frac{2x^{5/2}}{5}\right]_0^b$$
$$I_x = \frac{2}{15}bh^3 \qquad \textbf{Answer}$$

PROBLEMS

9.57 through 9.64. Use direct integration to determine the centroids or moments of inertia of the area indicated.

9.57 Determine the centroids of the rectangle shown in Fig. 9.30. Use vertical strips parallel with the y axis for the differential area.

9.58 Determine the centroids for the parabolic spandrel shown in Fig. 9.32. Use horizontal strips parallel with the x axis for the differential area.

9.59 Determine the centroids for the parabolic area shown in Fig. 9.35.

9.60 Determine the centroids for the cubical parabolic spandrel shown.

9.61 Determine the moments of inertia I_x and I_y for the cubical parabolic spandrel shown.

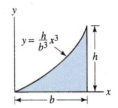

PROB. 9.60 and PROB. 9.61

9.62 Determine the moment of inertia I_y for the rectangular area shown in Fig. 9.33.

9.63. Determine the moment of inertia I_y for the triangular area shown in Fig. 9.34, and use the parallel-axis theorem to find the centroidal moment of inertia I_{y_c}.

9.64. Determine the moment of inertia I_y for the parabolic area shown in Fig. 9.35, and use the parallel-axis theorem to find the centroidal moment of inertia I_{y_c} where $\bar{x} = 3b/5$.

INTERNAL REACTIONS: STRESS FOR AXIAL LOADS

CHAPTER OBJECTIVES

In this chapter, we begin our study of the effects of forces on objects made from real materials. Before proceeding to Chapter 11, you should be able to

- Determine the internal reactions—axial forces, shear forces, and bending moments—created within objects that are acted on by external loads.

- Compute the normal and shear stresses developed by these internal reactions at various points in an object.

- Define and calculate the factor of safety for an object based on its level of stress and specified material properties.

10.1 INTRODUCTION

Strength of materials, or mechanics of materials, is concerned with methods for finding internal forces, or stresses, and deflections, or deformations, in deformable bodies when subjected to loads. It provides a means for determining the load-carrying capacity, or strength, and the induced deflections, or deformations, in the body. The methods we learn here apply equally to all sizes of deformable bodies, from the disk drive of the personal computer to massive earth-moving equipment.

The behavior of a material under load determines its range of useful applications. Often, the properties of a material can be varied either by the manufacturing process used or by composition of the material itself. The 58-in.-diameter turbine casings of Fig. 10.1, for example, achieve the high strength-to-weight ratio required for use on jet engines through a forming process known as press forging. By comparison, properties of the concrete and ceramic objects shown in Fig. 10.2 are highly dependent on the ratios of those components that comprise the material.

In statically determinate problems, the internal reaction will be found by cutting through a body with a section, isolating the free body on either side of the section, and applying the equilibrium equations. Forces at the cut section are internal reactions and represent the resultant of forces that are distributed over the section within the body. These dis-

tributed forces per unit of area are called *stresses.* However, in statically indeterminate problems, it will be necessary to consider changes in the shape or size of the body in addition to the equilibrium equations.

10.2 INTERNAL REACTIONS

In our study of trusses, we found internal forces in members of a truss by cutting the truss with a section and constructing a free-body diagram of either part of the truss. The forces in the members cut became external forces and were found by the equilibrium equations. In the case of the truss, the members cut were two-force members, and the magnitude of the internal force was unknown, but its direction was the same as the direction of the member. In the more general case, when a member is cut by a section, both an unknown internal force and couple act at the section, and both their magnitude and direction are unknown.

Consider the free-body diagram of a member involving loads and reactions in a plane [Fig. 10.3(a)]. The three reactions A_x, A_y, and B are found from the equations of equilibrium. To find the internal reactions or stress resultants at a section A–A, we cut the member at right angles to the axis of the member, separate the member at that section, and draw a free-body diagram of either part of the member. The *axial force F*, the *shear force V*, and the *bending moment M* that act on the left-hand part of the member are shown in Fig. 10.3(b). The equal and oppositely directed forces and couple acting on the right-hand part are shown in Fig. 10.3(c). Notice that if the two parts are put back together, the internal reactions add to zero, and we have a free-body diagram of the entire member. For convenience, the axial force and shear force are assumed to act through the centroid of the cross section of the member.

We will be interested in one other kind of internal reaction. It cannot be produced by forces in the plane of the member and is called the torque or twisting moment. The *torque T* is the moment about the axis of a member. Loads that produce a torque are shown in Fig. 10.4(a). The member is cut at section A–A, and the internal torques that occur are shown in Fig. 10.4(b) and (c). The torque is considered in more detail in Chapter 12.

FIGURE 10.1. Material properties can often be varied or enhanced by the manufacturing process used to generate a finished product. *Press forging* helped to maximize the toughness and strength of these 58-in.-diameter turbine casings for use on jet engines. (*Used with permission of the Wyman-Gordon Company, North Grafton, Mass.*)

FIGURE 10.2. Material properties are frequently related to the composition of a material. For the concrete sample (*left*), proportions of water, sand, cement, and stone used in the mixture affect the compressive strength, durability, and resistance to freeze–thaw cycles of the finished product. Because the ceramic bracket (*right*) is used in a precision optical system, its thermal properties must lie within a narrow range of values. To obtain these characteristics, the bracket is impregnated with other elements that produce the desired behavior.

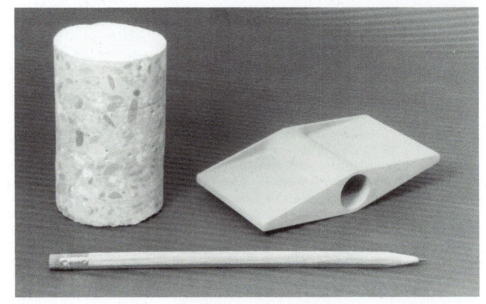

THOMAS YOUNG

SB.14(a)

SB.14(b)

The numerical constant relating stress and strain, known as *modulus of elasticity* or *Young's modulus,* was actually one of Thomas Young's (1773–1829) minor contributions to science. An English physician, Young (SB.14(a)), was an authority on vision and optics, and in this capacity he described the eye defect known as an astigmatism; developed a theory of color vision now called the Young–Helmholtz theory; and presented his belief that light itself behaved as a wave. In the field of natural science, he also made major contributions to the understanding of ocean tides. However, Young may be best known for helping to unravel the mystery of the *Rosetta stone,* a basalt slab discovered in northern Egypt in 1799 by Napoleon's troops. The stone, shown in SB.14(b), was inscribed with identical texts in three different languages: Greek, demotic, and hieroglyphic. Researchers worked for years trying to correlate the three texts to decipher the symbolic alphabet of the Egyptian hieroglyphic language. The task was a daunting one: In today's society, it would be equivalent to learning a symbolic language, such as Chinese, by studying both the English and Chinese versions of the operating instructions that accompany many of today's electronic products, including DVD players, televisions, and computers. Although final credit for decoding the hieroglyphic alphabet went to another researcher, most historians acknowledge that the two or three symbols initially identified by Young were instrumental in solving what was then one of the world's most intriguing mysteries.

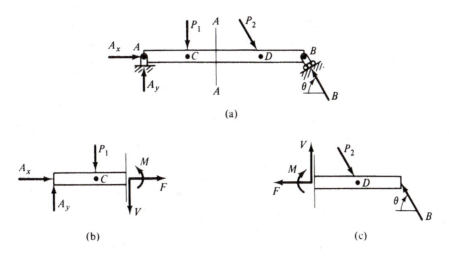

FIGURE 10.3

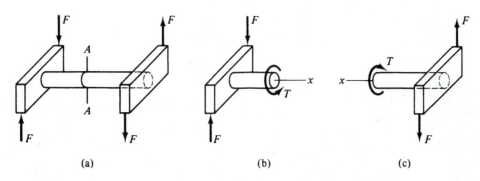

FIGURE 10.4

Example 10.1   Calculate the internal reactions for the bent bar shown in Fig. 10.5(a) at sections *M–M* and *N–N*.

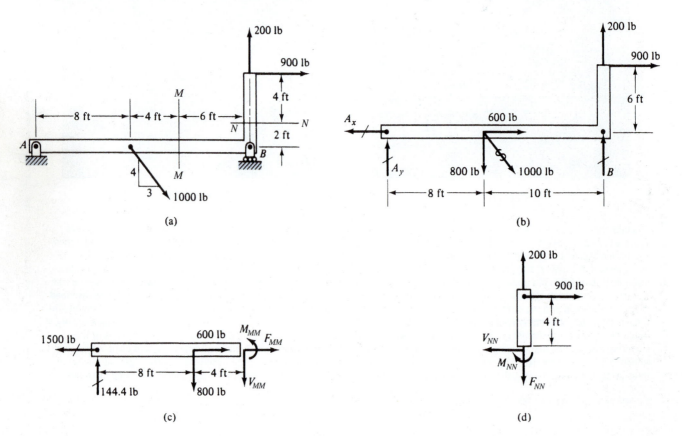

FIGURE 10.5

Solution: The free-body diagram of the bent bar is shown in Fig. 10.5(b). Components of the 1000-lb force are calculated and shown on the diagram. From equilibrium,

$$\Sigma M_A = 0 \quad -800(8) + B(18) + 200(18) - 900(6) = 0$$
$$B = 455.6 \text{ lb}$$
$$\Sigma M_B = 0 \quad -A_y(18) + 800(10) - 900(6) = 0$$
$$A_y = 144.4 \text{ lb}$$
$$\Sigma F_x = 0 \quad -A_x + 600 + 900 = 0$$
$$A_x = 1500 \text{ lb}$$

Check:

$$\Sigma F_y = 0 \quad A_y + B - 800 + 200 = 0$$
$$144.4 + 455.6 - 800 + 200 = 0 \quad \text{OK}$$

For Section M–M: The free-body diagram for the bent bar to the left of the section M–M is shown in Fig. 10.5(c). The bar to the left of the section M–M is selected because it results in a simple free-body diagram. The axial force, shear force, and bending moment for section M–M are shown. For convenience, we assume that the axial force and shear force act through the centroid C of the cross section. This portion of the beam is in equilibrium under the action of the forces, reactions, and internal reactions and couple. Therefore,

$$\Sigma F_x = 0 \qquad 600 - 1500 + F_{MM} = 0$$
$$F_{MM} = 900 \text{ lb} \quad \text{(tension)} \qquad \textbf{Answer}$$

$$\Sigma F_y = 0 \qquad 144.4 - 800 - V_{MM} = 0$$
$$V_{MM} = -655.6 \text{ lb}$$
$$= 656 \text{ lb} \uparrow \qquad \textbf{Answer}$$

$$\Sigma M_c = 0 \qquad -144.4(12) + 800(4) + M_{MM} = 0$$
$$M_{MM} = -1467 \text{ lb-ft}$$
$$= 1467 \text{ lb-ft} \, \circlearrowleft \qquad \textbf{Answer}$$

For Section N–N: The free-body diagram for the bent bar above section N–N is shown in Fig. 10.5(d). The bar above section N–N is selected because it results in a simple free-body diagram. As before, the part of the beam shown is in equilibrium. Therefore,

$$\Sigma F_y = 0 \qquad 200 - F_{NN} = 0$$
$$F_{NN} = 200 \text{ lb} \quad \text{(tension)} \qquad \textbf{Answer}$$

$$\Sigma F_x = 0 \qquad 900 - V_{NN} = 0$$
$$V_{NN} = 900 \text{ lb} \leftarrow \qquad \textbf{Answer}$$

$$\Sigma M_c = 0 \qquad -900(4) - M_{NN} = 0$$
$$M_{NN} = -3600 \text{ lb-ft}$$
$$= 3600 \text{ lb-ft} \, \circlearrowright \qquad \textbf{Answer}$$

PROBLEMS

10.1 through 10.14   Calculate the internal reactions for the member shown at the section(s) indicated.

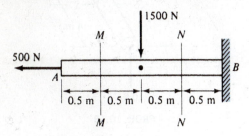

PROB. 10.1

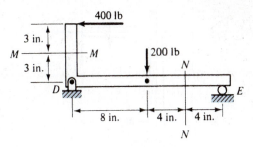

PROB. 10.3

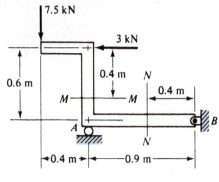

PROB. 10.2

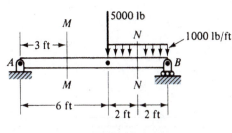

PROB. 10.4

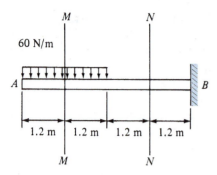

60 N/m

PROB. 10.5

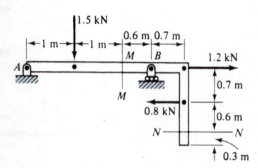

200 lb/ft

PROB. 10.6

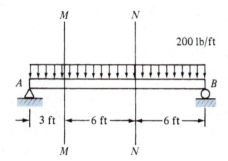

PROB. 10.7

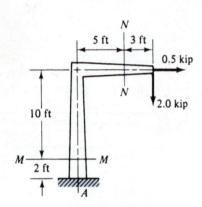

PROB. 10.8

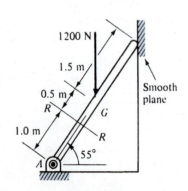

1200 N

Smooth plane

PROB. 10.9

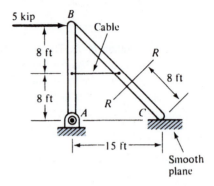

5 kip

Cable

Smooth plane

PROB. 10.10

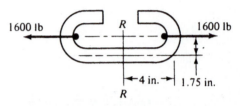

1600 lb — R — 1600 lb

4 in. — 1.75 in.

PROB. 10.11

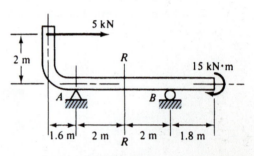

5 kN

15 kN·m

PROB. 10.12

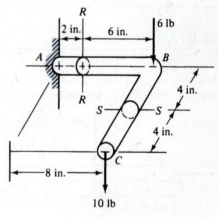

6 lb

10 lb

PROB. 10.13

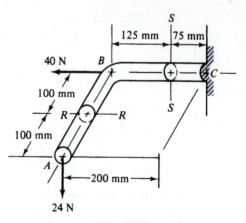

PROB. 10.14

10.3 STRESS

The internal reactions for a section are the resultant of distributed forces ΔT, which act on each small element of the cross-sectional area. In general, they vary in direction and magnitude, as shown in Fig. 10.6(a), for each small element of area. We are interested in intensity or force per unit area of these forces at various points of the cross section because their intensity affects the ability of the material to support loads and resist changes in shape. We will resolve the forces into components perpendicular (or normal) and parallel (or tangent) to the area on which they act, as shown in Fig. 10.6(b) and (c), respectively.

The intensity of the force normal to the area is called the *normal stress* and will be represented by the Greek lowercase letter σ (sigma). The average value of the normal stress over the area ΔA is defined by the equation

$$\sigma = \frac{\Delta F}{\Delta A}$$

where ΔF is the component of the force normal to the area ΔA. To find the stress at a point, the area ΔA must be decreased in size until it approaches a point. A normal stress is called *tensile stress* when it stretches the material on which it acts and *compressive stress* when it shortens the material on which it acts.

The intensity of the force parallel to the area is called the *shear stress* and will be represented by the Greek lowercase letter τ (tau). The average value of the shear stress over the area ΔA is defined by the equation

$$\tau = \frac{\Delta V}{\Delta A}$$

where ΔV is the component of the force parallel to the area ΔA. The shear stress at a point is found by decreasing the area ΔA in size until it approaches a point.

In cases where we have *uniform stress*, that is, stress that does not vary over the cross section, or we want to find an *average stress*, the normal stress can be found from the equation

$$\sigma = \frac{F}{A} \qquad \textbf{(10.1)}$$

and the shear stress from the equation

$$\tau = \frac{V}{A} \qquad \textbf{(10.2)}$$

The force F is the sum of the normal forces ΔF, the force V the sum of the parallel (tangential) forces ΔV, and A the area of the cross section or the sum of the elements of area ΔA of the cross section. The normal stress is caused by force components normal or perpendicular to the area of the cross section, and the shear stress is caused by force components parallel or tangent to the cross section. You should note several other points concerning stress:

1. The stress developed in an object is independent of the material from which that object is made. As Eqs. (10.1) and (10.2) demonstrate, stress is a computed quantity that is related only to the internal reactions and the geometry (area) of an object. Many students confuse the stress existing in an object with the *strength* of the material from which that object is formed. But *strength* is defined as the maximum allowable stress that a material can sustain; it is a characteristic, or property, of the material and is determined by experimental measurement.

2. Using stress to describe the internal loads in an object is much like the method of "unit pricing" found today in most supermarkets. If a 12-oz package of soap, for instance, costs $3.24 and a 15-oz package of the same product costs $4.35, then which package offers the lowest average cost to the consumer? To help shoppers evaluate these choices, the *unit* cost of each package must also be given: 27 cents *per ounce* for the small package, and 29 cents *per ounce* for the large. Unit pricing, then, eliminates the size effect when comparing two products. Similarly, if Material A can safely carry an axial load of 50,000 lb on a cross-sectional area of 2 square inches, and Material B can carry a load of 75,000 lb on an area of 3.75 square inches, which material is stronger, A or B? How much of B's increase in total load-carrying capability is due to its larger size and how much, if any, is due to the strength of the material itself? By comparing the stress, or unit load, in each material, we see that Material A is actually stronger at 25,000 lb *per square inch of area* as opposed to 20,000 lb *per square inch of area* for Material B. Stress, then, is used to eliminate size effects when comparing the loads applied to several different objects.

Units

Stresses are measured by units of force divided by units of area. The usual units for stress in the U.S. customary system are pounds per square inch, abbreviated as *psi*, or kip per square inch, abbreviated *ksi*. In metric, or SI units, stress is in newtons per square meter, abbreviated as N/m², or also

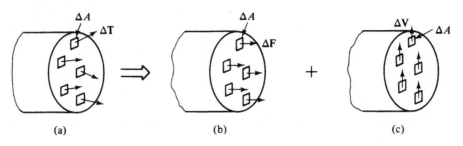

(a)   (b)   (c)

FIGURE 10.6

designated as a pascal (Pa). The pascal is a small unit of stress (1 psi = 6892.4692 Pa), and it may be more convenient to use the kilopascal (1 kPa = 10^3 Pa) or megapascal (1 MPa = 10^6 Pa = 1 N/mm^2).

10.4 STRESS IN AN AXIALLY LOADED MEMBER

Consider a straight two-force member of *uniform* cross section. The line of action of the loads passes through the centroid of the cross section as shown in Fig. 10.7(a). Near the *middle* of the bar at section B–B, a plane perpendicular to the line of action of the loads (axis of the bar), the internal reaction is a single axial force equal to P, as shown in Fig. 10.7(b) and (c). The normal stress distributed on the cross section is a *uniform* tension. From Eq. (10.1), the tensile stress

$$\sigma = \frac{P}{A} \quad \text{or} \quad \frac{\text{force}}{\text{area}} \left[\frac{\text{lb}}{\text{in.}^2} = \text{psi} \right] \quad \text{or} \quad \left[\frac{\text{N}}{\text{m}^2} = \text{Pa} \right]$$

$$(10.3)$$

where P is the axial force and A is the cross-sectional area. Because the stress distribution is uniform, Eq. (10.3) gives the actual stress for an axially loaded member. The distribution of stress is shown in Fig. 10.7(d) and (e). Notice that the stress multiplied by the cross-sectional area is equal to the axial force or stress resultant ($\sigma A = P$). No shear stress acts on the cross section.

If the directions of the forces in Fig. 10.7(a) are reversed, the member will be in compression. For a short member, where the compressive forces do not produce

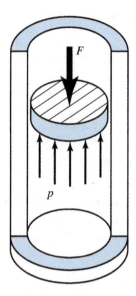

FIGURE 10.8. Axial *stress* is similar to the *pressure* created in a confined fluid. In this cutaway view, concentrated load F applied to a movable piston of area A (*shaded*) creates a pressure, p, in the liquid or gas beneath the piston. The formula for pressure, $p = F/A$, has the same form and units as Eq. (10.3) for stress.

buckling or collapse, the compressive stress can also be calculated by Eq. (10.3).

Note that the axial stress developed in a solid object is similar to the pressure created in a confined fluid. For instance, when concentrated load F is applied to a movable piston in the cylinder of Fig. 10.8, piston area A (shaded) distributes the load uniformly to the gas or liquid beneath the piston. This creates a pressure, p, in the fluid where, by definition, $p = F/A$ and has units that are identical to stress. Axial stress, then, can be thought of as a mechanical "pressure" caused by the internal reactions that develop in a solid object.

The three forms of Eq. (10.3) may be used in both the *analysis* and *design* of objects made from real materials:

1. When force P and area A are both known, this equation may be used directly to compute the stress created in an *existing object* (analysis), or to *select a material* whose allowable stress (see Secs. 10.8 and 11.2) is either equal to, or greater than, the computed value (design).

2. When area A and stress σ are known, load P carried by a member may be calculated as $P = \sigma \times A$. If σ is the maximum allowable stress for a particular material, this form of Eq. (10.3) yields the *maximum axial load* that should be applied (analysis).

3. When load P and stress σ are both known, the *required area* A can be determined using $A = P/\sigma$. Such information allows us to *select* or *specify* members of appropriate *size* for a given application (design).

Whichever form of Eq. (10.3) is used, you should be familiar with the formulas needed to compute values of A

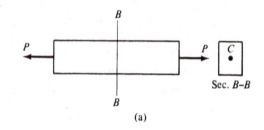

(a)

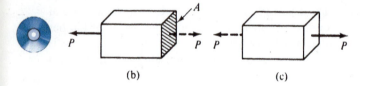

(b)   (c)

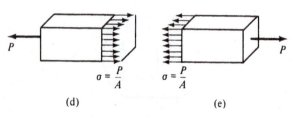

(d)   (e)

FIGURE 10.7

for simple *geometric shapes* (Table A.1 of the Appendix), as well as the sources of such values for standard *commercial shapes* such as pipe (Table A.8 of the Appendix), lumber (Table A.9 of the Appendix), and structural steel (Tables A.3–A.7 of the Appendix). The following examples demonstrate several common applications involving Eq. (10.3).

Example 10.2   The tensile link shown in Fig. 10.9(a) is flat, is 5 mm thick, and carries an axial load of $P = 1350$ N.

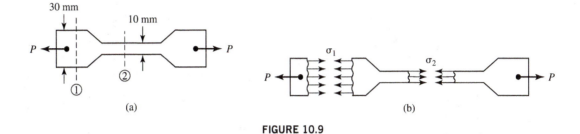

FIGURE 10.9

a.   Compute the tensile stress at sections 1 and 2 of the link.
b.   This link is to be made of plastic. From the list of common plastics listed as follows, select an appropriate material.

Plastic	Material Acronym	Allowable Stress (MPa)
High-density polyethylene	HDPE	15
Polycarbonate	PC	55
Polypropylene	PP	25
High-impact polystyrene	HIPS	30

c.   For your choice of material from (b), what is the maximum load, P_{max}, that can be carried by the link?

Solution: a.   Force P exists at all points along the length of this link and creates different stresses at cross sections having different areas. By making imaginary cuts at sections 1 and 2, we can envision the uniform stresses created at these sections by the concentrated load P [Fig. 10.9(b)]. Note that, like internal forces, the stresses on each pair of cut surfaces are equal in magnitude but opposite in direction. Because both cross sections are rectangular in shape and are equal to the product of the link's width and thickness at the appropriate section:

$$A_1 = 30 \text{ mm} \times 5 \text{ mm} = 150 \text{ mm}^2 = 150 \times 10^{-6} \text{ m}^2$$
$$A_2 = 10 \text{ mm} \times 5 \text{ mm} = 50 \text{ mm}^2 = 50 \times 10^{-6} \text{ m}^2$$

Then, from Eq. (10.3),

$$\sigma_1 = \frac{P}{A_1} = \frac{1350 \text{ N}}{150 \times 10^{-6} \text{ m}^2}$$
$$= 9 \times 10^6 \text{ Pa} = 9 \text{ MPa} \qquad \textbf{Answer}$$

$$\sigma_2 = \frac{P}{A_2} = \frac{1350 \text{ N}}{50 \times 10^{-6} \text{ m}^2}$$
$$= 27 \times 10^6 \text{ Pa} = 27 \text{ MPa} \qquad \textbf{Answer}$$

Note that maximum tensile stress occurs at section 2 and that both stresses are independent of the material from which this link is fabricated.

b.   The plastic selected must be able to endure a stress *at least as great* as the maximum value computed in part (a). Only PC and HIPS have allowable stresses equal to or greater than the 27 MPa created at section 2. Although either of these materials would be satisfactory, let us select HIPS because its apparent *strength* (30 MPa) is closest to the required value.

Choice: High-impact polystyrene   **Answer**

c. Our polystyrene link is stronger than necessary for the specified load of 1350 N. To find the maximum safe load, P_{max}, we simply compute the value of P required to create a stress in the link that is equal to the maximum allowable strength of the material:

$$P_{max} = \sigma_{max} \times A$$
$$= 30\,\text{MPa} \times (50 \times 10^{-6}\,\text{m}^2)$$
$$= 1500\,\text{N} \qquad \qquad \textbf{Answer}$$

Example 10.3 Southern pine has an allowable stress in compression of 6500 psi. If a short piece of this lumber must support an axial compressive load of 130,000 lb, as shown in Fig. 10.10(a),

a. Compute the minimum cross-sectional area required for this member and use the data in Table A.9 of the Appendix to specify an acceptable lumber size.
b. For your selection from (a), compute the actual stress developed in the piece.

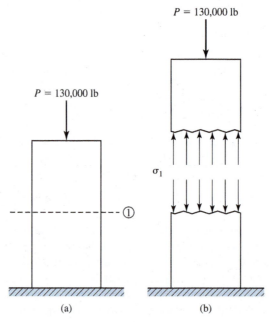

FIGURE 10.10

Solution: a. Making an imaginary cut at section 1 anywhere along the member, we see the uniform compressive stresses created on each internal surface [Fig. 10.10(b)]. To compute the *exact* area on which a force of 130,000 lb will create a stress of 6500 psi:

$$A = \frac{P}{\sigma} = \frac{130{,}000\,\text{lb}}{6500\,\text{lb/in.}^2} = 20.0\,\text{in.}^2 \qquad \qquad \textbf{Answer}$$

This is the *minimum* value for A; any smaller A would create a stress larger than 6500 psi. From Table A.9 of the Appendix, all lumber sizes larger than 4×6 have an acceptable cross-sectional area. We select the smallest (and most economical) size for which $A \geq 20.0\,\text{in.}^2$:

$$\text{Choice:} \quad 4 \times 8 (\text{nominal size}) @ A = 25.4\,\text{in.}^2 \qquad \qquad \textbf{Answer}$$

b. Because commercial lumber is available only in specific sizes, we are unlikely to find a size that has a cross-sectional area of exactly 20.0 in.^{2} Our 4×8 is larger than required; therefore, the actual stress developed will be less than 6500 psi:

$$\sigma = \frac{P}{A} = \frac{130{,}000\,\text{lb}}{25.4\,\text{in.}^2} = 5120\,\text{psi} \qquad \qquad \textbf{Answer}$$

Although Eq. (10.3) defines the theoretical value for average tensile or compressive stress, abrupt changes in cross-sectional area can create maximum stresses that are much higher than the average. These *stress concentrations* are discussed in Sec. 11.10. Also, although a stress computed using Eq. (10.3) may be allowable for static loads, a fluctuating load can cause progressive deterioration, and ultimate failure, in many materials. This phenomenon, known as the *fatigue* effect, is discussed in Sec. 11.11. For such conditions, experimental rather than theoretical values of allowable stress are often used.

10.5 AVERAGE SHEAR STRESS

There are many problems in which applied forces are transmitted from one body to another by developing internal reactions on planes that are parallel to the applied force. The planes are in shear, and the *average* shear stress can be found by dividing the shear force on the plane by the shear area of the plane. That is, applying Eq. (10.2), we have

$$\tau_{av} = \frac{V}{A_s} \quad \text{or} \quad \frac{\text{force}}{\text{area}}\left[\frac{\text{lb}}{\text{in.}^2} = \text{psi}\right] \quad \text{or} \quad \left[\frac{N}{m^2} = \text{Pa}\right]$$

(10.4)

An application of the concept of the average shear stress is shown in Fig. 10.11(a). Two I-bars are connected by a shear pin. We separate the bars by cutting the shear pin along two separate planes of area $A = \pi d^2/4$, as shown in Fig. 10.11(b) and (c). From equilibrium, each plane must have an internal reaction equal to $F/2$, which is parallel to the plane and therefore in shear. The shear stress for each plane is far from uniform, but an average value can be calculated from Eq. (10.4):

$$\tau_{av} = \frac{V}{A} = \frac{F/2}{\pi d^2/4}$$

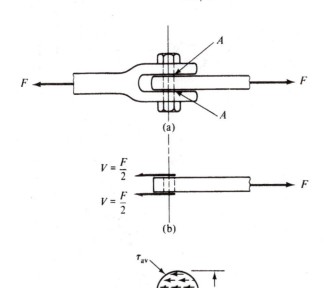

FIGURE 10.11

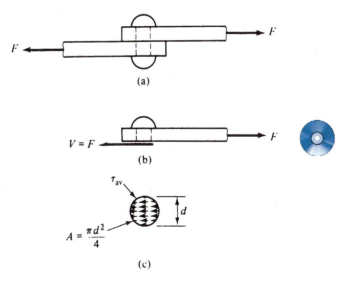

FIGURE 10.12

Because the shear pin has two planes that resist the force, it is said to be in *double shear*.

In another example, shown in Fig. 10.12(a), two plates are joined by a rivet. Separation of the plates requires that the rivet be cut along a single plane as shown in Fig. 10.12(b) and (c). The internal shear reaction from the equilibrium equation is equal to F, and the *average* value of the shear stress is

$$\tau_{av} = \frac{V}{A} = \frac{F}{\pi d^2/4}$$

The rivet is said to be in *single shear*.

10.6 BEARING STRESS

In certain structural and mechanical problems, one body is supported by another as shown in Fig. 10.13. The force intensity or normal stress between the two bodies in contact can be calculated by Eq. (10.1), if the resultant of the applied loads acts through the centroid C of the contact area. This

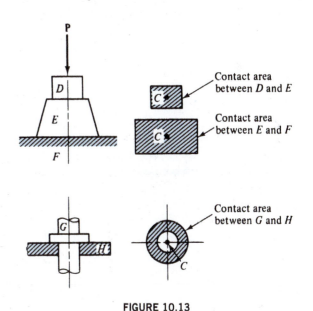

FIGURE 10.13

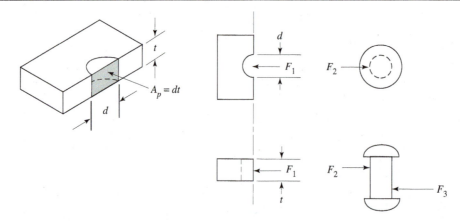

FIGURE 10.14

normal stress—called *bearing stress*—is found by dividing the resultant of the applied loads by the contact area. That is,

$$\sigma_b = \frac{P}{A_{contact}} \qquad (10.5)$$

When members are joined by connectors such as pins, rivets, or bolts, the connectors exert forces on the members to produce compressive or bearing stresses on the contact surface of the members.

Pins, Rivets, or Bolts

Consider again the single shear connection of Fig. 10.12. To separate the plates, we cut the top plate normal to the axis of

the member at the rivet hole, as shown in Fig. 10.14. The rivet exerts a force F_1 on the plate that is equal and opposite to the force F_2, which the plate exerts on the rivet. The stress produced is complex. However, in practice, a kind of nominal bearing or compressive stress is calculated based on an area equal to the *projection* of the rivet on the plate. This projected area $A_p = dt$, where d is the diameter of the rivet and t is the thickness of the plate. Accordingly, for a rivet, pin, or bolt the bearing stress

$$\sigma_b = \frac{F}{A_p} = \frac{F}{dt} \qquad (10.6)$$

Example 10.4   The collar bearing supports a load $P = 200$ kN as shown in Fig. 10.15(a). Find (a) the tensile stress in the shaft, (b) the shearing stress between the collar and the shaft, and (c) the bearing stress between the collar and the support.

Solution: a.   *Tensile stress:* The area of the shaft is shown in Fig. 10.15(b). The area $A = \pi d^2/4 = \pi(50)^2/4 = 1.963 \times 10^3$ mm^2. The tensile stress

$$\sigma = \frac{P}{A} = \frac{200 \times 10^3 \text{ N}}{1.963 \times 10^3 \text{ mm}^2} = 101.9 \text{ N/mm}^2 = 101.9 \text{ MPa} \qquad \textbf{Answer}$$

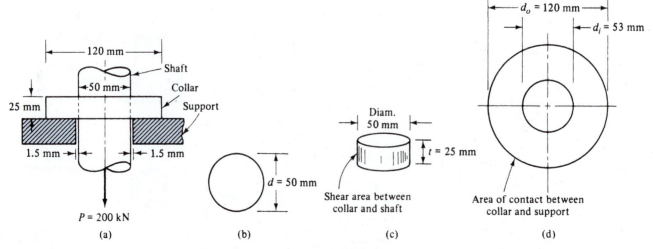

FIGURE 10.15

b. *Shear stress:* The shear area between the collar and the shaft is shown in Fig. 10.15(c). The area $A = \pi dt = \pi(50)(25) = 3.927 \times 10^3 \, \text{mm}^2$. The average shear stress

$$\tau_{av} = \frac{V}{A} = \frac{200 \times 10^3 \, \text{N}}{3.927 \times 10^3 \, \text{mm}^2} = 50.93 \, \text{N/mm}^2 = 50.9 \, \text{MPa} \qquad \textbf{Answer}$$

c. *Bearing stress:* The bearing area between the collar and the support is shown in Fig. 10.15(d). The area $A = \pi(d_o^2 - d_i^2)/4 = \pi[(120)^2 - (53)^2]/4 = 9.104 \times 10^3 \, \text{mm}^2$. The bearing stress

$$\sigma_b = \frac{P}{A} = \frac{200 \times 10^3 \, \text{N}}{9.104 \times 10^3 \, \text{mm}^2} = 21.97 \, \text{N/mm}^2 = 22.0 \, \text{MPa} \qquad \textbf{Answer}$$

Example 10.5   A steel column rests on a steel base plate that is supported by a concrete pier that is supported by a concrete footing, as shown in Fig. 10.16(a). Find the bearing stress between (a) the steel base plate and the pier, (b) the pier and footing, and (c) the footing and soil. Concrete weighs 150 lb/ft³.

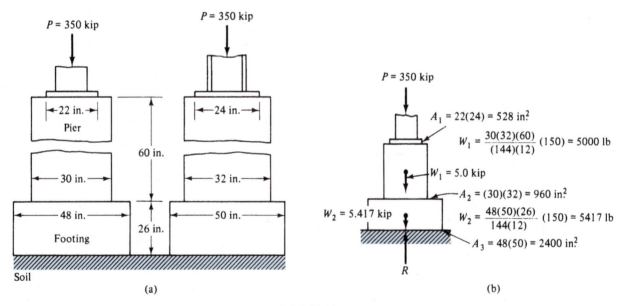

FIGURE 10.16

Solution: The free-body diagram is shown in Fig. 10.16(b). The weight of the pier and footing and the contact areas have been calculated and shown on the figure.

a. The force between the base plate and pier is $P = 350 \, \text{kip}$. The bearing stress

$$\sigma_b = \frac{P}{A} = \frac{350}{528} = 0.663 \, \text{ksi} \qquad \textbf{Answer}$$

b. The force between the pier and footing is $P = P + W_1 = 350 + 5 = 355 \, \text{kip}$. The bearing stress

$$\sigma_b = \frac{P}{A} = \frac{355}{960} = 0.370 \, \text{ksi} \qquad \textbf{Answer}$$

c. The force between the footing and the soil is $R = P + W_1 + W_2 = 350 + 5 + 5.4 = 360.4 \, \text{kip}$. The bearing stress

$$\sigma_b = \frac{P}{A} = \frac{360.4}{2400} = 0.150 \, \text{ksi} \qquad \textbf{Answer}$$

Example 10.6   A punch press is used to punch a 1.5-in.-diameter hole in a 0.35-in.-thick steel plate. Determine the force exerted by the press on the plate when the average shear resistance to punching of the steel plate is 58,000 psi.

Solution: The punch shears out a cylinder. The shear area is the cylindrical surface area of the cylinder, which has a diameter of 1.5 in. and a height equal to the thickness of the plate, which is 0.35 in., as shown in Fig. 10.17.

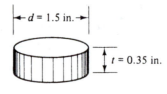

FIGURE 10.17

The area is given by

$$A = \pi \, dt = \pi(1.5)(0.35) = 1.650 \text{ in.}^2$$

From Eq. (10.4),

$$P_{\text{punch}} = V = A\tau = 1.650(58,000)$$
$$= 95,700 \text{ lb} \qquad\qquad \textbf{Answer}$$

Example 10.7   A hollow cylinder is to be designed to support a compressive load of 650 kN. The allowable compressive stress σ_a = 69.2 MPa. Compute the outside diameter of the cylinder if the wall thickness is 50 mm.

Solution: Solving for the required area from Eq. (10.7), we have

$$A_{\text{req}} = \frac{P}{\sigma_a} = \frac{650 \times 10^3}{69.2} = 9.393 \times 10^3 \text{ mm}^2$$
$$= 9393 \text{ mm}^2$$

The compression area is shown in Fig. 10.18. The area is determined by subtracting the inside circle from the outside circle as shown in the figure.

$$A = \frac{\pi}{4}(d_o^2 - d_i^2) = 9393 \text{ mm}^2$$

$$d_o^2 - d_i^2 = \frac{4}{\pi}(9393) = 11,960$$

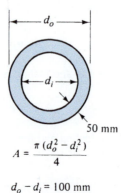

$$A = \frac{\pi(d_o^2 - d_i^2)}{4}$$

$$d_o - d_i = 100 \text{ mm}$$

FIGURE 10.18

Factoring the left-hand side of this equation yields

$$(d_o - d_i)(d_o + d_i) = 11{,}960 \qquad \text{(a)}$$

Because the wall thickness is 50 mm,

$$d_o - d_i = 100.0 \qquad \text{(b)}$$

Dividing Eq. (a) by Eq. (b), we have

$$d_o + d_i = \frac{11{,}960}{100} = 119.60 \qquad \text{(c)}$$

Adding Eqs. (b) and (c) yields

$$2d_o = 219.60$$
$$d_o = 109.8\,\text{mm} \qquad \textbf{Answer}$$

10.7 PROBLEMS INVOLVING NORMAL, SHEAR, AND BEARING STRESS

As we saw in the preceding examples, the equations for normal stress and shear stress are quite simple. For more involved problems, difficulty may arise in applying the equations either in finding the internal reaction from the equation of equilibrium or in visualizing the area that is acted on by normal or shear stress. We will consider several examples that will help clarify the problem areas.

Example 10.8   A timber frame or truss shown in Fig. 10.19(a) supports a load P of 50 kN. Find (a) the normal stress in all the members, (b) the horizontal shearing stress at the notched ends of timber AC, and (c) the bearing stress of C on plate D that measures 0.25 m × 0.25 m × 0.025 m.
The shear stress in part (b) is not to be confused with the horizontal shear stress due to beam bending. That stress is considered in Chap. 14.

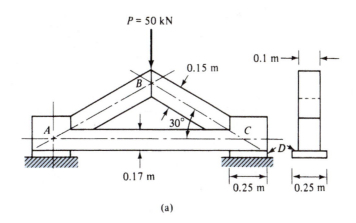

(a)

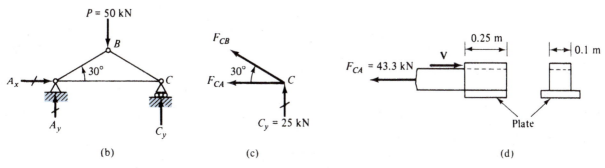

(b)   (c)   (d)

FIGURE 10.19

Solution: We idealize the problem by replacing the timber frame with a pin-connected simple truss, as shown in Fig. 10.19(b). From the equilibrium equations, $A_x = 0$, $A_y = 25.0$ kN, and $C_y = 25.0$ kN. The truss will be solved by the method of joints. A free-body diagram of joint C is shown in Fig. 10.19(c). For equilibrium of concurrent forces,

$$\Sigma F_y = 0 \quad 25.0 + F_{CB}\sin 30° = 0$$

$$F_{CB} = -\frac{25}{\sin 30} = -50\,\text{kN}$$

and

$$\Sigma F_x = 0 \quad -F_{CA} - F_{CB}\cos 30° = 0$$

$$F_{CA} = +50\cos 30° = 43.3\,\text{kN}$$

The force in AB is the same as the force in BC from symmetry. Therefore,

$$F_{AB} = -50\,\text{kN}$$

a. The cross-sectional area of members AB and BC are $0.15 \times 0.1 = 0.015\,\text{m}^2$ and member AC is $0.17 \times 0.1 = 0.017\,\text{m}^2$. The normal stress in members AB and BC is

$$\sigma = \frac{F_{AB}}{A} = \frac{-50\,\text{kN}}{0.015\,\text{m}^2} = -3333\,\text{kPa} = 3.33\,\text{MPa} \quad \text{(compression)} \qquad \textbf{Answer}$$

The normal stress in member AC is

$$\sigma = \frac{F_{AC}}{A} = \frac{43.3\,\text{kN}}{0.017\,\text{m}^2} = 2547\,\text{kPa} = 2.55\,\text{MPa} \quad \text{(tension)} \qquad \textbf{Answer}$$

b. The horizontal shear force acting at A and C must be equal to 43.3 kN and the shear area $0.25\,\text{m} \times 0.1\,\text{m} = 0.025\,\text{m}^2$, as shown in Fig. 10.19(d). Therefore,

$$\tau_{av} = \frac{V}{A} = \frac{43.3\,\text{kN}}{0.025\,\text{m}^2} = 1732\,\text{kPa} = 1.73\,\text{MPa} \qquad \textbf{Answer}$$

c. The force on the plate is equal to the reaction $C_y = 25$ kN, and the area in contact between the plate and member AC is $0.25\,\text{m} \times 0.1\,\text{m} = 0.025\,\text{m}^2$. Therefore, the bearing stress

$$\sigma_b = \frac{P}{A} = \frac{25\,\text{kN}}{0.025\,\text{m}^2} = 1000\,\text{kPa} = 1.0\,\text{MPa} \qquad \textbf{Answer}$$

Example 10.9 A horizontal force $P = 1.9$ kip is applied at pin B of the two pin-joined bars shown in Fig. 10.20(a). The pins have a diameter $d = 0.5$ in. at all connections. Determine the maximum average (a) normal stress in bars AB and BC, (b) the bearing stress at the pins A and C, and (c) the shear stress in the pins at A and C.

Solution: A free-body diagram of joint B, together with the components of the forces, is shown in Fig. 10.20(b). From equilibrium,

$$\Sigma F_y = 0 \quad \frac{F_{AB}}{\sqrt{2}} + \frac{F_{BC}}{\sqrt{5}} = 0 \qquad \textbf{(a)}$$

$$\Sigma F_x = 0 \quad \frac{2F_{BC}}{\sqrt{5}} - \frac{F_{AB}}{\sqrt{2}} = 1.9 \qquad \textbf{(b)}$$

Adding Eqs. (a) and (b), we have

$$\frac{3}{\sqrt{5}}F_{BC} = 1.9 \quad \text{or} \quad F_{BC} = \frac{\sqrt{5}}{3}(1.9) = 1.416\,\text{kip}$$

Substituting for F_{BC} in Eq. (a), we have

$$F_{AB} = \frac{-\sqrt{2}}{3}(1.9) = -0.896\,\text{kip}$$

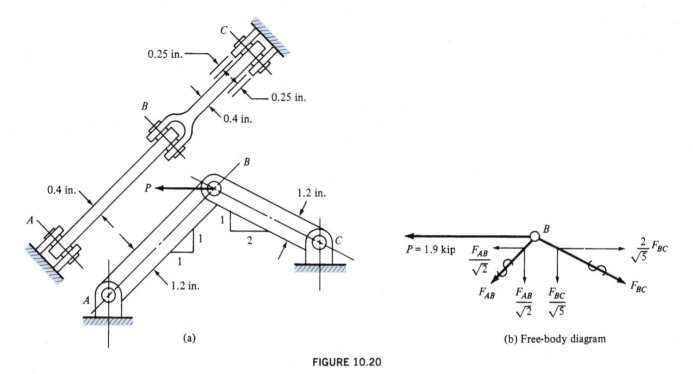

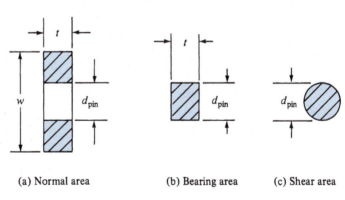

(a)   (b) Free-body diagram

FIGURE 10.20

a. The cross section of members AB and BC at pins A and C are shown in Fig. 10.21(a). The cross-sectional area $A = (w - d)t = (1.2 - 0.5)(0.4) = 0.280$ in.^{2} Therefore, the normal stress

$$\sigma_{AB} = \frac{F_{AB}}{A} = \frac{-0.896}{0.280} = -3.20 \text{ ksi} \qquad \sigma_{BC} = \frac{F_{BC}}{A} = \frac{1.416}{0.280} = 5.06 \text{ ksi} \qquad \textbf{Answer}$$

(a) Normal area   (b) Bearing area   (c) Shear area

FIGURE 10.21

b. The bearing area at pins A and C is shown in Fig. 10.21(b). The area $A = dt = (0.5)(0.4) = 0.2$ in.^{2} and the bearing stress

$$(\sigma_b)_A = \frac{F_{AB}}{A} = \frac{0.896}{0.2} = 4.48 \text{ ksi} \qquad (\sigma_b)_C = \frac{F_{BC}}{A} = \frac{1.416}{0.2} = 7.08 \text{ ksi} \qquad \textbf{Answer}$$

c. A single shear area for the pins at A and C is shown in Fig. 10.21(c). Because the pins are in double shear, the area $A = 2\pi d^2/4 = 2\pi(0.5)^2/4 = 0.393$ in.^{2}, and the shear stresses at A and C are

$$\tau_A = \frac{F_{AB}}{A} = \frac{0.896}{0.393} = 2.28 \text{ ksi} \qquad \tau_C = \frac{F_{BC}}{A} = \frac{1.416}{0.393} = 3.60 \text{ ksi} \qquad \textbf{Answer}$$

PROBLEMS

10.15 The loads shown are applied axially on the round steel bar that has a diameter of $d = 50$ mm. Determine the maximum normal stress in each section of the bar if $P_1 = 200$ kN, $P_2 = 150$ kN, $P_3 = 70$ kN, and $P_4 = 120$ kN.

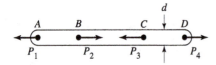

PROB. 10.15

10.16 Two round bars are welded together and loaded axially as shown. Determine the tensile stress at the middle of each bar if $d_1 = 2$ in., $d_2 = 1.2$ in., $P_1 = 15$ kip, and $P_2 = 5$ kip.

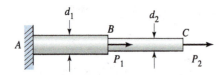

PROB. 10.16

10.17 Two square bars are welded together and loaded axially as shown. Determine the maximum normal stress in each bar if $w_1 = 40$ mm, $w_2 = 50$ mm, $P_1 = 130$ kN, $P_2 = 65$ kN, and $P_3 = 260$ kN.

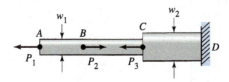

PROB. 10.17

10.18 A round aluminum tube has a wall thickness of 0.150 in. and an outside diameter of 1.5 in. The tube is used as a short compression member. If the axial load on the member is 15.5 kip, determine the axial stress.

10.19 A short wooden post 320 mm by 320 mm supports a compressive load of 450 kN. (a) Determine the axial stress in the post. (b) If the allowable compressive stress for the wood is 5.76 MPa, what is the maximum load that this post can support?

10.20 The failure tensile load for a steel wire 1.0 mm in diameter is 540 N. What is the failure tensile stress?

10.21 A 7/8-in.-diameter steel rod carries a load of 22,500 lb in tension. (a) What is the tensile stress in this rod? (b) If the maximum allowable tensile stress for the steel is 36,000 psi, what is the minimum required diameter for the rod?

10.22 A tension member supports an axial force P. What is the stress in the member if the member and load are as follows: (a) S10 × 35, $P = 200$ kip; (b) two L6 × 4 × 5/8, $P = 110$ kip?

10.23 A 400-N weight is supported by three wires joined together at C, as shown. If the diameter of the wire is 5 mm, determine the tensile stress in each wire.

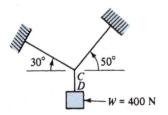

PROB. 10.23

10.24 A portable storage bin whose total weight is 18,000 lb is to be supported on four short legs made of L2 × 2 × 1/8 steel angle (see Table A.6 of the Appendix) as shown in Prob. 10.26(a).
a.   Assuming that one-fourth of the total bin weight is carried by each leg, find the compressive stress developed in the steel.
b.   If the maximum allowable compressive stress in this steel is 12 ksi, what is the maximum allowable bin weight?
c.   The bin will rest on a concrete floor for which the allowable bearing stress is 950 psi. To achieve that stress, a square metal plate is to be welded to the bottom of each leg, changing its "footprint" from that of Prob. 10.26(b) to that of Prob. 10.26(c). Determine the minimum side dimension, d, of these plates for the bin weight computed in (b).

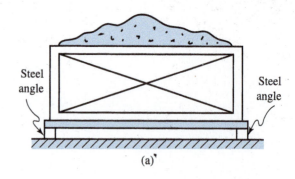

(a)

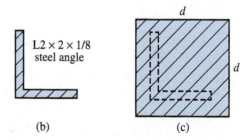

(b)   (c)

PROB. 10.24

10.25 A short T-shaped beam *AB* is supported at *C* by an eyebar and at *B* by a pin. The eyebar has a diameter $d = 40$ mm, and the pin at *D* has a diameter $d_p = 25$ mm. The beam supports a load $P = 45$ kN. Determine (a) the axial stress in *CD* and (b) the average shear stress in the pin at *D*.

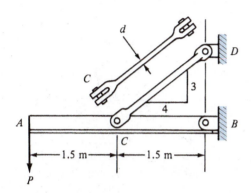

PROB. 10.25

10.26 A 1-1/4-in.-square steel punch is to push through an aluminum plate that is supported on a die, thus forming a square opening in the plate.
  a. If the allowable compressive stress in the steel is 24 ksi, what is the maximum punching force that this die can exert on the plate?
  b. If the aluminum has an ultimate shear strength of 6500 psi, find the maximum plate thickness, *t*, through which a hole may be punched.

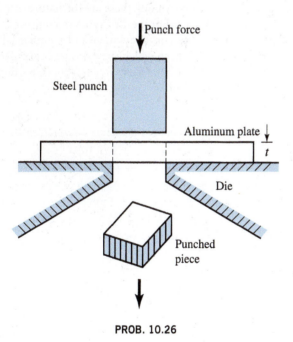

PROB. 10.26

10.27 A punch with a diameter $d = 25$ mm punches a hole in a steel plate of thickness $t = 15$ mm with a force $P = 330$ kN. Determine (a) the compressive stress in the punch and (b) the shear stress in the plate.

10.28 A concrete column with cross-sectional area $A = 120$ in.^{2} supports an axial compressive load $P = 9.0$ kip, including the weight of the column. The column is supported by a square spread footing with a depth of $d = 6$ in. and measuring $w = 25$ in. on each side. Determine (a) the compressive stress in the column and (b) the bearing stress in the soil. Concrete weighs 150 lb/ft^3.

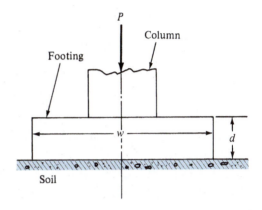

PROB. 10.28

10.29 The timber truss shown supports a load of $P = 10$ kip. Find (a) the normal stress in all the members, (b) the horizontal shear stress in timber *AC* at *C*, and (c) the bearing stress of *AC* on plate *D*, which measures 5 in. by 5 in. and is 0.25 in. thick.

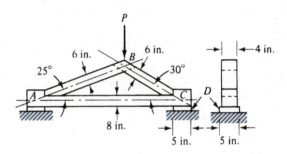

PROB. 10.29

10.30 A plate is supported by a 9/16-in.-diameter bolt and bracket shown. If the load on the plate is 4150 lb, determine the average shear stress in the bolt.

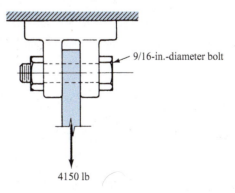

PROB. 10.30

10.31 The round hanger rod illustrated is supported by a floor plank. If the load on the rod is 27,600 lb, find (a) the tensile stress in the rod and (b) the bearing stress between the washer and the floor plank.

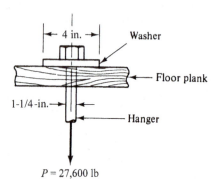

$$P = 27{,}600 \text{ lb}$$

PROB. 10.31

10.8 ALLOWABLE STRESS, FACTOR OF SAFETY

The *allowable stress* is the maximum stress that is considered safe for a material to support under certain loading conditions. The stress may be used to design load-supporting members of structures and machines. Allowable stress values are determined by tests and from experience gained from the performance of previous designs under service conditions. Allowable stress is also sometimes called the *working* or *design* stress. With the allowable normal stress σ_a known, we can solve Eq. (10.3) for the required area in tension or compression

$$A_{\text{req}} = \frac{P}{\sigma_a} \qquad (10.7)$$

and with the allowable shear stress τ_a known, we can solve Eq. (10.4) for the required area in shear

$$A_{\text{req}} = \frac{V}{\tau_a} \qquad (10.8)$$

Tests are performed on material to determine the material's resistance to stress. In one of the usual tests, a round bar is subjected to an axial tensile load. (See Sec. 11.2.) As the load is increased the bar stretches (increases in length). Finally, an *ultimate* or maximum load is applied, and the bar breaks or begins to support a smaller load. The *ultimate stress* is found by dividing the ultimate load by the *original* cross-sectional area of the bar. The ultimate stress is also sometimes called the *ultimate strength*. The *factor of safety* is defined as the ratio of some load that represents the strength for the member to the actual load existing in the member. That is,

$$\text{F.S.} = \frac{allowable\ load\ for\ the\ member}{actual\ load\ in\ the\ member} \qquad (10.9)$$

For tensile or compressive members, where the load is equal to stress multiplied by area, the ratio of the loads is identical to the ratio of stresses. Accordingly, for a tension member, a *factor of safety* that is based on the ultimate stress is equal to the ratio of the ultimate stress to the actual stress. Thus,

$$\text{F.S.} = \frac{\sigma_u}{\sigma_a} \qquad (10.10)$$

For ductile materials such as low-carbon steel, a factor of safety could be based on a stress value called the *yield point stress* (see Sec. 11.2), whereas a factor of safety based on the ultimate stress could be used for a brittle material such as gray cast iron. For long columns, the factor of safety is based on the buckling stress (see Chapter 17).

Values of the factor of safety used to design members depend on many factors. Among these are the nature of the loads, variation in material properties, types of failures possible, uncertainty in analysis, consequence of member failure, and the environment to which the member is exposed. Factors of safety range in value from over 1 to 20, with values between 3 and 15 common.

Example 10.10   The bell crank shown in Fig. 10.22(a) is in equilibrium with loads P and Q. The ultimate tensile stress is 400 MPa, and the ultimate shear stress is 250 MPa. The pins at A, B, and C are in double shear. If the factor of safety is 2.8, determine (a) the allowable load P, (b) the allowable load Q, (c) the required diameter of the rod d_2 at C, and (d) the diameter of the pins at A, B, and C.

Solution: The allowable stresses in tension and shear are given by Eq. (10.10):

$$\sigma_a = \frac{\sigma_u}{\text{F.S.}} = \frac{400}{2.8} = 142.9 \text{ MPa} \qquad \tau_a = \frac{\tau_u}{\text{F.S.}} = \frac{250}{2.8} = 89.3 \text{ MPa}$$

$$= 142.9 \text{ N/mm}^2 \qquad\qquad\qquad = 89.3 \text{ N/mm}^2$$

a.   At A, the area of the rod $A = \pi d_1^2/4 = \pi(15)^2/4 = 176.7 \text{ mm}^2$. The allowable load P_a from Eq. (10.3) is given by

$$P_a = \sigma_a A = 142.9(176.7) = 25.25 \times 10^3 \text{ N}$$

$$= 25.3 \text{ kN} \qquad\qquad\qquad\qquad\qquad\qquad \textbf{Answer}$$

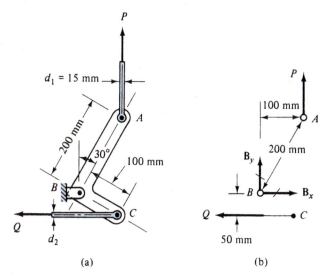

FIGURE 10.22

b. From the free-body diagram of the bell crank shown in Fig. 10.22(b),

$$\Sigma M_B = 0 \qquad 100P - 50Q = 0$$
$$Q = 2P = 2(25.25) = 50.5 \text{ kN} \qquad \textbf{Answer}$$

$$\Sigma F_x = 0 \qquad B_x - Q = 0$$
$$B_x = 50.5 \text{ kN}$$

$$\Sigma F_y = 0 \qquad B_y = P = 25.25 \text{ kN}$$
$$B = (B_x^2 + B_y^2)^{1/2} = [(25.25)^2 + (50.5)^2]^{1/2}$$
$$= 56.46 \text{ kN}$$

c. From Eq. (10.7), $A_{req} = F/\sigma_a$; therefore, the required area of the rod at C is given by

$$A_{req} = \frac{\pi d_2^2}{4} = \frac{Q}{\sigma_a} = \frac{50.5 \times 10^3}{142.9} = 3.534 \times 10^2 \text{ mm}^2$$

$$d_2^2 = \frac{4}{\pi}(3.534 \times 10^2) = 4.50 \times 10^2 \text{ mm}^2$$

$$d_2 = 21.2 \text{ mm} \qquad \textbf{Answer}$$

d. For the diameter of the pins, from Eq. (10.8),

$$A_{req} = \frac{2\pi d^2}{4} = \frac{V}{\tau_a} = \frac{V}{89.3}$$
$$d^2 = 7.129 \times 10^{-3} V$$

At A:
$$d_A^2 = 7.129 \times 10^{-3}(25.25 \times 10^3) = 1.800 \times 10^2 \text{ mm}^2$$
$$d_A = 13.4 \text{ mm} \qquad \textbf{Answer}$$

At C:
$$d_C^2 = 7.129 \times 10^{-3}(50.5 \times 10^3) = 3.600 \times 10^2 \text{ mm}^2$$
$$d_C = 18.97 = 19.0 \text{ mm} \qquad \textbf{Answer}$$

At B:
$$d_B^2 = 7.129 \times 10^{-3}(56.46 \times 10^3) = 4.025 \times 10^2 \text{ mm}^2$$
$$d_B = 20.1 \text{ mm} \qquad \textbf{Answer}$$

PROBLEMS

10.32 A circular bar supports an axial force of 35 kN. If the compressive stress in the bar is not to exceed 75.8 MN/m², determine the diameter of the bar.

10.33 A circular tube has an outer diameter of 1.2 in. and an inner diameter of 1.0 in. The allowable stress is 24,000 psi in tension and 16,000 psi in compression. Determine the allowable tensile and compression loads.

10.34 A steel wire must support a tensile load of 2000 N. The allowable stress $\sigma_a = 165$ MN/m^2. Determine the required diameter of the wire.

10.35 The steel column shown supports an axial compressive load of 105 kip. The allowable compressive stress in the steel is 20.5 ksi, and the allowable bearing stress in the concrete footing is 700 psi. (a) Determine the required cross-sectional area of the column. (b) If this column will be made of standard-weight steel pipe, select the smallest nominal diameter that may be used. (c) For the pipe size selected in (b), compute the *actual* stress developed in the column. (d) Calculate the required length of the steel base plate if the width of this plate is 10 in.

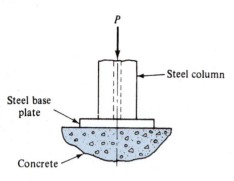

PROB. 10.35

10.36 A short compression member is fabricated from two steel pipes as shown. The 4- and 6-in. pipes have cross-sectional areas $A_4 = 3.17$ in.^{2} and $A_6 = 5.53$ in.^{2}, and the allowable compressive stress in the steel is 21.6 ksi. (a) What is the allowable load P_6 if $P_4 = 22$ kip? (b) What is the allowable load P_4 if $P_6 = 40$ kip?

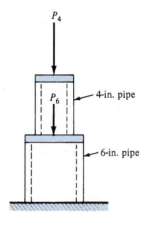

PROB. 10.36

10.37 Two wooden tension members are spliced together by wooden cover plates that are glued over the entire surface in contact. If $P = 7450$ lb and the average shear stress in the glue is 160 psi, what is the length d of the glued surfaces?

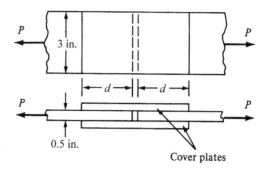

PROB. 10.37

10.38 Two $3/8 \times 6$-in. plates are joined by $3/4$-in.-diameter rivets as shown. Determine the maximum allowable load P on the plates if the allowable stress in tension is 22 ksi, the allowable stress in shear is 29 ksi, and the allowable stress in bearing is 58 ksi.

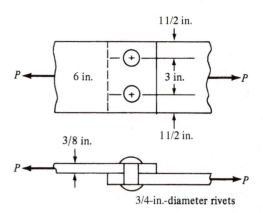

PROB. 10.38

10.39 The knuckle joint shown carries a load, P, of 62 kN and is assembled using a pin made of 1060-0 aluminum alloy. Find the minimum pin diameter, d, if the pin must have a factor of safety of 1.3 based on the material's ultimate shear strength of 42.5 MPa.

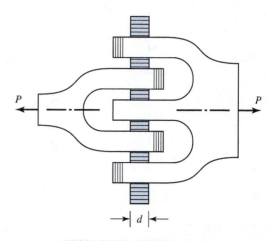

PROB. 10.39 and PROB. 10.40

10.40 Same as problem 10.39, but $P = 12,800$ lb, and ultimate shear strength is 6200 psi.

10.41 The bracket shown below supports a load $P = 4.5$ kN. The allowable stress is 165 MN/m^2 in tension and 110 MN/m^2 in compression. Determine the required cross-sectional areas of members AB and BC.

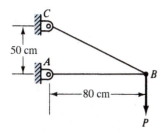

PROB. 10.41

10.42 The hoist shown supports a load $P = 10$ kip. Member BC is a round steel rod, and member AB is a square wooden post. The allowable stress is 20 ksi for steel in tension and 800 psi for wood in compression parallel to the grain. Determine the diameter of the steel rod and the dimensions of the square wooden post.

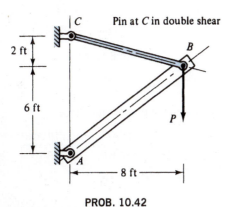

PROB. 10.42

10.43 The short beam ABC is supported by a pin at C and the link BD that is pinned at B and D. All pins are in single shear and have diameters $d = 15$ mm. If the steel for the link and supports has an ultimate normal stress of 400 MPa and the steel in the pins has an ultimate shear stress of 240 MPa, determine the largest allowable load P on the beam for which the factor of safety is 3.0.

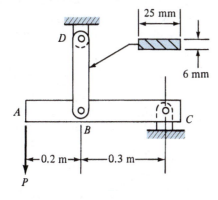

PROB. 10.43

10.44 The bell-crank lever shown is used to change the direction and magnitude of a force. The pin at B has a diameter of 0.65 in., and the pins at A and D have diameters of 0.5 in. The ultimate normal stress is 36 ksi in the link AD, and the ultimate shear stress is 14.5 ksi in the pins. What is the allowable load P if the factor of safety is 3.5?

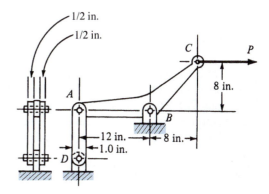

PROB. 10.44

10.45 The timber beam shown is supported by bearing plates at A and B, and the beam supports a load P on a bearing plate at C. The ultimate bearing stress for the beam perpendicular to the grain is 6.03 MPa and parallel to the grain 26.5 MPa. If the factor of safety is 3.5 and the load on the beam is limited by bearing stress, what is the allowable load?

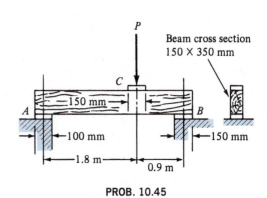

PROB. 10.45

10.46 A steel eyebar with a cross section measuring 1/2 in. by 4 in. supports a load $P = 44$ kip and is held at the end by a pin as shown. If the ultimate normal stress for the eyebar is 68 ksi and the ultimate shear stress for the pin is 48 ksi, what is the required diameter of the pin to make its factor of safety in shear equal to the factor of safety for the eyebar in tension?

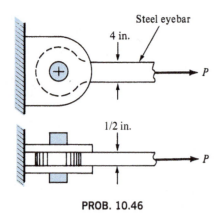

Steel eyebar

4 in.

P

1/2 in.

P

PROB. 10.46

10.47 A concrete wall footing with a width of 0.9 m and a depth of 0.6 m supports a centrally placed 0.6-m-thick wall. The ultimate bearing stress for the soil under the footing is 475 kN/m². What is the height of the wall if a factor of safety of 5.0 is used? Concrete weighs 23.5 kN/m³.

10.9 FURTHER ANALYSIS OF AXIAL LOADS: STRESSES ON OBLIQUE SECTIONS

In Sec. 10.4, we considered the stress on a cross section perpendicular to the axis of the bar. In this section, we will consider the stress on a plane that is perpendicular to the plane of the bar but oblique (inclined) with the axis. A straight two-force member of uniform cross-sectional area A is shown in Fig. 10.23(a)

To find the internal reactions on the oblique section B–B, we isolate the bar to the left of the section [Fig. 10.23(b)]. The resultant force on the oblique section must be equal to P for equilibrium. The force is inclined at an angle of θ with the normal to the oblique section. Therefore, the normal force $F = P\cos\theta$ and the shear force $V = P\sin\theta$. The area of the oblique cross-sectional area is equal to the cross-sectional area divided by the cos θ. That is,

$$A_\theta = \frac{A}{\cos\theta}$$

Therefore, the normal stress on the oblique plane [Fig. 10.23(c)]

$$\sigma_\theta = \frac{F}{A_\theta} = \frac{P\cos\theta}{A/\cos\theta} = \frac{P\cos^2\theta}{A} \quad (10.11)$$

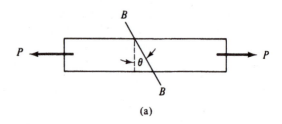

(a)

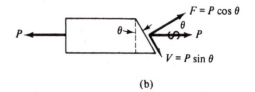

(b)

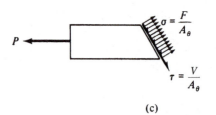

(c)

FIGURE 10.23

and the shear stress on the oblique plane

$$\tau_\theta = \frac{V}{A_\theta} = \frac{P\sin\theta}{A/\cos\theta} = \frac{P\sin\theta\cos\theta}{A}$$

Because $\sin\theta\cos\theta = (\sin 2\theta)/2$,

$$\tau_\theta = \frac{P\sin 2\theta}{2A} \quad (10.12)$$

The maximum *normal* stress from Eq. (10.11) occurs when $\theta = 0°$. That is, $\sigma_\theta = P/A$. The maximum value of the *shear* stress from Eq. (10.12) occurs when $\theta = 45°$ and is equal to

$$\tau_{45°} = \tau_{max} = \frac{P}{2A}$$

that is, *half* the value of the *maximum tensile or compressive* stress. Notice that the tensile or compressive stress on a plane when $\theta = 45°$ has the same value as the shear stress on the same plane.

Example 10.11 A steel bar 0.75 in. in diameter has an axial load of 12,000 lb. Determine (a) the normal and shearing stresses on a plane through the bar that forms an angle of 30° with the cross-section and (b) the maximum normal and shearing stresses in the bar.

Solution: The bar is shown in Fig. 10.24(a). A plane whose normal makes an angle of 30° with the axis of the bar cuts through the bar, and a free-body diagram of the bar to the left of the cut is shown in Fig. 10.24(b). Included are the normal and shear components of the internal reactions. The cross-sectional area perpendicular to the

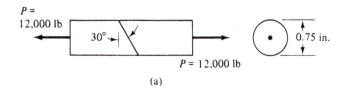

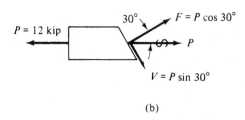

(a)

(b)

FIGURE 10.24

axis of the bar $A = \pi d^2/4 = \pi(0.75)^2/4 = 0.4418\,\text{in.}^2$ The area along an oblique plane of 30° is $A_\theta = A/\cos\theta$ or $A_{30} = 0.4418/\cos 30° = 0.5101\,\text{in.}^2$ The normal internal reactions $F = P\cos 30° = 12{,}000\cos 30° = 10{,}392\,\text{lb}$ and $V = P\sin 30° = 12{,}000\sin 30° = 6000\,\text{lb}$. Therefore,

a. The normal stress

$$\sigma_{30} = \frac{F}{A_{30}} = \frac{10{,}390}{0.5101} = 20{,}400\,\text{psi} \qquad\qquad \textbf{Answer}$$

$$\tau_{30} = \frac{V}{A_{30}} = \frac{6000}{0.5101} = 11{,}760\,\text{psi} \qquad\qquad \textbf{Answer}$$

b. The maximum normal and shearing stresses can be calculated by a formula.

$$\sigma_{max} = \sigma_0 = \frac{P}{A} = \frac{12{,}000}{0.4418} = 27{,}200\,\text{psi} \qquad\qquad \textbf{Answer}$$

$$\tau_{max} = \tau_{45°} = \frac{P}{2A} = \frac{12{,}000}{2(0.4418)} = 13{,}580\,\text{psi} \qquad\qquad \textbf{Answer}$$

Example 10.12   A short compression block with a cross section that measures 0.15 m × 0.20 m is loaded with an axial load P. The allowable compressive stress is $\sigma = 10{,}000\,\text{kN/m}^2$, and the allowable shear stress is $\tau = 830\,\text{kN/m}^2$. Find the maximum allowable load.

Solution: The formula for the maximum normal and shear stress may be used. The cross-sectional area

$$A = 0.15(0.20) = 0.03\,\text{m}^2$$

For compression,

$$\sigma_{max} = \frac{P}{A} \qquad P_a = \sigma_a A = 10{,}000(0.03) = 300\,\text{kN}$$

For shear,

$$\tau_{max} = \frac{P}{2A} \qquad P_a = \tau_a 2A = 830(2)(0.03) = 49.8\,\text{kN}$$

The allowable load is therefore based on the maximum shear stress. The allowable load

$$P_a = 49.8\,\text{kN} \qquad\qquad \textbf{Answer}$$

If the load on the compression block is increased until failure occurs, it will occur in shear along a plane that makes an angle of 45° with the axis of the compression block.

Example 10.13 The plate shown in Fig. 10.25(a) supports stresses $\sigma_x = 800\ \text{kN/m}^2$ and $\sigma_y = 400\ \text{kN/m}^2$. Determine the normal and shear stress on an oblique plane from A to D.

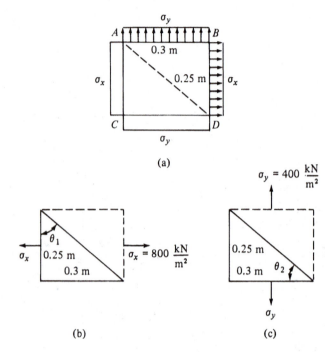

(a)

(b)   (c)

FIGURE 10.25

Solution: We use superposition and apply Eqs. (10.11) and (10.12) first in the x direction and then in the y direction.

The x direction: From Fig. 10.25(b), $\tan\theta_1 = 0.3/0.25$; therefore, $\theta_1 = 50.2°$. From Eqs. (10.11) and (10.12),

$$\sigma_{\theta_1} = \frac{P\cos^2\theta_1}{A} = 800(\cos 50.2)^2 = 327.8\ \text{kN/m}^2$$

$$\tau_{\theta_1} = \frac{P\sin 2\theta_1}{2A} = \frac{800\sin[2(50.2)]}{2} = 393.4\ \text{kN/m}^2$$

The y direction: From Fig. 10.25(c), $\tan\theta_2 = 0.25/0.3$; therefore, $\theta_2 = 39.8°$. From Eqs. (10.11) and (10.12),

$$\sigma_{\theta_2} = \frac{P\cos^2\theta_2}{A} = 400(\cos 39.8)^2 = 236.1\ \text{kN/m}^2$$

$$\tau_{\theta_2} = \frac{P\sin 2\theta_2}{2A} = \frac{400\sin[2(39.8)]}{2} = 196.7\ \text{kN/m}^2$$

Adding the normal stresses, we have

$$\sigma_n = \sigma_{\theta_1} + \sigma_{\theta_2} = 327.8 + 236.1 = 563.9\ \text{kN/m}^2 \qquad \textbf{Answer}$$

The shear stresses have opposite directions; therefore, we have

$$\tau = \tau_{\theta_1} - \tau_{\theta_2} = 393.4 - 196.7 = 196.7\ \text{kN/m}^2 \qquad \textbf{Answer}$$

PROBLEMS

10.48 A steel bar 0.875 in. in diameter has an axial tensile load of 20,000 lb. Determine (a) the normal and shearing stresses on a plane that makes an angle of 35° with

the direction of the load, and (b) the maximum normal and shearing stresses in the bar.

10.49 A short compression block with a cross section that measures 6 in. by 8 in. is loaded with an axial load P. The allowable compressive stress is 790 psi, and the allowable shear stress is 200 psi. Find the maximum allowable load.

10.50 The 3/4-in.-thick plate shown supports uniform loads with resultants $P = 800$ lb and $Q = 400$ lb. Deter-

mine (a) the normal and shear forces on the oblique plane from A to C, and (b) the normal and shear stresses on the oblique plane from B to D.

10.51 The uniform plate 25 mm thick is acted on by the stresses illustrated in the figure. Determine the normal and shear stresses on the oblique plane from A to C.

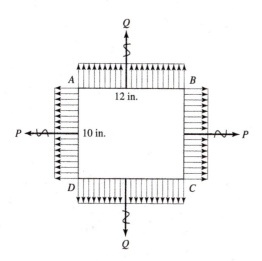

PROB. 10.50

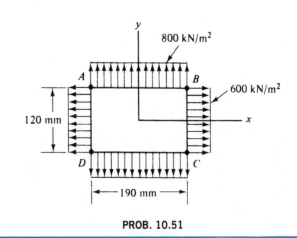

PROB. 10.51

CHAPTER ELEVEN

STRAIN FOR AXIAL LOADS: HOOKE'S LAW

CHAPTER OBJECTIVES

In this chapter, we examine the deformation of a material due to axial stress and correlate this deformation to various properties of the material itself. After completing the chapter, you should be able to

■ Define axial strain and compute numerical values of this quantity for simple loading conditions.

■ Describe a typical tension test, identify the important physical properties obtained from the resulting stress–strain curves, and explain the differences in these curves for specific types of materials.

■ Use the deformation of real materials, including thermal expansions and contractions, to solve problems that are otherwise statically indeterminate.

■ Define Poisson's ratio, and describe the material behavior on which this quantity is based.

■ Explain Saint-Venant's principle and its relevance to the values of average axial stress computed for an object.

■ Identify loading conditions under which theoretical values of computed axial stress might be modified to account for fatigue effects and stress concentrations, and apply appropriate stress concentration factors to calculations involving objects that have nonuniform cross sections.

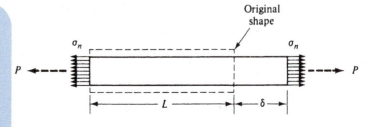

FIGURE 11.1

as shown in Fig. 11.1. The increase in length, or change in length, represented by the Greek lowercase letter δ (delta), is called *deformation*. The change in length δ is for a bar of length L. The change in length per unit of length, represented by the Greek lowercase letter ε (epsilon), is called *strain*. Strain is defined by the equation

$$\varepsilon = \frac{\delta}{L} \tag{11.1}$$

Strain is usually expressed dimensionally as inches per inch or meters per meter, even though it is a dimensionless quantity. Equation (11.1) gives the average strain over a length L. To obtain strain at a point, we let the length L approach zero. Lengths as small as 3 mm (1/8 in.) can be realized with electrical strain gauges.

In the case of an axial load, the stress in the direction of the load is called *axial stress*, and the strain in the direction of the load is called *axial strain*. With axial strain, we also have a smaller normal or lateral strain perpendicular to the load. When the axial stress is tensile, the axial strain is associated with an increase in length, and the lateral strain is associated with a decrease in width. The reverse is true for a compression stress. Tensile strain is called *positive strain*, and compressive strain is called *negative strain*.

11.1 AXIAL STRAIN

We have seen that when an axial load is applied to a bar, normal stresses are produced on a cross section perpendicular to the axis of the bar. In addition, the bar increases in length,

---

Example 11.1 A concrete test cylinder 200 mm in diameter by 400 mm high is tested to failure. The compressive strain at failure is 0.0012. Determine the total amount the cylinder shortens before failure.

Solution: From the definition for strain,

$$\varepsilon = \frac{\delta}{L} \quad \text{or} \quad \delta = \varepsilon L$$

The total shortening or deformation of the cylinder is given by

$$\delta = 0.0012(400) = 0.48\,mm \qquad \textbf{Answer}$$

11.2 TENSION TEST AND STRESS–STRAIN DIAGRAM

One of the most common tests of material is the tension test. In the usual tension test, the cross section of the specimen is round, square, or rectangular. If a large enough piece of the material to be tested is available and can be machined, a round cross section could be used. For a thin plate, a rectangular or square section would be used. The profile for a typical round test specimen is shown in Fig. 11.2. The fillets are provided to reduce the stress concentration caused by the abrupt change in section. The deformation or change in length of the specimen is measured for a specified distance known as the *gauge length*. The strain is therefore the deformation divided by the gauge length. The ends of the specimen must be properly shaped to fit the gripping device on the testing machine used to apply the deformation or load. (For an interesting application involving gauge length and percent strain, see Application Sidebar SB.11, "Flat Belts," in Chapter 8.)

The tension specimen is placed in a testing machine such as the one shown in Fig. 11.3. Strain can be measured either by sensors built into the testing machine or by separate gauges attached directly to the specimen. Until about 1930, the most common type of external gauge used for this purpose was the *extensometer*, a mechanical device that is

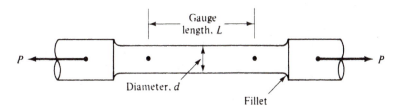

FIGURE 11.2

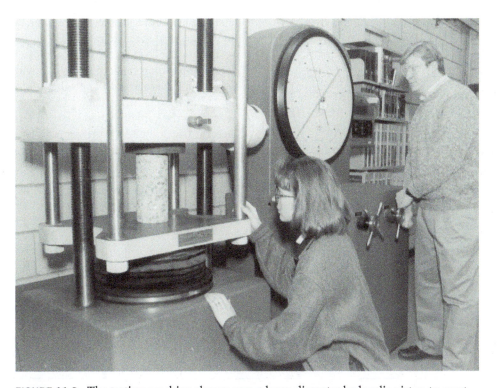

FIGURE 11.3. The testing machine shown uses a large-diameter hydraulic piston to exert tensile or compressive forces as high as 120,000 lb. Here, technicians are positioning a cylindrical concrete specimen in the loading fixture at left to perform a compressive test similar to that of Example 11.1.

FIGURE 11.4. Extensometer, which separates when its range has been exceeded. Such devices typically operate on a 2-in. gauge length. *(Courtesy of Tinius Olsen Testing Machine Co., Inc.)*

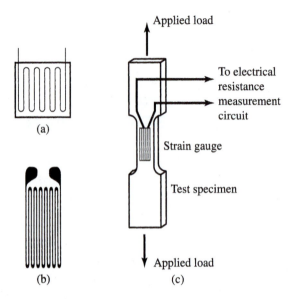

FIGURE 11.5. Electrical resistance strain gauges made of fine-diameter wire (a) or metallic foil (b) are generally mounted on a backing material of plastic film or paper. Gauge lengths from 1/64th in. to 6 in. are common for such devices. When glued to a test specimen (c), the gauge's "no load" resistance is first measured; changes in resistance due to deformation of the foil or wire can be correlated to strain in the specimen.

clamped to the specimen (Fig. 11.4) and that operates using a mechanical or optical lever system.

Perhaps the most common type of gauge in use today, however, is the *electrical resistance strain gauge.* This consists of a length of metallic foil or fine-diameter wire that is generally formed into a looped configuration [Fig. 11.5(a) and (b)]. The gauge is epoxied to the surface of a test specimen [Fig. 11.5(c)] and connected via lead wires to a sensitive electrical circuit that measures the resistance of the foil or wire used in the gauge. With no load applied, the electrical resistance has a certain value; when a load is applied, the foil or

wire deforms, thus producing a measurable change in resistance that can be directly correlated to the specimen's strain.

Deformations or loads are *gradually* applied to the specimen, and simultaneous readings of the load and deformation are taken at specified intervals. Values of stress are found by dividing the load by the original cross-sectional area and the corresponding value of strain by dividing the deformation by the gauge length. The values obtained can then be plotted in a stress–strain curve. The *shape* of the curve will depend on the kind of material tested. (The temperature and speed at which the test is performed also affect the results.)

The stress–strain curves shown in Fig. 11.6 are for three different kinds of material:

1. Low-carbon steel, a ductile material with a yield point [Fig. 11.6(a)]

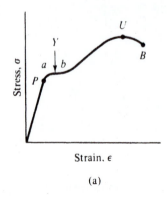

(a)

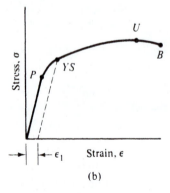

(b)

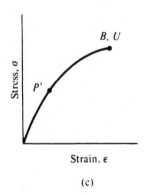

(c)

FIGURE 11.6

2. A ductile material, such as aluminum alloy, which does not have a yield point [Fig. 11.6(b)]

3. A brittle material, such as cast iron or concrete, in compression [Fig. 11.6(c)]

We are concerned here with the shape of the curves only. The actual magnitude of the stresses and strains for various materials would differ widely. For example, the stress at point U for concrete and cast iron could be 4000 psi (28 MPa) or 20,000 psi (140 MPa), respectively.

Several points of interest can be identified on the curves:

1. *Proportional limit:* the maximum stress for which stress is proportional to strain. (Stress at point P.)

2. *Yield point:* stress for which the strain increases without an increase in stress. (Horizontal portion of the curve *ab*. Stress at point Y.)

3. *Yield strength:* the stress that will cause the material to undergo a certain specified amount of permanent strain after unloading. (Usual permanent strain $\varepsilon_1 = 0.2$ percent. Stress at point YS.)

4. *Ultimate strength:* maximum stress material can support up to failure. (Stress at point U.)

5. *Breaking strength:* stress in the material based on original cross-sectional area at the time it breaks. Also called fracture or rupture strength. (Stress at point B.)

Compression tests are made in a manner similar to the tension test just described. The cross section of the compression specimen is preferably of a uniform circular shape, although a rectangular or square shape is often used. For most compression tests, the recommended ratio of length to major cross-sectional dimension (diameter or side length) of the specimen is 2:1. This ratio allows a uniform state of stress to develop on the cross section, while reducing the tendency of the specimen to buckle sideways. In Fig. 11.7, for example, the right-hand sample (a nominal 4×4) is 8 in. long and exhibits a typical compressive failure. The left-hand specimen, 12 in. in length, failed by a combination of compression and buckling.

For ductile materials, values of yield-point stress are commonly used as the *allowable stress* for in-service applications. A good example of this is the metal used in structural members (see Fig. 9.11), most of which are made of *A36 steel,* an industry designation based on the material's yield-point strength of 36,000 psi. Although properties for ductile materials are generally taken to be the same in tension and compression, such materials are rarely tested in compression because they tend to "wad up" and do not exhibit a clear point of failure. In Fig. 11.8, for instance, the aluminum (top right) and copper (bottom right) compression specimens simply deformed without failure until the testing machine's maximum capacity was reached. However, the steel tensile specimens shown on the left (untested specimen at top, "reassembled" failed specimen in middle, separated failed specimen at bottom) show the effects of *necking* (reduction in diameter caused by elongation), which precedes the unmistakable, and easily measurable, point of failure.

FIGURE 11.7. The nominal 4×4 at right has a length ratio of 2:1 and exhibits a typical compression fracture. The specimen at left has a 3:1 length ratio and failed through a combination of buckling and compression.

Brittle materials, such as cast iron or concrete, often have little or no strength in tension and are used primarily for compressive loads; allowable stresses for these materials are generally set at some percentage of the material's ultimate strength. Typical stress values for some common materials are given in Table A.10 of the Appendix.

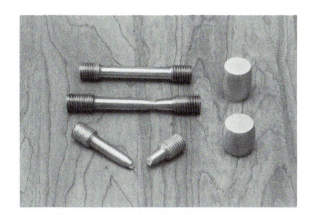

FIGURE 11.8. The steel tensile specimens at left (*top,* untested; *middle,* failed but "reassembled;" *bottom,* failed with pieces left separated) exhibit the typical elongation and necking that precedes an unmistakable point of fracture. However, the copper (*bottom right*) and aluminum (*top right*) compression specimens simply deformed under load, showing no clear signs of failure. Because of this behavior, ductile materials are not generally tested in compression.

METAL FORMING

SB.15(a)

SB.15(b)

Many common metal products are mass produced at room temperature from sheets of ductile metals such as steel and aluminum. These include automobile body panels, motorcycle fenders and gas tanks, the exterior surfaces of aircraft, and everyday items such as pots and pans. Most of these items are shaped by a *die forming* process that exerts tensile forces on the metal while deforming it beyond its elastic limit or yield point. The extent to which the metal may be shaped without failing by thinning, cracking, or rupture is defined by its *formability limit*.

In contrast, *hand forming* methods, some of which date back several centuries, use compressive forces in the shaping process and are particularly effective for *strain-sensitive* materials such as aluminum. Such methods also allow the creation of intricate shapes, such as compound curves, by exceeding the formability limits for tensile processes. In SB.15(a), for instance, a 1930s vintage power hammer is being used to develop a complex stainless-steel prototype for the exhaust-discharge nozzles of a modern helicopter. The primary disadvantages of hand forming are that it is time consuming and expensive. Nevertheless, these techniques are alive and well due to a renewed interest in the restoration of antique vehicles and aircraft, creation of custom-built automobiles and motorcycles, and fabrication of decorative architectural shapes. The craftsman shown in SB.15(b), for example, is using an air-driven planishing hammer from the 1950s to smooth the newly formed fender of a 1932 street rod. *(Courtesy of Fay Butler Fab/Metal Shaping, Wheelwright, Massachusetts, www.faybutler.com)*

11.3 HOOKE'S LAW

We can see in Fig. 11.6(a) and (b) and to a lesser degree in Fig. 11.6(c) that stress is directly proportional to strain (the curve is a straight line) on the lower end of the stress–strain curve. Based on tests of various materials and on the idealized behavior of those materials, *Hooke's law states that stress is proportional to strain*. In Fig. 11.9, we show a stress–strain curve for a material that follows Hooke's law. The slope of the stress–strain curve is the *elastic modulus* or modulus of elasticity, *E*. The elastic modulus is also known as Young's modulus (see Application Sidebar SB.14, "Thomas Young," in Chapter 10). The elastic modulus, *E*, is equal to the slope of the stress–strain curve.

$$E = \frac{\text{stress}}{\text{strain}} = \frac{\sigma}{\varepsilon} \quad \text{or} \quad \sigma = E\varepsilon \qquad (11.2)$$

This is the mathematical statement of Hooke's law. Hooke's law only applies up to the proportional limit of the material.

Because strain is dimensionless, the elastic modulus, *E*, has the same units as stress. The modulus is a measure of the

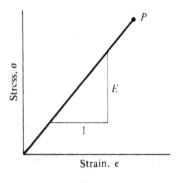

FIGURE 11.9

TABLE 11.1   Sources of Information on Properties and Uses of Materials in Manufacturing

Aluminum Association
1525 Wilson Boulevard, Suite 600
Arlington, VA 22209
703-358-2960
www.aluminum.org

American Ceramic Society
600 N. Cleveland Ave, Suite 210
Westerville, OH 43082
866-721-3322
www.ceramics.org

American Iron and Steel Institute
1140 Connecticut Avenue NW, Suite 705
Washington, DC 20036
202-452-7100
www.steel.org

American National Standards Institute
25 W. 43rd Street, 4th Floor
New York, NY 10036
212-642-4900
www.ansi.org

American Society of Mechanical Engineers
Three Park Avenue
New York, NY 10016
800-843-2763
www.asme.org

American Society for Nondestructive Testing
1711 Arlingate Lane
P.O. Box 28518
Columbus, OH 43228
800-222-2768/614-274-6003
www.asnt.org

American Society for Testing and Materials
100 Barr Harbor Drive, P.O. Box C700
West Conshohocken, PA 19428
610-832-9500
www.astm.org

Copper Development Association
(formerly Copper and Brass Research
Association)
260 Madison Avenue
New York, NY 10016
212-251-7200
www.copper.org

International Magnesium Association
100 N. Rand Road, Suite 214
Wauconda, IL 60084
847-526-2010
www.intlmag.org

International Titanium Association
2655 West Midway Boulevard, Suite 300
Broomfield, CO 80020
303-404-2221
www.titanium.org

Society of Manufacturing Engineers
One SME Drive
Dearborn, MI 48121
800-733-4763/313-425-3000
www.sme.org

Society of Plastics Engineers
14 Fairfield Drive
P.O. Box 403
Brookfield, CT 06804
203-775-0471
www.4spe.org

stiffness or resistance of a material to loads. Except for brittle materials, high values of E generally correspond to stiffer materials, while low values of E are consistent with more elastic materials. Steel, for instance, has an approximate value of $E = 30 \times 10^6$ psi (207×10^3 MPa), whereas wood has an average value of $E = 1.5 \times 10^6$ psi (10.3×10^3 MPa). Average values of the elastic modulus are given in Table A.10 of the Appendix. For more detailed information on specific materials and their uses, readers are encouraged to contact the professional organizations listed in Tables 11.1 and 11.2.

TABLE 11.2   Sources of Information on Properties and Uses of Materials in Construction

American Concrete Institute International
P.O. Box 9094
Farmington Hills, MI 48333
248-848-3700
www.concrete.org

American Institute of Steel Construction
One E. Wacker Drive, Suite 700
Chicago, IL 60601
312-670-2400
www.aisc.org

American Institute of Timber Construction
7012 S. Revere Parkway, Suite 140
Centennial, CO 80112

303-792-9559
www.aitc-glulam.org

American Society of Civil Engineers
1801 Alexander Bell Drive
Reston, VA 20191
800-548-2723
www.asce.org

American Wood Council (of the American Forest
and Paper Association)
1111 19th Street NW, Suite 800
Washington, DC 20036
800-878-8878/202-463-2766
www.awc.org, www.afandpa.org

Continued

TABLE 11.2	Sources of Information on Properties and Uses of Materials in Construction (*cont'd.*)

APA: The Engineered Wood Association
(formerly the American Plywood Association)
7011 So. 19th
Tacoma, WA 98466
253-565-6600
www.apawood.org

Forest Products Society
(formerly Forest Product Research Society)
2801 Marshall Court
Madison, WI 53705
608-231-1361
www.forestprod.org

International Code Council
500 New Jersey Avenue, NW, 6th Floor
Washington, DC 20001
888-422-7233
www.iccsafe.org

National Institute of Building Sciences
1090 Vermont Avenue NW, Suite 700
Washington, DC 20005
202-289-7800
www.nibs.org

Portland Cement Association
5420 Old Orchard Road
Skokie, IL 60077
847-966-6200
www.cement.org

Precast/Prestressed Concrete Institute
209 W. Jackson Boulevard, # 500
Chicago, IL 60606
312-786-0300
www.pci.org

Southern Forest Products Association
2900 Indiana Avenue
Kenner, LA 70065
504-443-4464
www.sfpa.org

Structural Building Components Association
(Formerly Wood Truss Council of America)
6300 Enterprise Lane
Madison, WI 53719
608-274-4849
www.sbcindustry.com, www.woodtruss.org

USDA Forest Service, Forest Products Laboratory
One Gifford Pinchot Drive
Madison, WI 53726
608-231-9200
www.fpl.fs.fed.us

Western Wood Products Association
522 SW Fifth Avenue, Suite 500
Portland, OR 97204
503-224-3930
www.wwpa.org

Example 11.2   The axial load at the proportional limit of a 0.505-in.-diameter bar was 4900 lb. If a 2-in. gauge length on the bar has increased by 0.0032 in. at the proportional limit, find (a) the stress at the proportional limit, (b) the strain at the proportional limit, and (c) the elastic modulus.

Solution:

a. The cross-sectional area

$$A = \frac{\pi d^2}{4} = \frac{\pi(0.505)^2}{4} = 0.2003 \text{ in.}^2$$

and the stress at the proportional limit

$$\sigma = \frac{P_P}{A} = \frac{4900}{0.200} = 24{,}500 \text{ psi} \qquad \textbf{Answer}$$

b. The strain at the proportional limit

$$\varepsilon_P = \frac{\delta}{L} = \frac{0.0032 \text{ in.}}{2.0 \text{ in.}} = 0.0016 \qquad \textbf{Answer}$$

c. The elastic modulus

$$E = \frac{\sigma}{\varepsilon} = \frac{24{,}500}{0.0016} = 15.3 \times 10^6 \text{ psi} \qquad \textbf{Answer}$$

Example 11.3   A flat bar with a cross section of 5.0 mm × 50 mm elongates 2.1 mm in a length of 1.5 m as a result of an axial load of 45 kN. The proportional limit of the material is 240,000 kN/m². Determine (a) the axial stress in the bar and (b) the elastic modulus.

Solution:

a. Stress is given by

$$\sigma = \frac{P}{A} = \frac{45 \times 10^3}{5(50)} = 180 \frac{\text{N}}{\text{mm}^2} = 180 \text{ MPa} \qquad \textbf{Answer}$$

b. The length 1.5 m = 1500 mm. Therefore, the strain

$$\varepsilon = \frac{\delta}{L} = \frac{2.1 \text{ mm}}{1500 \text{ mm}} = 0.0014$$

Because the stress is below the proportional limit, the elastic modulus can be found from

$$E = \frac{\sigma}{\varepsilon} = \frac{180}{0.0014} = 128.6 \times 10^3 \, \text{MPa}$$ **Answer**

11.4 AXIALLY LOADED MEMBERS

From Hooke's law, Eq. (11.2),

$$\sigma = E\varepsilon$$

When the stress and strain are caused by axial loads, we have

$$\sigma = \frac{P}{A} = E\frac{\delta}{L}$$

or

$$\delta = \frac{PL}{AE} = \sigma\frac{L}{E} \qquad (11.3)$$

Example 11.4 Determine the total change in length of the 25-mm-diameter rod when it is acted on by three forces shown in Fig. 11.10(a). Let $E = 70 \times 10^3 \, \text{MPa} = 70 \times 10^3 \, \text{N/mm}^2$.

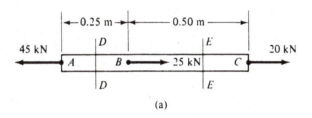

(a)

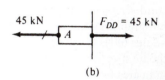

(b)

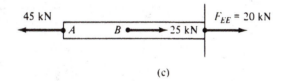

(c)

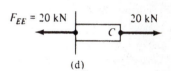

(d)

FIGURE 11.10

Solution: The total change in length must be equal to the sum of the changes in length for each section of the rod. Cut the bar at any section *D–D* between points *A* and *B* and draw the free-body diagram of the bar to the left of the section as shown in Fig. 11.10(b). We see from the equilibrium equation that the internal reaction must be a tensile force of 45 kN. Similarly, cut the bar at any section *E–E* between points *B* and *C* and draw a free-body diagram of the bar either to the left, as shown in Fig. 11.10(c), or to the right, as shown in Fig. 11.10(d), of the section. The internal reaction, as given by the equilibrium equation, must be a tensile force of 20 kN. The cross-sectional area $A = \pi d^2/4 = \pi(25)^2/4 = 491 \, \text{mm}^2$. The change in length for the section of the rod from *A* to *B* is found from Eq. (11.3).

$$\delta_1 = \frac{PL}{AE} = \frac{45 \times 10^3 \, \text{N} \, (250 \, \text{mm})}{491 \, \text{mm}^2 \, (70 \times 10^3 \, \text{N/mm}^2)} = 0.327 \, \text{mm}$$

Similarly, the change in length for the section of the rod from *B* to *C* is

$$\delta_2 = \frac{PL}{AE} = \frac{20 \times 10^3 \, \text{N} \, (50 \, \text{mm})}{491 \, \text{mm}^2 \, (70 \times 10^3 \, \text{N/mm}^2)} = 0.291 \, \text{mm}$$

Therefore, the total change in length

$$\delta = \delta_1 + \delta_2 = 0.327 + 0.291 = 0.618 \, \text{mm}$$ **Answer**

PROBLEMS

11.1 A concrete test cylinder 6 in. in diameter and 12 in. high is loaded in compression. As a result of the load, the cylinder shortened by 0.01 in. Determine the strain.

11.2 A steel rod 2.5 mm in diameter is tested in tension. The strain at the proportional limit is 0.0016. Determine the deformation at the proportional limit for a gauge length 20 mm long.

11.3 An axial load is applied to a bar. The length $L = 16$ in. increases by 0.022 in. The width $w = 3$ in. decreases by 0.00112 in., and the thickness $t = 0.75$ in. decreases by 0.00028 in. Determine the three components of strain.

11.4 A steel cable is used on a hoist. If the maximum strain in the cable is 0.004, what is the deformation for 30 m of this cable?

In Probs. 11.5 and 11.6, tabulated data from tension tests of steel are as follows. Calculate stress and the corresponding strain and draw a complete stress–strain curve. Draw a second stress–strain curve up to the yield point on the same graph. Determine the (a) proportional limit, (b) modulus of elasticity, (c) upper and lower yield points, (d) ultimate strength, (e) breaking strength, (f) percent reduction in area (see Sec. 11.8), and (g) percent elongation (see Sec. 11.8).

11.5 For the complete stress–strain curve, scales of 1 in. = 10 ksi and 1 in. = 0.05 in./in. may be used. For the stress–strain curve up to the yield point, use a scale of 1 in. = 0.0005 in./in.

11.6 For the complete stress–strain curve, scales of 10 mm = 25 MPa and 10 mm = 0.020 mm/mm may be used. For the stress–strain curve up to the yield point, use a scale of 10 mm = 0.00020 mm/mm.

(Data for Prob. 11.5)
Tension Test of Steel (Gauge Length = 2 in., Original Diameter = 0.501 in.)

Load (kip)	Deformation (in.)	Load (kip)	Deformation (in.)
0	0.0000	8.4	0.0050
1.0	0.0003	7.8	0.0095
2.0	0.0007	9.65	0.1000
3.0	0.0011	11.8	0.2000
4.0	0.0014	12.25	0.3000
5.0	0.0018	12.5	0.5000
6.0	0.0021	12.2	0.6000
7.0	0.0025	10.2	0.7000
8.0	0.0028	8.4	0.7200
8.6	0.0035		
Final diameter = 0.331 in.			

(Data for Prob. 11.6)
Tension Test of Steel (Gauge Length = 40 mm, Original Diameter = 10 mm)

Load (kN)	Deformation (mm)	Load (kN)	Deformation (mm)
0	0.0000	20.8	0.0920
2.0	0.0050	19.5	0.3400
4.0	0.0090	20.0	0.7600
6.0	0.0148	23.0	1.6000
8.0	0.0197	25.5	2.0000
10.0	0.0247	28.4	4.0000
12.0	0.0296	29.4	6.0000
14.0	0.0346	29.8	8.0000
16.0	0.0395	29.6	10.0000
18.0	0.0444	29.0	12.0000
20.0	0.0512	27.9	14.0000
21.36	0.0610	23.2	15.2000
Final diameter = 5.52 mm			

11.7 A copper bar with a cross-sectional area of 1.5 in.^{2} is loaded as shown. What is the total change in length of the bar? $E_C = 16 \times 10^6$ psi.

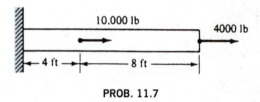

PROB. 11.7

11.8 An aluminum bar with a cross-sectional area of 750 mm^2 is loaded as shown. What is the total change in length of the bar? $E_A = 70 \times 10^6$ kN/m^2.

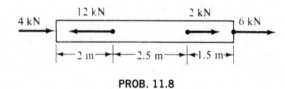

PROB. 11.8

11.9 A nominal 2 × 8 board 12 ft long is made of MSR 2250f-1.9E lumber and carries a compressive load of 15.5 kip parallel to the grain. (a) Find the factor of safety for this board. (b) Compute the total axial strain experienced by the board.

11.10 Same as Prob. 11.9, but board is a nominal 50 × 200, has a length of 3.65 m, and carries a compressive load of 68.5 kN.

11.11 A piece of standard-weight steel pipe is to be used as a 7-ft-long column and will carry a compressive load of 65,000 lb.
   a. Determine the smallest nominal-size pipe that can be used if the following conditions must be met:
     • Compressive stress does not exceed 36,000 psi.
     • Total deformation (strain) does not exceed 0.05 in.
   b. For the size pipe selected in (a), find the actual stress and actual deformation of this column.

11.12 An 8-m-long piece of L 102 × 76 × 6.35 angle is used in the construction of a communication tower.
   a. If a tensile force of 220 kN is applied at each end of this member, find the stress developed.
   b. The angle is made of ASTM-A36 steel. For the given load, what is the factor of safety for this member based on yield-point stress?
   c. Find the total deformation (strain) of this angle for the given load.

11.13 The aluminum bar AB, shown in the figure, has a rectangular cross section 6 in. by 1.75 in. Determine (a) the axial stress in the bar and (b) the lengthening of the bar due to the loads. $E_A = 10 \times 10^6$ psi.

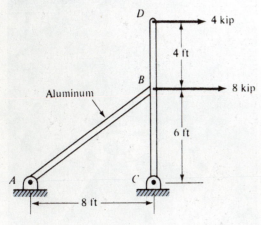

PROB. 11.13

11.14 The round steel rod shown in the figure has a diameter of 25 mm. Determine (a) the axial stress in the rod and (b) the lengthening of the rod due to the load. $E_S = 200 \times 10^6$ kN/m².

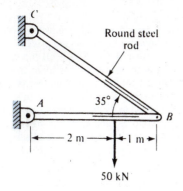

PROB. 11.14

11.15 A short square steel structural tube is used as a compression member as shown. The allowable compressive stress is 24,000 psi, and the allowable axial deformation 0.008 in. Determine the allowable load. $E_S = 30 \times 10^6$ psi.

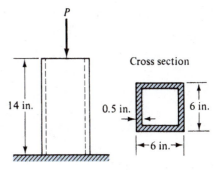

PROB. 11.15

11.16 A nominal 4 × 4 wooden post 10 ft tall carries a compressive load of 20 kip. (a) Select the lowest MSR and MEL lumber grades that provide a factor of safety of at least 1.25 for the post. (b) Compute the total deformation for each post.

11.17 Same as Prob. 11.16, but the post is a nominal 100 × 100 that is 3.0 m tall and carries a load of 88.5 kN.

11.18 Two round bars are joined and used as a tension member as illustrated. Determine the magnitude of the force P that will lengthen the two bars 0.25 mm. $E_S = 200 \times 10^6$ kN/m², $E_A = 70 \times 10^6$ kN/m².

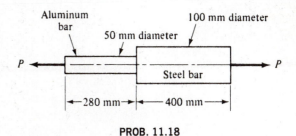

PROB. 11.18

11.19 The axial load at the proportional limit of an 0.8-in.-diameter bar was 12,300 lb. If a 4-in. gauge length on the bar has increased by 0.0060 in. at the proportional limit, determine (a) the stress at the proportional limit, (b) the strain at the proportional limit, and (c) the elastic modulus.

11.20 A flat bar of cross section 15 mm by 70 mm elongates 2.59 mm in a length of 1.3 m as the result of an axial load of 230 kN. The proportional limit of the material is 240 kN/m². Determine the axial stress in the bar and the elastic modulus.

RECYCLED METALS IN MANUFACTURING

SB.16(a)

SB.16(b)

SB.16(c)

SB.16(d)

Many industries recycle metals as part of their manufacturing processes. For the casting process described here, scrap steel (mostly in the form of used brake rotors, as shown in SB.16(a)), iron ingots, and pieces of cast iron from previous pours are placed in a furnace and melted. To determine the actual composition of the mix, samples of molten metal are drawn off (see SB.16(b)) and formed into small wafers for testing. Individual wafers are placed in a machine called a *spectrometer*, which discharges a spark on the sample's surface and analyzes the spectrum of light emitted by this spark. Because each chemical element has its own light signature, the machine is able to determine the exact composition of the melted mixture. Figure SB.16(c) shows a tested sample with spark marks on its surface and another wafer positioned in the machine for testing. The mix is then adjusted to its desired composition, primarily through the addition of carbon, silica, and nickel as necessary. The resulting mixture of molten cast iron is then poured into sand molds to form castings, such as the valve housing and cover shown in Figure SB.16(d). Note that the housing in the background weighs almost 4600 lb, while the cover in the foreground is approximately 60 in. in diameter. *(Courtesy of Rodney Hunt Company, Orange, Massachusetts, www.rodneyhunt.com)*

11.5 STATICALLY INDETERMINATE AXIALLY LOADED MEMBERS

If a machine or structure is made up of one or more axially loaded members, the equations of statics may not be sufficient to find the internal reactions in the members. The problem is said to be *statically indeterminate*. In such cases, equations for the geometric fit of the members are required. Eq. (11.3), $\delta = PL/AE$, is used to write these equations.

Example 11.5 Two steel plates 1 in. × 4 in. × 15 in. are attached to a pine block 4 in. × 4 in. × 15 in., as shown in Fig. 11.11(a). A rigid bearing plate on top transmits a compressive load of 200 kip to the composite member. Find the internal reaction and stress in the pine and steel. The elastic modulus for steel is $E_S = 30 \times 10^6$ psi and for pine is $E_P = 1.5 \times 10^6$ psi.

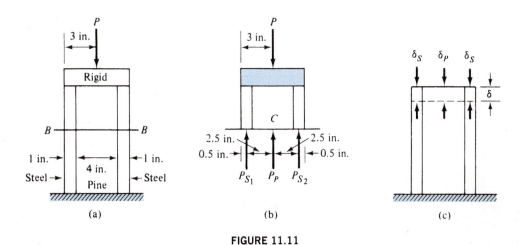

FIGURE 11.11

Solution: A section B–B is cut through the member, and a free-body diagram is drawn for the member above the section, as shown in Fig. 11.11(b). The forces in the steel plates are P_{S_1} and P_{S_2}. The force in the pine block is P_P. From equilibrium equations,

$$\Sigma M_C = 0 \qquad P_{S_2}(2.5) - P_{S_1}(2.5) = 0$$
$$P_{S_2} = P_{S_1} = P_S$$

$$\Sigma F_y = 0 \qquad P_S + P_S + P_P - P = 0$$
$$2P_S + P_P = P = 200 \text{ kip}$$

Fig. 11.11(c) shows the deformations for the steel plates and the pine block. Because the force on each steel plate is the same, the deformations of the steel plates are equal. The rigid bearing plate under the load ensures that the deformation of the pine block must equal the deformation of the steel plates. That is,

$$\delta_S = \delta_P$$

From Eq. (11.3),

$$\frac{P_S L_S}{A_S E_S} = \frac{P_P L_P}{A_P E_P} \quad \text{or} \quad P_S = \frac{A_S E_S L_P}{A_P E_P L_S} P_P$$

where the subscript S refers to the steel plate and the subscript P refers to the pine block. The length of both the steel and the pine are equal. Substituting values of the cross-sectional area and elastic modulus, we have

$$P_S = \frac{4(30)(10^6)L_P}{16(1.5)(10^6)L_S} P_P = 5P_P$$

Substituting $P_S = 5P_P$ in the equilibrium equation yields

$$2(5P_P) + P_P = 200$$
$$11P_P = 200$$
$$P_P = 18.18 \text{ kip}$$

Then,
$$P_S = 5P_P = 5(18.18) = 90.9 \text{ kip}$$
$$P_P = 18.2 \text{ kip} \qquad P_S = 90.9 \text{ kip}$$ **Answer**

The stress in the steel and pine is given by

$$\sigma_S = \frac{P_S}{A_S} = \frac{90.9}{4} = 22.7 \text{ ksi}$$ **Answer**

$$\sigma_P = \frac{P_P}{A_P} = \frac{18.2}{16} = 1.14 \text{ ksi}$$ **Answer**

Example 11.6   The bar shown in Fig. 11.12(a) is attached at its ends to unyielding supports. There is no initial stress in the bar. A force P is then applied as shown. (a) What are the forces in the steel and aluminum parts of the bar? (b) If the allowable working stress in the steel is 150 MN/m² and in the aluminum it is 60 MN/m², what safe load P can the bar support? $E_S = 200 \times 10^6$ kN/m², $E_A = 70 \times 10^6$ kN/m².

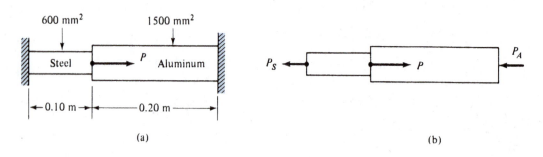

(a)   (b)

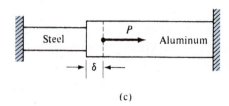

(c)

FIGURE 11.12

Solution: a.   In Fig. 11.12(b) a free-body diagram of the bar is shown. From equilibrium,

$$\Sigma F_x = 0 \qquad -P_S + P - P_A = 0$$
$$P_S + P_A = P$$

Figure 11.12(c) shows the deformation for the steel and aluminum. The tensile deformation for the steel is equal to the compressive deformation of the aluminum. That is,

$$\delta_S = \delta_A$$

From Eq. (11.3),

$$\frac{P_S L_S}{A_S E_S} = \frac{P_A L_A}{A_A E_A} \quad \text{or} \quad P_S = \frac{A_S E_S L_A}{A_A E_A L_S} P_A$$

Substituting numerical values, we have

$$P_S = \frac{600(200)(10^9)(0.20)}{1500(70)(10^9)(0.10)} P_A = 2.286 P_A$$

Substituting $P_S = 2.286P_A$ in the equilibrium equation,

$$2.286P_A + P_A = P$$

$$3.286P_A = P \qquad P_A = 0.304P \qquad \text{Answer}$$

$$P_S = 2.286P_A = 2.286(0.304P)$$

$$= 0.695P \qquad \text{Answer}$$

b.   The allowable load in the steel and aluminum part of the bar is

$$P_A = \sigma_A A_A = 60(1500) = 90 \times 10^3\,\text{N}$$
$$P_S = \sigma_S A_S = 150(600) = 90 \times 10^3\,\text{N}$$

The *total* allowable load based on the allowable steel stress is

$$P = \frac{P_S}{0.695} = \frac{90,000}{0.695} = 129.5 \times 10^3\,\text{N}$$

and on the allowable aluminum stress

$$P = \frac{P_A}{0.304} = \frac{90,000}{0.304} = 269 \times 10^3\,\text{N}$$

Therefore, the maximum allowable load is

$$P = 129.5\,\text{kN} \qquad \text{Answer}$$

PROBLEMS

11.21 A bar is supported at B and C as illustrated. A force of 10,000 lb is applied at A. Determine the reactions at the supports.

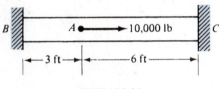

PROB. 11.21

11.22 As shown in the figure, the bar is supported at C and D. A 240-kip force is applied at E. Determine the reactions at the supports and the stress and strain in each part of the bar. Assume that $E = 30 \times 10^6$ psi.

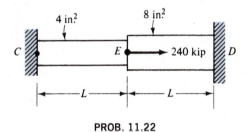

PROB. 11.22

11.23 A steel pipe is filled with concrete and used as a column with a cross section as shown. The column is loaded in compression. If the total compressive force is 150 kip, determine the stress in each material. $E_S = 30 \times 10^6$ psi and $E_C = 3 \times 10^6$ psi.

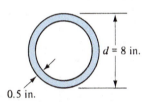

PROB. 11.23

11.24 A 14-in. square concrete post is 6 ft tall and reinforced by nine 3/4-in.-diameter steel rods as shown in the cross-sectional end view. Assume that the materials are high-strength concrete and ASTM-A36 structural steel.
a.   If a compressive load of 250,000 lb is applied to this column, find the stress in the steel and the stress in the concrete.
b.   Find the total axial deformation (strain) of the column.

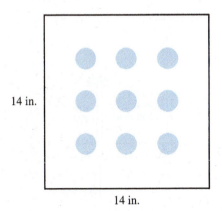

PROB. 11.24 and PROB. 11.25

11.25 Repeat Prob. 11.24 if the reinforcing rods are 1 in. in diameter.

11.26 A square wooden post of 12-in. × 12-in. nominal size is braced by four pieces of L 2 × 2 × 1/4 steel angle, as shown. Assume that the wood is Douglas fir and that the pieces of angle are exactly the same length as the wood. If this reinforced post carries a total compressive load of 200,000 lb, find (a) the axial stress in the steel and in the wood, and (b) the factor of safety based on yield-point stress for the steel and the wood.

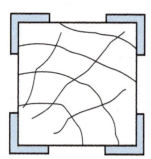

PROB. 11.26

11.27 A power company uses cable that consists of an A36 steel core (shaded in figure) that is 9/32 in. in diameter and is surrounded by seven strands of 6061 annealed aluminum wire, each of which is 1/8 in. in diameter. This cable carries a total tensile load of 2850 lb. Find (a) the stress in the steel and stress in the aluminum, and (b) the deformation in 500 ft of cable.

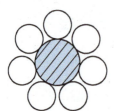

PROB. 11.27

11.28 A rigid bar AB is supported by three rods as shown. If a load $P = 200$ kN is applied to the bar, determine the deformation and stress in each rod. $E_S = 200 \times 10^6$ kN/m² and $E_C = 110 \times 10^6$ kN/m².

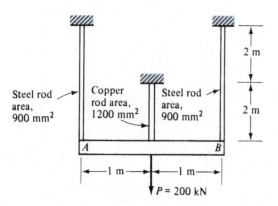

PROB. 11.28

11.29 In the figure shown, the steel bar AB has a cross section 0.75 in. × 0.5 in., and the aluminum bar CD has a cross section of 1.0 in. × 0.5 in. Assume that bar BC is

rigid. A load of $P = 4000$ lb is applied to bar BC so it remains horizontal. Determine (a) the axial stress in the steel and aluminum bar, (b) the vertical displacement of BC, and (c) the location of the load measured from point B. $E_S = 30 \times 10^6$ psi and $E_A = 10 \times 10^6$ psi.

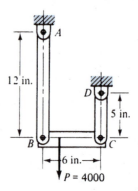

PROB. 11.29

11.30 Bars M and O are made of brass, and bar N is made of steel as illustrated. Each bar has a cross-sectional area of 2000 mm². What load P will cause axial stress in the steel to be one-half of the axial stress in the brass? $E_S = 200 \times 10^6$ kN/m² and $E_B = 100 \times 10^6$ kN/m².

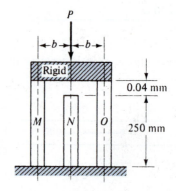

PROB. 11.30

11.31 Two wires are used to lift a 10-kip weight as shown. One wire is 60.0 ft long, and the other wire is 60.08 ft long. If the cross-sectional area of each wire is 0.15 in.², determine the load supported by each wire. $E = 30 \times 10^6$ psi.

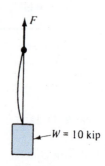

PROB. 11.31

11.32 A 1/4-in.-diameter steel bolt made of A633 Grade A steel is bolted through a hollow tube made of 6061 annealed aluminum as shown. The tube has an inside diameter of

3/8 in. and an outside diameter of 5/8 in. For bolt threads designated as UNC 1/4-20 ("Unified Coarse" thread form, 1/4-in. diameter, 20 threads per inch), the nut advances 1/20 in. for each revolution on the bolt. If the nut on this assembly is finger tightened as far as possible and then turned another quarter turn using a wrench, find the stress developed in the bolt and in the tube. (Hint: The tensile force developed in the bolt is exactly equal to the compressive force developed in the tube.)

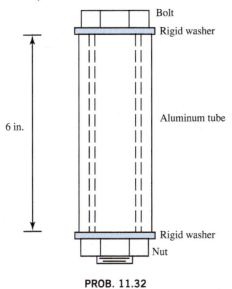

PROB. 11.32

11.6 POISSON'S RATIO

When a load is applied along the axis of a bar, axial strain is produced. At the same time, a lateral (perpendicular to the axis) strain is also produced. If the axial force is in tension, the length of the bar increases, and the cross section contracts or decreases. That is, a *positive axial stress produces a positive axial strain and a negative lateral strain*. For a negative axial stress, the axial strain is negative and the lateral strain is positive.

The ratio of lateral strain to axial strain is called *Poisson's ratio*. It is constant for a given material provided that the material is not stressed above the proportional limit, is homogeneous, and has the same physical properties in all directions. Poisson's ratio, represented by the Greek lowercase letter v (nu), is defined by the equation

$$v = \frac{-\text{lateral strain}}{\text{axial strain}} \qquad \textbf{(11.4)}$$

The negative sign ensures that Poisson's ratio is a positive number.

The value of Poisson's ratio, v, varies from 0.25 to 0.35 for different metals. For concrete, it may be as low as $v = 0.1$, and for rubber, as high as $v = 0.5$.

Example 11.7   A test conducted on a steel bar whose cross section is 0.5 in. × 4 in. is performed. An axial load of $P = 30{,}500$ lb produces a deformation of 0.00103 in. over a gauge length of 2 in. and a decrease of 0.000078 in. in the 0.5-in. thickness of the bar. Determine Poisson's ratio, v, the elastic modulus, E, and the decrease in the 4.0-in. cross-sectional dimension. The proportional limit for this steel is 40.6 ksi.

Solution: The lateral strain is

$$\varepsilon_t = \frac{\delta_t}{d} = \frac{-0.000078}{0.50} = -0.000156$$

Because the thickness decreases, the lateral strain is negative. The axial strain is

$$\varepsilon_a = \frac{\delta_a}{L} = \frac{0.00103}{2.0} = 0.000515$$

Therefore, Poisson's ratio is given by

$$v = \frac{-\varepsilon_t}{\varepsilon_a} = \frac{-(-0.000156)}{0.000515} = 0.303 \qquad \textbf{Answer}$$

The cross-sectional area of the rod $A = (0.5)(4.0) = 2.0\text{ in.}^2$ From Eq. (11.3),

$$\delta = \frac{PL}{AE} \qquad E = \frac{PL}{A\delta} = \frac{30{,}500\text{ lb (2 in.)}}{2.0\text{ in.}^2 (0.00103\text{ in.})} = 29.6 \times 10^6\text{ psi} \qquad \textbf{Answer}$$

The lateral strain is $\varepsilon_t = -0.000156$. Then,

$$\delta_t = \varepsilon_t L_t = -0.000156(4.0) = -0.000624\text{ in.} \qquad \textbf{Answer}$$

The axial stress will be calculated to verify that the stress is below the proportional limit.

$$\sigma = \frac{P}{A} = \frac{30{,}500\text{ lb}}{2.0\text{ in.}^2} = 15{,}250\text{ psi} < \sigma_P = 40{,}600\text{ psi} \qquad \text{OK}$$

11.7 THERMAL DEFORMATION: THERMALLY INDUCED STRESS

When the temperature of a material is raised (or lowered), the material expands (or contracts) unless it is prevented from doing so by external forces acting on the material. If the temperature of a body of length L is changed by a temperature Δt, the linear deformation is given by

$$\delta_t = L\alpha(\Delta t) \qquad (11.5)$$

where α is the coefficient of linear expansion and has units of per Fahrenheit (Celsius) degree. Note that Eq. (11.5) is in-dependent of the body's cross-sectional area. Average values of α for selected materials are given in Table A.10 of the Appendix. Associated with the linear thermal deformation δ_t is a *thermal* strain $\varepsilon_t = \delta_t/L$. From Eq. (11.5),

$$\varepsilon_t = \alpha(\Delta t) \qquad (11.6)$$

When the material is free to expand (or contract), there is no stress associated with the thermal strain.

Indeterminate problems involving temperature changes follow the same methods as the problems in Sec. 11.4.

Example 11.8 A steel rod having a cross-sectional area of 0.002 m² and length of 1.5 m is attached to supports. Find (a) the force exerted by the rod on the supports if the temperature falls 60°C and the supports are unyielding, and (b) the force exerted by the rod on the supports if they yield 0.0003 m while the temperature rises 50°C.

$$E_S = 200 \times 10^9\,\text{N/m}^2 \qquad \alpha_S = 11.5 \times 10^{-6}\,\text{per °C}$$

Solution: a. Remove one end support, decrease the temperature of the steel rod by 60°C, and let it shorten. Calculate the decrease in length as shown in Fig. 11.13(a). Then, apply a load that will restore the bar to its original length as shown in Fig. 11.13(b). From Eq. (11.5),

$$\delta_t = L\alpha(\Delta t) = 1.5(11.5)(10^{-6})(60)$$
$$= -0.001035\,\text{m}$$

From Eq. (11.3),

$$\delta = \frac{PL}{AE}$$

or

$$P = \frac{\delta AE}{L} = \frac{0.001035(0.002)(200)(10^9)}{1.5}$$

$$= 276{,}000\,\text{N} = 276\,\text{kN} \qquad \textbf{Answer}$$

The rod is in tension. Thus the force of the rod on the supports will tend to bring the supports closer together.

b. Remove one support and increase the temperature of the steel rod. Calculate the increase in length due to a temperature rise of 50°C as shown in Fig. 11.13(c). Then apply a force that will restore the

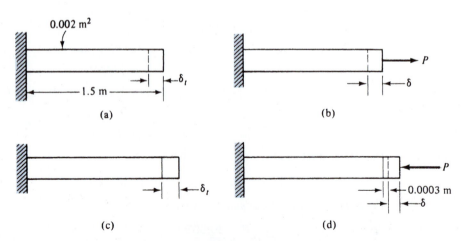

FIGURE 11.13

rod to its original length less 0.0003 m, which the supports are permitted to yield, as shown in Fig. 11.13(d). From Eq. (11.5),

$$\delta_t = L\alpha(\Delta t) = 1.5(11.5)(10^{-6})(50)$$
$$= 0.0008625\,m$$

Because
$$\delta = \delta_t - 0.0003 = 0.0008625 - 0.0003$$
$$= 0.0005625\,m$$

From Eq. (11.3),

$$P = \frac{\delta AE}{L} = \frac{0.0005625(0.002)(200)(10^9)}{1.5}$$
$$= 150,000\,N = 150\,kN \qquad \textbf{Answer}$$

The rod is in compression. Thus, the force of the rod on the supports will tend to move the supports farther apart.

Example 11.9   A steel bolt through a 25-in.-long aluminum tube has the nut turned on until it just touches the end of the tube. No stress is introduced in the bolt or the tube. The temperature of the assembly is raised from 60°F to 70°F. If the cross-sectional area of the bolt is 0.08 in.^{2} and the cross-sectional area of the tube is 0.25 in.^{2}, find the stress induced in the bolt and tube by the temperature change. $E_S = 30 \times 10^6$ psi, $E_A = 10 \times 10^6$ psi, $\alpha_S = 6.5 \times 10^{-6}$ per °F, $\alpha_A = 13.0 \times 10^{-6}$ per °F.

Solution: Remove the steel bolt from the aluminum tube; increase their temperature. Let them expand freely, and calculate the thermal expansion of the steel bolt, δ_{t_S}, and the thermal expansion of the aluminum tube, δ_{t_A}, as shown in Fig. 11.14(a).

$$\delta_{t_S} = L\alpha(\Delta t) = 25(6.5)(10^{-6})(10) = 0.001625\,in.$$
$$\delta_{t_A} = L\alpha(\Delta t) = 25(13.0)(10^{-6})(10) = 0.00325\,in.$$

To place the steel bolt back into the aluminum tube, the steel bolt must be stretched δ_S by a force P_S, and the aluminum tube must be compressed δ_A by a force P_A so the tube and the bolt are the same length, as shown in Fig. 11.14(b). That is,

$$\delta_A + \delta_S = \delta_{t_A} - \delta_{t_S}$$
$$\frac{P_A L_A}{A_A E_A} + \frac{P_S L_S}{A_S E_S} = 0.00325 - 0.001625$$
$$\frac{P_A(25)}{0.25(10)(10^6)} + \frac{P_S(25)}{0.08(30)(10^6)} = 0.001625$$
$$10.0P_A + 10.41P_S = 1625$$

From equilibrium,

$$P_A = P_S = P$$

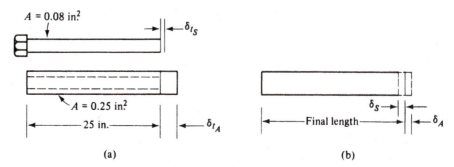

A = 0.08 in.^{2}

A = 0.25 in.^{2}

25 in.

(a)

Final length

(b)

FIGURE 11.14

THERMAL EXPANSION

SB.17(a)

SB.17(b)

One application that uses the *thermal expansion* of metals is the fabrication of casks used for dry storage of commercial spent nuclear fuel. The main body of each cask consists of a 90,000-lb steel forging that contains a cylindrical cavity whose diameter is approximately 74 in. and is machined to an accuracy of 0.005 in. This forging, with a wall thickness of 8 in., provides most of the shielding required by the spent fuel. A steel liner, used for protection against brittle fracture, is fabricated by rolling a 2-in.-thick flat plate into a cylinder and welding and then machining it to form a finished piece that is 179 in. long. To *heat-shrink* the forging and liner into a single assembly, the forging [seen covered with white insulation in the lower portion of SB.17(a) and (b)] is heated for 24 hours to a temperature of 750°F. With the liner at room temperature, this creates a diametral clearance between forging and liner of 0.375 in. The liner, filled with water for additional weight and to help maintain the necessary temperature gradient, is then lowered into the forging. This entire process, from start [SB.17(a)] to finish [SB.17(b)] is completed in about 30 sec. Once the forging has returned to room temperature, the mechanical bond between the two pieces is so strong that the liner can only be removed by machining. *(Courtesy of RANOR Inc., Westminster, Massachusetts, www.ranor.com, and Transnuclear, Inc., Hawthorne, New York, www.cogema.com)*

Therefore,

$$20.416P = 1625$$

$$P = 79.6 \text{ lb}$$

and the stresses in the steel bolt and the aluminum tube are

$$\sigma_S = \frac{P}{A_S} = \frac{79.6}{0.08} = 995 \text{ psi} \quad \text{(tension)}$$ **Answer**

$$\sigma_A = \frac{P}{A_A} = \frac{79.6}{0.25} = 318 \text{ psi} \quad \text{(compression)}$$ **Answer**

11.8 ADDITIONAL MECHANICAL PROPERTIES OF MATERIALS

In addition to the properties discussed in Sec. 11.2, the following are of interest:

Elastic Limit: The highest stress that can be applied without permanent strain when the stress is removed. To determine the elastic limit would require the application of larger and larger loadings and unloadings of the material until permanent strain is detected. The test would be difficult. It is rarely done. Numerical values of the proportional limit are often used in its place.

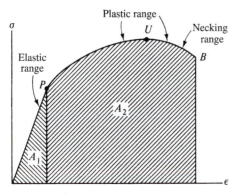

FIGURE 11.15

Elastic Range: Response of the material as shown on the stress–strain curve from the origin up to the proportional limit P (Fig. 11.15).

Plastic Range: Response of the material as shown on the stress–strain curve from the proportional limit P to the breaking strength B (Fig. 11.15).

Necking Range: Response of the material as shown on the stress–strain curve from the ultimate strength U to the breaking strength B (Fig. 11.15). Beyond the ultimate strength, the cross-sectional area of a localized part of the specimen decreases rapidly until rupture occurs. This phenomenon is referred to as *necking* and is a characteristic of low-carbon steel. Brittle materials do not exhibit it at usual temperatures. The necking range is part of the plastic range.

Percentage Reduction in Area: When a ductile material is stretched beyond its ultimate strength, the cross section "necks down," and the area reduces appreciably. It is defined by the equation

$$\text{Percentage reduction in area} = \frac{A_o - A_f}{A_o} 100$$

where A_o is the original and A_f the final minimum cross-sectional area. It is a measure of ductility.

Percentage Elongation: The percentage elongation represents a comparison of the increase in the length of the gauge length to the original gauge length. It is defined by the equation

$$\text{Percentage elongation} = \frac{L_f - L_o}{L_o} 100$$

where L_o is the original and L_f the final gauge length. It is also a measure of ductility.

Modulus of Resilience: The work done on a unit volume of material from a zero force up to the force at the proportional limit. This is equal to the area under the stress–strain curve from zero to the proportional limit. (Area A_1 of Fig. 11.15. Units of in.-lb/in.^{3} or N $\cdot$ (m/m^3.)

Modulus of Toughness: The work done on a unit volume of material from a zero force up to the force at the breaking point. This is equal to the area under the stress–strain curve from zero to the breaking strength. (Areas A_1 and A_2 of Fig. 11.15. Units of in.-lb/in.^{3} or N $\cdot$ (m/m^3.)

Recent Developments in Materials Technology

Research in the development and applications of new materials is a continual process. Today, much of this work takes place at the microscopic level and involves *nanotechnology* (see Section 1.5). Several of the most recent discoveries are listed next, and although some are still in the laboratory stage, all offer great promise for new and exciting products that could benefit society.

- The strongest material ever tested is a carbon material called *graphene.* Said to be 200 times stronger than structural steel, it consists of a single layer of graphite atoms that may be rolled into tiny tubes (called nanotubes). These tubes can then be used as the basis of graphite fibers found in a variety of commercial products requiring high strength and light weight.

- A process has been developed in which ceramic particles are added to molten aluminum, and a gas is blown into the mixture. When solidified, a honeycombed *cellular aluminum* material is formed that is lightweight yet strong. However, the metal's most important property is its ability to absorb energy, sound, and vibration, making it ideal for use in automobile "collision crumple zones" and in various protective devices exposed to bombs and bullets.

- The use of *nano-sized additives* has been found to double the effective life of concrete by preventing penetration of chloride and sulfate ions from sources such as road salt, seawater, and soils. It has long been known that infiltration by these ions can cause internal damage and cracking, but previous attempts to reduce such effects have concentrated on developing denser, less-porous concretes. Instead, the small-molecule additives serve as a *diffusion barrier* to slow the influx of ions.

- A new structural material similar to concrete has been developed that uses a mixture of *fly ash* and organic materials. (Fly ash is a waste product consisting of small particles removed from the combustion gases produced by the burning of coal.) This new material has good insulating properties and fire resistance, as well as high strength and light weight. Unlike concrete, the fly-ash mixture does not use cement or aggregate (sand and rock) and emerges from curing ovens in its final form and at full strength.

- Several forms of *enhanced steels* have recently been developed. Sometimes referred to as "super steels," these materials are stronger than their traditional counterparts, can withstand extreme levels of heat and radiation, and generally have much higher resistance to corrosion. The optimized properties are obtained primarily by reducing irregularities in the steel during its production. Some of these advanced high-strength steels have yield points near 700 MPa, compared to a 250 MPa yield strength for standard ASTM-A36 structural steel.

- A *structural coating* has been developed using a combination of carbon nanotubes and polymers. When applied to a bridge, the coating can detect internal cracks long before they become visible. Said to cost about $150 per square foot of coverage, the new material has an expected life of several decades.

- The U.S. Environmental Protection Agency recently approved public health claims that *copper*, *brass*, and *bronze* are capable of killing potentially deadly bacteria, including methicillin-resistant *Staphylococcus aureus* (MRSA). In U.S. hospitals alone, approximately 2 million people acquire MRSA each year, resulting in nearly 100,000 deaths. Although this bacteria is generally resistant to antibiotics, copper-alloy surfaces destroy more than 99.9% of MRSA germs within 2 hours at room temperature and continue to kill these bacteria even after repeated contamination.

PROBLEMS

11.33 Determine the Poisson's ratio for the bar material in Prob. 11.3.

11.34 A member 1.5 m long has a cross section 75 mm by 75 mm. The member becomes 0.7 mm longer and 0.01 mm narrower after loading. Determine Poisson's ratio for the material.

11.35 A load of 210 kip is applied to a 20-ft-long rod with a diameter of 2.25 in. The rod stretches 0.42 in., and the diameter decreases by 0.001 in. Determine Poisson's ratio and the elastic modulus for the material.

11.36 A flat bar of cross section 20 mm by 70 mm elongates 4.0 mm in a length of 2.0 m as a result of axial load. If the elastic modulus is 199×10^6 kN/m^2 and Poisson's ratio is 0.25 for the material, find the axial load and the total change in each cross-sectional dimension.

11.37 An aluminum rod with a diameter of 2.0 in. and a length of 1.5 ft is attached to supports at its ends. Determine (a) the force exerted by the supports on the rod if the temperature drops 50°F and the supports are unyielding, and (b) the force exerted by the supports on the rod if they yield 0.0001 in. while the temperature increases 60°F. $E_A = 10 \times 10^6$ psi and $\alpha_A = 13.2 \times 10^{-6}(°F)^{-1}$.

11.38 If an object "just fits" between rigid supports at a given temperature, then the mechanical deformation provided by these supports must exactly equal any thermal deformation caused by a change in temperature. In other words, from Eqs. (11.3) and (11.5),

$$\sigma \frac{L}{E} = L\alpha(\Delta t)$$

which reduces to

$$\sigma = E\alpha(\Delta t)$$

a. If a 40-ft length of steel pipe is set in place between two rigid supports at an ambient temperature of 45°F, find the axial stress developed in the pipe when the ambient temperature rises to 105°F.

b. Find the force exerted on each support if the pipe is 6-in. nominal ("in name only") standard weight pipe. (See Table A.8 of the Appendix.)

11.39 At a temperature of 0°C, a precast wall made of low-strength concrete just fits between two rigid supports. If $L = 3$ m, $h = 1$ m, and $w = 10$ cm,
a. Find the *stress* developed in this wall if the temperature rises to 35°C.
b. What *force* does the wall exert on each support when the temperature is 35°C?

11.40 Same as Prob. 11.39, but the installation temperature is 34°F, the final temperature is 88°F, $L = 12$ ft, $h = 48$ in., and $w = 6$ in.

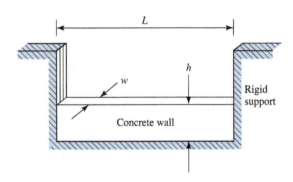

PROB. 11.39 and PROB. 11.40

11.41 The composite aluminum and steel member illustrated is attached to supports at its ends. Determine (a) the force exerted by the supports on the member if the temperature drops 60°F and the supports are unyielding, and (b) the force exerted by the supports on the member if they yield 0.01 in., while the temperature increases 70°F. $E_S = 30 \times 10^6$ psi, $E_A = 10 \times 10^6$ psi, $\alpha_S = 6.5 \times 10^{-6}(°F)^{-1}$, and $\alpha_A = 13.2 \times 10^{-6}(°F)^{-1}$.

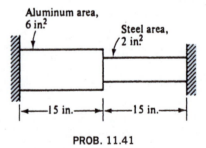

PROB. 11.41

11.42 An electric microswitch contains a thin strip of A633 steel and a thin strip of C26800 copper placed side by side as shown. At an ambient temperature of 20°C, the steel strip is 50 mm long and the copper strip is 0.0254 mm shorter than the steel. At what temperature will both strips be the same length?

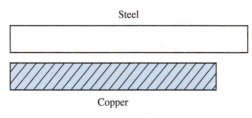

Steel

Copper

PROB. 11.42

11.43 A machinist plans to shrink-fit a thin copper ring onto a steel shaft. At a temperature of 68°F, the outside diameter of the shaft is 2.500 in., the inside diameter of the copper ring is 2.495 in., and the ring is assumed to be stress free.
  a. To what temperature must the ring be heated so it just fits over the shaft?
  b. If the assembly is then cooled to 68°F, what stress develops in the ring? Do you think the ring will crack?

11.44 An aluminum bar is attached at its ends to rigid supports. The compressive stress in the bar is 2500 psi when the temperature is 100°F. What is the stress in the bar when the temperature is lowered to 50°F? $E_A = 10 \times 10^6$ psi and $\alpha_A = 13.2 \times 10^{-6} (°F)^{-1}$.

11.45 A bar with a composite cross section as shown has its temperature raised by 30°C. If no slippage occurs between the bronze and aluminum, determine the axial stress in each material. $E_A = 70 \times 10^6$ kN/m^2, $E_B = 110 \times 10^6$ kN/m^2, $\alpha_A = 23.8 \times 10^{-6}(°C)^{-1}$, and $\alpha_B = 18.0 \times 10^{-6}(°C)^{-1}$.

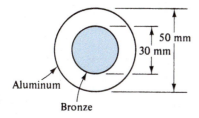

Aluminum

Bronze

50 mm
30 mm

PROB. 11.45

11.46 The cross section of a concrete column consisting of steel reinforcing bars embedded in a concrete cylinder is shown in the figure. If no slippage occurs between the concrete and the steel, determine the axial stress introduced into the steel and concrete due to a temperature increase of 100°F. $E_S = 30 \times 10^6$ psi, $E_C = 3 \times 10^6$ psi, $\alpha_S = 6.5 \times 10^{-6}(°F)^{-1}$, and $\alpha_C = 6.2 \times 10^{-6}(°F)^{-1}$.

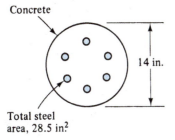

Concrete

14 in.

Total steel area, 28.5 in.^{2}

PROB. 11.46

11.9 STRAIN AND STRESS DISTRIBUTIONS: SAINT-VENANT'S PRINCIPLE

In our discussion thus far of the uniform axially loaded bar, we have assumed a uniform distribution of normal stress on any plane section near the middle of the bar away from the load. We ask here what effect a concentrated compressive load has on the stresses and strains near the load. To visualize the answer, we consider a short rubber bar of rectangular cross section. A grid or network of uniformly spaced horizontal and perpendicular lines is drawn on the side of the bar as in Fig. 11.16(a) to form square elements. (Half of the bar is shown in the figure.) Then, compressive loads are applied to the bar [Fig. 11.16(b)]. Close to the load, the elements are subjected to large deformations or strain, while other elements on the end of the bar, away from the load, remain virtually free of deformations. Moving axially away from the load, we see a gradual smoothing of the deformations of the elements, and thus a uniform distribution of strain and the resulting stress along a cross section of the bar. Our observations are also verified by results from the theory of elasticity.[†] From the theory, we show in Fig. 11.17 the distribution of stress for various cross sections of an axially loaded short compression member. The sections are taken at distances of $b/4, b/2,$ and b from the load where b is the width of the member. It appears from the figure that the concentrated load produces a highly nonuniform stress distribution and large local stresses near the load. However, notice how quickly the stress smoothes out to a nearly uniform distribution. Away from the load at a distance equal to the width of the bar, the maximum stress differs from the average stress, $\sigma_{ave} = P/A$, by only 2.7 percent.

This smoothing out of the stress distribution is an illustration of Saint-Venant's principle. Barre de Saint-Venant, a French engineer and mathematician, observed that near loads, high localized stresses may occur, but away from the load at a distance equal to the width or depth of the member, the localized effect disappears and the value of the stress can be determined from an elementary formula such as $\sigma = P/A$. The principle is important because it applies to

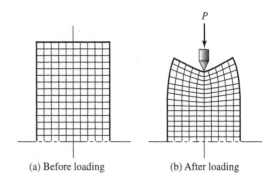

(a) Before loading   (b) After loading

FIGURE 11.16. Deformation of a grid pattern on an axially loaded rubber bar. (Top half of the bar shown in the figure.)

[†]S. Timoshenko and J. N. Goodier, *Theory of Elasticity*, 3rd ed. (New York: McGraw-Hill, Inc., 1970), 60.

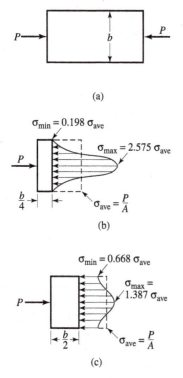

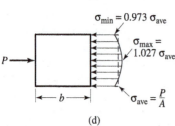

FIGURE 11.17

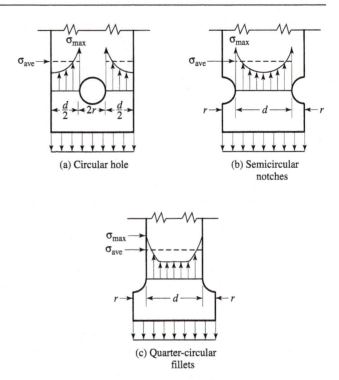

(a) Circular hole (b) Semicircular notches

(c) Quarter-circular fillets

FIGURE 11.18. Stress distribution in flat bars with axial loading.

Values of the stress-concentration factor are available in the technical literature in the form of graphs and tables. In Fig. 11.19, we show stress-concentration factors for axially loaded bars with the three different geometric discontinuities of Fig. 11.18. The following example illustrates the use of the graphs.

almost every other type of member and load as well. It permits us to develop simple relationships between loads and stresses and loads and deformations. The determination of the local effect of loads is then considered as a separate problem usually by experiment or the theory of elasticity.

11.10 STRESS CONCENTRATIONS

We see that the stress near a concentrated load is several times larger than the average stress in the member. A similar condition exists at discontinuities in a member. Various types of members under different loading conditions have been studied experimentally and by the theory of elasticity. In Fig. 11.18, we show the stress distribution in flat bars under an axial load with a circular hole, semicircular notches, and quarter-circular fillets.

In the study of geometric discontinuities, a ratio called the *stress-concentration factor* K is defined. It is the ratio of the maximum local stress to the average stress. That is,

$$K = \frac{\sigma_{max}}{\sigma_{ave}} \qquad (11.7)$$

where the average stress is determined for the net area (the smallest cross-sectional area of the discontinuity).

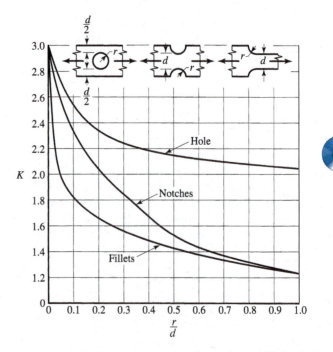

FIGURE 11.19. Stress-concentration factor for axially loaded flat bars.[†] (Average stress is computed for minimum cross section.)

[†]M. M. Frocht, "Photolastic Studies in Stress Concentration," *Mechanical Engineering* 58, no. 8 (August 1936), pp. 485–89.

Example 11.10   A tension member with a cross section measuring w by t with a hole of radius r supports an axial load P as shown in Fig. 11.20. Determine the minimum required thickness t if the member is made of material with (a) an ultimate strength of 25 ksi if $P = 1.5$ kip, $w = 1.2$ in., and $r = 0.25$ in. (using a factor of safety of 2), and (b) an ultimate strength of 340 MPa if $P = 14.0$ kN, $w = 40$ mm, and $r = 10$ mm (using a factor of safety of 2.5).

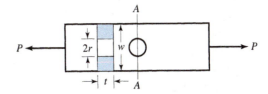

FIGURE 11.20

Solution: a.   To find the stress-concentration factor, we computed the ratio

$$\frac{r}{d}\frac{\text{(radius of the hole)}}{\text{(net or minimum width)}} = \frac{0.25 \text{ in.}}{(1.2 - 0.5) \text{ in.}} = 0.357$$

From the curve in Fig. 11.19 for $r/d = 0.357$, the stress-concentration factor $K = 2.2$. The allowable stress

$$\sigma_a = \frac{\sigma_u}{\text{F.S.}} = \frac{25}{2} = 12.5 \text{ ksi}$$

and the maximum normal stress including the stress-concentration factor

$$\sigma = K\frac{P}{A} = \frac{2.2(1.5)}{(1.2 - 0.5)t} \text{ ksi}$$

Equating the maximum normal stress σ and the allowable stress $\sigma_a = 12.5$ ksi, we have

$$\frac{2.2(1.5)}{(1.2 - 0.5)t} = 12.5$$

and solving for t,

$$t = \frac{2.2(1.5)}{0.7(12.5)} = 0.381 \text{ in.} \qquad\qquad \textbf{Answer}$$

b.   We first compute the ratio

$$\frac{r}{d} = \frac{\text{(radius of the hole)}}{\text{(net or minimum width)}} = \frac{10}{(40 - 20)} = 0.5$$

From the curve in Fig. 11.19 for $r/d = 0.5$, the stress-concentration factor $K = 2.13$. The allowable stress

$$\sigma_a = \frac{\sigma_u}{\text{F.S.}} = \frac{340}{2.5} = 136 \text{ Mpa} = 136 \text{ N/mm}^2$$

The maximum normal stress including the stress-concentration factor

$$\sigma = K\frac{P}{A} = \frac{2.13(14 \times 10^3)}{(40 - 20)t} \text{ N/mm}^2$$

Equating the maximum normal stress σ and the allowable stress $\sigma_a = 136$ N/mm², we have

$$\frac{2.13(14 \times 10^3)}{20t} = 136$$

and solving for t,

$$t = \frac{2.13(14 \times 10^3)}{20(136)} = 10.96 \text{ mm} \qquad\qquad \textbf{Answer}$$

STRESS CONCENTRATION

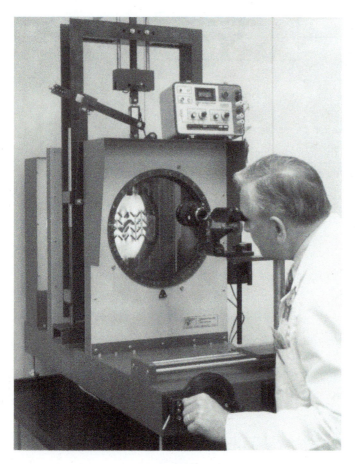

SB.18(a)

SB.18(b)

Numerical values for *stress-concentration factors* are generally determined experimentally. In one common technique, a model or prototype of any specific two-dimensional object is constructed of a transparent *photoelastic* material (typically polycarbonate, polyurethane, or epoxy sheet). This model is then placed in a loading fixture SB.18(a), and a source of polarized light is positioned behind the assembly. If a load is now applied to the model, the transmitted light reveals a series of colored fringes SB.18(b) that appear when the model is viewed through appropriate filters. Similar to the lines of constant elevation found on a topographic map, these fringes represent lines of constant stress. The configuration of the fringes and their proximity to each other yield valuable information, both qualitative and quantitative, about how the actual stresses in the object differ from theoretical values. Three-dimensional objects may be coated with a photoelastic film and viewed under load using reflected polarized light. *(Courtesy of Vishay Intertechnology, Inc., Malvern, Pennsylvania, www.vishay.com)*

Whether stress concentrations are important in design depends on the nature of the loads and the material used for the member. If the loads are applied statically on a ductile material (defined in Sec. 11.2), stress concentrations are usually not significant. However, for impact or repeated loads on ductile material or static loading on brittle material (defined in Sec. 11.2), stress concentration cannot be ignored.

Stress concentrations are usually not important in conventional building design. They may, however, be important in the design of supports for machinery and equipment and for crane runways. Stress concentrations should always be considered in the design of machines. In most machine part failures, cracks form at points of high stress. The cracks continue to grow under repeated loading until the section can no longer support the loads. The failure is usually sudden and dangerous.

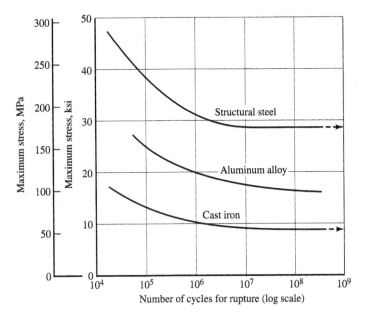

FIGURE 11.21. $\sigma - n$ Curve for ferrous and nonferrous materials.
(For illustration purposes only.)

11.11 REPEATED LOADING, FATIGUE

Many structural and most machine members are subjected to repeated loading and the resulting variations of the stresses in the members. These stresses may be significantly less than the static breaking strength of the member, but if repeated enough times, failure can occur. The failure is due to *fatigue.* The mechanism of a fatigue failure is a progressive cracking that leads to fracture. If a crack forms from repeated loads, the crack usually forms at a point of maximum stress. Such a point may form on the outside of the member at scratches, machine marks, etc., or inside the member from flaws in member material or from stress concentrations due to abrupt changes in the geometry of the member.

In a fatigue test, a specimen of the material is loaded and unloaded until failure occurs. The repeated loading produces stress reversals or large stress changes in either tension or compression. The lower the stress level the greater the number of cycles before failure. In the test, the stress level is lowered in steps until a level is reached where failure does not occur. That stress level is the *fatigue strength* or *endurance limit.* From the test, a $\sigma - n$ curve relating σ (stress required for failure) to n (the number of cycles of stress for failure) is obtained as in Fig. 11.21. From the figure, we see that a specimen of ferrous material such as structural steel may be loaded at a stress level of 39 ksi (269 MPa) with failure in 10^5 cycles, but at a level of 29 ksi (200 MPa) failure occurs at 10^7 cycles without stress concentrations.

Nonferrous metals such as aluminum and magnesium do not exhibit a fatigue strength and must be tested until the number of cycles (service life) that the metal will be subjected to is reached. For example, in Fig. 11.21, if an aluminum member has a service life of 10^7 cycles, the maximum stress would be 17.5 ksi (121 MPa) without stress concentrations.

Because the fatigue test is difficult and time consuming and the tension test is quick and easy, it is important to know the relationship between the fatigue strength and the tensile strength. If handbook data are not available and stress concentrations are not present, the fatigue strength for ferrous materials may be taken at approximately 50 percent of the tensile strength. For nonferrous metals, it must be lowered to approximately 35 percent.

PROBLEMS

11.47 A long slot is cut from a steel rod as shown. If $w = 4$ in., $t = 0.75$ in., $r = 0.7$ in., $L_1 = 1.5$ ft, $L_2 = 8$ ft, $P = 24$ kip, and $E = 30 \times 10^6$ psi, find (a) the maximum stress in the rod and (b) the total elongation of the rod. Assume that the stress-concentration factor for a round hole is applicable. When determining the elongation, assume that stress concentrations can be neglected and the slot extends the entire length L_1.

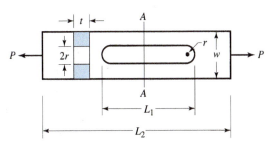

PROB. 11.47 and PROB. 11.48

11.48 Solve Prob. 11.47 for $w = 100\,\text{mm}, t = 20\,\text{mm}$, $r = 18\,\text{mm}$, $L_1 = 0.5\,\text{m}$, $L_2 = 2.5\,\text{m}$, $P = 110\,\text{kN}$, and $E = 200\,\text{Gpa}$.

11.49 A hole is drilled through a flat bar as shown. Dimensions for the bar are $w = 50\,\text{mm}$, $t = 12.5\,\text{mm}$, and $P = 25\,\text{kN}$. Determine the maximum stress in the bar if the diameter D of the hole is equal to (a) 3.5 mm, (b) 7.0 mm, and (c) 10.5 mm.

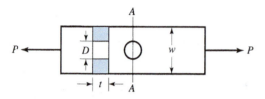

PROB. 11.49 and PROB. 11.50

11.50 Solve Prob. 11.49 if the dimensions of the bar are $w = 2\,\text{in.}$, $t = 0.5\,\text{in.}$, $P = 5.5\,\text{kip}$, and the diameter D of the hole is equal to (a) 1/8 in., (b) 1/4 in., and (c) 3/8 in.

11.51 Circular notches are cut in a flat bar as shown. If $w = 1\,\text{in.}$, $t = 0.5\,\text{in.}$, $r = 1/8\,\text{in.}$, and the allowable stress is 20 ksi, determine (a) the allowable load, (b) the allowable load for a similar bar without the notches, and (c) the percent reduction in the allowable load caused by the notches.

11.52 Solve Prob. 11.51 if the dimensions of the bar are $w = 25\,\text{mm}$, $t = 12.5\,\text{mm}$, $r = 3.5\,\text{mm}$, and the allowable stress is 140 MPa.

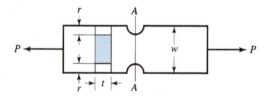

PROB. 11.51 and PROB. 11.52

11.53 A standard steel test specimen is under load, and the total elongation of the gauge length g is 0.002 in. If $w = 3/4\,\text{in.}$, $t = 3/8\,\text{in.}$, $r = 1/8\,\text{in.}$, and $g = 2\,\text{in.}$, determine (a) the axial load P on the bar, (b) the corresponding stress in the middle of the specimen, and (c) the maximum stress at the fillets. ($E = 30 \times 10^6\,\text{psi}$)

11.54 Solve Prob. 11.53 if the elongation of the gauge length g is 0.05 mm when $w = 20\,\text{mm}$, $t = 10\,\text{mm}$, $r = 4\,\text{mm}$, and $g = 50\,\text{mm}$. ($E = 200\,\text{GPa}$)

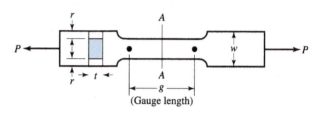

PROB. 11.53 and PROB. 11.54

11.55 The bar shown is made from a material with an allowable stress of 12 ksi. If $w = 4\,\text{in.}$ and $r = 3/4\,\text{in.}$, determine the minimum allowable thickness t of the bar that can support 15 kip.

11.56 The bar shown has a thickness of 15 mm. If $w = 60\,\text{mm}$, $r = 15\,\text{mm}$, and the allowable stress is 140 MPa, determine the maximum allowable axial force that can be applied to the bar.

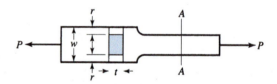

PROB. 11.55 and PROB. 11.56

SHEAR STRESSES AND STRAINS: TORSION

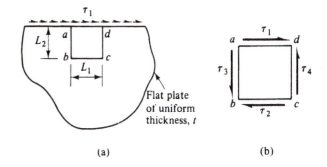

(a)   (b)

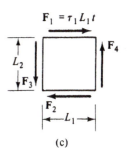

(c)

FIGURE 12.1

12.1 INTRODUCTION

In Chapters 10 and 11, we considered axially loaded members. In this chapter, we begin by discussing shearing stresses and strains. Then, we consider slender circular members that are subjected primarily to torsion or twisting. Such members are commonly called shafts, and we will learn that the circular shaft acted on by torsion is in a state of pure shear.

One of the main uses of the shaft is to transmit mechanical power from one location to another. In the automobile, a shaft is used to transmit power in the steering column and in the drive shaft and axle. A shaft may also be used to transmit power from an electric motor to various mechanical devices. An application of a shaft that does not involve the transmission of power is the torsion spring.

12.2 SHEARING STRESS ON PLANES AT RIGHT ANGLES

Consider the shear stress acting tangent to a plane edge of a flat plate of uniform thickness, t [Fig. 12.1(a)]. The average shear stress over a part of the edge from a to d is τ_1. The average shear stress is shown on the four sides of an element of the plate $abcd$ in Fig. 12.1(b). The plate is in equilibrium; therefore, the element $abcd$ of the plate is in equilibrium. We draw a free-body diagram of the element in Fig. 12.1(c).

A force $F_1 = \tau_1 A = \tau_1 L_1 t$ acts to the right on the top of the element. The shear force on the bottom of the element is equal to the average shear stress τ_2 multiplied by the shear area $F_2 = \tau_2 A = \tau_2 L_1 t$. Forces on the left and right sides of the element are $F_3 = \tau_3 L_2 t$ and $F_4 = \tau_4 L_2 t$, respectively. From equilibrium,

$$\Sigma F_x = 0 \qquad F_1 - F_2 = 0$$
$$\tau_1 L_1 t - \tau_2 L_1 t = 0 \quad \text{or} \quad \tau_1 = \tau_2$$
$$\Sigma F_y = 0 \qquad -F_3 + F_4 = 0$$
$$-\tau_3 L_2 t + \tau_4 L_2 t = 0 \quad \text{or} \quad \tau_3 = \tau_4$$
$$\Sigma M_b = 0 \qquad -F_1 L_2 + F_4 L_1 = 0$$
$$-\tau_1 L_1 t L_2 + \tau_4 L_2 t L_1 = 0 \quad \text{or} \quad \tau_1 = \tau_4$$

Therefore, $\qquad \tau_1 = \tau_2 = \tau_3 = \tau_4 = \tau$

Thus, all four average shear stresses acting on the element are equal. Notice that the shear stresses meet tip to tip and tail to tail at the corners of the element. If we let the element shrink to a point, the average stresses become stresses at a point. Hence, *the shearing stresses at a point acting on mutually perpendicular planes are equal.*

12.3 SHEARING STRAINS

The shearing stresses shown in Fig. 12.1(b) cause distortion of the element *abcd*. The diagonal *b–d* is lengthened, and the diagonal *a–c* is shortened. The distortion is illustrated in Fig. 12.2. The shearing strain is equal to the distance *aa'* divided by the distance *ab*. Because the angle represented by the Greek lowercase letter γ (gamma) is small,

$$\tan\gamma = \gamma(\text{radians}) = \frac{aa'}{ab}$$

Hence, the shearing strain is numerically equal to γ measured in radians.

12.4 HOOKE'S LAW FOR SHEAR

Experiments show that as for normal stress and strain, shearing stress is proportional to shearing strain as long as stress does not exceed the proportional limit. Hooke's law for shear may be expressed as

$$\tau = G\gamma \qquad (12.1)$$

where *G* is the *shear modulus* or the *modulus of rigidity*. The shearing modulus, *G*, is a constant for a given material. It is

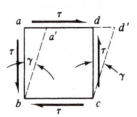

FIGURE 12.2

expressed in the same units as stress because γ is measured in radians, which are dimensionless quantities. The shear modulus, *G*, has a value about 40 percent of the value of the elastic modulus *E*.

It can be shown that the three elastic constants—modulus of elasticity, *E*, modulus of rigidity, *G*, and Poisson's ratio, ν—are not independent of each other for an isotropic material. An *isotropic material* has the same properties in all directions. The relationship between elastic constants is given by the equation

$$G = \frac{E}{2(1 + \nu)} \qquad (12.2)$$

Example 12.1   In the test on a steel bar, $E = 29.57 \times 10^6$ psi and $\nu = 0.303$. Find the shear modulus, *G*.

Solution: From Eq. (12.2),

$$G = \frac{29.57(10^6)}{2(1 + 0.303)} = 11.35 \times 10^6 \text{ psi} \qquad \textbf{Answer}$$

PROBLEMS

12.1 Determine the shear modulus for a magnesium alloy that has a modulus $E = 6.5 \times 10^6$ psi and Poisson's ratio $\nu = 0.340$.

12.2 Determine the shear modulus for a steel that has a modulus $E = 206 \times 10^6$ kN/m² (kPa) and Poisson's ratio $\nu = 0.25$.

12.3 Determine Poisson's ratio for a copper alloy that has a modulus $E = 105 \times 10^6$ kN/m² (kPa) and a shear modulus of $G = 37.2 \times 10^6$ kN/m² (kPa).

12.4 Determine Poisson's ratio for a cast iron that has a modulus of $E = 12 \times 10^6$ psi and a shear modulus $G = 4.8 \times 10^6$ psi.

12.5 TORSION OF A CIRCULAR SHAFT

We consider here the torsion of a circular shaft. Let the bottom end of the shaft be fixed and a torque *T* be applied at the top end as shown in Fig. 12.3(a).

Geometry of Deformation

Cutting through the shaft, we draw a free-body diagram of the part of the shaft between plane cross sections *J–J* and *K–K* in Fig. 12.3(b). Both cross sections are normal to the axis of the member. We assume that these plane cross sections remain plane after the torque is applied. The plane *OABD*, which was parallel to the axis of the shaft, moves to a new position *OEBD* as the top cross section rotates through an angle, φ (phi), after application of the torque *T*. The line *OA* remains straight as it rotates through the angle φ to the new position *OE*. The magnitude of the shearing strain at a distance *r* from the center of the shaft is given by the angle *FHG* = γ expressed in radians. The angle γ is small and can be expressed closely in radian measure as

$$\gamma = \frac{FG}{FH} = \frac{FG}{L} \qquad (a)$$

The angle *FOG* = φ can be expressed in radians as $\varphi = FG/FO = FG/r$ or $FG = r\varphi$. Combining this with Eq. (a), we obtain

$$\gamma = \frac{r\varphi}{L} \qquad (12.3)$$

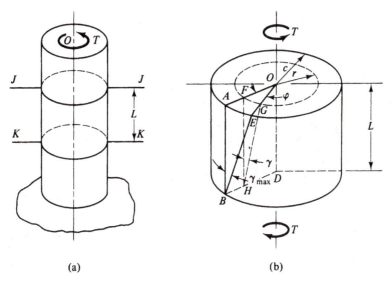

FIGURE 12.3

Similarly, the magnitude of the maximum shearing strain is given by the angle $ABE = \gamma_{max}$. The angle γ_{max} is also small and can be expressed closely in radian measure as

$$\gamma_{max} = \frac{AE}{AB} = \frac{AE}{L} \qquad \textbf{(b)}$$

The angle AOE or φ can be expressed in radians as $\varphi = AE/AO = AE/c$ or $AE = c\varphi$. Combining this with Eq. (b) yields

$$\gamma_{max} = \frac{c\varphi}{L} \qquad \textbf{(12.4)}$$

Hooke's Law

In our consideration of the geometry of deformation, no restrictions have been placed on the material from which the shaft was made except that it must be isotropic (have the same physical properties in all directions). We assume now that the material follows Hooke's law for shear. Combining Hooke's law, Eq. (12.1), with Eqs. (12.3) and (12.4), we have

$$\tau = G\gamma = \frac{G\varphi}{L} r \qquad \textbf{(12.5)}$$

and

$$\tau_{max} = G\gamma_{max} = \frac{G\varphi}{L} c \qquad \textbf{(12.6)}$$

Dividing Eq. (12.6) by Eq. (12.5),

$$\frac{\tau_{max}}{\tau} = \frac{c}{r} \qquad \textbf{(c)}$$

Thus, we see that the shear stress on the cross section of the shaft is proportional to the distance from the center of the cross section as shown in Fig. 12.4.

Equilibrium

Consider (Fig. 12.5) a narrow ring of mean radius r and area ΔA concentric with the center of the cross section of the shaft. The magnitude of the shearing stress at every point in the ring is equal to τ, and it is directed normal to a radius drawn from the center of the cross section O to that point. The moment about O or torque due to the stresses on the ring is, therefore,

$$\underset{\substack{\text{(torque}\\\text{for}\\\text{ring)}}}{\Delta T} = \underset{\substack{\text{(stress)(area)}\\\text{(force)}}}{\tau \ \ \Delta A} \qquad \underset{\text{(moment arm)}}{r}$$

Adding the torque for all rings to include the entire cross-sectional area, we obtain the total torque.

$$T = \underset{\substack{\text{total}\\\text{area}}}{\Sigma \, \tau r \, \Delta A}$$

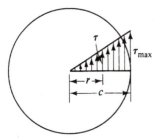

FIGURE 12.4

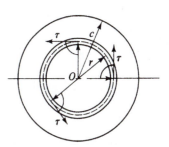

FIGURE 12.5

Deformation of a Circular Shaft

Substituting the value of the shear stress, τ, from Eq. (12.5), we have

$$T = \sum_{\substack{\text{total} \\ \text{area}}} \frac{G\varphi}{L} r^2 \Delta A$$

Because the shear modulus G, angle of twist φ, and the length L are constant for each ring,

$$T = \frac{G\varphi}{L} \sum_{\substack{\text{total} \\ \text{area}}} r^2 \Delta A \qquad \textbf{(d)}$$

However, $\sum r^2 \Delta A$ is also a constant for a particular cross section called the *polar moment of inertia, J*. The polar moment of inertia was discussed in Sec. 9.10. The formula for the polar moment of inertia of a circle of radius c and diameter d is found from the equation $J = I_x + I_y$ and Table A.2 of the Appendix. Therefore,

$$J = \frac{\pi c^4}{2} = \frac{\pi d^4}{32} \qquad \textbf{(12.7)}$$

Substituting the polar moment of inertia, J, in Eq. (d), we obtain

$$T = \frac{G\varphi J}{L} \quad \text{or} \quad \varphi = \frac{TL}{GJ} \qquad \textbf{(12.8)}$$

where φ is the angle of twist in radians, a dimensionless quantity. The internal torque T is expressed in pound-inches or newton·meters (N·m), L in inches or meters (m), G in psi or newtons/meter^2 (N/m^2), and J in in.^{4} or m^4. Note that for a hollow shaft,

$$J = \frac{\pi(d_o^4 - d_i^4)}{32} \qquad \textbf{(12.9)}$$

where d_o and d_i represent the outside and inside shaft diameters, respectively.

Stresses in a Circular Shaft

The relationship of shear stress to torque may be obtained by substituting Eq. (12.8) in Eq. (12.5), that is,

$$\tau = \frac{Gr\varphi}{L} = \frac{Gr}{L}\frac{TL}{GJ} \quad \text{or} \quad \tau = \frac{Tr}{J} \qquad \textbf{(12.10)}$$

The maximum value of the shear stress occurs when r has its maximum value, $r = c$.

$$\tau_{\max} = \frac{Tc}{J} \qquad \textbf{(12.11)}$$

This equation is known as the *torsion formula*. It gives the maximum stress in terms of the torque and the dimensions of the member. Torque, T, will be expressed in pound-inches or newton·meters (N·m), c in inches or meters, and J in in.^{4} or m^4. The torsion shearing stress will have units of

$$\frac{(\text{lb-in.})(\text{in.})}{(\text{in.}^4)} = \frac{\text{lb}}{\text{in.}^2} = \text{psi}$$

or in SI units

$$\frac{(\text{N·m})(\text{m})}{(\text{m}^4)} = \frac{\text{N}}{\text{m}^2} \quad \text{or} \quad \text{Pa} \quad \text{(pascals)}$$

For solid shafts, Eq. (12.11) can be put into a form that is especially useful in designing a shaft when maximum torque and allowable shearing stress are known. Letting $c = r = d/2$, and $J = \pi d^4/32$, minimum required shaft diameter, $d_{\min}$, becomes

$$d_{\min} = \sqrt[3]{\frac{16T}{\pi\tau}} = \left(\frac{16\,T}{\pi\tau}\right)^{1/3} \qquad \textbf{(12.12)}$$

12.6 FURTHER COMMENTS ON THE TORSION OF A CIRCULAR SHAFT

In our analysis of the torsion of a circular shaft, we have only considered the shear stress acting on a cross section of the shaft. In Fig. 12.6, we consider an element *abcd* on the outside of a circular shaft before and after torque is applied. Side *ad* of the element lies along a cross section, and sides *ab* is normal to the cross section. Thus, the shear stress on sides *ad* and *ab* must be equal, from Sec. 12.2. The value of the stress is given by Eq. (12.11).

We may also apply the same argument to an element on any concentric cylinder of radius r inside the circular shaft.

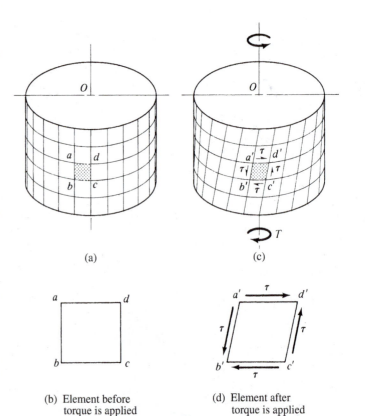

(a)

(b) Element before torque is applied

(c)

(d) Element after torque is applied

FIGURE 12.6

Therefore, the distribution of shear stress on the cross section and on a plane normal to the cross section are the same, as shown in Fig. 12.7.

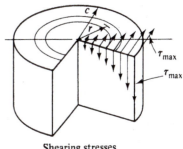

12.7 PROBLEMS INVOLVING DEFORMATION AND STRESS IN A CIRCULAR SHAFT

In this section, we solve several problems involving the deformation and stress in a circular shaft.

Shearing stresses
on mutually perpendicular
planes are equal

FIGURE 12.7

Example 12.2   Find the torque that a solid circular shaft with a diameter of 100 mm can transmit if the maximum shearing stress is 50 MPa. What is the angle of twist per meter of length? ($G = 80 \times 10^3$ Mpa)

Solution: The polar moment of inertia J is given by

$$J = \frac{\pi d^4}{32} = \frac{\pi (100)^4}{32} = 9.82 \times 10^6 \, mm^4$$

From Eq. (12.11), $\tau_{max} = Tc/J$; therefore,

$$T = \frac{\tau_{max} J}{c} = \frac{50(9.82 \times 10^6)}{50} = 9.82 \times 10^6 \, N \cdot mm \qquad \textbf{Answer}$$

The angle of twist is given by Eq. (12.8):

$$\varphi = \frac{TL}{GJ} = \frac{9.82 \times 10^6 (10^3)}{80 \times 10^3 (9.82 \times 10^6)} = 1.250 \times 10^{-2} \, rad$$

$$= \frac{180}{\pi}(1.250 \times 10^{-2}) = 0.716° \qquad \textbf{Answer}$$

Example 12.3   Find the shearing stress on the outside and inside of a long hollow circular shaft with an outside diameter of 6 in. and an inside diameter of 4 in. if a torque of 20,000 lb-ft is applied. What will the angle of twist be in a length of 15 ft? ($G = 12 \times 10^6$ psi)

Solution: The ratio of J/c for a hollow circular shaft with the cross section shown in Fig. 12.8 is given by

$$\frac{J}{c} = \frac{\pi(d_o^4 - d_i^4)}{32}\left(\frac{2}{d_o}\right)$$

If we let $d_i/d_o = n$, then

$$\frac{J}{c} = \frac{\pi(1 - n^4)d_o^3}{16} \qquad \textbf{(a)}$$

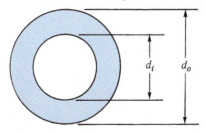

$$J = \frac{\pi}{32}(d_o^4 - d_i^4) = \frac{\pi}{32}(1 - n^4)d_o^4$$
where $n = d_i/d_o$

FIGURE 12.8

For $d_o = 6$ in. and $d_i = 4$ in., $n = 4/6 = 0.667$, and from Eq. (a),

$$\frac{J}{c} = \frac{\pi\left[1 - (0.667)^4\right](6)^3}{16} = 34.0 \text{ in.}^3$$

From Eq. (12.11), the stress on the outside of the shaft

$$\tau_o = \frac{T}{J/c} = \frac{20,000(12)}{34.0} = 7060 \text{ psi} \qquad \textbf{Answer}$$

The stress is proportional to the distance from the center of the shaft; therefore, the stress on the inside of the hollow shaft

$$\tau_i = \frac{d_o}{d_i}\tau_o = \frac{4}{6}(7060) = 4710 \text{ psi} \qquad \textbf{Answer}$$

The ratio $J/c = 34.0$ in.^{3}; therefore, $J = 34c = 34(3) = 102$ in.^{4} From Eq. (12.8), the angle of twist

$$\varphi = \frac{TL}{GJ} = \frac{20,000(12)(15)(12)}{12 \times 10^6(102)} = 0.0353 \text{ rad}$$

$$= \frac{180}{\pi}(0.0353) = 2.02° \qquad \textbf{Answer}$$

Example 12.4   Torques are applied at pulleys A, B, and C as shown in Fig. 12.9(a). The hollow shaft has an outside diameter $d_o = 2.5$ in. and an inside diameter $d_i = 2.0$ in. Determine the maximum shearing stress on the outside of the shaft.

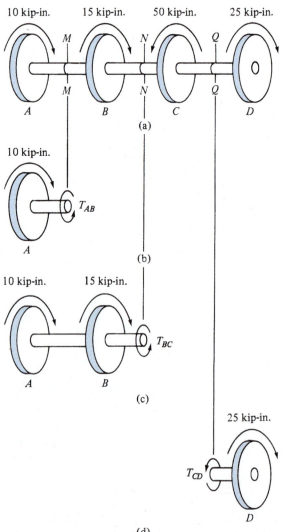

FIGURE 12.9

Solution: To determine the maximum torque in the shaft, we cut the shaft at sections M–M, N–N, and Q–Q, and draw free-body diagrams of the section to the left for sections M–M and N–N as shown in Fig. 12.9(b) and (c). For the section Q–Q, the free-body diagram is drawn for the section to the right [Fig. 12.9(d)]. (The section to the right results in a simpler free-body diagram.) For sections M–M, N–N, and Q–Q, from equilibrium

$$\Sigma M_x = 0 \qquad T_{AB} - 10 = 0 \qquad T_{AB} = 10 \, \text{kip-in.}$$

$$\Sigma M_x = 0 \qquad T_{BC} - 10 - 15 = 0 \qquad T_{BC} = 25 \, \text{kip-in.}$$

$$\Sigma M_x = 0 \qquad T_{CD} - 25 = 0 \qquad T_{CD} = 25 \, \text{kip-in.}$$

The maximum torque $T_{max} = T_{BC} = T_{CD} = 25$ kip-in. For $d_o = 2.5$ and $d_i = 2.0$, $n = d_i/d_o = 2.0/2.5 = 0.8$. From Eq. (a) of Example 12.3,

$$\frac{J}{c} = \frac{\pi}{16}(1 - n^4)d_o^3 = \frac{\pi}{16}\left[1 - (0.8)^4\right](2.5)^3 = 1.811 \text{ in.}^3$$

The maximum shear stress from Eq. (12.11):

$$\tau_{max} = \frac{T_{max}}{J/c} = \frac{25.0}{1.811} = 13.80 \, \text{ksi} \qquad\qquad\qquad \textbf{Answer}$$

Example 12.5   The electric motor exerts a torque of 3.2 kN·m on the pulley system shown in Fig. 12.10(a). If the shaft is solid and the allowable shear stress is 70 MPa, determine the required diameter of each shaft.

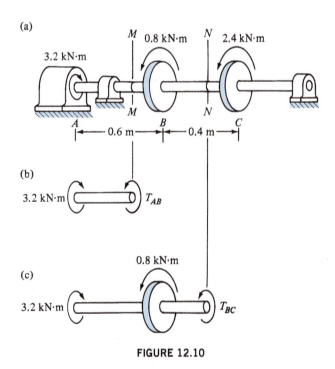

FIGURE 12.10

Solution: We cut the shaft at sections M–M and N–N between AB and BC and draw free-body diagrams in Fig. 12.10(b) and (c). For sections M–M and N–N, from equilibrium,

$$\Sigma M_x = 0 \qquad -3.2 + T_{AB} = 0 \qquad T_{AB} = 3.2 \, \text{kN·m}$$

$$\Sigma M_x = 0 \qquad -3.2 + 0.8 + T_{BC} = 0 \qquad T_{BC} = 2.4 \, \text{kN·m}$$

Shaft AB: From Eq. (12.12),

$$d_{min} = \sqrt[3]{\frac{16 \times 3.2 \times 10^3 \, \text{N·m}}{\pi \times 70 \times 10^6 \, \text{N/m}^2}} = \sqrt[3]{0.0002328 \, \text{m}^3}$$

$$d_{min} = 0.0615 \, \text{m} = 61.5 \, \text{mm} \qquad\qquad\qquad\qquad \textbf{Answer}$$

Shaft BC: Again, from Eq. (12.12),

$$d_{min} = \sqrt[3]{\frac{16 \times 2.4 \times 10^3 \text{ N·m}}{\pi \times 70 \times 10^6 \text{ N/m}^2}} = \sqrt[3]{0.0001746 \text{ m}^3}$$

$$d_{min} = 0.0559 \text{ m} = 55.9 \text{ mm} \qquad\qquad \textbf{Answer}$$

Example 12.6   If the solid shaft of Example 12.5 [Fig. 12.10(a)] has a uniform diameter of 65 mm, determine the angle through which (a) pulley B rotates with respect to motor A and (b) pulley C rotates with respect to motor A. Use $G = 80 \times 10^3$ MPa for the shaft.

Solution: For the shafts, the polar moment of inertia $J = \pi d^4/32 = \pi(64)^4/32 = 1.752 \times 10^6 \text{ mm}^4$.

Shaft AB:

$$\varphi_{AB} = \frac{TL}{GJ} = \frac{3.2 \times 10^6(600)}{80 \times 10^3(1.752 \times 10^6)} = 13.70 \times 10^{-3} \text{ rad}$$

$$= \frac{180}{\pi}(13.70 \times 10^{-3}) = 0.785°$$

Shaft BC:

$$\varphi_{BC} = \frac{TL}{GJ} = \frac{2.4 \times 10^6(400)}{80 \times 10^3(1.752 \times 10^6)} = 6.84 \times 10^{-3} \text{ rad}$$

$$= \frac{180}{\pi}(6.84 \times 10^{-3}) = 0.392°$$

a.   The rotation of pulley B from motor A:

$$\varphi = \varphi_{AB} = 0.785° \qquad\qquad \textbf{Answer}$$

b.   The rotation of pulley C from motor A:

$$\varphi = \varphi_{AB} + \varphi_{BC} = 0.785 + 0.392 = 1.777° \qquad\qquad \textbf{Answer}$$

12.8 TORSION TEST

In a test of a solid circular shaft in torsion, the outside of the shaft is more highly stressed than the interior of the shaft; therefore, the outside of the shaft reaches the proportional limit or yield point before it is reached in the interior of the shaft. As a result, the onset of yielding is covered up by the fact that the inside of the shaft continues to support loads with stresses proportional to strain. Only after considerable yielding has occurred does the effect show in the test data. To overcome this difficulty, a test on a thin hollow circular

specimen can be performed. The stress across the thin wall of the specimen will have approximately the same value, and the onset of yielding will show in the test data soon after it first occurs. From Eq. (12.8), $\varphi = TL/GJ$ or

$$\frac{T}{\varphi/L} = GJ \qquad\qquad \textbf{(a)}$$

In the torsion test, a torque is applied to a solid or hollow circular shaft by a torsion testing machine, and for each specified value of the torque, we measure the angle of twist with a mechanical device called a *troptometer*. The values obtained can then be plotted in a torque-angle of twist per unit of length curve. A typical curve is shown in Fig. 12.11.

From Eq. (a), the slope of such a curve will be equal to the shear modulus multiplied by the polar moment of inertia. The shear modulus is approximately 40 percent of the elastic modulus ($G \approx 0.40E$).

12.9 POWER TRANSMISSION

To design a power transmission shaft, we require a relationship between power and torque. A cable is wound around the pulley shown in Fig. 12.12, and a force F is applied at the

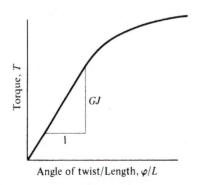

FIGURE 12.11

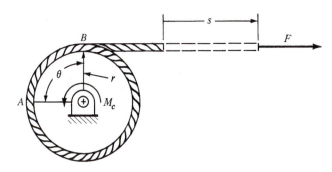

FIGURE 12.12

where P is the power produced by a constant torque T acting on a body rotating at a constant angular velocity θ/t.

U.S. Customary Units

The unit of power is the horsepower (hp); (1 hp = 550 ft-lb/sec or 33,000 ft-lb/min). The angular velocity θ/t is usually expressed in revolutions per minute (rpm). Expressing Eq. (c) in customary units, we have

$$P = \frac{TN}{63,000} \qquad (12.13)$$

where T = torque (lb-in.)
　　　N = revolutions per minute (rpm)
　　　P = horsepower (hp)

end of the cable. Motion is resisted by a friction couple M_c in the bearing. As a result of the force, the pulley rotates through an angle θ, and the end of the cable moves a distance s. The work Wk done by the force is equal to the product of the force F and the distance s, that is,

$$Wk = Fs \qquad (a)$$

The distance s that the force moves is equal to the arc length AB; thus, $s = r\theta$, where θ is in radians. Accordingly, from Eq. (a), $Wk = Fr\theta$. However, the torque $T = Fr$. Therefore,

$$Wk = T\theta \qquad (b)$$

Power is defined as the work per time. That is, $P = Wk/t$ or from Eq. (b)

$$P = \frac{T\theta}{t} \qquad (c)$$

SI Units

The unit of power is the watt (W); (1 W = 1 N·m/sec or 1 kW = 1000 N·m/sec). The angular velocity θ/t is usually expressed in revolutions per minute (rpm) or revolutions per second [hertz (Hz)]. Expressing Eq. (c) in SI units, we have

$$P = \frac{TN}{9550} \qquad (12.14)$$

where T = torque (N·m)
　　　N = revolutions per minute (rpm)
　　　P = kilowatts (kW)

For conversions: 1 hp = 0.746 kW and 1 kW = 1.341 hp.
$$(12.15)$$

Example 12.7　A steel shaft with a diameter d = 1.75 in. and length 3.5 ft transmits 80 hp from an electric motor to a compressor. If the allowable shear stress is 8000 psi and the allowable angle of twist is 2°, what is the minimum allowable speed of rotation? Use $G = 12 \times 10^6$ psi.

Solution:　For the shaft, $J = \pi d^4/32 = \pi(1.75)^4/32 = 0.921$ in.^{4} The torque as limited by stress is given by Eq. (12.11).

$$T = \frac{\tau J}{c} = \frac{8000(0.921)}{0.875} = 8.42 \times 10^3 \text{ lb-in.}$$

The angle of twist of 2° = $2\pi/180 = 3.49 \times 10^{-2}$ rad. The torque as limited by the angle of twist is given by Eq. (12.8).

$$T = \frac{\varphi GJ}{L} = \frac{3.49 \times 10^{-2}(12 \times 10^6)(0.921)}{3.5(12)} = 9.18 \times 10^3 \text{lb-in.}$$

The torque due to stress is the control. From Eq. (12.13),

$$N = \frac{63,000P}{T} = \frac{63,000(80)}{8.42 \times 10^3} = 598.6 = 599 \text{ rpm} \qquad \textbf{Answer}$$

Example 12.8　A motor through a set of gears at B drives a shaft at 28.9 rpm as shown in Fig. 12.13. The power required at A and C are 75 hp and 25 hp, respectively. If the allowable shearing stress in the shaft is 41 MPa, what are the required diameters? What are the angles of twist between B and A and between B and C? Use $G = 83 \times 10^3$ MPa, and assume torsion only in the shafts.

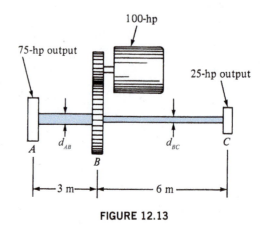

FIGURE 12.13

Solution: For shaft AB, 75 hp = 75(0.746) = 55.95 kW. From Eq. (12.14),

$$T_{AB} = \frac{9550P}{N} = \frac{9550(55.95)}{28.9} = 18.49 \times 10^3 \text{ N·m} = 18.49 \text{ kN·m}$$

For shaft BC, because torque is proportional to power,

$$T_{BC} = \frac{25}{75}(18.49) = 6.16 \text{ kN·m}$$

From Eq. (12.12), we have for shaft AB

$$d_{AB} = \sqrt[3]{\frac{16 \times 18.49 \times 10^3 \text{ N·m}}{\pi \times 41 \times 10^6 \text{ N/m}^2}} = \sqrt[3]{0.002297 \text{ m}^3}$$

$$d_{AB} = 0.1319 \text{ m} = 131.9 \text{ mm} \qquad \textbf{Answer}$$

From Eq. (12.12), we have for shaft BC

$$d_{BC} = \sqrt[3]{\frac{16 \times 6.16 \times 10^3 \text{ N·m}}{\pi \times 41 \times 10^6 \text{ N/m}^2}} = \sqrt[3]{0.0007652 \text{ m}^3}$$

$$d_{BC} = 0.09147 \text{ m} = 91.5 \text{ mm} \qquad \textbf{Answer}$$

For shaft AB, $J = \pi d^4/32 = \pi(131.9)^4/32 = 29.7 \times 10^6 \text{ mm}^4$, and from Eq. (12.8), we have

$$\varphi_{AB} = \frac{TL}{GJ} = \frac{18.49 \times 10^6(3 \times 10^3)}{83 \times 10^3(29.7 \times 10^6)} = 0.225 \times 10^{-3} \text{ rad}$$

$$= \frac{180}{\pi}(0.225 \times 10^{-3}) = 1.29° \qquad \textbf{Answer}$$

For shaft BC, $J = \pi d^4/32 = \pi(91.5)^4/32 = 6.88 \times 10^6 \text{ mm}^4$, and from Eq. (12.8), we have

$$\varphi_{BC} = \frac{TL}{GJ} = \frac{6.16 \times 10^6(6 \times 10^3)}{83 \times 10^3(6.88 \times 10^6)} = 0.647 \times 10^{-3} \text{ rad}$$

$$= \frac{180}{\pi}(0.647 \times 10^{-3}) = 3.66° \qquad \textbf{Answer}$$

PROBLEMS

Where appropriate, use the following values of the shear modulus: $G_{steel} = 12 \times 10^6$ psi (83×10^3 MPa) and $G_{aluminum} = 4 \times 10^6$ psi (28×10^3 MPa).

12.5 A solid circular shaft has a diameter of 50 mm. If the torque applied is 2.25 kN·m, what is the maximum shearing stress?

12.6 A hollow aluminum shaft has an outside diameter of 2.50 in. and an inside diameter of 2.25 in. What is the maximum shearing stress if a torque of 21,500 lb-in. is applied?

12.7 Find the torque required to produce a maximum shearing stress of 16,500 psi in a solid circular steel shaft with a diameter of 0.875 in.

DYNAMOMETERS

SB.19(a)

SB.19(b)

A device called a *dynamometer* is used to obtain the torque of a rotating shaft as required by Eqs. (12.11) and (12.12). Power sources such as internal combustion engines and electric motors are placed on a test stand [SB.19(a)], loads are applied to the output shaft using mechanical or electrical devices, and the torques produced by the power source under these loads are measured. The most common mechanical loading techniques use either a *drum brake* (similar to those found on automobiles and trucks) or a *water brake* consisting of a simple water pump that is partially filled with liquid. Two of the more popular electrical loading devices are the *eddy-current brake* and the *dc generator*. Both use magnetic fields to vary the resistance encountered by a metal rotor as it spins within the brake. For those applications where use of a dynamometer is impractical, sensors such as the miniature *wireless transducer* shown in [SB.19(b)] are often used. This device slides onto the rotating shaft and monitors torques produced during actual operation of a machine. The sensor receives torque measurements from shaft-mounted strain gauges, converts the signals to digital form, and transmits the data to a computer for recording or display. *(Courtesy of Land & Sea, Inc., North Salem, New Hampshire, www.land-and-sea.com)*

12.8 A circular tube has an outside diameter of 150 mm and an inside diameter of 100 mm. If the allowable shearing stress is 55 MPa, what is the allowable torque?

12.9 Torques are applied to the solid shaft shown. What is the maximum shearing stress in the shaft, and where does it occur?

12.10 The solid shaft shown is made of steel; what is the angle of twist in degrees between A and D?

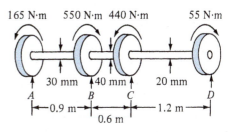

PROB. 12.9 and PROB. 12.10

12.11 A circular aluminum alloy tube has an outside diameter of 4 in. and an inside diameter of 3.4 in. What is the angle of twist in degrees if the torque applied is 30,000 lb-in. and the shaft is 4.5 ft long?

12.12 The electric motor exerts a torque of 18,000 lb-in. on the pulley system shown. If the shaft is solid and the allowable shear stress is 10,000 psi, determine the diameter of shafts AB and BC.

12.13 If the solid steel shaft shown has a diameter of 2.5 in., determine the angle through which (a) pulley B rotates with respect to motor A and (b) pulley C rotates with respect to motor A.

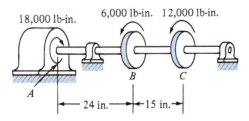

PROB. 12.12 and PROB. 12.13

12.14 The standard 1/2-hp electric motor found on most washing machines and dishwashers operates at 1750 rpm. If such a motor is connected to a water pump by a 1/4-in.-diamter steel shaft,
a. Find the torque carried by the shaft.
b. Compute the maximum shear stress developed in the shaft.
c. Find the factor of safety of the shaft for an allowable shear stress of 8500 psi.

12.15 A length of extra-strong 2-in. pipe (see Table A.8 of the Appendix) is used to connect the output shaft of an internal combustion engine to the spindle of a large circular sawblade used to cut logs into boards. If the pipe has an allowable shear stress of 15,600 psi and will operate at 800 rpm, what is the maximum horsepower engine that can be used?

12.16 A power source at C actuates valves at A and B through a long round steel control rod. A torque of 8.5 kN·m is required to actuate each valve. If the allowable shear stress in the control rod is 80 MPa, (a) what diameters are required for each section of the rod? (b) what is the angle of twist in the rod between the power source at C and the valve at A?

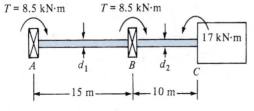

PROB. 12.16

12.17 The motor at A through a shaft AB and a set of gears drives the shaft CD. The shear stress in shaft AB is 55 MPa. If the shear stress in CD is also 55 MPa, (a) what is the diameter of shaft CD? (b) what is the initial angle of twist between the motor at A and the pulley at D? (Hint: From the shaft at B to the shaft at C through the set of gears, the torques transmitted are proportional to the pitch diameters ($T_B/T_C = D_B/D_C$), and the angles of rotation are inversely proportional to the pitch diameters ($\varphi_B/\varphi_c = D_C/D_B$). The torques and angles of rotation are in opposite directions.)

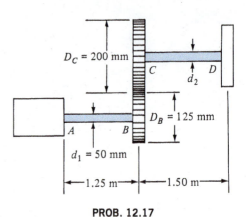

PROB. 12.17

12.18 A solid steel shaft delivers 5 hp at 30 revolutions per second. If the allowable shearing stress is limited to 95 MPa, find the required diameter of the shaft.

12.19 The allowable stress in a steel shaft with an outside diameter of 1.5 in. and an inside diameter of 1.1 in. is 12,000 psi. What hp can the shaft deliver at 1750 rpm?

12.20 A steel shaft with a diameter of 40 mm and length of 1.1 m transmits 60 hp from an electric motor to a compressor. If the allowable shear stress is 50 MPa and the allowable angle of twist is 1.5°, what is the minimum allowable speed of rotation?

12.21 A motor through a set of gears delivers power to a hollow steel shaft with an outside diameter of 40 mm and an inside diameter of 30 mm. The shaft is rotating at 900 rpm. Twenty-five percent of the power supplied by the motor is transmitted to A, and 75 percent of the power supplied is transmitted to C. If the allowable shear stress in the shaft is 105 MPa, what is the maximum power that the motor can deliver to the shaft?

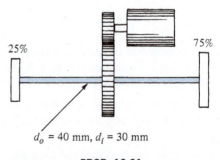

PROB. 12.21

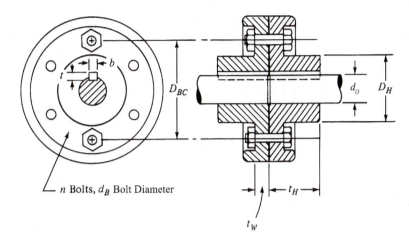

FIGURE 12.14

12.10 FLANGE COUPLINGS

Couplings are used to join sections of shafts or to connect the shaft of a motor to the shaft of a driven machine. To join sections together, *rigid* couplings may be used. With rigid couplings, alignment between the ends of the joined shafts is required to prevent bending in the shafts or loads on the bearings. The *flexible* coupling is commonly used to join the shaft of a motor to the shaft of a driven machine where misalignment of the shafts is common.

Standard couplings are available from manufacturers. However, the analysis of the coupling provides a good illustration of various types of stress.

Here, we only analyze the rigid coupling. Consider the two couplings that are formed with a flange and hub as shown in Fig. 12.14. The couplings are used to join or couple two sections of shaft together. The shafts are keyed to the hubs, and the flanges are then bolted together. This coupling is analyzed in the following example.

Example 12.9 The coupling shown in Fig. 12.14 is used to join two shafts that transmit 38 kW at 200 rpm. The shaft, couplings, and keys have dimensions and specifications as follows:

$$d_o = 50 \text{ mm} \qquad t_H = 80 \text{ mm} \qquad n = \text{no. bolts} = 6$$
$$D_H = 90 \text{ mm} \qquad t_W = 15 \text{ mm} \qquad d_B = 10 \text{ mm}$$
$$D_{BC} = 135 \text{ mm} \qquad b = t = 15 \text{ mm}$$

Find the stresses in the (a) shaft, (b) key, (c) bolts, and (d) flange. (e) If the yield stresses for the couplings, shafts, and bolts are $\sigma_y = 410$ Mpa and $\tau_y = 205$ MPa, what are the factors of safety based on yielding for the elements in parts (a) through (d)?

Solution: From Eq. (12.13), we have

$$T = \frac{9550P}{N} = \frac{9550(38)}{200} = 1814 \text{ N·m}$$

a. From Eq. (12.11), the shear stress in the shaft

$$\tau = \frac{Tc}{J} = \frac{16T}{\pi d^3} = \frac{16(1.814 \times 10^6)}{\pi(50)^3} = 73.9 \text{ N/mm}^2 \text{ (MPa)} \qquad \textbf{Answer}$$

b. The force on the key multiplied by the radius of the shaft is equal to the torque. Therefore,

$$F_K = \frac{2T}{d_o} = \frac{2(1.814 \times 10^6)}{50} = 72.6 \times 10^3 \text{ N}$$

The area in shear for the key $A_S = bt_H = 15(80) = 1200 \text{ mm}^2$ [Fig. 12.15(a)]; therefore, the shear stress in the key

$$\tau = \frac{F_K}{A_S} = \frac{72.6 \times 10^3}{1.2 \times 10^3} = 60.5 \text{ N/mm}^2 \text{ (MPa)} \qquad \textbf{Answer}$$

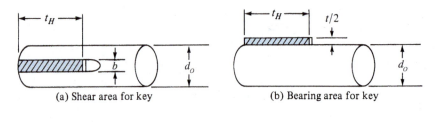

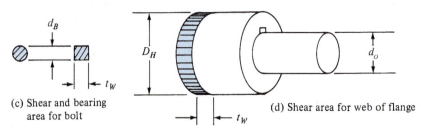

FIGURE 12.15

The area in bearing for the key $A_B = tt_H/2 = 15(80)/2 = 600\,\text{mm}^2$ [Fig. 12.15(b)]; therefore, the bearing stress in the key

$$\sigma_B = \frac{F_K}{A_B} = \frac{72.6 \times 10^3}{600} = 121.0\,\text{N/mm}^2\,(\text{MPa}) \qquad \textbf{Answer}$$

c. The force on each bolt times the number of bolts times the radius of the bolt circle is equal to the torque. Therefore,

$$F_B = \frac{2T}{nD_{BC}} = \frac{2(1.814 \times 10^6)}{6(135)} = 4.48 \times 10^3\,\text{N}$$

The shear area for the bolts $A_S = \pi d_B^2/4 = \pi(10)^2/4 = 78.5\,\text{mm}^2$. Therefore, the shear stress in the bolts

$$\tau = \frac{F_B}{A_S} = \frac{4.48 \times 10^3}{78.5} = 57.0\,\text{N/mm}^2\,(\text{MPa}) \qquad \textbf{Answer}$$

The area in bearing for the bolt on the web of the flange $A_B = d_B t_W = 10(15) = 150\,\text{mm}^2$ [Fig. 12.15(c)]. Therefore,

$$\sigma_B = \frac{F_B}{A_B} = \frac{4.48 \times 10^3}{150} = 29.9\,\text{N/mm}^2\,(\text{MPa}) \qquad \textbf{Answer}$$

d. The flange may shear from the hub along a cylindrical surface at the junction of the hub and the web of the flange. The force on this cylindrical surface multiplied by the radius of the surface is equal to the torque, that is,

$$F_H = \frac{2T}{D_H} = \frac{2(1.814 \times 10^6)}{90} = 40.3 \times 10^3\,\text{N}$$

The area of the cylindrical surface is equal to the circumferential length of the hub multiplied by the thickness of the web of the flange $A_H = \pi D_H t_W = \pi(90)(15) = 4.24 \times 10^3\,\text{mm}^2$ [Fig. 12.15(d)]. Therefore,

$$\tau = \frac{F_H}{A_H} = \frac{40.3 \times 10^3}{4.24 \times 10^3} = 9.50\,\text{N/mm}^2\,(\text{MPa}) \qquad \textbf{Answer}$$

e. Factors of safety are as follows:

Shaft: F.S. (shear) $= \dfrac{205}{73.9} = 2.77$

Key: F.S. (shear) $= \dfrac{205}{60.5} = 3.39,$ F.S. (bearing) $= \dfrac{410}{121.0} = 3.39$

Bolt: F.S. (shear) $= \dfrac{205}{57.0} = 3.60,$ F.S. (bearing) $= \dfrac{410}{29.9} = 13.71$

Web: F.S. (shear) $= \dfrac{205}{9.50} = 21.6$ **Answers**

For rigid couplings that contain several concentric rows of bolts, it is customary to assume that the force in any bolt is proportional to its distance from the shaft axis. In other words,

$$\frac{F_{B1}}{r_1} = \frac{F_{B2}}{r_2} \qquad (12.16)$$

This relationship, shown graphically in Fig. 12.16, is based on the potential deformations caused at different radii by a rigid coupling. Because each bolt force causes a "mini-torque" about the shaft axis, the total torque, T, that the coupling can carry is

$$T = n_1 r_1 F_{B1} + n_2 r_2 F_{B2} \qquad (12.17)$$

Here, n_1 and n_2 are the number of bolts at radius r_1 and r_2, respectively.

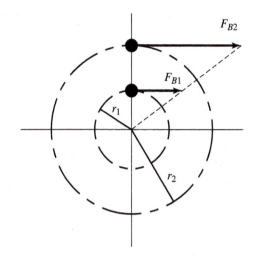

FIGURE 12.16

THREADED FASTENERS

SB.20(a)

ROLLED THREAD

SB.20(b)

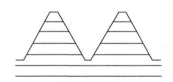

CUT THREAD

SB.20(c)

SB.20(d)

External threads, such as those found on bolts and screws, may be formed in several ways using various types of dies. The two most common methods of mass-producing threaded fasteners today are by *cutting*, in which material is *removed* from a cylindrical metal blank to form the desired thread shape, or by *rolling*, in which the blank material is *compressed* into the shape of a thread. In the rolling process shown [SB.20(a)], a metal blank is rolled between one moving and one stationary flat die. Such dies, which are normally housed in an automated machine, may be used to manufacture screws and bolts having diameters between 0.060 in. and 1.125 in. at production rates up to 1000 pieces per minute. Rolling has several advantages over cutting. Because no material is removed, the grain structure of the metal within the fastener [SB.20(b)] remains continuous and flows up into the thread form. This grain structure flow results in a much stronger thread form compared with the interrupted grain structure of the cut thread [SB.20(c)]. A fastener with rolled thread will have improved tensile strength, increased surface hardness, and superior surface finish on the thread profile. The rolling process may also be used to produce knurled and splined forms, as well as special grooves, worms, pinions, and annular or helical nails [SB.20(d)]. *(Courtesy of Reed-Rico, Holden, Massachusetts, www.pccspd.com)*

Example 12.10   A flange coupling contains ten 1/2-in.-diameter steel bolts whose allowable shear stress is 14,000 psi. Four of these bolts are located on a 6-in. bolt circle, and six are on a 9-in. bolt circle. Based on shear in the bolts, what is the maximum torque that this coupling can carry?

Solution: For a 1/2-in.-diameter bolt, $A = 0.196$ in.^{2}, and the maximum force that can be exerted by *any* of the bolts is

$$F_{max} = \tau \times A = 14,000 \text{ lb/in.}^2 \times 0.196 \text{ in.}^2 = 2740 \text{ lb}$$

From Fig. 12.16, we see that the outside row of bolts must carry the highest force, so $F_{B2} = 2740$ lb, and F_{B1} is obtained from Eq. (12.16) as

$$F_{B1} = r_1 \times \frac{F_{B2}}{r_2} = 3 \text{ in.} \times \frac{2740 \text{ lb}}{4.5 \text{ in.}} = 1830 \text{ lb}$$

Then, from Eq. (12.17),

$$T = 4 \times (3 \text{ in.} \times 1830 \text{ lb}) + 6 \times (4.5 \text{ in.} \times 2740 \text{ lb})$$
$$= 95,900 \text{ in.-lb.}$$ **Answer**

PROBLEMS

12.22 A solid shaft whose diameter is 3.5 in. carries a torque of 950 ft-lb. This shaft is connected to a coupling by the rectangular key shown. Find the required length of the key if its allowable shear stress is 11,200 psi.

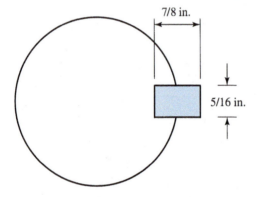

7/8 in.

5/16 in.

PROB. 12.22

12.23 a.   A hollow shaft whose outside diameter is $d_o = 3.00$ in. has a uniform wall thickness of $t_S = 1/8$ in. What is the maximum torque that this shaft can carry without exceeding a torsional shear stress of 20,000 psi?

    b.   The shaft from part (a) is attached to a coupling containing three bolts on a bolt circle whose diameter is $D_{BC} = 5$ in. If the bolt material has an allowable shear stress of 18,500 psi, what is the minimum diameter (to the nearest 1/16th in.) for these bolts?

12.24 a.   Same as Prob. 12.23(a), except that $d_o = 75$ mm, $t_S = 3$ mm, and the allowable shear stress for the shaft is 138 MPa.

    b.   Same as Prob. 12.23(b), except that $D_{BC} = 125$ mm, and the allowable shear stress for the bolts is 128 MPa. (Find d_B to the nearest 1 mm.)

12.25 Five bolts are made of low-carbon steel having a yield point in shear of 19,200 psi. These bolts are 3/8-in. in diameter and are to be used in a flange coupling that transmits 285 hp at 1250 rpm. If the bolts are to have a factor of safety in shear of 1.5, find the distance, r, from the center of this coupling at which the bolts should be placed. Is this a minimum or maximum distance?

12.26 A stationary power plant consists of a diesel engine connected to an electric generator. The diesel produces 875 hp at 550 rpm.

    a.   If the diesel's solid output shaft is to be made of steel whose allowable shear stress is 10,500 psi, find, to the nearest 1/16 in., the minimum shaft diameter required.

    b.   If a 4-1/4-in.-diameter solid output shaft is used, find the maximum shear stress in the shaft.

    c.   The 4-1/4-in. shaft is locked to a bolted flange coupling by a steel key that is 3-1/8 in. long and has an allowable shear stress of 19,500 psi. Compute the minimum width of this key.

    d.   The coupling contains four 13/16-in.-diameter bolts on a 7-in. bolt circle. Find the average shear stress in these bolts.

12.27 The coupling shown in Fig. 12.14 is used to join two shafts that transmit 26 hp at 150 rpm. The shaft, coupling, and key have dimensions and specifications as follows:

$$d_o = 45 \text{ mm} \qquad t_H = 70 \text{ mm}$$
$$D_H = 80 \text{ mm} \qquad t_W = 15 \text{ mm}$$
$$D_{BC} = 130 \text{ mm} \qquad b = t = 10 \text{ mm}$$
$$n = \text{no. bolts} = 5$$

Bolt diameter = 15 mm

Find the stresses in the (a) shaft, (b) keys, (c) bolts, and (d) flanges. (e) If the yield stresses for all parts are $\sigma_y = 360$ MPa and $\tau_y = 250$ MPa, what are the factors of safety for the elements in parts (a) through (d)?

12.28 Two steel shafts are joined by couplings that are forged integrally with the shaft as shown. The diameters of the shaft and bolt circle are $d_o = 75$ mm and $D_{BC} = 115$ mm, the flange has a thickness $t_{FL} = 10$ mm, and there are six bolts. If the allowable shear stress in the shafts, flanges, and bolts is 93 MPa, (a) determine the bolt diameter required to develop the full shear capacity of the shaft. (b) Determine the average shear and bearing stresses in the flanges. (c) If the shaft is rotating at 100 rpm, determine the horsepower the shaft can transmit.

12.29 Two solid steel shafts are joined by couplings that are forged integrally with the shaft as shown. The diameter of the shaft and bolt circle are $d_o = 5$ in. and $D_{BC} = 10$ in., the flange has a thickness $t_{FL} = 5/8$ in., and there are eight 1-in.-diameter bolts. If the allowable shear stress in the shaft is 8000 psi, (a) determine the full torque capacity of the shafts based on shear stress. (b) What horsepower can the shafts transmit at 100 rpm? (c) What is the shear stress in the bolts? (d) Determine the bearing stress and maximum shear stress in the flanges.

12.30 A flange coupling assembly contains eight 3/4-in. steel bolts, four located on a 7-in. bolt circle and four on a 10-in. bolt circle. If the allowable shearing stress in these bolts is 14,500 psi, find the maximum horsepower that the coupling can transmit at a speed of 175 rpm.

12.31 The design for a new bolted flange coupling calls for six steel bolts, three located on a 6-in. bolt circle and

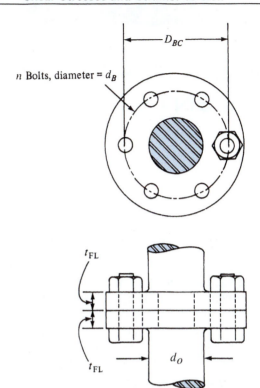

PROB. 12.28 and PROB. 12.29

three on a 4-in. bolt circle. If the coupling is to transmit 105 hp at 1200 rpm, and the allowable shear stress in the bolts is 9750 psi, find, to the nearest 1/16 inch, the required bolt diameter.

SHEAR FORCES AND BENDING MOMENTS IN BEAMS

CHAPTER OBJECTIVES

Here, we examine the internal forces and moments created at any point along the length of a beam. Before proceeding to Chapter 14, you should, with some proficiency, be able to

- Use the methods of Chapter 5 to compute the forces provided by supports for beams subjected to various combinations of external loads.
- Quickly and accurately construct shear and moment diagrams that describe the internal forces and moments existing at all points in a beam.
- From shear and moment diagrams, determine magnitudes and locations for the maximum values of internal force and moment existing in a beam.

13.1 INTRODUCTION

We previously considered four kinds of internal reactions in members. They were axial forces, shear forces, bending moments, and torsions. In Chapters 10 and 11, we examined the axial forces and the stresses and strains they produce. The ability of torque to produce stress and strain was explored in Chapter 12. In this chapter, we examine the beam, that is, a member acted on by loads that produce bending. The internal reactions in the beam are shear forces and bending moments and, to a lesser extent, axial forces. Because we considered axial forces in some detail in Chapters 10 and 11, they are not included in this chapter. The combined effects of axial forces and bending moments are discussed in Chapter 16.

Besides the values of the shear forces and bending moments at selected cross sections of a beam, we may also be interested in their values at cross sections throughout the beam. Such information is represented graphically on shear force and bending moment diagrams, where the horizontal axis of the diagram represents the location of the beam cross section and the vertical axis represents the value of shear force or bending moment at each cross-sectional location.

13.2 TYPES OF BEAMS

As previously indicated, beams are structural and machine members acted on by loads that produce bending. To produce bending, the loads must be applied transversely, that is, perpendicular to the axis of the beam.

Beams may be classified according to the kinds of supports and loading. If a beam is freely supported at its ends with either pins or rollers, it is called a *simply supported beam* or *simple beam*. Simple beams are shown in Fig. 13.1(a) and (b). The beam that is fixed at one end and free at the other is called a *cantilever beam*. Figure 13.1(c) and (d) show cantilever beams. Beams with overhanging ends are shown in Fig. 13.1(e) and (f). The supports for such beams are either pins or rollers. The simple beam, cantilever beam, and beam with overhanging

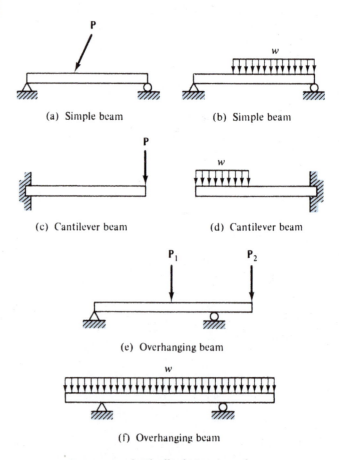

(a) Simple beam (b) Simple beam

(c) Cantilever beam (d) Cantilever beam

(e) Overhanging beam

(f) Overhanging beam

FIGURE 13.1. Statically determinate beams.

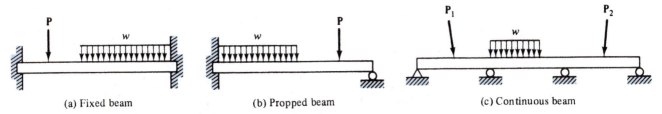

(a) Fixed beam   (b) Propped beam   (c) Continuous beam

FIGURE 13.2. Statically indeterminate beams.

ends are all *determinate* because the three unknown reactions for each beam can be determined by the equations of static equilibrium. Examples of *statically indeterminate* beams are shown in Fig. 13.2, where we have a *fixed beam* in which both ends are fixed, a *propped beam* in which one end is fixed and the other end supported by a roller, and a *continuous beam*, which is supported by a pin and two or more rollers.

13.3 BEAM REACTIONS

Before an analysis of the internal reactions can be made, the beam reactions must be calculated. Because the internal reactions depend on the beam reactions, care should be taken to ensure their accuracy. In the following examples, the procedure for finding beam reactions is reviewed.

Example 13.1   Find the reactions for the simply supported beam [Fig. 13.3(a)] loaded as shown. Neglect the weight of the beam.

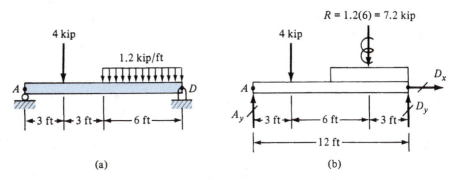

(a)   (b)

FIGURE 13.3

Solution: A free-body diagram is shown in Fig. 13.3(b). The uniformly distributed load has been replaced by its resultant. From equilibrium,

$$\Sigma M_A = 0 \qquad -4(3) - 7.2(9) + D_y 12 = 0 \qquad D_y = 6.4 \text{ kip} \qquad \textbf{Answer}$$

$$\Sigma M_D = 0 \qquad -A_y 12 + 4(9) + 7.2(3) = 0 \qquad A_y = 4.8 \text{ kip} \qquad \textbf{Answer}$$

$$\Sigma F_x = 0 \qquad\qquad\qquad\qquad\qquad\qquad\qquad D_x = 0 \qquad\qquad \textbf{Answer}$$

Check:

$$\Sigma F_y = 0 \qquad A_y - 4 - 7.2 + D_y = 0$$
$$4.8 - 4 - 7.2 + 6.4 = 0 \qquad \text{OK}$$

Example 13.2   Find the reactions for the cantilever beam shown in Fig. 13.4(a). Neglect the weight of the beam.

Solution: The triangular load has been replaced in Fig. 13.4(b) by its resultant. (Note that the moment of a couple is the same about any point in the plane of the couple.) From equilibrium,

$$\Sigma F_x = 0 \qquad\qquad\qquad\qquad\qquad A_x = 0 \qquad\qquad\qquad \textbf{Answer}$$

$$\Sigma F_y = 0 \qquad\qquad A_y - 4 - 1.8 = 0 \qquad A_y = 5.8 \text{ kN} \qquad\qquad \textbf{Answer}$$

$$\Sigma M_A = 0 \qquad M_A - 4(0.9) + 3 - 1.8(3.3) = 0 \qquad M_A = 6.54 \text{ kN·m} \qquad \textbf{Answer}$$

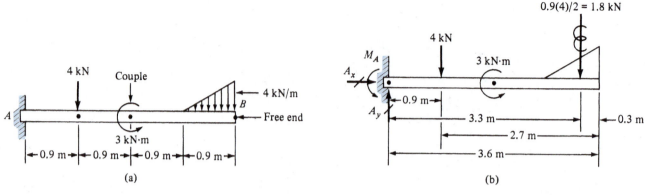

FIGURE 13.4

Check:

$$\Sigma M_B = 0 \qquad M_A - A_y 3.6 + 4(2.7) + 3 + 1.8(0.3) = 0$$
$$6.54 - 5.8(3.6) + 4(2.7) + 3 + 1.8(0.3) = 0 \qquad \text{OK}$$

13.4 SHEAR FORCES AND BENDING MOMENTS IN BEAMS

To determine the shear force and bending moment at a given location in a beam, we use the method of sections to cut the beam at that location and then draw a free-body diagram of either end of the beam. Both ends of the beam are in equilibrium; therefore, on either end, the resultant external forces and moments are balanced by the internal shear forces and bending moments. We will establish a sign convention for the internal reactions in the following articles.

Shear Forces

In Fig. 13.5(a), we show the direction of a positive internal shear force. For positive shear $+V$ at any section of a beam, the sum of the external forces ΣF to the left of the section act upward and the sum of the external forces ΣF to the right of the section act downward. This tends to cause the beam on the left to move up and the beam on the right to move down. The direction of negative internal shear forces is reversed [see Fig. 13.5(b)]. We will assume a positive shear force, and if the equilibrium equation gives a negative value, the shear is negative.

We see from Fig. 13.5(a) and (b) and the equilibrium equations that *the shear force at any section of a beam is equal to the algebraic sum of the normal forces to the left of the section. Forces directed up produce positive shear, and forces directed down produce negative shear.* Or, in equation form,

$$V(\text{at section}) = +\uparrow \Sigma F(\text{left of section}) \qquad \textbf{(13.1)}$$

We also see from Fig. 13.5(a) and (b) that the sign convention is reversed when considering the beam to the right of the section. We modify Eq. (13.1) accordingly.

$$V(\text{at section}) = +\downarrow \Sigma F(\text{right of section}) \qquad \textbf{(13.2)}$$

Bending Moments

In Fig. 13.6(a), we show the direction of positive internal bending moments. For a positive bending moment $+M$ at any section of a beam, the sum of the moments ΣM of the external forces to the left of the section act in a clockwise direction, and the sum of the moments ΣM of the external forces to the right of the section act in a counterclockwise direction. This tends to cause compression on the top of the beam and tension on the bottom of the beam. (To visualize this phenomenon, lie on your stomach and arch your back.

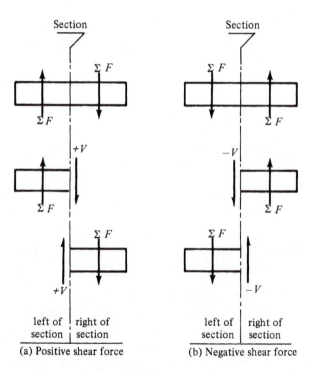

(a) Positive shear force   (b) Negative shear force

FIGURE 13.5

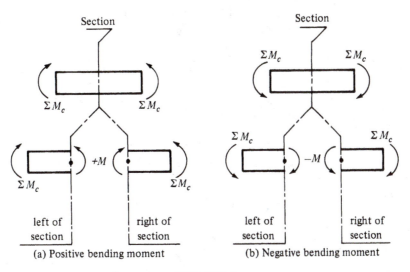

FIGURE 13.6

Are your stomach and back in tension or compression?) The direction of negative bending moments is reversed [see Fig. 13.6(b)]. Negative bending moments tend to cause tension on the top of the beam and compression on the bottom of the beam. We will assume a positive bending moment and if the equilibrium equation gives a negative value, the bending moment is negative.

From Fig. 13.6(a) and (b) and the equilibrium equations we see that *the bending moment at any section of a beam is equal to the algebraic sum of the moments of the forces about the centroid of the cross section for forces left of the section.*

Clockwise moments are positive, and counterclockwise moments are negative. Or, in equation form,

$$M(\text{at section}) = \circlearrowleft \Sigma M_c \text{ (left of section)} \quad (13.3)$$

From Fig. 13.6(a) and (b), we also see that the sign convention is reversed when the beam to the right is considered. Equation (13.3) is modified accordingly.

$$M(\text{at section}) = \circlearrowright \Sigma M_c \text{ (right of section)} \quad (13.4)$$

In the following examples, we calculate the shear force and bending moment at designated cross sections of various beams.

Example 13.3 For the beam shown in Fig. 13.7(a), find the shear force and bending moment at $x = 4$ ft and $x = 10$ ft. For convenience, we repeat Fig. 13.3(a) and the beam reactions in Fig. 13.7(a).

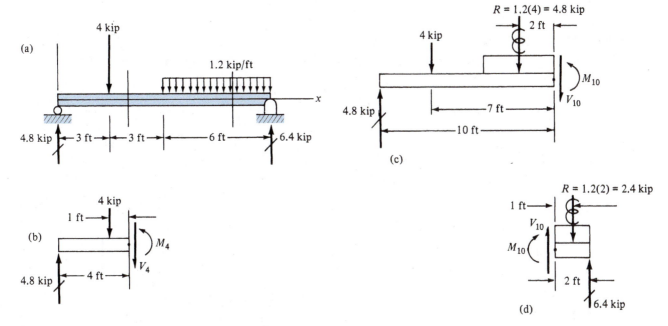

FIGURE 13.7

Solution: Free-body diagrams for the beam to the left of $x = 4$ ft are shown in Fig. 13.7(b) and for the beam to the left and right of $x = 10$ ft in Fig. 13.7(c) and (d). Notice that the section at $x = 10$ ft cuts through the distributed load. The resultant for the part of the distributed load to the left of the section [Fig. 13.7(c)] has a magnitude $R = 1.2(4) = 4.8$ kip and acts at the middle of the distributed load 2 ft to the left of the section. The resultant for the part to the right of the section [Fig. 13.7(d)] has a magnitude $R = 1.2(2) = 2.4$ kip and acts at the middle of the distributed load 1 ft to the right of the section.

For the beam to the left of $x = 4$ ft [Fig. 13.7(b)],

$$\Sigma F_y = 0 \qquad 4.8 - 4 - V_4 = 0 \qquad V_4 = 0.8 \text{ kip} \qquad \textbf{Answer}$$

$$\Sigma M_c = 0 \qquad -4.8(4) + 4(1) + M_4 = 0 \qquad M_4 = 15.2 \text{ kip-ft} \qquad \textbf{Answer}$$

For the beam to the left of $x = 10$ ft [Fig. 13.7(c)],

$$\Sigma F_y = 0 \qquad 4.8 - 4 - 4.8 - V_{10} = 0 \qquad V_{10} = -4.0 \text{ kip} \qquad \textbf{Answer}$$

$$\Sigma M_c = 0 \qquad -4.8(10) + 4(7) + 4.8(2) + M_{10} = 0 \qquad M_{10} = 10.4 \text{ kip-ft} \qquad \textbf{Answer}$$

The same results are obtained from the simpler free-body diagram for the section to the right of $x = 10$ ft [Fig. 13.7(d)]:

$$\Sigma F_y = 0 \qquad V_{10} - 2.4 + 6.4 = 0 \qquad V_{10} = -4.0 \text{ kip} \qquad \textbf{Answer}$$

$$\Sigma M_c = 0 \qquad 6.4(2) - 2.4 - M_{10} = 0 \qquad M_{10} = 10.4 \text{ kip-ft} \qquad \textbf{Answer}$$

Alternative Solution

For the beam to the left of $x = 4$ ft [Fig. 13.7(b)]: From Eqs. (13.1) and (13.3),

$$V(\text{at section}) = +\uparrow\Sigma F \text{ (left of section)}$$

$$V_4 = 4.8 - 4.0 = 0.8 \text{ kip} \qquad \textbf{Answer}$$

$$M(\text{at section}) = \Sigma M_c \text{ (left of section)}$$

$$M_4 = 4.8(4) - 4(1) = 15.2 \text{ kip-ft} \qquad \textbf{Answer}$$

For the beam to the right of $x = 10$ ft [Fig. 13.7(d)]: From Eqs. (13.2) and (13.4),

$$V(\text{at section}) = +\downarrow\Sigma F \text{ (right of section)}$$

$$V_{10} = 2.4 - 6.4 = -4.0 \text{ kip} \qquad \textbf{Answer}$$

$$M(\text{at section}) = \Sigma M_c \text{ (right of section)}$$

$$M_{10} = 6.4(2) - 2.4(1) = 10.4 \text{ kip-ft} \qquad \textbf{Answer}$$

Example 13.4 For the cantilever beam shown in Fig. 13.8(a), find the shear force and bending moment at $x = 1.0$ m and at $x = 2.7$ m. Fig. 13.4(a) and the beam reactions are repeated in Fig. 13.8(a).

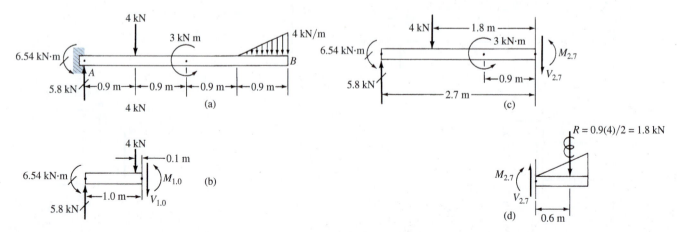

FIGURE 13.8

Solution: For the beam to the left of $x = 1.0\,$m [Fig. 13.8(b)]: From Eqs. (13.1) and (13.3),

$$V(\text{at section}) = +\!\uparrow \Sigma F(\text{left of section})$$

$$V_{1.0} = 5.8 - 4 = 1.8\,\text{kN} \qquad\qquad \textbf{Answer}$$

$$M(\text{at section}) = \boldsymbol{\zeta}\Sigma M_c(\text{left of section})$$

$$M_{1.0} = 5.8(1.0) - 6.54 - 4(0.1) = -1.14\,\text{kN·m} \qquad\qquad \textbf{Answer}$$

For the beam to the right of $x = 2.7\,$m [Fig. 13.8(d)]: From Eqs. (13.2) and (13.4),

$$V(\text{at section}) = +\!\downarrow \Sigma F(\text{right of section})$$

$$V_{2.7} = 1.8\,\text{kN} \qquad\qquad \textbf{Answer}$$

$$M(\text{at section}) = \boldsymbol{\jmath}\Sigma M_c(\text{right of section})$$

$$M_{2.7} = -1.8(0.6) = -1.08\,\text{kN·m} \qquad\qquad \textbf{Answer}$$

Example 13.5 For the beam with overhang shown in Fig. 13.9(a), find the shear forces and bending moments at sections to the left and right of B ($x = 2\,$m), to the left and right of C ($x = 6\,$m), halfway between C and D ($x = 7.5\,$m), and at the free end D ($x = 9\,$m).

Solution: The free-body diagram for the entire beam is shown in Fig. 13.9(b). The resultant of the distributed load has been calculated and is shown in the figure. Equilibrium requires that

$$\Sigma F_x = 0 \qquad A_x = 0$$

$$\Sigma M_A = 0 \qquad -300(2) + C_y6 - 200(7.5) = 0$$

$$C_y = 350\,\text{kN}$$

$$\Sigma M_C = 0 \qquad -A_y6 + 300(4) - 200(1.5) = 0$$

$$A_y = 150\,\text{kN}$$

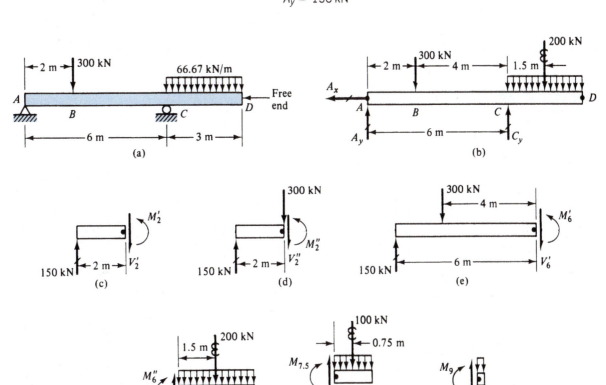

FIGURE 13.9

Check:

$$\Sigma F_y = 0 \quad A_y - 300 + C_y - 200 = 0 \quad \text{OK}$$

In Fig. 13.9(c) and (d), free-body diagrams are shown for sections immediately to the left and right of point B. Point B must be excluded from the first free-body diagram because the 300-kN load acts at point B. The primed value of shear and bending moment (V' and M') are reserved for a section to the left of a point, and the double-primed values (V'' and M'') for a section to the right.

From Eqs. (13.1) and (13.3),

$$V_2' = 150 \text{ kN} \qquad\qquad\qquad \textbf{Answer}$$
$$M_2' = 150(2) = 300 \text{ kN·m} \qquad\qquad \textbf{Answer}$$

and

$$V_2'' = 150 - 300 = -150 \text{ kN} \qquad\qquad \textbf{Answer}$$
$$M_2'' = 150(2) = 300 \text{ kN·m} \qquad\qquad \textbf{Answer}$$

The free-body diagrams are shown in Fig. 13.9(e) and (f) for portions of the beam to the left and right of point C. (Note that C_y does not appear on either diagram.)
From Eqs. (13.1) and (13.3),

$$V_6' = 150 - 300 = -150 \text{ kN} \qquad\qquad \textbf{Answer}$$
$$M_6' = 150(6) - 300(4) = -300 \text{ kN·m} \qquad \textbf{Answer}$$

From Eqs. (13.2) and (13.4),

$$V_6'' = 200 \text{ kN} \qquad\qquad\qquad \textbf{Answer}$$
$$M_6'' = -200(1.5) = -300 \text{ kN·m} \qquad\qquad \textbf{Answer}$$

For the section at $x = 7.5$ m [Fig. 13.9(g)], we use Eqs. (13.2) and (13.4) to find the shear and bending moment.

$$V_{7.5} = 100 \text{ kN} \qquad\qquad\qquad \textbf{Answer}$$
$$M_{7.5} = -100(0.75) = -75 \text{ kN·m} \qquad\qquad \textbf{Answer}$$

For the free end of the beam, point D, $x = 9$ m [Fig. 13.9(h)], we use Eqs. (13.2) and (13.4) to find the shear and bending moment.

$$V_9 = 0 \qquad\qquad\qquad \textbf{Answer}$$
$$M_9 = 0 \qquad\qquad\qquad \textbf{Answer}$$

PROBLEMS

13.1 and 13.2 For the simply supported beam shown, determine (a) the shear force for sections to the right of A, B, and C, and (b) the bending moment for sections at A, B, C, and D.

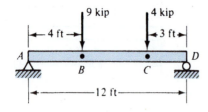

PROB. 13.1

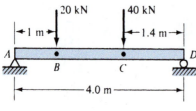

PROB. 13.2

13.3 and 13.4 Determine (a) the shear force for sections to the right of A and B, and (b) the bending moment for sections at A, B, and C for the cantilever beam shown.

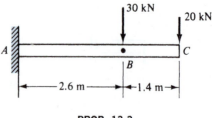

PROB. 13.3

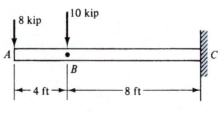

PROB. 13.4

13.5 For the beam with the overhang shown, determine (a) the shear force for sections to the right of A, B, and C, and (b) the bending moment for sections at A, C, and D and to the right of B.

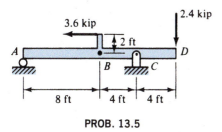

PROB. 13.5

13.6 Determine (a) the shear force for sections to the right of A and B, and (b) the bending moment for sections to the right of A and at B for the cantilever beam shown.

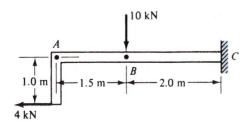

PROB. 13.6

13.7 For the beam with the overhang shown, determine (a) the shear force for sections to the right of A and B and to the left of C, and (b) the bending moment for sections at A, B, and C.

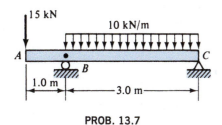

PROB. 13.7

13.8 Determine (a) the shear force for sections to the left of B and to the right of B and C, and (b) the bending moments for sections at A, B, C, and D for the beam with the overhang shown.

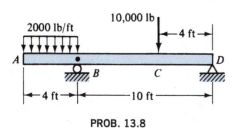

PROB. 13.8

13.9 Determine (a) the shear force for sections to the right of A and B, and (b) the bending moment for sections at A, B, and C for the cantilever beam shown.

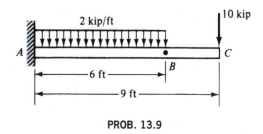

PROB. 13.9

13.10 For the cantilever beam shown, determine (a) the shear force for sections at B and to the right of C, and (b) the bending moment for sections at A, B, C, and D.

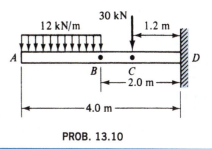

PROB. 13.10

13.5 SHEAR-FORCE AND BENDING-MOMENT DIAGRAMS

In many problems, the value of the internal reactions at selected cross sections along the beam may be sufficient. In other problems, however, the values throughout the beam are required. To draw the shear-force and bending-moment diagrams, we need their values at selected cross sections. In Fig. 13.10, we have repeated Fig. 13.9(a) and have drawn the various diagrams from the data obtained in Example 13.5. The moment-diagram curve crosses the axis halfway between $x = 2$ m and $x = 6$ m, that is, at $x = 4$ m.

In reviewing the various diagrams, we see that if we are dealing with *concentrated loads or reactions,* the shear diagram consists of straight horizontal lines that are broken only at the load or reaction. The moment diagram consists of straight sloping lines broken at the loads or reactions. When dealing with *uniform loads,* the shear diagram consists of straight sloping lines. The moment diagram consists of curved lines (the curved lines are second-degree curves).

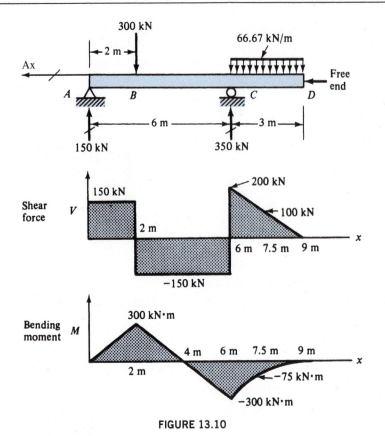

FIGURE 13.10

Example 13.6   Draw the shear-force and bending-moment diagrams for the beam shown in Fig. 13.11(a). The free-body diagram for the entire beam is shown in Fig. 13.11(b).

Solution: From equilibrium, we write

$$\Sigma M_A = 0 \qquad -10(2) - 20(6) + 2(10) + B_y 12 = 0$$
$$B_y = 10 \text{ kip}$$
$$\Sigma M_B = 0 \qquad -A_y 12 + 10(10) + 20(6) - 2(2) = 0$$
$$A_y = 18 \text{ kip}$$

Check:

$$\Sigma F_y = 0 \qquad A_y - 10 - 20 + 2 + B_y = 0$$
$$18 - 10 - 20 + 2 + 10 = 0 \qquad \text{OK}$$

     We must calculate the shear to the left and right of $x = 2$ ft, at $x = 4$ ft, at $x = 8$ ft, and to the left and right of $x = 10$ ft. The bending moment must be calculated at $x = 2$ ft, 4 ft, 6 ft, 8 ft, and 10 ft. Free-body diagrams for the various sections are drawn in Fig. 13.11(c)–(k). Applying Eqs. (13.1) and (13.3) to free-body diagrams shown in Fig. 13.11(c)–(g), we have

$$V_2' = 18 \text{ kip} \qquad\qquad M_4 = 18(4) - 10(2) = 52 \text{ kip-ft}$$
$$M_2' = 18(2) = 36 \text{ kip-ft} \qquad V_6 = 18 - 10 - 10 = -2 \text{ kip}$$
$$V_2'' = 18 - 10 = 8 \text{ kip} \qquad M_6 = 18(6) - 10(4) - 10(1) = 58 \text{ kip-ft}$$
$$M_2'' = 18(2) = 36 \text{ kip-ft} \qquad V_8 = 18 - 10 - 20 = -12 \text{ kip}$$
$$V_4 = 18 - 10 = 8 \text{ kip} \qquad M_8 = 18(8) - 10(6) - 20(2) = 44 \text{ kip-ft}$$

Applying Eqs. (13.2) and (13.4) to free-body diagrams shown in Fig. 13.11(h)–(k), we have

$$V_8 = -2 - 10 = -12 \text{ kip} \qquad V_{10}'' = -10 \text{ kip}$$
$$M_8 = 2(2) + 10(4) = 44 \text{ kip-ft} \qquad M_{10}'' = 10(2) = 20 \text{ kip-ft}$$

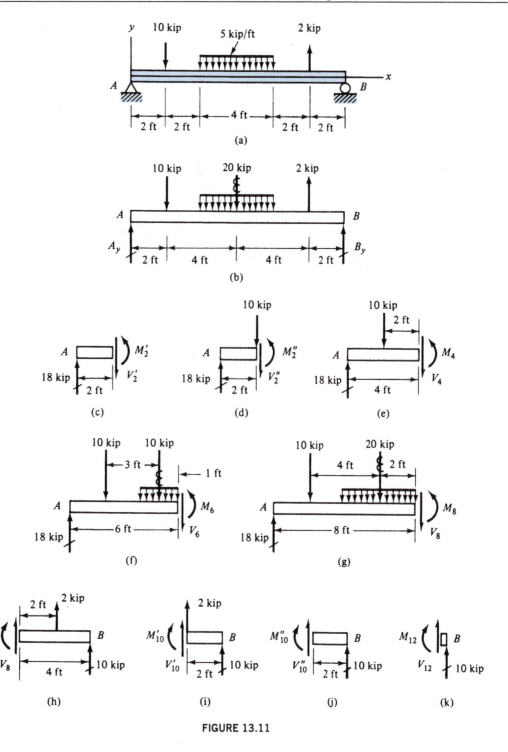

FIGURE 13.11

$$V'_{10} = -2 - 10 = -12 \text{ kip} \qquad V_{12} = -10 \text{ kip}$$
$$M'_{10} = 10(2) = 20 \text{ kip-ft} \qquad M_{12} = 0$$

From the data, we now draw the shear and moment diagrams in Fig. 13.12(b) and (c). The maximum moment occurs when the shear diagram crosses the axis, that is, when the shear is zero. The shear is zero at a distance x_1 along the distributed load such that the beam load $5x_1$ is equal to 8, that is, $5x_1 = 8$ or $x_1 = 1.6$ ft. Therefore, the shear is zero at $x = 4 + 1.6 = 5.6$ ft. A free-body diagram of the beam is shown for $x = 5.6$ ft in Fig. 13.12(d). From rotational equilibrium,

$$\Sigma M_c = 0 \qquad -18(5.6) + 10(3.6) + 8(0.8) + M_{5.6} = 0$$
$$M_{5.6} = 58.4 \text{ kip-ft}$$

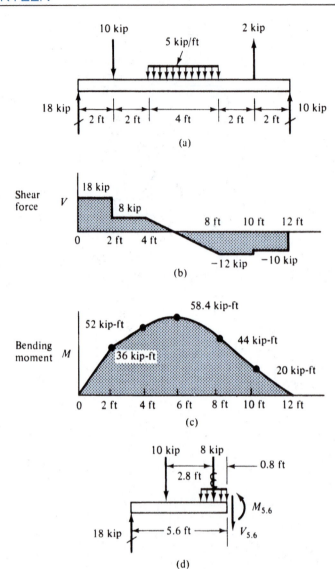

FIGURE 13.12

13.6 RELATIONS AMONG LOADS, SHEAR FORCES, AND BENDING MOMENTS

We will consider the free-body diagram of a small element of a beam of length Δx [Fig. 13.13(b)] that lies between sections A–A and B–B of the beam in Fig. 13.13(a). The shear and bending moment on section A–A will be taken as V and M, respectively, and are shown in a positive sense in the figure. Let ΔV represent the change in shear from sections A–A to B–B and ΔM represent the change in bending moment from sections A–A to B–B. Then, the shear at section B–B is equal to $V + \Delta V$, and the bending moment at section B–B is equal to $M + \Delta M$. In the figure, they are shown in a positive sense. The resultant of the distributed load is the area of the distributed load $w(\Delta x)$, and it acts at the middle of the element as shown. Loads directed up are considered positive.

From equilibrium,

$$\Sigma F_y = 0 \qquad V + w(\Delta x) - (V + \Delta V) = 0$$

$$\Delta V = w(\Delta x) \qquad \textbf{(a)}$$

$$\Sigma M_c = 0$$

$$-M - w(\Delta x)\frac{(\Delta x)}{2} + (M + \Delta M) - V(\Delta x) = 0$$

$$-w\frac{(\Delta x)^2}{2} + \Delta M - V(\Delta x) = 0$$

Now, the element Δx is small, and the term $w(\Delta x)^2/2$ is even smaller and can be dropped from the equation. We have, therefore,

$$\Delta M = V(\Delta x) \qquad \textbf{(b)}$$

Equations (a) and (b) have important physical interpretations. Equation (a) indicates that the change in the shear from

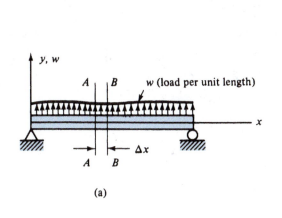

(a)

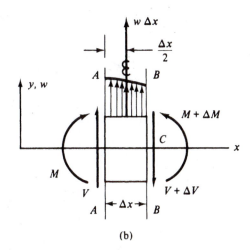

(b)

FIGURE 13.13

sections A–A to B–B is equal to the external load on the beam between those two sections. If we add all elements between any sections 1 and 2, we see that the change in shear between 1 and 2 is equal to the external load on the beam between sections 1 and 2 (area under load curve from 1 to 2). In equation form,

$$V_2 - V_1 = \text{external load between 1 and 2}$$

or $$\Delta V_{1 \to 2} = \text{load}_{1 \to 2} \tag{13.5}$$

Rewriting Eq. (a), we have

$$\frac{\Delta V}{\Delta x} = w \tag{13.6}$$

Therefore, *the slope of the shear diagram* $\Delta V / \Delta x$ *at any point is equal to the value of the load at that point.*

Equation (b) can be interpreted in a similar way. *The change in bending moment between sections 1 and 2 is equal to the area under the shear curve between 1 and 2.* In equation form,

$$M_2 - M_1 = \text{area under shear curve between 1 and 2}$$

or $$\Delta M_{1 \to 2} = \text{area shear curve}_{1 \to 2} \tag{13.7}$$

Rewriting Eq. (b), we have

$$\frac{\Delta M}{\Delta x} = V \tag{13.8}$$

Therefore, *the slope of the bending moment diagram* $\Delta M / \Delta x$ *at any point is equal to the value of the shear force V at that point.* For values of x at which $V = 0$, slope of the bending diagram is also zero. Graphically, this means that the moment diagram has formed a relative "peak" in the positive or negative direction. We see, then, that *the maximum moment must occur at a point where the value of shear is zero.*

Example 13.7   Use the relationships among loads, shears, and bending moments to sketch the shear and moment diagrams for the simply supported beam shown in Fig. 13.14(a).

Solution: The reactions are calculated and shown in Fig. 13.14(a). From Eq. (13.5) and Fig. 13.14(a), we calculate the shears. The value of shear to the left of A is zero.

$$V_A = 0$$
$$\Delta V(\text{left to right of } A) = 10.5 \, \text{kip}$$
$$V(\text{right of } A) = 10.5 \, \text{kip}$$
$$\Delta V(\text{left to right of } B) = -12 \, \text{kip}$$
$$V(\text{right of } B) = 10.5 - 12 = -1.5 \, \text{kip}$$
$$\Delta V(\text{right of } B \text{ to } C) = 0$$
$$V(\text{at } C) = -1.5 \, \text{kip}$$
$$\Delta V(C \text{ to left of } D) = 1(-6) = -6.0 \, \text{kip}$$
$$V(\text{left of } D) = -1.5 - 6.0 = -7.5 \, \text{kip}$$
$$\Delta V(\text{left of } D \text{ to right of } D) = 7.5 \, \text{kip}$$
$$V(\text{right of } D) = -7.5 + 7.5 = 0 \quad \text{OK}$$

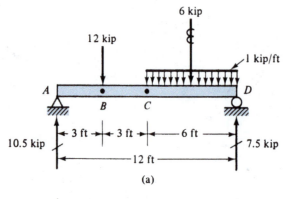

(a)

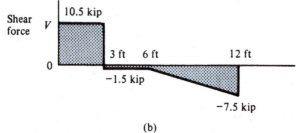

(b)

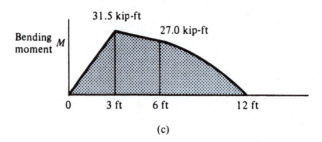

(c)

FIGURE 13.14

The shear diagram is drawn in Fig. 13.14(b). The moment diagram will be constructed from the shear diagram and Eq. (13.6) as follows:

$$M_A = 0$$
$$\Delta M_{A \to B} = 10.5(3) = 31.5 \text{ kip-ft}$$
$$M_B = 31.5 \text{ kip-ft}$$
$$\Delta M_{B \to C} = -1.5(3) = -4.5 \text{ kip-ft}$$
$$M_C = 31.5 - 4.5 = 27 \text{ kip-ft}$$
$$\Delta M_{C \to D} = \frac{-(7.5 + 1.5)(6)}{2} = -27 \text{ kip-ft}$$
$$M_D = 27 - 27 = 0 \quad \text{OK}$$

The moment diagram is drawn in Fig. 13.14(c).

Example 13.8   Use the relationships among loads, shears, and bending moments to draw the shear and moment diagrams for the beam shown in Fig. 13.15(a).

Solution: The free-body diagram is shown in Fig. 13.15(b). From the equations of equilibrium,

$$\Sigma M_B = 0 \quad 20(2.7) + 36(0.9) - M_C = 0 \quad M_C = 86.4 \text{ kN·m}$$
$$\Sigma F_y = 0 \quad -20 - 36 + C_y = 0 \quad C_y = 56 \text{ kN}$$

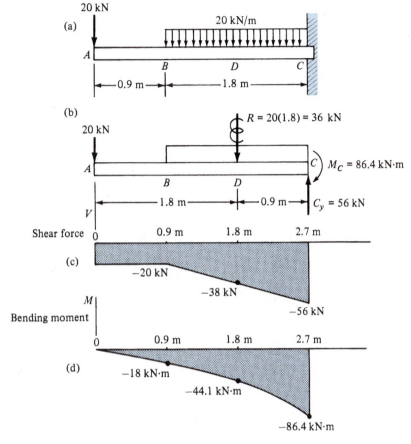

FIGURE 13.15

Check:

$$\Sigma M_A = 0 \quad -36(1.8) + C_y 2.7 - M_C = 0$$
$$-36(1.8) + 56(2.7) - 86.4 = 0 \quad \text{OK}$$

From Eq. (13.5) and Fig. 13.15(b), we calculate the shears. The value of the shear to the left of A is zero. That is,

$$V(\text{left of } A) = 0$$
$$\Delta V(\text{left of } A \text{ to right of } A) = -20 \text{ kN}$$
$$V(\text{right of } A) = -20 \text{ kN}$$
$$\Delta V(\text{right of } A \text{ to } B) = 0$$
$$V(\text{at } B) = -20 \text{ kN}$$
$$\Delta V(\text{from } B \text{ to left of } C) = -36 \text{ kN}$$
$$V(\text{left of } C) = -20 - 36 = -56 \text{ kN}$$
$$\Delta V(\text{left of } C \text{ to right of } C) = 56 \text{ kN}$$
$$V(\text{right of } C) = -56 + 56 = 0 \quad \text{OK}$$

The shear diagram is plotted in Fig. 13.15(c). The shear at D (midpoint between B and C) $V_D = (-20 - 56)/2 = -38$ kN. The moment diagram shown in Fig. 13.15(d) can be constructed from the shear diagram and Eq. (13.6) as follows:

$$M_A = 0$$
$$\Delta M_{A \to B} = -20(0.9) = -18 \text{ kN·m}$$
$$M_B = -18 \text{ kN·m}$$
$$\Delta M_{B \to D} = 0.9\left(\frac{-20 - 38}{2}\right) = -26.1 \text{ kN·m}$$

$$M_D = -18 - 26.1 = -44.1 \text{ kN·m}$$

$$\Delta M_{D \to C} = 0.9\left(\frac{-38 - 56}{2}\right) = -42.3 \text{ kN·m}$$

$$M_C = -44.1 - 42.3 = -86.4 \text{ kN·m} \quad \text{OK}$$

Example 13.9 Use the relations among loads, shears, and bending moments to sketch the shear and moment diagrams for the beam shown in Fig. 13.16(a).

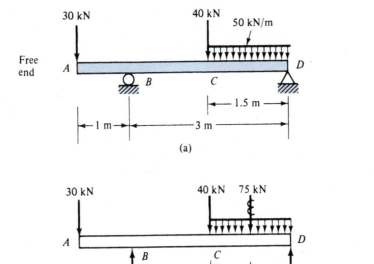

(a)

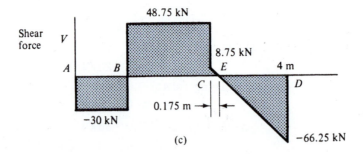

(b)

(c)

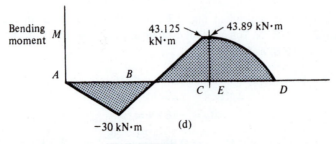

(d)

FIGURE 13.16

Solution: The free-body diagram is shown in Fig. 13.16(b). From the usual equations of equilibrium,

$$\Sigma M_B = 0 \quad 30(1) - 40(1.5) - 75(2.25) + D_y3 = 0$$
$$D_y = 66.25 \text{ kN}$$
$$\Sigma M_D = 0 \quad 30(4) - B_y3 + 40(1.5) + 75(0.75) = 0$$
$$B_y = 78.75 \text{ kN}$$

Check:

$$\Sigma F_y = 0 \quad -30 + B_y - 40 - 75 + D_y = 0$$
$$-30 + 78.75 - 40 - 75 + 66.25 = 0 \quad \text{OK}$$

From Eq. (13.5) and the free-body diagram, we calculate the shears. The value of the shear to the left of point A is zero. That is,

$$V(\text{left of point } A) = 0$$
$$\Delta V(\text{left of } A \text{ to right of } A) = -30 \text{ kN}$$
$$V(\text{right of } A) = -30 \text{ kN}$$
$$\Delta V(\text{from } A \text{ to } B) = 0$$
$$V(\text{left of } B) = -30 \text{ kN}$$
$$\Delta V(\text{left of } B \text{ to right of } B) = +78.75 \text{ kN}$$
$$V(\text{right of } B) = 78.75 - 30 = 48.75 \text{ kN}$$
$$\Delta V(\text{from } B \text{ to } C) = 0$$
$$V(\text{left of } C) = 48.75 \text{ kN}$$
$$\Delta V(\text{left of } C \text{ to right of } C) = -40 \text{ kN}$$
$$V(\text{right of } C) = -40 + 48.75 = 8.75 \text{ kN}$$
$$\Delta V(\text{right of } C \text{ to left of } D) = -75 \text{ kN}$$
$$V(\text{left of } D) = 8.75 - 75 = -66.25 \text{ kN}$$
$$\Delta V(\text{left of } D \text{ to right of } D) = +66.25$$
$$V(\text{right of } D) = -66.25 + 66.25 = 0 \quad \text{OK}$$

To find the location of E where the shear is zero, we write

$$V(\text{from right of } C \text{ to } E) = -50x' = -8.75 \quad \text{or} \quad x' = 0.175 \text{ m}$$

The section where the shear is zero is at $x = 2.5 + 0.175 = 2.675$ m. The shear diagram is drawn in Fig. 13.16(c). From Eq. (13.6) and the shear diagram, we calculate the following values for bending moments:

$$M_A = 0$$
$$\Delta M_{A \to B} = -30(1) = -30 \text{ kN·m}$$
$$M_B = -30 \text{ kN·m}$$
$$\Delta M_{B \to C} = 48.75(1.5) = 73.125 \text{ kN·m}$$
$$M_C = -30 + 73.125 = 43.125 \text{ kN·m}$$
$$\Delta M_{C \to E} = \frac{(0.175)(8.75)}{2} = +0.766 \text{ kN·m}$$
$$M_E = 43.125 + 0.766 = 43.891 \text{ kN·m}$$
$$\Delta M_{E \to D} = \frac{(1.5 - 0.175)(-66.25)}{2} = -43.891 \text{ kN·m}$$
$$M_D = 43.891 - 43.891 = 0 \quad \text{OK}$$

The moment diagram is drawn in Fig. 13.16(d).

Example 13.10 Use the relations among loads, shears, and bending moments to sketch the shear and moment diagrams for the beam shown in Fig. 13.17(a).

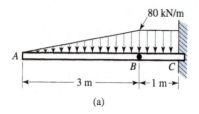

(a)

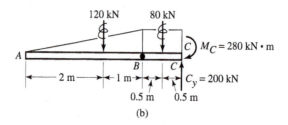

(b)

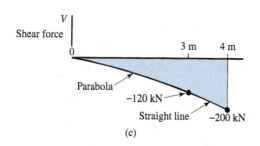

(c)

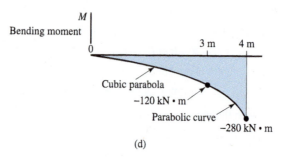

(d)

FIGURE 13.17

Solution: The free-body diagram of the beam is shown in Fig. 13.17(b). From the equations of equilibrium,

$$\Sigma M_C = 0 \qquad 120(2) + 80(0.5) - M_C = 0 \qquad M_C = 280 \text{ kN·m}$$
$$\Sigma F_y = 0 \qquad -120 - 80 + C_y = 0 \qquad C_y = 200 \text{ kN}$$

Check:

$$\Sigma M_A = 0 \qquad -120(2) - 80(3.5) + C_y(4) - M_C = 0$$
$$-120(2) - 80(3.5) + 200(4) - 280 = 0 \qquad \text{OK}$$

From Eq. (13.5) and the free-body diagram, we calculate the shears.

$$V_A = 0$$

$$\Delta V(\text{from } A \text{ to } B) = \frac{-80(3)}{2} = -120 \text{ kN}$$

$$V_B = -120 \text{ kN}$$

$$\Delta V(\text{from } B \text{ to } C) = -80(1) = -80 \text{ kN}$$

$$V_C = -120 - 80 = -200 \text{ kN} \qquad \text{OK}$$

The shear diagram is drawn in Fig. 13.17(c) and will be used together with Eq. (13.6) to construct the moment diagram of Fig. 13.17(d). (The parabolic area of the shear diagram from A to B is $bh/3$, where b is the base and h the height of the parabola.)

$$M_A = 0$$

$$M_{A \to B} = \frac{-120(3)}{3} = -120\,\text{kN·m}$$

$$M_B = -120\,\text{kN·m}$$

$$M_{B \to C} = -\frac{(120 + 200)}{2} = -160\,\text{kN·m}$$

$$M_C = -120 - 160 = -280\,\text{kN·m} \quad \text{OK}$$

In reviewing the shear and moment diagrams in Figs. 13.14 through 13.17, we observe the following:

1. *Concentrated loads or reactions* produce a shear diagram of straight horizontal lines that are broken only at a concentrated load or reaction, and a moment diagram of straight sloping lines broken at a load or reaction.

   a. On shear diagrams, a concentrated load or reaction appears as a straight vertical line that is equal in direction (up or down) and numerical value to the load or reaction itself.

   b. On moment diagrams, the change in moment between any two points along a beam is equal to the rectangular area under the shear diagram between those same two points.

2. *Uniform loads* produce a shear diagram of straight sloping lines and a moment diagram that varies as the square of the distance along the beam (as a parabola).

   a. Slope of the shear diagram equals the numerical value (lb/ft or kN/m) of the uniform load.

   b. For beams loaded only by uniform loads, the change in moment between any two points along the beam is equal to the triangular area under the shear diagram between those same two points.

3. *Triangular loads* produce a shear diagram that varies as the square of the distance along the beam (as a parabola) and a moment diagram that varies as the cube of the distance along the beam (as a cubic parabola).

4. For *all types of beam loading:*

   a. The value of shear at any point along a beam is equal to the slope of the moment diagram at that point.

   b. The maximum moment must occur at a beam location for which the value of shear is zero.

PROBLEMS

13.11 through 13.35 Sketch and label the shear-force diagram for the beam shown.

13.36 through 13.60 Sketch and label the bending moment diagram for the beam shown.

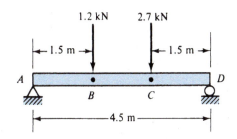

PROB. 13.11 and PROB. 13.36

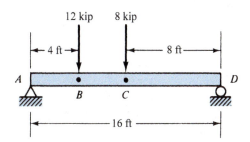

PROB. 13.12 and PROB. 13.37

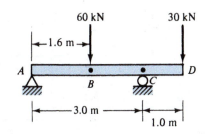

PROB. 13.13 and PROB. 13.38

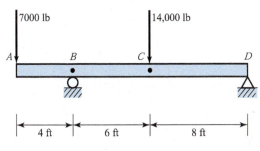

PROB. 13.14 and PROB. 13.39

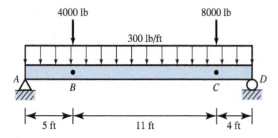

PROB. 13.15 and PROB. 13.40

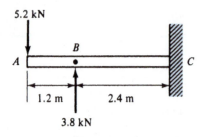

PROB. 13.16 and PROB. 13.41

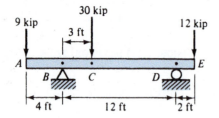

PROB. 13.17 and PROB. 13.42

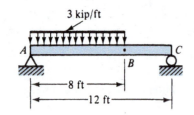

PROB. 13.18 and PROB. 13.43

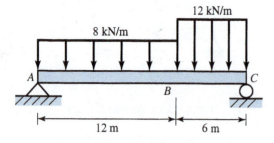

PROB. 13.19 and PROB. 13.44

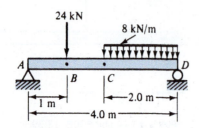

PROB. 13.20 and PROB. 13.45

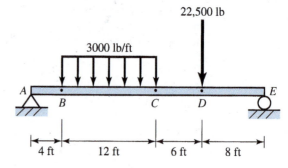

PROB. 13.21 and PROB. 13.46

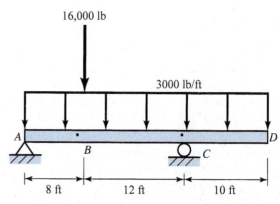

PROB. 13.22 and PROB. 13.47

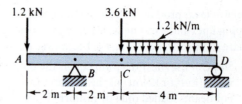

PROB. 13.23 and PROB. 13.48

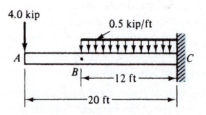

PROB. 13.24 and PROB. 13.49

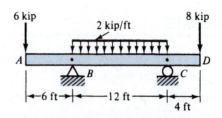

PROB. 13.25 and PROB. 13.50

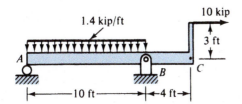

PROB. 13.26 and PROB. 13.51

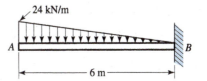

PROB. 13.32 and PROB. 13.57

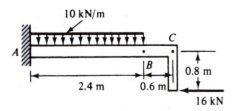

PROB. 13.27 and PROB. 13.52

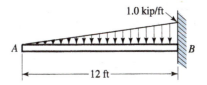

PROB. 13.33 and PROB. 13.58

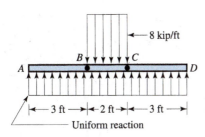

PROB. 13.28 and PROB. 13.53

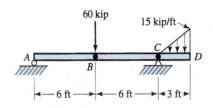

PROB. 13.34 and PROB. 13.59

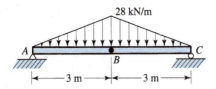

PROB. 13.29 and PROB. 13.54

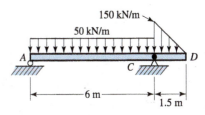

PROB. 13.35 and PROB. 13.60

13.61 to 13.63   For the given shear-force diagram, sketch and label the beam-loading diagram and the bending-moment diagram.

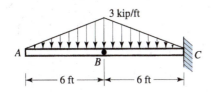

PROB. 13.30 and PROB. 13.55

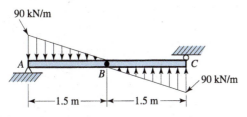

PROB. 13.31 and PROB. 13.56

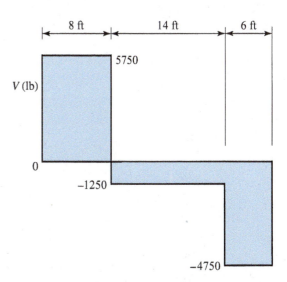

PROB. 13.61

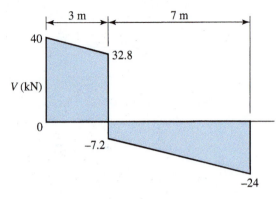

PROB. 13.62

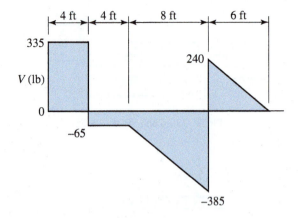

PROB. 13.63

BENDING AND SHEARING STRESSES IN BEAMS

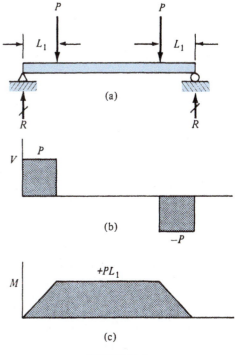

FIGURE 14.1

14.1 INTRODUCTION

If no axial force is present in a beam, the bending moment or bending couple at a cross section of the beam is the resultant of internal stresses that act normal to the cross section. The shear force is the resultant of shear stresses that act tangent to the cross section. In this chapter, we will be concerned with the magnitude and distribution of these stresses. We begin by considering the nature of the deformations that are assumed to occur when a beam is subjected to bending.

14.2 PURE BENDING OF A SYMMETRIC BEAM

The beam that we consider is subjected to *pure bending:* That is, a bending moment without either shearing or axial forces. In Fig. 14.1(a), the beam has been loaded by couples to produce pure bending. Notice the absence of a shearing force in the shear diagram of Fig. 14.1(b) and a constant bending moment in the interval of the beam

between the loads in Fig. 14.1(c). Furthermore, the beam has a uniform cross section, the cross section is symmetric about a vertical axis, the loads act in the plane of symmetry, and bending takes place in the same plane.

For purposes of illustration, consider a beam with a rectangular cross section. Horizontal and vertical lines have been drawn on the side of the beam [Fig. 14.2(a)]. In addition, we imagine horizontal and vertical lines drawn on a cross section of the beam [Fig. 14.2(b)]. Two loads are

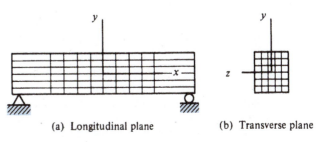

(a) Longitudinal plane   (b) Transverse plane

FIGURE 14.2. Beam before bending.

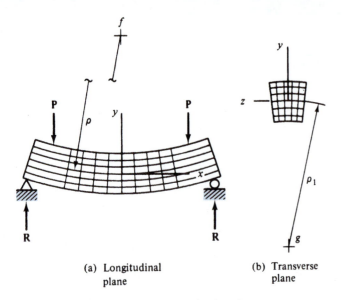

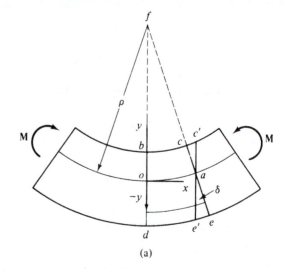

(a) Longitudinal plane

(b) Transverse plane

FIGURE 14.3. Beam after bending.

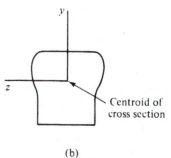

(b)

FIGURE 14.4

applied to the beam [Fig. 14.3(a) and (b)] to produce pure bending in the middle interval of the beam. As a result of the loads, horizontal lines on the side of the beam form arcs of circles that have a common center at point f, a distance of $y = \rho$ (the Greek lowercase letter rho) *above* the beam. The vertical lines on the side of the beam remain straight and rotate so they are directed toward the center of the circles at point f. At the same time, imaginary horizontal lines on the cross section form arcs of circles that have a common center at point g, a distance of $y = -\rho_1$ (rho) *below* the beam. The imaginary vertical lines on the cross section remain straight and rotate so they are directed toward the center of the circle at point g. Thus, we see that when pure bending produces longitudinal curvature, in the xy plane, it is accompanied by transverse curvature, also called *anticlastic curvature,* in the yz plane. The existence of anticlastic curvature may be observed by bending a rubber eraser between the thumb and forefinger. The deformations of the cross section have negligible effect on most beams and will not be considered in the following discussions.

14.3 DEFORMATION GEOMETRY FOR A SYMMETRIC BEAM IN PURE BENDING

In our discussion of the bending of a rectangular beam, we saw that the vertical straight lines on the side of the beam remained straight after bending. We also saw that they rotated through an angle so they were directed toward the center of the arcs of the circles that were formed by the horizontal lines. Generalizing these observations for a beam of any symmetrical cross section, we can say that *cross sections of a beam that are plane and normal to the axis of the beam before bending remain plane and normal to the axis after bending.* Consider part of such a beam, as shown in Fig. 14.4. Let plane cross section $c'e'$ rotate as a result of bending through an angle to a new position ce. No deformation

occurs in the longitudinal fiber oa. Fiber oa is the edge of a surface formed by fibers that do not stretch. The surface extends from the front to the back of the beam and is called the *neutral surface.* Below the neutral surface, longitudinal fibers stretch and, above the neutral surface, longitudinal fibers compress.

The intersection of the neutral surface and the cross section of the beam forms an axis called the *neutral axis.* The stretching of a horizontal fiber a distance $-y$ below the neutral axis at o is given by δ. Therefore, the strain of the fiber is

$$\epsilon = \frac{\delta}{oa} \qquad \text{(a)}$$

From similar triangles $e'ae$ and ofa,

$$\frac{\delta}{y} = \frac{oa}{\rho} \quad \text{or} \quad \frac{\delta}{oa} = \frac{y}{\rho}$$

Substituting into Eq. (a), we have

$$\epsilon = \frac{-y}{\rho} \qquad \text{(14.1)}$$

where the negative sign accounts for the fact that tensile strain occurs when y is negative. We see from Eq. (14.1) that the strains of the longitudinal fibers are proportional to the distance from the neutral axis and inversely proportional to the radius of curvature.

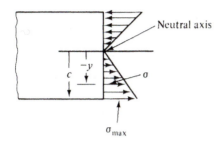

FIGURE 14.5

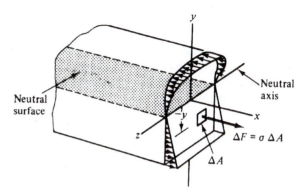

FIGURE 14.6

14.4 HOOKE'S LAW: DISTRIBUTION OF BENDING STRESS

The deformation geometry that we have considered up to this point depended only on the fact that the beam had a constant cross section and that the cross section and bending were symmetrical with respect to the xy plane. We will now assume that the material follows Hooke's law for longitudinal stresses and strains; that is, $\sigma = E\epsilon$. Combining Hooke's law and Eq. (14.1), we have

$$\sigma = \frac{-Ey}{\rho} \qquad (14.2)$$

Thus, we see that the bending stress is proportional to the distance y from the neutral axis and inversely proportional to the radius of curvature ρ.

The distribution for bending stress is shown in Fig. 14.5. The maximum stress occurs at a point that is the farthest distance c from the neutral axis. The stress at any distance y is related to the maximum stress by the expression obtained from similar triangles.

$$\frac{\sigma}{-y} = \frac{\sigma_{max}}{c} \quad \text{or} \quad \sigma = \frac{-\sigma_{max}}{c}y \qquad (14.3)$$

14.5 BENDING STRESS FORMULA: FLEXURE FORMULA

The position of the neutral axis can be found from the condition that the resultant axial force for the cross section of the beam must be equal to zero. The element of force ΔF acting on an element of area ΔA of the cross section is equal to the product of the stress σ and the area ΔA, as shown in Fig. 14.6. With the value of σ given by Eq. (14.3), we have

$$\Delta F = \sigma(\Delta A) = \frac{-\sigma_{max}}{c}y(\Delta A) \qquad (a)$$

From equilibrium in the axial direction, the sum of the elements of force ΔF must vanish, that is,

$$\Sigma \frac{-\sigma_{max}}{c}y(\Delta A) = \frac{-\sigma_{max}}{c}\Sigma y(\Delta A) = 0 \qquad (b)$$

Because σ_{max} and c are nonzero constants for a given cross section, $\Sigma y\Delta A$ must be equal to zero. The summation

$\Sigma y\Delta A$ represents the moment of the elements of area ΔA of the cross section with respect to the neutral axis. Therefore,

$$\Sigma y\Delta A = \bar{y}A = 0$$

where $\bar{y}$ is the coordinate of the centroid measured from the neutral axis and A is the area of the cross section. The centroidal distance $\bar{y} = 0$ because the area A is not zero. Therefore, *the neutral axis passes through the centroid of the cross section of the beam*. Thus, the z axis is the neutral axis.

From rotational equilibrium, the bending moment must be equal to the sum of the moments about the neutral axis of the elements of force ΔF, that is,

$$M = \Sigma y(\Delta F) \qquad (c)$$

Substituting for ΔF from Eq. (a), we have

$$M = \Sigma \frac{-\sigma_{max}}{c}y^2\Delta A = \frac{-\sigma_{max}}{c}\Sigma y^2\Delta A \qquad (d)$$

Recall that

$$I = \Sigma y^2\Delta A \qquad (e)$$

is the moment of inertia of the cross-sectional area with respect to the neutral axis (z axis). It has a definite value for a given cross section. (See Sec. 9.7 for a discussion of moments of inertia.) Substituting Eq. (e) in Eq. (d), we have

$$M = \frac{-\sigma_{max}}{c}I \quad \text{or} \quad \sigma_{max} = \frac{-Mc}{I} \qquad (f)$$

The negative sign in this equation is usually dropped, and the sign of the stress is determined by inspection. The stress distribution on the cross section must produce a resultant couple that has the same direction as the bending moment. Thus, a positive bending moment produces compression on the top of the beam and tension on the bottom of the beam. The reverse is true for a negative bending moment. Rewriting Eq. (f) without the negative sign, we have

$$\sigma_{max} = \frac{Mc}{I} \qquad (14.4)$$

This equation is known as the *bending stress formula* or the *flexure formula*. It gives the *maximum* normal stress on a cross section that has a bending moment M. By eliminating σ_{max} between Eqs. (14.3) and (14.4), we have

$$\sigma = \frac{My}{I} \qquad (14.5)$$

This formula gives the normal stress at any point on a cross section a distance y from the neutral axis in terms of the bending moment and the dimensions of the member. The sign of the stress is determined by inspection. Bending moment M will be expressed in inch-pounds or newton·meters (N·m), y in inches or meters, and I in in.^{4} or m^4. The normal stress will have units of

$$\frac{\text{(in.-lb)(in.)}}{\text{(in.}^4)} = \frac{\text{lb}}{\text{in.}^2} = \text{psi}$$

or, in SI units,

$$\frac{\text{(N·m)(m)}}{\text{(m}^4)} = \frac{\text{N}}{\text{m}^2} = \text{Pa (pascals)}$$

The bending stress formula was developed for a beam subject to pure bending only. However, it can be shown that the formula can also be used when shear as well as bending exists on the cross section.

14.6 ELASTIC SECTION MODULUS

In many problems involving the bending of a beam, the maximum normal stress is the quantity desired. Both I and c in Eq. (14.4) are constants for a given cross section. There-

fore, we define a new constant, $I/c = S$, which is called the *elastic section modulus*. Equation (14.4) can be rewritten in terms of the elastic section modulus as follows:

$$\sigma_{max} = \frac{M}{I/c} \quad \text{or} \quad \sigma_{max} = \frac{M}{S} \tag{14.6}$$

The bending moment will be expressed in inch-pounds or newton·meters (N·m) and the elastic section modulus in in.^{3} or m^3. As before, the normal stress will be expressed in lb/in.^{2} (psi) or N/m^2 (Pa). The moment of inertia or the elastic section modulus of the cross section with respect to the neutral axis may be found by the methods of Sec. 9.7 or for rolled steel shapes from Tables A.3 through A.8 of the Appendix, and for wooden members in Tables A.9, A.12, and A.13 of the Appendix.

14.7 PROBLEMS INVOLVING THE BENDING STRESS FORMULA

In this section, we solve several problems involving the normal stress due to the bending of a beam.

Example 14.1   A simply supported rectangular wooden beam with a cross section 8 in. × 10 in. and a span of 14 ft supports a 1.5-kip load [Fig. 14.7(a)]. Determine the *maximum* bending stress (a) at a section 2 ft from the left end and (b) for any section. Neglect the weight of the beam.

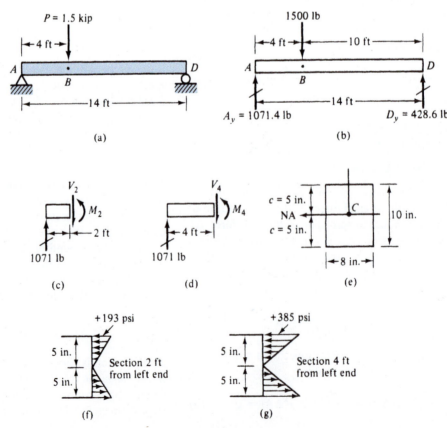

FIGURE 14.7

Solution: The bending moment and the moment of inertia with respect to the neutral axis are required to find the bending stress.

Bending Moment: A free-body diagram for the beam is drawn in Fig. 14.7(b). From equilibrium, $A_y = 1071$ lb and $D_y = 429$ lb. The bending moment for $x = 2$ ft is determined from the free-body diagram in Fig. 14.7(c). From equilibrium,

$$\Sigma M_c = 0 \qquad -1071(2) + M_2 = 0$$
$$M_2 = 2142 \text{ lb-ft} = 25,700 \text{ lb-in.}$$

The maximum bending moment will occur under the concentrated load at $x = 4$ ft. From the free-body diagram shown in Fig. 14.7(d), we have

$$\Sigma M_c = 0 \qquad -1071(4) + M_4 = 0$$
$$M_4 = 4284 \text{ lb-ft} = 51,400 \text{ lb-in.}$$

Moment of Inertia: From Table A.2 of the Appendix, the formula for the moment of inertia with respect to the centroidal or neutral axis is $I = bh^3/12$. For the cross section shown in Fig. 14.7(e),

$$I = \frac{bh^3}{12} = \frac{8(10)^3}{12} = 667 \text{ in.}^4$$

a. The maximum stress at a section 2 ft from the left end of the beam is given by the bending stress formula, Eq. (14.4).

$$\sigma_{max} = \frac{Mc}{I} = \frac{25,700(5)}{667} = 192.7 \text{ psi} \qquad \textbf{Answer}$$

b. The maximum stress for any section occurs 4 ft from the left end of the beam, where the bending moment is maximum. The maximum stress is given by the bending stress formula.

$$\sigma_{max} = \frac{Mc}{I} = \frac{51,400(5)}{667} = 385 \text{ psi} \qquad \textbf{Answer}$$

The stresses are shown in Fig. 14.7(f) and (g).

Example 14.2   A cantilever beam 3.5 m long supports a load of 20 kN on its free end. The beam and cross section are shown in Fig. 14.8(a) and (b). Find the maximum bending stress in the beam.

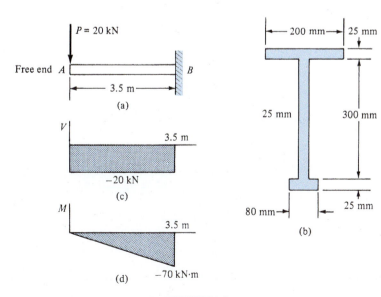

FIGURE 14.8

Solution: The maximum bending moment and moment of inertia of the cross section about the neutral axis (centroidal axis) are required to find the bending stress.

Bending Moment: We plot the shear force and bending moment diagram in Fig. 14.8(c) and (d). From the bending moment diagram, the maximum moment $M_B = -70\,\text{kN·m}$.

Moment of Inertia: The moment of inertia will be calculated by the method of Sec. 9.11.

The *I* with unequal flanges can be divided into three rectangular areas, as in Fig. 14.9. The solution for the centroid is shown in Table (A). From the sums in the table,

$$\bar{y} = \frac{\Sigma Ay}{\Sigma A} = \frac{3.025 \times 10^6}{14.5 \times 10^3} = 208.6\,\text{mm}$$

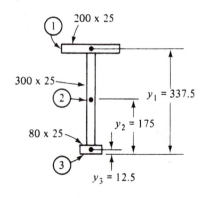

(All dimensions in mm)

FIGURE 14.9

TABLE (A)	Example 14.2		
	A	y	Ay
①	5000	337.5	1.688×10^6
②	7500	175	1.312×10^6
③	2000	12.5	0.025×10^6
	$\Sigma A = 14{,}500\,\text{mm}^2$		$\Sigma Ay = 3.025 \times 10^6\,\text{mm}^3$

The transfer distances for each area are shown in Fig. 14.10. The moment of inertia for each area about its *own* centroidal axis $I_c = bh^3/12$. For areas ①, ②, and ③,

$$(I_c)_1 = \frac{200(25)^3}{12} = 260.4 \times 10^3\,\text{mm}^4$$

$$(I_c)_2 = \frac{25(300)^3}{12} = 56.25 \times 10^6\,\text{mm}^4$$

$$(I_c)_3 = \frac{80(25)^3}{12} = 104.2 \times 10^3\,\text{mm}^4$$

Thus, I_c for areas ① and ③ are negligible compared with area ②. The solution for the moment of inertia about the neutral axis is shown in Table (B). From the sums in the table,

$$I_{NA} = \Sigma I_c + \Sigma(Ad^2) = 56.2 \times 10^6 + 168.5 \times 10^6$$
$$= 224 \times 10^6\,\text{mm}^4 = 224 \times 10^{-6}\,\text{m}^4$$

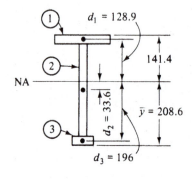

(All dimensions in mm)

FIGURE 14.10

TABLE (B)	Example 14.2			
	A	d	I_c	Ad^2
①	5000	128.9	—	83.08×10^6
②	7500	33.6	56.25×10^6	8.47×10^6
③	2000	196.1	—	76.91×10^6
			$\Sigma I_c = 56.2 \times 10^6\,\text{mm}^4$	$\Sigma(Ad^2) = 168.5 \times 10^6\,\text{mm}^4$

The maximum bending stress will occur at the bottom of the beam at a maximum distance $c = 208.6 \times 10^{-3}$ m from the neutral axis. The stress will be compressive because the bending moment is negative. From Eq. (14.4),

$$\sigma_{max} = \frac{Mc}{I} = \frac{70(208.6 \times 10^{-3})}{224 \times 10^{-6}} = 65 \times 10^3 \, \text{kN/m}^2$$

$$= 65 \, \text{MPa} \qquad\qquad \textbf{Answer}$$

Example 14.3 A W16 × 67 steel beam with an overhang supports a uniform load, including its own weight of 4 kip/ft, as shown in Fig. 14.11(a). Find the maximum bending stress.

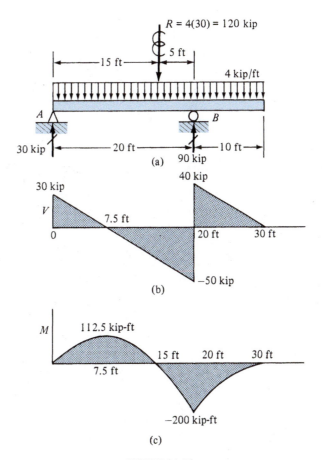

FIGURE 14.11

Solution: We plot the shear force and bending moment diagrams in Fig. 14.11(b) and (c). The bending moment $M_{max} = -200$ kip-ft $= -2.4 \times 10^6$ lb-in. occurs over the support at B.

Section properties for the W16 × 67 beam are given in Table A.3 of the Appendix. $S = I/c = 117$ in.^{3} The maximum stress from Eq. (14.6) is given by

$$\sigma_{max} = \frac{M}{S} = \frac{2.4 \times 10^6}{117} = 20,500 \, \text{psi} \qquad\qquad \textbf{Answer}$$

Example 14.4 A simply supported uniform beam with a span L supports a uniformly distributed load w. The beam and cross section are shown in Fig. 14.12(a) and (b). If the beam has a span $L = 4$ m and the allowable bending stress is 70 MPa, find the allowable uniform load w—including the weight of the beam—in units of kN/m.

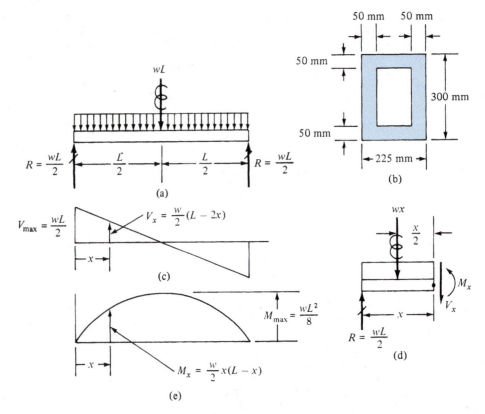

FIGURE 14.12

Solution: The reactions for the beam are both equal to one-half of the resultant of the distributed load, or $R = wL/2$. To determine the shear force and bending moment for a section at a distance x along the beam, we draw the free-body diagram shown in Fig. 14.12(d). From the diagram and Eqs. (13.1) and (13.3),

$$V(\text{at section}) = +\uparrow \Sigma F(\text{left of section})$$

$$V_x = \frac{wL}{2} - wx = \frac{w}{2}(L - 2x) \qquad \textbf{(a)}$$

$$M(\text{at section}) = \Subset + \Sigma M_c(\text{left of section})$$

$$M_x = \frac{wL}{2}x - \frac{w}{2}x^2 = \frac{w}{2}x(L - x) \qquad \textbf{(b)}$$

From Eqs. (a) and (b),

$$V_{\text{max}} = \frac{wL}{2} \quad \text{and} \quad M_{\text{max}} = \frac{wL^2}{8} \qquad \textbf{(c)}$$

The maximum moment occurs at the middle of the beam. For $L = 4\,\text{m}$, the maximum bending moment from Eq. (c) $M_{\text{max}} = wL^2/8 = w(4)^2/8 = 2w(\text{kN·m})$ *where w has units of* kN/m.

The moment of inertia for the hollow rectangular cross section [Fig. 14.12(b)] is found by subtracting the moment of inertia of the inside rectangle from the moment of inertia of the outside rectangle. The formula for the moment of inertia of a rectangle about the centroid (neutral) axis is $I = bh^3/12$, and for the hollow rectangle $I = [225(300)^3 - 125(200)^3]/12 = 422.9 \times 10^6\,\text{mm}^4$. Accordingly, the section modulus $S = I/c = 422.9 \times 10^6/150 = 2.819 \times 10^6\,\text{mm}^3 = 2.819 \times 10^{-3}\,\text{m}^3$.

From the bending stress formula [Eq. (14.6)], we have

$$M = \sigma S$$

$$2w(\text{kN·m}) = 70 \times 10^3 \frac{\text{kN}}{\text{m}^2}(2.819 \times 10^{-3}\,\text{m}^3)$$

$$w = 98.7\,\text{kN/m} \qquad \qquad \textbf{Answer}$$

Example 14.5   A simply supported W14 × 38 steel beam with a span L supports two concentrated loads P at its third points [Fig. 14.13(a)]. If the loads are $P = 35$ kip, the span $L = 9$ ft, and the maximum allowable bending stress is 24 ksi, is the beam satisfactory? Neglect the weight of the beam.

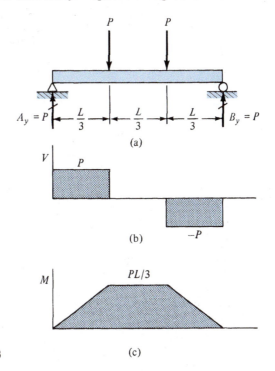

FIGURE 14.13

Solution: We plot the shear and moment diagrams in Fig. 14.13(b) and (c). The maximum bending moment $M_{max} = PL/3 = 35(9)/3 = 105$ kip-ft occurs throughout the middle third of the beam. The section modulus for the W14 × 38 from Table A.3 of the Appendix is $S = I/c = 54.7$ in.³ Accordingly, the maximum stress from Eq. (14.6)

$$\sigma_{max} = \frac{M}{S} = \frac{105(12)}{54.7} = 23.0 \text{ ksi} < \sigma_a = 24 \text{ ksi}$$

Therefore, the beam is satisfactory.   **Answer**

PROBLEMS

14.1 A beam of rectangular cross section is 125 mm wide and 200 mm deep. If the maximum bending moment is 28.5 kN·m, determine (a) the maximum tensile and compressive bending stress, and (b) the bending stress 25 mm from the top of the section.

14.2 A rectangular beam 50 mm wide and 100 mm deep is subjected to bending. What bending moment will cause a maximum bending stress of 137.9 MN/m² (MPa)?

14.3 Determine the bending moment in a rectangular beam 3 in. wide and 6 in. deep if the maximum bending stress is 15,000 psi.

14.4 A rectangular tube has outside dimensions that are 2 in. wide and 4 in. deep and a wall thickness of 0.25 in. The tube is acted on by a bending moment of 12,000 lb-in. Determine (a) the maximum tensile and compressive bending stress, and (b) the bending stress 0.25 in. from the bottom of the tube.

14.5 A rectangular tube has an outside width of 100 mm, an outside depth of 150 mm, and a wall thickness of 10 mm. If the bending moment is 1.4 kN·m, determine (a) the maximum bending stress in the tube, and (b) the bending stress 10 mm from the top of the tube.

14.6 What bending moment will cause a maximum stress of 72.0 MN/m² (MPa) in the beam with the cross section shown?

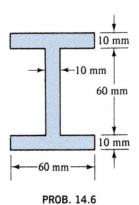

PROB. 14.6

14.7 What bending moment will cause a maximum bending stress of 10,000 psi in the beam with the cross section shown?

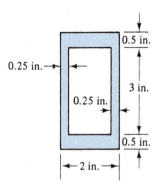

PROB. 14.7

14.8 A beam with the cross section shown is acted on by a bending moment of 10,000 lb-in. Determine (a) the maximum tensile and compressive bending stress, and (b) the bending stress 1 in. from the bottom of the beam.

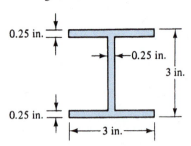

PROB. 14.8

14.9 Determine the maximum tensile and compressive bending stress for the simply supported beam with a concentrated load $P = 12$ kip and cross section as shown.

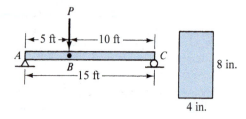

PROB. 14.9

14.10 A simply supported beam has a uniform load $w = 3.5$ kip/ft and cross section as shown. Determine (a) the maximum tensile and compressive bending stress, and (b) the bending stress 3 ft to the right of point A and 3 in. from the bottom of the beam.

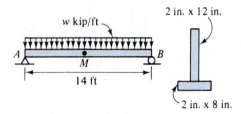

PROB. 14.10

14.11 Compute the maximum tensile and compressive bending stress for the simply supported beam with uniform load $w = 85$ kN/m and cross section as shown.

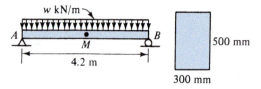

PROB. 14.11

14.12 A cantilever beam has a cross section and concentrated loads $P_1 = 30$ kN and $P_2 = 10$ kN, as shown. Compute (a) the maximum tensile and compressive bending stress, and (b) the bending stress 2 m to the right of point A and 50 mm from the top of the beam.

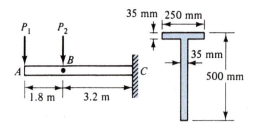

PROB. 14.12

14.13 Determine the maximum tensile and compressive bending stress for the cantilever beam with concentrated loads $P_1 = 3$ kip and $P_2 = 9$ kip and cross section shown.

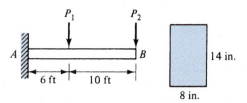

PROB. 14.13

14.14 A cantilever beam has a uniform load $w = 1.10$ kip/ft and cross section as shown. Determine (a) the maximum tensile and compressive bending stress, and (b) the bending stress 10 ft to the left of point B and 2 in. from the top of the beam.

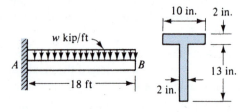

PROB. 14.14

14.15 Compute the maximum tensile and compressive bending stress for the cantilever beam with uniform load $w = 90$ kN/m and cross section shown.

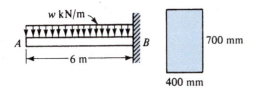

PROB. 14.15

14.16 The allowable tensile and compressive bending stress for the beam shown in the figure for Prob. 14.10 is 20 ksi. Calculate the allowable uniform load w.

14.17 Compute the allowable uniform load w for the beam shown in the figure for Prob. 14.11 if the allowable tensile bending stress is 42.0 MN/m² (MPa) and the allowable compressive bending stress is 168.0 MN/m² (MPa).

14.18 Determine the allowable load P_2 for the beam shown in the figure for Prob. 14.12 if the allowable tensile and compressive bending stress is 68.7 MN/m² (MPa). Assume that $P_1 = P_2$.

14.19 The allowable tensile bending stress is 6700 psi and the allowable compressive bending stress is 27,000 psi for the beam shown in the figure for Prob. 14.13. Find the allowable load P_1 if we assume that $P_2 = 3P_1$.

14.20 Compute the allowable uniform load w for the beam shown in the figure for Prob. 14.14 if the allowable tensile and compressive bending stress is 15.0 ksi.

14.21 Calculate the allowable uniform load w for the beam shown in the figure for Prob. 14.15 if the allowable tensile and compressive bending stress is 100 MN/m² (MPa).

For Probs. 14.22 through 14.26, see the Appendix for properties of rolled steel sections.

14.22 The W36 × 160 steel cantilever beam is acted on by a concentrated load and uniform load as shown. Determine (a) the maximum tensile and compressive bending stress if $P = 24$ kip and $w = 2$ kip/ft, and (b) the allowable uniform load w if the bending stress is 13.8 ksi and $P = 12w$.

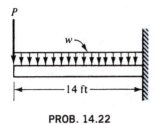

PROB. 14.22

14.23 The W33 × 221 steel simply supported beam is loaded with concentrated loads and uniform load as shown. Determine (a) the maximum tensile and compressive bending stress if $P = 150$ kip and $w = 10$ kip/ft, and (b) the allowable uniform load w if the allowable bending stress is 24 ksi and $P = 15w$.

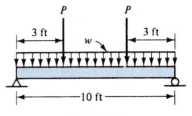

PROB. 14.23

14.24 A W18 × 35 steel beam supports a uniform load with overhanging ends as shown. Determine (a) the maximum tensile and compressive bending stress if $w = 3$ kip/ft, and (b) the allowable uniform load w if the allowable bending stress is 12 ksi.

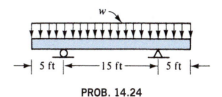

PROB. 14.24

14.25 A W16 × 50 steel beam supports a uniform load and a concentrated load as shown. Determine (a) the maximum tensile and compressive bending stress if $w = 1.5$ kip/ft and $P = 15$ kip, and (b) the allowable load w if the allowable bending stress is 17 ksi and $P = 10w$.

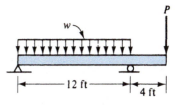

PROB. 14.25

14.26 A W18 × 55 steel beam supports a uniform load with an overhanging end as shown. Determine (a) the maximum bending stress if $w = 2$ kip/ft, and (b) the allowable uniform load w if the allowable bending stress is 15.8 ksi.

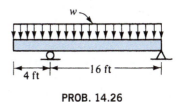

PROB. 14.26

For Probs. 14.27 through 14.30, see Tables A.12 and A.13 of the Appendix for properties of engineered wood products.

14.27 A simply supported beam is fabricated from two 44 × 241 LVL boards as shown. This beam has a length $L = 6$ m, carries an unknown distributed load of w N/m, and has an allowable bending stress of 19.0 MPa. Find the maximum permissible value for w.

14.28 Same as Prob. 14.27, but the beam consists of two 1-3/4 × 9-1/2 LVL boards, $L = 18$ ft, and the allowable bending stress is 2750 psi.

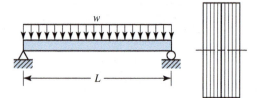

PROB. 14.27 and PROB. 14.28

14.29 A simply-supported 44 × 356 wooden I-joist has a length $L = 5$ m and carries a uniformly distributed load $w = 1460$ N/m as shown. (a) Find the maximum bending stress developed in this beam. (b) If the maximum allowable stress in the beam is 20.3 MPa, find the maximum permissible value of L.

14.30 Same as Prob. 14.29, but the beam is a 1-3/4 × 9-1/2 I-joist, $w = 100$ lb/ft, and for (a), $L = 16$ ft, while for (b), the allowable stress is 2950 psi.

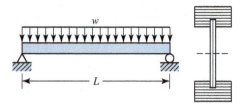

PROB. 14.29 and PROB. 14.30

14.8 SHEARING STRESS IN BEAMS

The shear force at a section is the resultant of the shear stresses that act tangent to the section. We assume that the shear stresses are parallel to the y axis as shown in Fig. 14.14(a) for a beam that is symmetric with respect to the xy plane. In addition to the shear stress on the cross section, shear stress of equal magnitude exists on horizontal planes that are perpendicular to the plane of the cross section. That is, as shown in Fig. 14.14(b), the shear stresses that exist at points on line qq' in the plane $qq'r'r$ are at every point equal in magnitude to the shear stresses that exist at points on line qq' in the plane $qq'p'p$. The existence of horizontal shear stress and its equality to the shear stress on the cross section was to be expected from the results of Sec. 12.2, where we found that shearing stresses at a point acting on mutually perpendicular planes are equal.

As an aid in visualizing horizontal shear stress, let us consider two rectangular beams stacked one on top of the other [Fig. 14.15(a)]. If we neglect friction, the two beams—under load—bend independently with one sliding over the other [Fig. 14.15(b)]. If the beam is solid, this sliding action is resisted by shearing stresses. We may visualize a solid beam as made up of many horizontal layers. To bend as a solid beam, shear stresses are set up between each horizontal layer or on each horizontal plane. Accordingly, this stress is often

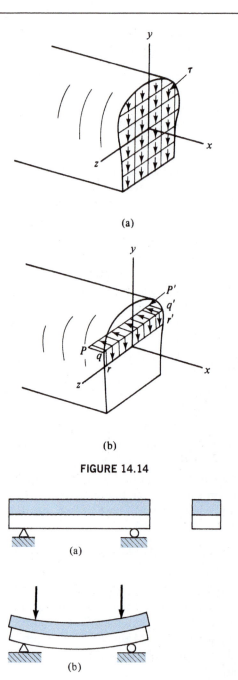

(a)

(b)

FIGURE 14.14

(a)

(b)

FIGURE 14.15

referred to as *horizontal* shearing stress. (A good example of this effect can be observed using an ordinary deck of playing cards. Holding one end of the deck lightly on the fingertips of each hand, it is relatively easy to bend the entire deck using your thumbs. However, if you grasp each end of the deck tightly between a thumb and forefinger, the cards are constrained from sliding and thus it becomes considerably more difficult to bend the deck.)

14.9 HORIZONTAL SHEARING STRESS FORMULA

In a beam subject to pure bending, no shearing force exists and therefore no horizontal shearing stress. The derivation of the bending stress formula was based on the deformation

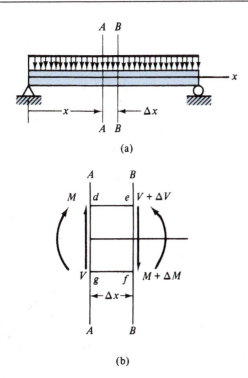

(a)

(b)

FIGURE 14.16

geometry of pure bending. The shearing stress formula is not based on deformation geometry. In fact, the formula does not satisfy compatible deformation geometry, and its use is more restricted than the bending stress formula. However, it gives satisfactory results for many engineering problems.

The shearing stress formula will be derived from the requirements of equilibrium. Consider an element of a beam of length Δx between two sections A–A and B–B as shown in Fig. 14.16(a). A free-body diagram for the element is drawn in Fig. 14.16(b), where the shearing force and bending moment on the sections are shown. In Fig. 14.17(a), the bending moments have been replaced by bending stresses. The bending stresses at any point on section A–A (plane $dd'g'g$) are given by

$$\sigma_1 = \frac{My}{I} \tag{a}$$

while the bending stresses at any point on section B–B (plane $ee'f'f$) are given by

$$\sigma_2 = \frac{(M + \Delta M)y}{I} \tag{b}$$

To find the shearing stress on a horizontal plane, we cut the element along a horizontal plane a distance y_1 above the neutral surface and draw a free-body diagram of the part of the element above the cut [Fig. 14.17(b)]. The force F_1 is the sum of the stresses σ_1 multiplied by the elements of area ΔA summed over the plane area $dd'i'i$. Thus, $F_1 = \Sigma \sigma_1 \Delta A$. The stress σ_1 is given by Eq. (a). Therefore,

$$F_1 = \Sigma \frac{My}{I} \Delta A = \frac{M}{I} \Sigma y \Delta A \tag{c}$$

where the summation must be carried out for the area above y_1 of the cross section [Fig. 14.17(c)]. Similarly, the force F_2 is the sum of the stress σ_2 multiplied by the elements of area ΔA summed over the plane area $ee'h'h$. Thus, $F_2 = \Sigma \sigma_2 \Delta A$. The stress is given by Eq. (b); therefore,

$$F_2 = \Sigma \frac{M + \Delta M}{I} y \Delta A = \frac{M + \Delta M}{I} \Sigma y \Delta A$$

or $$F_2 = \frac{M}{I} \Sigma y \Delta A + \frac{\Delta M}{I} \Sigma y \Delta A \tag{d}$$

where the summation must be carried out for the area above y_1 of the cross section [Fig. 14.17(c)]. The shear force along the horizontal plane $i'ihh'$ is given by the product of the average shear stress τ multiplied by the shear area $b(\Delta x)$, that is, $\tau b(\Delta x)$. From the equilibrium equation $\Sigma F_x = 0$, we write

$$F_1 - F_2 + \tau b(\Delta x) = 0$$

Substituting Eqs. (c) and (d) for F_1 and F_2, we have

$$\frac{M}{I} \Sigma y \Delta A - \frac{M}{I} \Sigma y \Delta A - \frac{\Delta M}{I} \Sigma y \Delta A + \tau b(\Delta x) = 0$$

or $$\tau b(\Delta x) = \frac{\Delta M}{I} \Sigma y \Delta A$$

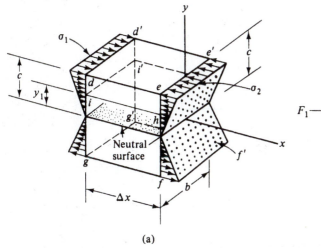

(a)

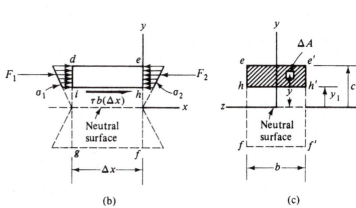

(b)   **(c)**

FIGURE 14.17

Dividing both sides of the equation by $b(\Delta x)$, this becomes

$$\tau = \frac{\Delta M \Sigma y \Delta A}{(\Delta x) Ib}$$

In Sec. 13.6, we found that $\Delta M/\Delta x = V$; therefore,

$$\tau = \frac{V \Sigma y \Delta A}{Ib} \qquad (14.7)$$

The summation, $\Sigma y \Delta A$, represents the first moment of the area of the cross section of the beam above y_1 with respect to the neutral or z axis. (The first moment of the area below y_1 gives the negative of the first moment of the area above y_1.) Let $\Sigma y \Delta A = Q$; then, Eq. (14.7) may be written as

$$\tau = \frac{VQ}{Ib} \qquad (14.8)$$

If the shearing force V is in pounds (lb) or newtons (N), the first moment of area Q is in in.^{3} or m^3, the moment of inertia in in.^{4} or m^4, and the width b in in. or m, then the average shear stress τ will be in lb/in.^{2} (psi) or newtons/meter^2 (N/m^2).

Example 14.6 A rectangular beam 400 mm × 800 mm supports a shear force of 5 kN. Find the horizontal shearing stress at the various levels shown in Fig. 14.18(a), and draw a graph showing the distribution of stress across the depth of the beam.

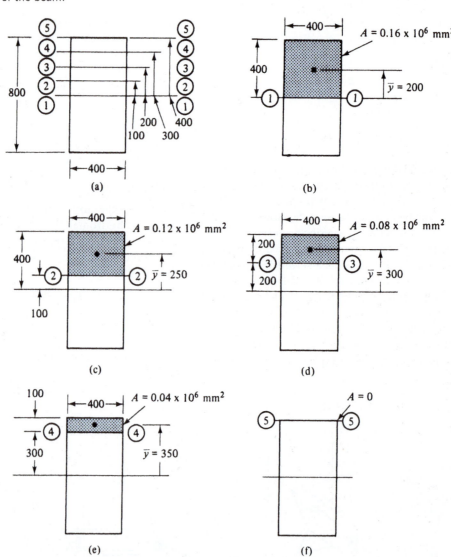

FIGURE 14.18 (All dimensions in mm)

Solution: The moment of inertia of the rectangular cross section with respect to the neutral axis is given by the formula

$$I_{NA} = \frac{bh^3}{12} = \frac{400(800)^3}{12} = 17.07 \times 10^9 \text{ mm}^4$$

The shear force $V = 5$ kN $= 5 \times 10^3$ N and the width of the section $b = 400$ mm. Accordingly, $V/Ib = 5 \times 10^3/[(17.07 \times 10^9)(400)] = 0.7323 \times 10^{-9}$ N/mm^5.

Values of A and $\bar{y}$ at each level are calculated and shown in Fig. 14.18(b) through (f). The values of A and $\bar{y}$, together with the corresponding shear stresses, are listed in Table (A).

	A $(10^6 \ mm^2)$	$\bar{y}$ (mm)	$Q = A\bar{y}$ $(10^6 \ mm^3)$	$\tau = \dfrac{VQ}{Ib}$ $(10^{-3} \ N/mm^2)$	τ (kPa)
TABLE (A) Example 14.6					
Level					
①–①	0.16	200	32	23.43	23.4
②–②	0.12	250	30	21.97	22.0
③–③	0.08	300	24	17.58	17.6
④–④	0.04	350	14	10.25	10.3
⑤–⑤	0	—	0	0	0

The distribution of horizontal shear stress from top to bottom of the beam is shown graphically in Fig. 14.19. The distribution below the neutral axis was calculated in the same way as the distribution above, except that for convenience the area used to calculate Q was taken below the shear plane instead of above it. In fact, the shearing stress distribution along the depth of a rectangular beam is parabolic. The maximum value occurs at the neutral axis and is zero at the top and bottom. This is *not* true for all beam shapes. For example, it is not true for a beam with a triangular cross section.

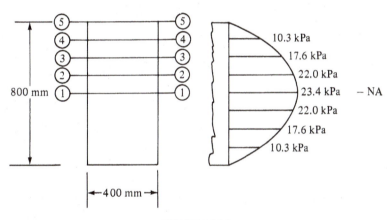

FIGURE 14.19

Example 14.7   Derive a formula for the maximum horizontal shearing stress in a beam of rectangular cross section.

Solution: The moment of inertia with respect to the neutral axis $I_{NA} = bh^3/12$. The moment of the area above the neutral axis (plane of maximum shearing stress) $Q = \bar{y}A = bh^2/8$, and the width of the shear plane $b = b$ as shown in Fig. 14.20. Therefore,

$$\tau = \frac{VQ}{Ib} = \frac{V(bh^2/8)}{(bh^3/12)b} = \frac{3}{2}\frac{V}{bh}$$

or

$$\tau_{max} = \frac{3}{2}\frac{V}{A} \tag{14.9}$$

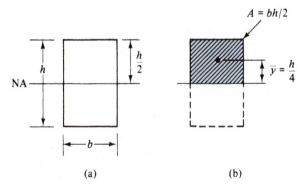

FIGURE 14.20

where A is the area of the cross section. Thus, the maximum horizontal shearing stress is 1-1/2 times the average value ($\tau_{av} = V/A$).

Example 14.8   A simply supported wooden beam with a span of 10 ft supports a load at the middle of its span [Fig. 14.21(a)]. The cross section is 10 in. wide and 14 in. deep. The allowable shear stress for the wood is 120 psi. Find P.

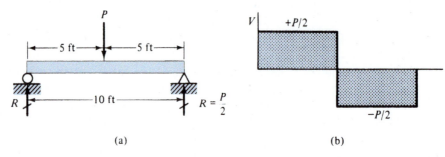

(a)   (b)

FIGURE 14.21

Solution: The shearing force diagram is shown in Fig. 14.21(b). The maximum shear force $V = P/2$, the cross-sectional area $A = 10(14) = 140 \text{ in.}^2$, and the allowable shearing stress is 120 psi. From Eq. (14.9),

$$\tau_{max} = \frac{3V}{2A} \quad \text{or} \quad V = \frac{\tau_{max}2A}{3}$$

Therefore,

$$\frac{P}{2} = \frac{120(2)(140)}{3} = 11,200 \text{ lb}$$

or

$$P = 22,400 \text{ lb}$$ **Answer**

Example 14.9   Find the maximum shearing stress for the beam and cross section in Fig. 14.22(a) and (b). (The maximum shearing stress occurs at the neutral axis.)

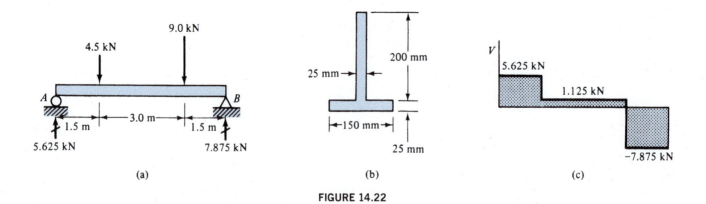

(a)   (b)   (c)

FIGURE 14.22

Solution: The reactions were calculated from the equations of equilibrium and the shear diagram drawn in Fig. 14.22(c). The maximum shear force at any section is 7.875 kN. The T-shaped cross section is divided into two rectangular areas, as in Fig. 14.23. The solution for the centroid is shown in Table (A). From the sums in the table,

$$\bar{y} = \frac{\Sigma Ay}{\Sigma A} = \frac{671.8 \times 10^3}{8.75 \times 10^3} = 76.8 \text{ mm}$$

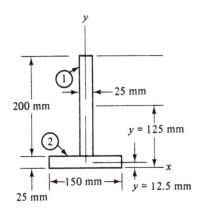

FIGURE 14.23

TABLE (A)	Example 14.9		
	A	y	Ay
①	5000	125	625.00×10^3
②	3750	12.5	46.88×10^3
	$\Sigma A = 8750\,\text{mm}^2$		$\Sigma Ay = 671.88 \times 10^3\,\text{mm}^3$

The transfer distances for each area are given in Fig. 14.24(a). The solution for the moment of inertia about the neutral axis is shown in Table (B). From the sums in the table,

$$I_{NA} = \Sigma I_c + \Sigma (Ad^2) = 16.86 \times 10^6 + 27.12 \times 10^6$$
$$= 43.98 \times 10^6\,\text{mm}^4 = 43.98 \times 10^{-6}\,\text{m}^4$$

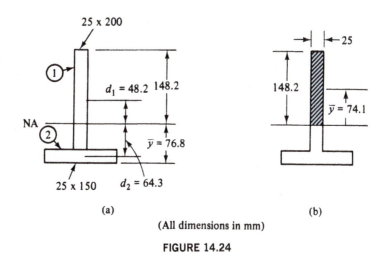

(a)   (b)

(All dimensions in mm)

FIGURE 14.24

The moment of the cross-sectional area above the neutral axis is determined from Fig. 14.24(b).

$$Q = \bar{y}A = 74.1(25)(148.2) = 0.2745 \times 10^6\,\text{mm}^3$$

The width of the beam at the shear plane (neutral axis) is $b = 25\,\text{mm}$.

The maximum shearing stress that occurs at the neutral axis is found from the horizontal shearing stress formula. Therefore,

$$\tau_{max} = \frac{VQ}{Ib} = \frac{7.875(0.2745)(10^6)}{(43.98)(10^6)(25)} = 1.966 \times 10^{-3}\,\text{kN/mm}^2$$
$$= 1.966\,\text{MN/m}^2\,\text{(MPa)}$$ **Answer**

TABLE (B)	Example 14.9			
	A	d	I_c	Ad^2
①	5000	48.2	ⓐ16.66×10^6	11.62×10^6
②	3750	64.3	ⓑ0.20×10^6	15.50×10^6
			$\Sigma I_c = 16.86 \times 10^6\,\text{mm}^4$	$\Sigma(Ad^2) = 27.12 \times 10^6\,\text{mm}^4$

ⓐ $I_c = 25(200)^3/12$
ⓑ $I_c = 150(25)^3/12$

Example 14.10   An I-beam with cross section as shown in Fig. 14.25(a) supports a shear force of 75 kip. Find the horizontal shearing stress at the various levels shown, and draw a graph showing the distribution of stresses across the depth of the beam.

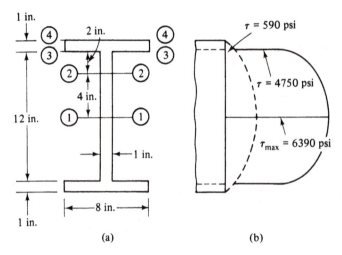

FIGURE 14.25

Solution: The moment of inertia may be found by calculating the moment of inertia of a rectangle 8 in. × 14 in. and subtracting the moment of inertia of two rectangles 3.5 in. × 12 in. That is,

$$I = \frac{8(14)^3}{12} - 2\left[\frac{3.5(12)^3}{12}\right] = 821.3 \text{ in.}^4$$

Therefore,

$$\frac{V}{I} = \frac{75,000}{821.3} = 91.3 \text{ lb/in.}^4$$

Values of Q and b and the corresponding shear stress for the various levels required are listed in Table (A). The distribution of shear stress is shown in Fig. 14.25(b). The stress is symmetric with respect to the neutral axis, and the maximum stress occurs at the neutral axis. At level ③–③, values of $b = 8$ in. and $b = 1$ in. are used to find the endpoints of the two curves. Both curves are parts of parabolas.

TABLE (A)   Example 14.10

Level	A ($in.^2$)	$\bar{y}$ (in.)	$Q = A\bar{y}$ ($in.^3$)	b (in.)	$\tau = \dfrac{VQ}{Ib}$ (psi)
①–①	8 × 1 = 8 6 × 1 = 6	6.5 3.0	52 18 } 70	1.0	6390
②–②	8 × 1 = 8 2 × 1 = 2	6.5 5.0	52 10 } 62	1.0	5660
③–③	8 × 1 = 8	6.5	52	1.0 8.0	4750 590
④–④	0	7.5	0	8.0	0

$\dfrac{V}{I} = 91.3$ lb/in.^{4}

The maximum shear stress is supported by the web of the beam; therefore, the maximum shear stress is often approximated by dividing the total shear force V by the area of the web to find the average stress.

$$\tau_{av} = \frac{V}{A_{web}} \tag{14.10}$$

$$= \frac{75,000}{12(1)} = 6250 \text{ psi}$$

We see that the average stress is lower than the maximum stress by approximately 2 percent. For the I-beam considered, the average value is a reasonable approximation of the maximum value. However, for standard rolled steel I-sections, the difference between the maximum and the average shear stress ranges in value from approximately 10 to 15 percent.

PROBLEMS

14.31 Determine the maximum shear stress in a rectangular beam 6 in. wide and 8 in. deep produced by a shear force of 38,400 lb.

14.32 A rectangular beam has a cross section 150 mm wide and 200 mm deep. Determine the maximum shear stress caused by a shear force of 165 kN.

14.33 What shear force will cause a maximum shear stress of 7.21 MPa in a rectangular beam 125 mm wide and 225 mm deep?

14.34 A rectangular beam has a cross section 5 in. wide and 9 in. deep. Determine the shear force that will cause a maximum shear stress of 1400 psi.

14.35 The beam shown has a rectangular cross section 4 in. wide and 8 in. deep. Determine the maximum shear stress.

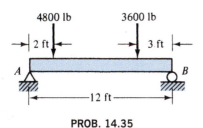

PROB. 14.35

14.36 Calculate the maximum shear stress in the beam shown. The rectangular cross section is 500 mm wide and 900 mm deep.

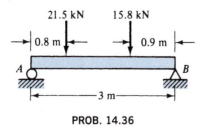

PROB. 14.36

14.37 For the simply supported beam with a uniform load and cross section as shown, calculate (a) the maximum shear stress, (b) the average shear stress [Eq. (14.10)], and (c) the percent difference between the

maximum shear stress and average shear stress. [Percent difference = $100(\tau_{max} - \tau_{av})/\tau_{max}$.]

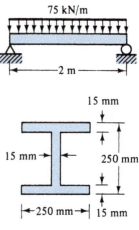

PROB. 14.37

14.38 A simply supported beam has a uniform load and cross section as shown. Determine (a) the maximum shear stress, (b) the average shear stress [Eq. (14.10)], and (c) the percent difference between the maximum shear stress and the average shear stress. [Percent difference = $100(\tau_{max} - \tau_{av})/\tau_{max}$.]

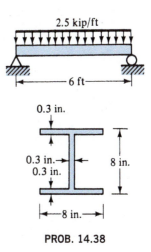

PROB. 14.38

14.39 Calculate the maximum shear stress for the cantilever beam with the cross section shown.

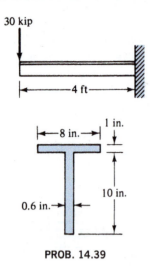

PROB. 14.39

14.40 Find the maximum shear stress for the cantilever beam with the cross section shown.

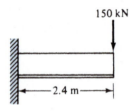

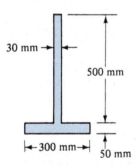

PROB. 14.40

14.41 A beam has a cross section as shown. At a section where the shear force is 12 kN, (a) determine the shear stress at 25-mm intervals along the depth, and (b) plot the shear stress distribution.

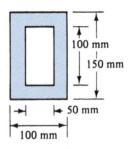

PROB. 14.41

14.42 A beam has a cross section as shown. If the shear force on the section is 12 kip, (a) calculate the shear stress at 1-in. intervals along the depth, and (b) plot the shear stress distribution.

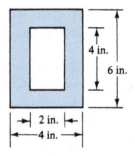

PROB. 14.42

14.10 SHEAR FLOW FORMULA

Shear flow q is defined as the force per unit of length along the shear plane. As such, it is equal to the product of the shearing stress and the width of the beam at the shear plane, that is,

$$q = \tau b$$

Substituting for the value of shearing stress from Eq. (14.8) in this equation, we have

$$q = \frac{VQ}{I} \qquad (14.11)$$

Shearing force V is in lb. or newtons (N), the first moment of the area Q is in in.^{3} or m^3, and the moment of inertia is in in.^{4} or m^4. Then, shear flow q will be lb/in. or newtons/meter (N/m).

The shear flow formula can be used to find the number or spacing of connectors used to fasten members together to form a beam cross section.

Example 14.11 The I-beam shown in Fig. 14.26(a) and (b) is made by bolting together three planks 14 ft long. What should be the maximum spacing of the bolts to resist the shearing stress? The bolts can safely resist a shearing force of 1100 lb. Neglect the weight of the beam.

Solution: The free-body diagram for the beam is drawn in Fig. 14.26(c). From equilibrium, $A_y = 2857$ lb and $C_y = 1143$ lb.

The shear diagram is constructed in Fig. 14.26(d). We see that the beam must transmit a maximum shear force $V = 2857$ lb. The cross section of the beam [Fig. 14.26(b)] is symmetric about a horizontal

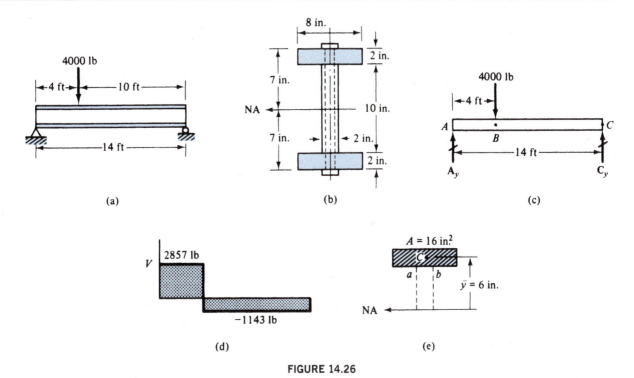

FIGURE 14.26

and vertical axis. Therefore, the neutral axis can be located by inspection. The neutral axis is shown in the figure.

The moment of inertia about the neutral axis may be determined by calculating the moment of inertia of a rectangle 8 in. × 14 in. and subtracting the moment of inertia of two rectangles 3 in. × 10 in. That is,

$$I = \frac{8(14^3)}{12} - (2)\frac{3(10^3)}{12} = 1329 \text{ in.}^4$$

The shear plane ab where the bolts must resist the shear force is shown in Fig. 14.26(e). The value of Q is equal to the moment of the area above plane ab (shaded area) about the neutral axis. The centroid for this area $\bar{y} = 6$ in., and the shaded area $A = 2(8) = 16$ in.^{2} Therefore, $Q = \bar{y}A = 6(16) = 96$ in.^{3} From Eq. (14.11), the shear flow

$$q = \frac{VQ}{I} = \frac{2857(96)}{1329} = 206.4 \text{ lb/in.}$$

Each bolt resists a shearing force $F_V = 1100$ lb and must resist the force developed by the shear flow q over a length s. Therefore,

$$F_V = qs \quad \text{or} \quad s = \frac{F_V}{q} = \frac{1100 \text{ lb}}{206.4 \text{ lb/in.}} = 5.329 \text{ in.}$$

For convenience, use a spacing $s = 5.25$ in. **Answer**

Example 14.12 Four planks are nailed together to form a box beam with a cross section, as shown in Fig. 14.27(a). Each nail can resist a shear force of 270 N. What should be the maximum spacing of the nails to resist a shear force of 5 kN?

Solution: The cross section of the beam [Fig. 14.27(a)] is symmetric about a horizontal and vertical axis, and the neutral axis is located by inspection. To determine the moment of inertia about the neutral axis, we subtract the moment of inertia of the inside rectangle from the outside rectangle. That is,

$$I = \frac{200(250)^3}{12} - \frac{100(150)^3}{12} = 232.3 \times 10^6 \text{ mm}^4$$

The shear planes ab and $a'b'$ where nails must resist the shear force are shown in Fig. 14.27(b). The value of Q is equal to the moment of the area between the shear planes (shaded area) about the neutral axis. The

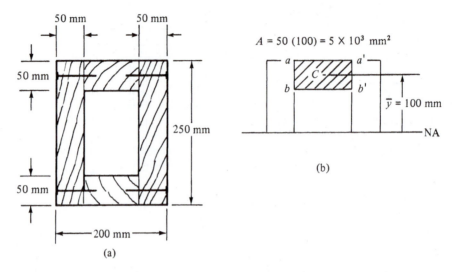

FIGURE 14.27

centroid for this area $\bar{y} = 100$ mm and the shaded area $A = 50(100) = 5 \times 10^3$ mm². Accordingly, $Q = \bar{y}A = 100(5 \times 10^3) = 0.5 \times 10^6$ mm³. From Eq. (14.11), the shear flow

$$q = \frac{VQ}{I} = \frac{5 \times 10^3(0.5 \times 10^6)}{232.3 \times 10^6} = 10.76 \text{ N/mm}$$

The beam and upper plane are both symmetric with the plane of loading; therefore, the shear flow for each shear plane is one-half of 10.76 or 5.38 N/mm. Each nail resists a shear force $F_V = 270$ N and must resist the shear force developed by the shear flow over a length s. Therefore,

$$F_V = qs \quad \text{or} \quad s = \frac{F_V}{q} = \frac{270}{5.38} = 50.2 \text{ mm}$$

Use a spacing $s = 50$ mm.   **Answer**

PROBLEMS

14.43 A box beam is made by nailing together four full-size planks to form a cross section as shown. What should be the maximum spacing of the nails to resist a shear force of 1100 N? Each nail can resist 270 N in shear.

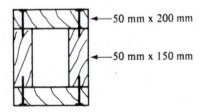

PROB. 14.43

14.44 Four full-size planks are nailed together to form a box beam with a cross section as shown. Each nail can resist a shear force of 80 lb. What should be the maximum spacing of the nails to resist a shear force of 300 lb?

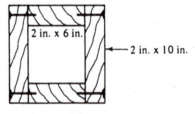

PROB. 14.44

14.45 A wooden beam is made up of three full-size planks that are joined together by screws to form a cross section as shown. Each screw can resist a force of 120 lb. What should be the maximum spacing of the screws to resist a shear force of 1200 lb? The moment of inertia of the cross section $I_{NA} = 776$ in.[4]

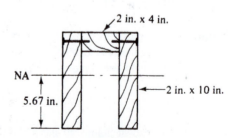

PROB. 14.45

14.46 Two planks are joined together by screws to form a T-shaped cross section as shown. Each screw has a shear strength of 525 N. What should be the maximum spacing of the screws to resist a shear force of 720 N? The moment of inertia of the cross section $I_{NA} = 113.5 \times 10^6$ mm^4.

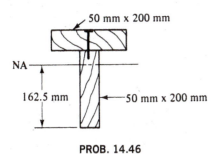

PROB. 14.46

14.47 Two W8 × 15 rolled steel shapes are joined by bolts to form a cantilever beam with loading and a cross section as shown. What should be the maximum spacing of the 1/2-in.-diameter bolts if the allowable shear stress for the bolts is 22 ksi?

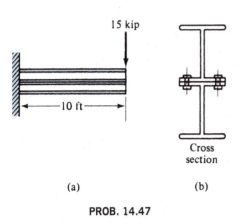

(a)

(b)

PROB. 14.47

14.48 The built-up shape shown in the figure is used as a beam. The cover plates are joined to the W10 × 45 by 3/4-in.-diameter rivets. What should be the maximum spacing (pitch) of the rivets to resist a shear force of 43.5 kip? The allowable shear stress in the rivets is 20 ksi. The moment of inertia of the cross section $I_{NA} = 644$ in.^{4}

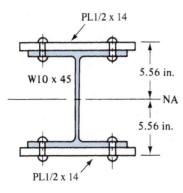

PROB. 14.48

14.49 For the beam cross section shown, determine the required spacing of the 1.0-in.-diameter rivets. The shear force on the section is 200 kip, and the allowable shear stress in the rivets is 20 ksi. The moment of inertia of the cross section $I_{NA} = 8225$ in.^{4}

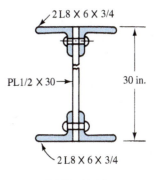

PROB. 14.49

14.11 DESIGN OF BEAMS FOR STRENGTH

In the phase of design called *detail design*, we determine the size and proportions of the members that make up a structure. The principles of statics and strength of materials, together with the appropriate specifications, are used to ensure that the member has adequate strength, stiffness, and stability.

In this section, the span, support conditions, and loading are known. The allowable stresses in bending and shear are specified, and we select a beam so the bending and shear stresses do not exceed the allowable values. Although we do not consider stiffness and stability here, all beams selected have adequate stiffness and stability.

Wood Beams of Rectangular Cross Section

Wood has a low allowable shear stress parallel to the grain. Therefore, horizontal shear stress is often the controlling factor in design. The bending or flexural stress must also be no greater than the allowable value.

Shear Stress Requirement   The maximum horizontal shear stress in a beam of rectangular cross section is given by Eq. (14.9):

$$\tau_{max} = \frac{3V_{max}}{2A}$$

where A is the area of the cross section. For wood, the maximum shear stress parallel to the grain of the wood must be equal to or less than the allowable value. Therefore,

$$\tau_a \geq \frac{3V_{max}}{2A} \quad \text{or} \quad A_{req} \geq \frac{3V_{max}}{2\tau_a} \qquad \textbf{(14.12)}$$

Bending Stress Requirement   The maximum bending stress in a beam is given by Eqs. (14.4) and (14.6):

$$\sigma_{max} = \frac{M_{max}c}{I} = \frac{M_{max}}{S}$$

where S is the section modulus of the cross section. Therefore, the maximum bending stress must be equal to or less than the allowable value.

$$\sigma_a \geq \frac{M_{max}}{S} \quad \text{or} \quad S_{req} \geq \frac{M_{max}}{\sigma_a} \qquad (14.13)$$

The appropriate rectangular timber beam is selected from Table A.9 of the Appendix, where the cross-sectional area and section modulus are tabulated for various timber sizes. In addition to the average allowable stresses for various wood species listed in Table A.10 of the Appendix, machine-graded lumber having more standardized properties is also available (see Sec. 9.6 and Table A.11). Like machine-graded lumber, LVL boards are often rated by their manufacturers in terms of allowable bending stress and modulus of elasticity. This information is generally stamped directly on the boards or is available from the manufacturer/supplier (see Sec. 9.6 and Table A.12).

Deep narrow beams have a large section modulus for a given area but may not have sufficient stability to prevent lateral buckling or twisting. Lateral buckling is buckling perpendicular to the plane of bending. Accordingly, unless bracing is used to prevent lateral buckling, rectangular beams are usually selected with the depth from two to three times the width.

Wood I-Joists or I-Beams

Bending stress and shear stress limits for wooden I-joists are specific to individual manufacturers. In practice, these structural members are often selected not for their strength, but rather for the amount of deflection they experience during use. As we will see in Chapter 15, deflections of 1/360th of the span for a joist or beam are common. However, worried that such floors may be uncomfortably "springy," architects, builders, and potential homeowners will often opt for larger I-joists, shorter spans, or closer joist spacing to achieve deflections of 1/480th or 1/960th of the span. As a result, many new structures have joists and beams that are sufficiently stiff but are overly strong in terms of stress. Also, because these members have little lateral stability, manufacturer guidelines for installation and bracing should be closely followed.

Structural Steel Beams

The American Institute of Steel Construction *Manual of Steel Construction—Allowable Stress Design* (9th ed., 1989) recommends an allowable bending stress $\sigma_a = 0.66\sigma_y$ and an allowable *average* shear stress $\tau_a = 0.4\sigma_y$ provided that the steel section considered meets certain specifications. All sections used here satisfy the specifications. In addition, possible lateral buckling is prevented by full lateral support of the compression flange of the beam. For the most commonly used structural steel, the yield stress $\sigma_y = 36$ ksi (250 MPa). (See Table A.10, material no. 1, in the Appendix.) Therefore, the allowable bending stress

$$\sigma_a = 0.66\sigma_y = 0.66(36) = 23.76 = 24\,\text{ksi}(165\,\text{MPa})$$

and the allowable *average* shear stress

$$\tau_a = 0.4\sigma_y = 0.4(36) = 14.4\,\text{ksi}\,(100\,\text{MPa})$$

Steel beams are usually selected from the allowable bending stress. The required section modulus is established from Eq. (14.13), and the *lightest* beam is selected from Table A.3 of the Appendix. The lightest beam section contains the least steel and is usually the most economical. The average shear stress in the web of the beam is then checked by Eq. (14.10) and found for most beams to be less than the allowable value. In Eq. (14.10), the area of the web is taken to be equal to the depth of the section, d, multiplied by the thickness of the web, t. That is,

$$\tau_{av} = \frac{V_{max}}{dt} \qquad (14.14)$$

Not all beams are selected from the bending stress. Some exceptions are beams with heavy loads near beam supports and beams that have a flange or flanges removed to accommodate a connection. When shear stress controls, a beam is selected with a web area large enough so the shear stress is less than the allowable value. Limitations on beam deflection may also control the size of a beam. Beam deflections are discussed in Chapter 15.

Example 14.13   The simply supported beam of Fig. 14.28(a) must span a 14.ft opening and carry a concentrated load of 3500 lb at its midpoint. A builder is considering use of a wooden *header* made by nailing together two pieces of S4S 2-in.-wide lumber as shown in Fig. 14.28(b), or a wide-flange steel member aligned in the standard position of Fig. 14.28(c). Considering only bending stress,

    a.   What size member is required for MSR lumber designated as 2400f—2.0E (see Sec. 9.6 and Table A.11) if the weight of the beam is neglected? For this size member, what is the maximum bending stress in the beam?

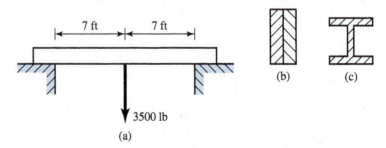

(a)

(b)   (c)

FIGURE 14.28

b. Repeat part (a) if the weight of the beam must be considered.

c. Select the most economical steel W-member that may be used if the beam's weight is neglected and a bending stress of 24,000 psi is allowed. What is the maximum bending stress in the chosen beam?

d. Repeat part (c) if the beam's weight must be considered.

Solution: a. The load, shear, and moment diagrams for this beam are shown in Fig. 14.29. Maximum moment is 147,000 lb·in., and the allowable bending stress is 2400 psi. From Eq. (14.6),

$$S = \frac{M}{\sigma} = \frac{147,000 \text{ lb·in.}}{2400 \text{ lb/in.}^2} = 61.25 \text{ in.}^3$$

Because two boards are used in this beam, each must have a section modulus of at least 30.625 in.^{3} From Table A.9 of the Appendix, we see that a 2 × 12 S4S has a section modulus of 31.6 in.^{3} Actual bending stress in the beam, then, is

$$\sigma = \frac{M}{S} = \frac{147,000 \text{ lb·in.}}{2 \times 31.6 \text{ in.}^3} = 2326 \text{ psi} \qquad \textbf{Answer}$$

b. Table A.9 of the Appendix tells us that a typical wooden 2 × 12 weighs 4.69 lb/ft. Taking this weight into account yields the load, shear, and moment diagrams of Fig. 14.30. For the new maximum moment of 149,760 lb·in., our required section modulus is

$$S = \frac{M}{\sigma} = \frac{149,760 \text{ lb·in.}}{2400 \text{ lb/in.}^2} = 62.4 \text{ in.}^3$$

Each board must have a section modulus of 31.2 in.^{3}, so 2 × 12 S4S lumber will still be sufficient. Actual bending stress in the beam is

$$\sigma = \frac{M}{S} = \frac{149,760 \text{ lb·in.}}{2 \times 31.6 \text{ in.}^3} = 2370 \text{ psi} \qquad \textbf{Answer}$$

Note that this bending stress is only about 2 percent higher than our computed value from part (a).

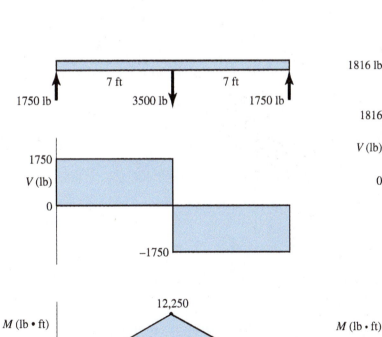

FIGURE 14.29

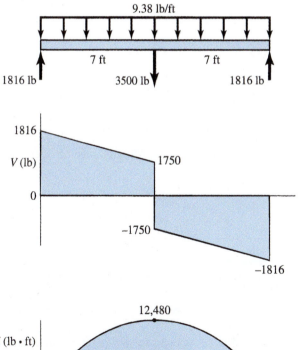

FIGURE 14.30

c. Neglecting beam weight, maximum moment is 147,000 lb·in., so the minimum required section modulus for a W-member becomes

$$S = \frac{M}{\sigma} = \frac{147{,}000 \text{ lb·in.}}{24{,}000 \text{ lb/in.}^2} = 6.125 \text{ in.}^3$$

From Table A.3 of the Appendix, two members having section moduli close to the required value are

$$W6 \times 12 @ S = 7.31 \text{ in.}^3$$
$$W8 \times 10 @ S = 7.81 \text{ in.}^3$$

Selecting the lightest (most economical) member (W8 × 10) yields a maximum bending stress of

$$\sigma = \frac{M}{S} = \frac{147{,}000 \text{ lb·in.}}{7.81 \text{ in.}^3} = 18{,}822 \text{ psi} \qquad \textbf{Answer}$$

d. Accounting for the beam's weight of 10 lb/ft, we obtain the load, shear, and moment diagrams of Fig. 14.31.

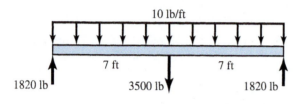

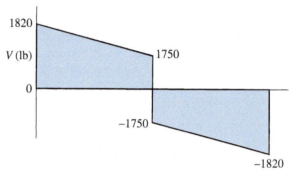

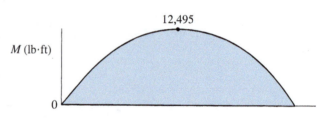

FIGURE 14.31

For the new maximum moment of 149,940 lb·in., the minimum required section modulus becomes

$$S = \frac{M}{\sigma} = \frac{149{,}940 \text{ lb·in.}}{24{,}000 \text{ lb/in.}^2} = 6.248 \text{ in.}^3$$

The W8 × 10 member is still satisfactory, and maximum bending stress is

$$\sigma = \frac{M}{S} = \frac{149{,}940 \text{ lb·in.}}{7.81 \text{ in.}^3} = 19{,}198 \text{ psi} \qquad \textbf{Answer}$$

Again, this represents a change of less than 2 percent from the stress value computed in part (c).

Example 14.14   Select a wooden beam of rectangular cross section to support loads as shown in Fig. 14.32(a) and determine the size of the bearing plates at A, B, and C. The allowable stress in bending is 1475 psi, in shear parallel to the grain 115 psi, and in bearing perpendicular to the grain of the wood 200 psi.

Solution: Solving for the reactions at A and C from the free-body diagram shown in Fig. 14.32(b), we have $A_y = 2160$ lb and $C_y = 1440$ lb. The shear force and bending-moment diagrams are plotted in Fig. 14.32(c) and (d). From Eq. (14.12), the required cross-sectional area

$$A_{req} \geq \frac{3V_{max}}{2\tau_a} = \frac{3(2160)}{2(115)} = 23.2 \text{ in.}^2$$

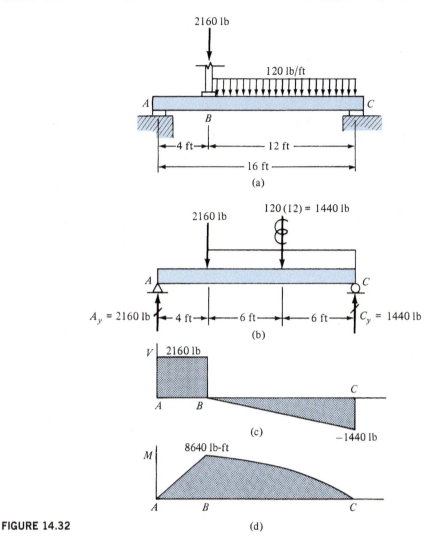

FIGURE 14.32 (d)

The maximum bending moment M_{max} = 8640 lb-ft = 103,680 lb-in., and from Eq. (14.13), the required section modulus

$$S_{req} \geq \frac{M_{max}}{\sigma_a} = \frac{103,680}{1475} = 70.3 \text{ in.}^3$$

We select two possible cross sections from Table A.9 of the Appendix.

4 in. × 12 in. $A = 39.4 \text{ in.}^2$ $S = 73.8 \text{ in.}^3$
6 in. × 10 in. $A = 52.2 \text{ in.}^2$ $S = 82.7 \text{ in.}^3$

Check of the 4 × 12-in. Wooden Beam: The 4 × 12-in. nominal size timber is the lightest section, and it weighs 10.9 lb/ft. It also has a depth three times the width. To check the section, we determine the additional shear force at A and the additional bending moment at B due to the weight of the beam. From Eqs. (b) and (c) of Example 14.4, the additional shear force at the reactions

$$V_{max} = \frac{wL}{2} = \frac{10.9(16)}{2} = 87 \text{ lb}$$

and the additional bending moment at $x = 4$ ft

$$M_{x=4\text{ ft}} = \frac{10.9(4)}{2}(16 - 4) = 262 \text{ lb-ft}$$

Therefore, the maximum bending moment and shear force including the weight of the beam M_{max} = 8640 + 262 = 8902 lb-ft = 106,824 lb-in. and V_{max} = 2160 lb + 87 lb = 2247 lb. From Eq. (14.12), the required cross-sectional area including the weight of the member

$$A_{req} \geq \frac{3V_{max}}{2\tau_a} = \frac{3(2247)}{2(115)} = 29.3 \text{ in.}^2 < A = 39.4 \text{ in.}^2 \quad \text{OK}$$

and from Eq. (14.13), the required section modulus including the weight of the member

$$S_{req} \geq \frac{M_{max}}{\sigma_a} = \frac{106,824}{1475} = 72.4 \text{ in.}^3 < S = 73.8 \text{ in.}^3 \quad \text{OK}$$

Use a 4 × 12-in. nominal size beam.   **Answer**

Bearing plates are provided at the concentrated load and the supports. From Eq. (9.4), the required bearing area $(A_b)_{req} = P/\sigma_{ba}$, where P is the load on the plate and σ_{ba} is the allowable bearing stress. The bearing area is equal to the width of the beam w times the length of the plate L, that is, $A_b = wL$; therefore,

$$(wL)_{req} \geq \frac{P}{\sigma_{ba}} \quad \text{or} \quad L_{req} \geq \frac{P}{w\sigma_{ba}} \tag{a}$$

The width of the beam is 3.5 in., and the allowable bearing stress perpendicular to the grain is 200 psi.

Bearing Plate at A: $P = R = 2160 \text{ lb} + 87 \text{ lb} = 2247 \text{ lb}$ (including the weight of the beam). From Eq. (a),

$$L_{req} \geq \frac{P}{w\sigma_{ba}} = \frac{2247}{3.5(200)} = 3.21 \text{ in.} \quad \text{(use 3.5 in.)}$$

Use PL3.5 in. × 3.5 in.   **Answer**

Bearing Plate at B: $P = 2160 \text{ lb}$. From Eq. (a),

$$L_{req} \geq \frac{P}{w\sigma_{ba}} = \frac{2160}{3.5(200)} = 3.09 \text{ in.} \quad \text{(use 3.5 in.)}$$

Use PL3.5 in. × 3.5 in.   **Answer**

Bearing Plate at C: $P = R = 1440 \text{ lb} + 87 \text{ lb} = 1527 \text{ lb}$ (including weight of beam). From Eq. (a),

$$L_{req} \geq \frac{P}{w\sigma_{ba}} = \frac{1527}{3.5(200)} = 2.18 \text{ in.} \quad \text{(use 2.5 in.)}$$

Use PL3.5 in. × 2.5 in.   **Answer**

Example 14.15   A simply supported beam has a span and loads as shown in Fig. 14.33. Select the lightest wide flange (W) section that can safely support the loads, including the weight of the beam. The allowable stresses are 24 ksi in bending and 14.5 ksi in shear.

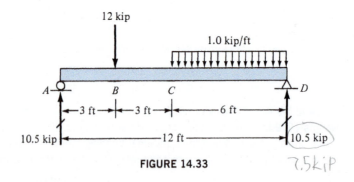

FIGURE 14.33

Solution: The reactions and the shear-force and bending-moment diagrams are shown in Example 13.7. The maximum bending moment is at point B and is equal to 31.5 kip-ft or 378 kip-in., and the maximum shear force acts between A and B and is equal to 10.5 kip.

We obtain a *tentative* beam selection by neglecting the weight of the beam. From Eq. (14.13),

$$S_{req} \geq \frac{M_{max}}{\sigma_a} = \frac{378}{24} = 15.75 \text{ in.}^3$$

The following shapes from Table A.3 of the Appendix have section moduli larger than 15.75 in.^{3} and are the lightest shape in each depth group that satisfies the weight requirement.

W16 × 26   38.4 in.^{3}

W14 × 22   29.0 in.^{3}

$$W12 \times 16 \quad 17.1 \text{ in.}^3 < \text{lightest section}$$
$$W10 \times 22 \quad 23.2 \text{ in.}^3$$
$$W8 \times 24 \quad 20.9 \text{ in.}^3$$
$$W6 \times 25 \quad 16.7 \text{ in.}^3$$

Check of W12 × 16 Beam: The beam weighs 16 lb/ft or 0.016 kip/ft. To find the additional bending moment at $B(x = 3 \text{ ft})$ from the weight of the beam, we use Eq. (b) from Example 14.4.

$$M_{x=3\,ft} = \frac{0.016(3)}{2}(12 - 3) = 0.216 \text{ kip-ft} = 2.59 \text{ kip-in.}$$

and the additional shear force at A from the weight of the beam is found from Eq. (c) of Example 14.4.

$$V_{max} = \frac{wL}{2} = \frac{0.016(12)}{2} = 0.096 = 0.1 \text{ kip}$$

Accordingly, $M_{max} = 378 + 2.59 = 380.6$ kip-in. and $V_{max} = 10.5 + 0.1 = 10.6$ kip. From Eq. (14.13),

$$S_{req} \geq \frac{M_{max}}{\sigma_a} = \frac{380.6}{24} = 15.9 \text{ in.}^3 < S_x = 17.1 \text{ in.}^3 \quad \text{OK}$$

and from Eq. (14.14),

$$\tau_{max} = \frac{V_{max}}{dt} = \frac{10.6}{11.99(0.220)} = 4.02 \text{ ksi} < \tau_a = 14.5 \text{ ksi} \quad \text{OK}$$

In this example, the weight of the beam added about 1.0 percent to the required section modulus. It is usually safe to neglect the weight of a steel beam unless the required section modulus is close to the section modulus of the selected beam. However, the weight of a beam should be included in the design of a member that supports that beam.

Use the W12 × 16 beam.   **Answer**

Example 14.16   Beams, girders, and columns support a reinforced 4-in.-thick concrete floor as shown in the plan view of Fig. 14.34. The beams, girders, and floor continue on all sides, and the concrete floor provides full lateral support for the members. Reinforced concrete weighs 150 lb/ft³. Design beam B_1 and girder G_1 if the floor supports a live load of 65 lb/ft² and the dead load of its own weight. *Assume that all members are simply supported* and use steel that has an allowable bending stress of 24 ksi and an allowable shear stress of 14.5 ksi. (*Girders* are major beams that support cross beams.)

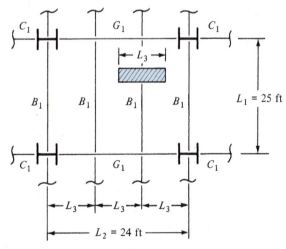

FIGURE 14.34

Solution: **Design of Beam B_1:** The weight of the 4-in.-thick concrete floor per square foot is $(4/12)(150) = 50$ lb/ft². The beam supports a strip of floor of width $L_3 = 8$ ft as shown in the figure; therefore, the dead load (weight) of the concrete floor per linear foot of the beam

$$w_1 = 50 \text{ lb/ft}^2 \, (8 \text{ ft}) = 400 \text{ lb/ft of beam}$$

and the live load per linear foot of the beam

$$w_2 = 65 \text{ lb/ft}^2(8 \text{ ft}) = 520 \text{ lb/ft of beam}$$

Therefore, the total load per foot of the beam, not including the weight of the beam,

$$w = w_1 + w_2 = 400 + 520 = 920 \text{ lb/ft of beam}$$

The beam is simply supported and has a span of 25 ft and a uniform load $w = 920 \text{ lb/ft} = 0.920 \text{ kip/ft}$. From Eq. (c) of Example 14.4

$$M_{max} = \frac{wL^2}{8} = \frac{0.92(25)^2}{8} = 71.88 \text{ kip-ft} = 862.5 \text{ kip-in.}$$

and from Eq. (14.13), the required section modulus

$$S_{req} \geq \frac{M_{max}}{\sigma_a} = \frac{862.5}{24} = 35.9 \text{ in.}^3$$

We select a W12 × 30 with $S = 38.6 \text{ in.}^3$ from Table A.3 of the Appendix. It is the lightest beam in any depth group that has a section modulus greater than 35.9 in.^{3} The uniform load including the weight of the beam $w = 0.92 + 0.03 = 0.95 \text{ kip/ft}$; therefore,

$$M_{max} = \frac{wL^2}{8} = \frac{0.95(25)^2}{8} = 74.2 \text{ kip-ft} = 890.6 \text{ kip-in.}$$

and the required section modulus

$$S_{req} \geq \frac{M_{max}}{\sigma_a} = \frac{890.6}{24} = 37.1 \text{ in.}^3 < S_x = 38.6 \text{ in.}^3 \quad \text{OK}$$

Check of shear stress: From Eq. (c) of Example 14.4,

$$R = V_{max} = \frac{wL}{2} = \frac{0.95(25)}{2} = 11.88 \text{ kip}$$

$$\tau_{max} = \frac{V_{max}}{dt} = \frac{11.88}{12.34(0.260)} = 3.70 \text{ ksi} < \tau_a = 14.5 \text{ ksi} \quad \text{OK}$$

Use W12 × 30 for beam B_1.　　　　　　　　　　　　　　　　　　　　**Answer**

Design of Girder G_1: Each girder supports the end reactions ($R_B = 11.88 \text{ kip}$) from four beams—two end reactions at each third point of the girder—and its own weight. The shear-force and bending-moment diagrams for a beam loaded at its third points are shown in Fig. 14.13 of Example 14.5. From that example, $P = 2R_B = 2(11.88) = 23.76 \text{ kip}$ and $L = 24 \text{ ft}$. The maximum shear force $V_{max} = P = 23.76 \text{ kip}$, and the maximum moment $M_{max} = PL/3 = 23.76(24)/3 = 190.1 \text{ kip-ft} = 2280 \text{ kip-in.}$ The required section modulus from Eq. (14.13)

$$S_{req} \geq \frac{M_{max}}{\sigma_a} = \frac{2280}{24} = 95 \text{ in.}^3$$

We select a W18 × 55 from Table A.3 of the Appendix with a section modulus $S = 98.3 \text{ in.}^3$ It is the lightest beam in any depth group with a section modulus greater than 95 in.^{3} The beam weighs 55 lb/ft = 0.055 kip/ft; therefore, the maximum moment including the weight of the beam

$$M_{max} = 190.1 + \frac{0.055(24)^2}{8} = 194.1 \text{ kip-ft} = 2329 \text{ kip-in.}$$

Accordingly, the required section modulus

$$S_{req} \geq \frac{M_{max}}{\sigma_a} = \frac{2329}{24} = 97.0 \text{ in.}^3 < S_x = 98.3 \text{ in.}^3 \quad \text{OK}$$

Check of shear stress: The maximum shear force including the weight of the beam

$$V_{max} = 23.76 + \frac{0.055(24)}{2} = 24.4 \text{ kip}$$

Therefore, the maximum shear stress

$$\tau_{max} = \frac{V_{max}}{dt} = \frac{24.4}{18.11(0.390)} = 3.46 \text{ ksi} < \tau_a = 14.5 \text{ ksi} \quad \text{OK}$$

Use W18 × 55 for girder G_1.   **Answer**

PROBLEMS

14.50 A steel beam has an allowable bending stress of 24,000 psi. If this beam carries a maximum moment of 80,000 ft-lb,
  a. What is the minimum section modulus required for the beam?
  b. Based on your answer from (a), select the most economical (lightest) W-shape member that is satisfactory for this application.

14.51 Same as Prob. 14.50, except that the allowable bending stress is 165 MPa and the maximum moment is 108 kN-m.

14.52 Floor joists in a particular residential structure are simply supported on a span of 17 ft. If the joists have a center-to-center spacing of 16 in., typical building codes require that each joist be able to safely carry a uniformly distributed load of 107 lb/ft along its entire length. Neglecting the beam's weight, use Table A.9 in the Appendix to select a suitable joist size for MSR boards that are marked
  a. 1950f—1.7E
  b. 2250f—1.9E

14.53 A W33 × 221 steel member acts as a simply supported beam. If the maximum allowable bending stress *due to the beam's own weight* is 12,000 psi, compute the maximum allowable length for this beam.

14.54 A simply supported rectangular beam with a span of 6 ft has a concentrated load of 4.1 kip at midspan. Select the lightest rectangular wood beam that can safely support the loads (neglect the weight of the beam). The allowable stress in bending is 1200 psi and in shear parallel to the grain 100 psi.

14.55 Select the lightest rectangular wooden beam that can safely support the loads shown, including its own

weight. Design bearing plates for the supports and the concentrated load. The allowable stress in bending is 1200 psi, in shear parallel to the grain 100 psi, and in bearing perpendicular to the grain of the wood 180 psi.

14.56 A simply supported rectangular beam with a span of 3 m has a distributed load of 15 kN/m, not including its own weight. Select the lightest rectangular wooden beam that can safely support the loads including its own weight. The allowable stress in bending is 8.28 MPa and in shear parallel to the grain 0.69 MPa.

14.57 Select the lightest rectangular wooden beam that can safely support the loads shown, including its own weight. Design bearing plates for the supports. The allowable stress in bending is 8.26 MPa, in shear parallel to the grain 0.69 MPa, and in bearing perpendicular to the grain of the wood 1.24 MPa.

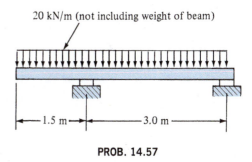

20 kN/m (not including weight of beam)

— 1.5 m — — 3.0 m —

PROB. 14.57

14.58 The cantilever timber beam has loads and span as shown. Select the most economical rectangular beam that can support the loads, including its own weight. The allowable stress in bending is 1300 psi and in shear parallel to the grain of the wood 130 psi.

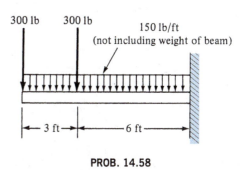

300 lb   300 lb   150 lb/ft
(not including weight of beam)

— 3 ft — — 6 ft —

PROB. 14.58

14.59 A timber beam having a rectangular cross section is used as a simply supported beam with a span of 8 ft. It supports two equal concentrated loads of $P = 6000$ lb, so there is a load 2 ft from each end (loads at quarter

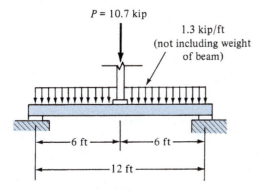

P = 10.7 kip

1.3 kip/ft
(not including weight of beam)

— 6 ft — — 6 ft —

— 12 ft —

PROB. 14.55

points). The allowable stress in bending is 800 psi, in shear parallel to the grain 80 psi, and bearing perpendicular to the grain of the wood 210 psi. Determine (a) the beam required to safely support the loads (neglect weight of beam), and (b) the bearing plate sizes required at the supports and concentrated loads.

14.60 A simply supported steel beam with a span of 16 ft has a uniform load of 5.9 kip/ft over the entire span. Select the lightest wide-flange beam that can safely support the loads. The allowable stresses in bending and shear are 24 ksi and 14.5 ksi, respectively.

14.61 through 14.64 For the beam shown, select the lightest wide-flange steel beam that can safely support the loads. For beams in U.S. customary units, the allowable stresses in bending and shear are 24 ksi and 14.5 ksi, respectively, and in SI units the allowable stresses in bending and shear are 165 MPa and 100 MPa, respectively.

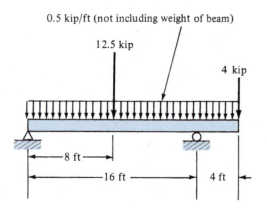

0.5 kip/ft (not including weight of beam)

12.5 kip

4 kip

8 ft

16 ft

4 ft

PROB. 14.61

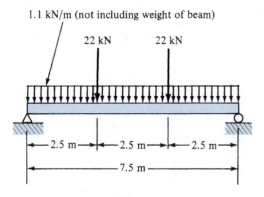

1.1 kN/m (not including weight of beam)

22 kN   22 kN

2.5 m   2.5 m   2.5 m

7.5 m

PROB. 14.62

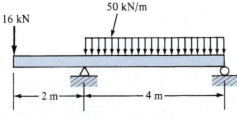

50 kN/m

16 kN

2 m   4 m

PROB. 14.63

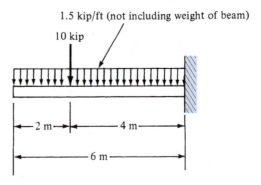

1.5 kip/ft (not including weight of beam)

10 kip

2 m   4 m

6 m

PROB. 14.64

14.65 Beams and girders are used to support a reinforced concrete floor as shown in Fig. 14.34. The beams, girders, and floor continue on all sides of the plan view, and the concrete floor provides full lateral support. Assume that all beams and girders are simply supported. Use steel with an allowable stress in bending and shear of 24 ksi and 14.5 ksi, respectively. Design beam B_1 and girder G_1 if the concrete floor supports a live load q_L (lb/ft^2) and its own dead weight q_D (lb/ft^2), and has spans L_1 (ft) and L_2 (ft) as follows:

a. $q_L = 70$ lb/ft^2, $q_D = 50$ lb/ft^2, $L_1 = 20$ ft, $L_2 = 21$ ft

b. $q_L = 105$ lb/ft^2, $q_D = 75$ lb/ft^2, $L_1 = 21$ ft, $L_2 = 24$ ft

*14.12 RESIDENTIAL DESIGN USING TABULATED VALUES

Many load-carrying members found in residential structures are treated as simple beams. These are generally made of wood and include girders, floor joists, ceiling joists, and roof rafters. Although the methods presented earlier in this chapter allow for the analysis and design of virtually any beam configuration and loading, most residential construction follows standardized practices. For this reason, the structural elements used in a residence are often selected from tabulated values rather than by individual analysis. In this section, we will examine some of the common methods used and compare them to the theoretical techniques already discussed.

Sources of Data

Tables of data for residential construction are available from several sources, including professional associations, institutes, and societies such those listed in Table 11.2. Many of these organizations publish handbooks, brochures, technical bulletins, and other support materials that describe recommended practices for specific structural elements.

*An asterisk denotes optional material that may be omitted without loss of continuity.

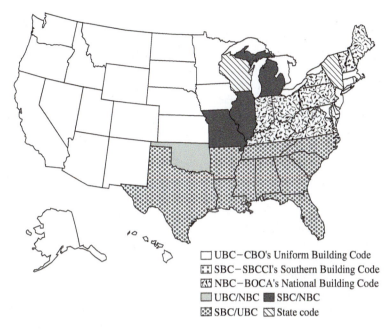

FIGURE 14.35. Through the mid-1990s, state use of the three model building codes was approximately as shown. *(Courtesy of PFS Research Foundation, Madison, Wisconsin, www.pfscorporation.com)*

Perhaps the best comprehensive sources of data are the state and local building codes for your own geographic area. These codes are generally adapted from one of three national building codes that are used as *models* by most states and municipalities. These three codes are as follows: the National Building Code (NBC) created by the Building Officials and Code Administrators International, Inc. (BOCA); the Uniform Building Code (UBC) established by the International Conference of Building Officials (ICBO); and the Standard Building Code (SBC) developed by the Southern Building Code Congress International, Inc. (SBCCI). Through the mid-1990s, adaptation of the various codes was approximately as shown in Fig. 14.35. Notice that some states used several codes, whereas other states had their own codes that differed from the three model codes. In 1994, BOCA, ICBO, and SBCCI joined together to form the International Code Council (ICC), whose purpose was to develop a single, comprehensive, national code for use as a model in all regions of the country. Although the International Building Code (IBC) is now in place, adoption is voluntary and is the responsibility of individual states and municipalities. It may be quite some time before the IBC becomes the single code used in all states; however, as of mid-2005, 45 states plus Washington, DC, had adopted at least some portion of this code.

You should note that the tables used in this section contain average values taken from a variety of sources and are used for demonstration purposes only. To find equivalent values for your own geographic area, look for specific local codes and building regulations in either your municipal library or town/city hall.

The General Method

All structures are designed to safely carry a combination of live and dead loads. *Live loads* are *temporary* and include the forces or weights exerted by people, furniture, wind, and snow. *Dead loads* are a *permanent* part of the structure and generally represent the weight of materials used in the structure. Live loads and some dead loads are expressed as *distributed loads per unit area* (typically lb/ft^2 in the U.S. customary system), average values of which are given in Table 14.1. In effect, if the appropriate combination of live and dead loads from this table were applied to each square foot of area on the floors, ceiling, and roof in a residential structure, the building is expected to remain standing without undergoing failure in any of its load-bearing elements. Use of these design loads is an *averaging* procedure, particularly when applied to floors. For example, although a water bed, piano, or billiards table will create a heavier load on some portion of a room's floor, the effect is averaged across the entire floor surface as part of the room's live load. Similarly, both *bearing*

TABLE 14.1	Typical Recommended Live and Dead Loads for One- and Two-Story Residential Structures	
Structural Element	**Live (psf)**	**Dead (psf)**
Each floor	40	10
Sleeping areas	30	14
Attics (limited storage)	20	10
Balconies and decks	60	12
Partitions (2 × 4 studs plastered on 2 sides)	0	20
Roofs (rises 4 in./ft to 12 in./ft)	42[†]	
Asphalt shingled		12
Felt and gravel		18
Slate (1/4 in.)		30

[†]Combined wind and snow load (northeastern United States).

partitions (used to support floor and ceiling joists) and *curtain* partitions (used to divide large living spaces into smaller rooms) are assumed to add a uniform dead load of 20 psf to the entire floor area in a residence, regardless of the number or location of these partitions. One of the reasons that this load-averaging procedure has proven satisfactory is because *bridging* materials (solid wood blocks or rigid metal strips) are used between adjacent floor joists. This bridging, as well as the flooring itself, tends to distribute any relatively concentrated loads over several joists, thus creating an average uniform load in the vicinity of the concentrated load.

In certain applications where it is necessary to determine specific dead loads, the values given in Table 14.2 may be used. These represent the average *weight per unit volume* for some common building materials and are used to compute the *total* dead load (such as the weight, in lbs, of a foundation wall) created by one portion of a structure.

To examine the distribution of live and dead loads in a residence, let us first discuss the structure's load-bearing elements in the same sequence that the building is constructed—from the bottom up. Referring to Fig. 14.36, we see that all our design loads are ultimately supported by the ground, specifically at the contact areas under the *foundation wall footings* and *column footings*. Although we generally assume the "ground" to be relatively indestructible, Table 14.3 indicates that the allowable *bearing stress* beneath these footings varies widely for different types of soil. The footings, which are always made of poured concrete, may be adjusted

TABLE 14.2	Average Weights of Common Building Materials
Material	**Weight (lb/ft³)**
Poured concrete	150
Concrete block with mortar	85
Brick with mortar	120
Seasoned timber	
Fir, Douglas	32
Hemlock, Eastern	26
Oak, white	45
Pine, Eastern white	23
Pine, longleaf	39
Spruce, red and Sitka	26

in size to reduce the actual bearing stress to an acceptable level for existing on-site conditions.

Next, foundation walls of concrete block or poured concrete are built directly above the wall footings and support one end of the *floor joists*. [A wooden member called a *sill* (not shown) is generally fastened to the top of wall; this provides a nailing surface for the outside ends of the joists.] The other ends of these joists are carried by a wooden *girder*, which is typically fabricated using three or more 2×8s, 2×10s, or 2×12s in its cross section. Girders are supported by steel or wooden *columns* set atop the column footings placed

---

CONCRETE

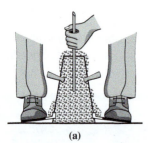 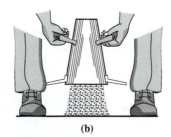 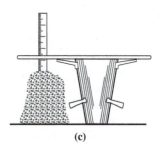

(a)   (b)   (c)

SB.21

Cement is a finely ground powder consisting of lime, silica, alumina, and iron. When combined with water and *aggregate* (sand and stones of various sizes), the mixture hardens into a solid, durable material called *concrete*. Although the development of modern concrete is generally attributed to the Romans, who used it in their buildings as early as 100 B.C., cement was first patented in 1824 by the English mason Joseph Aspdin. He named the material *portland cement* because of its resemblance to limestone quarried on the nearby Isle of Portland. The first cement plant in the United States was built in 1872 at Coplay, Pennsylvania, while the first Canadian plant appeared in Hull, Quebec, in 1889.

Although compressive strength is often the most important property of a concrete mixture, its *workability*—the ease with which it may be pumped to remote locations or placed in complex forms around steel reinforcing rods—must also be considered. The standard test for workability is the *slump test,* in which a steel cone is filled with fresh concrete and tamped with a steel rod; the cone is removed; and the slump of the mix, in inches, is measured (see SB.21(a)–(c), respectively). For most large commercial applications, the amount of slump, and the mixture's compressive strength at 30 days, are both specified by contract.

Despite its long history of use as a building material, concrete mixtures with enhanced properties continue to be developed (see Section 11.8). Within the past two decades, engineers have created *reactive-powder* concrete, whose compressive strength approaches that of ordinary steel; lightweight *autoclaved aerated* concrete, which is 80% air; and *translucent* concrete, which transmits light.

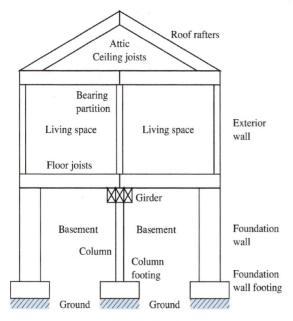

FIGURE 14.36

TABLE 14.4	Built-up Wood Girders: Allowable Spans Between Columns[†]		
Structure Width (ft)	Nominal Girder Size (in.)	Type of Structure	
		One Story	Two Story
24	6 × 8	6'-6"	4'-10"
	6 × 10	8'-4"	6'-3"
	6 × 12	10'-3"	7'-8"
26	6 × 8	6'-4"	4'-8"
	6 × 10	8'-0"	6'-0"
	6 × 12	9'-9"	7'-4"
28	6 × 8	6'-2"	4'-6"
	6 × 10	7'-9"	5'-9"
	6 × 12	9'-6"	7'-0"
30	6 × 8	6'-0"	4'-4"
	6 × 10	7'-6"	5'-6"
	6 × 12	9'-3"	6'-8"
32	6 × 8	5'-8"	4'-3"
	6 × 10	7'-3"	5'-4"
	6 × 12	8'-9"	6'-6"

[†]Typical spans for 1000 psi bending stress in structures having trussed roofs and load-bearing center walls on the first floor of two-story residences.

at specified distances along the girder. Table 14.4 lists the allowable distances between columns for various girder configurations and structure sizes, while Table 14.5 gives safe compressive loads for columns of different materials and lengths. Floor joists themselves may be sized using Table 14.6. [Although the actual unsupported span for a floor joist is the distance between the inside faces of sill and girder, it is conventional (and conservative) to use half of the structure's outside width as the required span for floor and ceiling joists.]

Exterior walls constructed on each floor support one end of the floor joists for the floor above. Walls on the top floor carry the outside ends of both *ceiling joists* and *roof rafters*, or support the ends of roof trusses. (The weights for these walls pass directly to the foundation and create no

bending stresses in the floor joists.) Bearing partitions, usually assumed to be located directly above the girder, are used to support the inside ends of floor and ceiling joists (nontrussed roofs). Roof rafters support each other along a board called the *ridge* (not shown) or are part of a roof truss. Safe spans for ceiling joists and roof rafters (horizontal span) are given in Tables 14.7 and 14.8, respectively.

TABLE 14.3	Safe Ground Loads Beneath Footings
Ground Condition	**Allowable Bearing Stress (psf)**
Rock	
Medium hard	80,000
Soft	28,000
Hardpan	20,000
Gravel	
Very dense	18,000
Dense	12,000
Medium dense	8000
Loose	6000
Clay	
Hard	8000
Stiff	4000
Medium	3000
Soft	2000
Sand	
Dense	8000
Medium dense	6000
Loose	4000

TABLE 14.5	Safe Loads for Columns (kip)			
	Column Material/Nominal Size			
Unbraced Column Length	Concrete-filled Steel Pipe		Spruce, Pine, Fir	
	3.5-in. diameter	4-in. diameter	6 × 6	8 × 8
6 ft	25.0	33.2	22.1	42.9
7 ft	22.1	30.2	21.3	42.4
8 ft	19.2	27.1	19.6	41.2
9 ft	16.3	24.0	18.3	40.4
10 ft	13.7	20.9	16.8	38.6

TABLE 14.6	Allowable Floor Joist Spans (40 psf Live Load)[†]		
Nominal Joist Size (in.)	Center-to-Center Joist Spacing (in.)		
	12	16	24
2 × 6	10'-2"	9'-3"	8'-2"
2 × 8	13'-6"	12'-2"	10'-8"
2 × 10	17'-2"	15'-6"	13'-6"
2 × 12	21'-0"	19'-0"	16'-8"

[†]For modulus of elasticity of 1,400,000 psi; maximum deflection equal to span divided by 360; maximum bending stress of 1350 psi.

TABLE 14.7	Allowable Spans for Ceiling Joists (20 psf Live Load)[†]		
Nominal Joist Size (in.)	Center-to-Center Joist Spacing (in.)		
	12	16	24
2 × 4	9'-4"	8'-6"	7'-6"
2 × 6	14'-8"	13'-4"	11'-8"
2 × 8	19'-6"	17'-8"	15'-6"
2 × 10	24'-9"	22'-6"	19'-8"

[†]For modulus of elasticity of 1,400,000 psi; drywall ceiling and maximum deflection of span divided by 240; maximum bending stress of 1650 psi.

TABLE 14.8	Allowable Rafter Spans (40 psf Live Load)[†]		
Nominal Rafter Size (in.)	Center-to-Center Rafter Spacing (in.)		
	12	16	24
2 × 6	11'-8"	10'-2"	8'-6"
2 × 8	15'-4"	13'-4"	10'-9"
2 × 10	19'-8"	17'-0"	13'-9"
2 × 12	24'-0"	20'-8"	16'-9"

[†]Spans represent horizontal projections of rafters; maximum bending stress of 1500 psi; maximum deflection equal to span divided by 180; minimum modulus of elasticity of 1,100,000 psi.

As mentioned in Sec. 9.6, modern residential construction has seen a tremendous increase in the use of engineered wood products. LVL boards are frequently used to fabricate traditional wooden girders, with typical values for allowable bending stress and modulus of elasticity of 3000 psi and 2,000,000 psi, respectively. I-joists are now common for floor and ceiling joists, and for roof rafters as well, with nontraditional o.c. spacings of 19.2 in. and 32 in. supplementing the standard spacings of 12, 16, and 24 in. Although design values for clear spans and allowable loads vary considerably between manufacturers, Table 14.9 lists some average clear-span values for common sizes of I-joists used in flooring systems.

As you can see from the previous discussion, selection of the beams in a residential structure—girder, floor joists, ceiling joists, and rafters—can be done directly from the tabulated values with little or no calculation. However, when analyzing or designing a structure in this manner, it is customary to also check the column loads, as well as bearing stresses at the wall and column footings. To do so, we make several assumptions concerning distribution of loads:

1. Roof loads are transmitted directly to the foundation walls and footings.

2. All live and dead loads on joists, girders, and rafters are applied in a direction perpendicular to the lengths of these beams.

3. All beams are considered to be simply supported, and share their loads equally between end supports.

TABLE 14.9	Typical Spans of Wooden I-Joists for Floors[†]				
Joist Size (in.)	Center-to-Center Joist Spacing (in.)				
	12	16	19.2	24	32
1-3/4 × 9-1/2	17'-2"	15'-8"	14'-10"	13'-10"	12'-6"
× 11-7/8	20'-6"	18'-9"	17'-8"	16'-6"	14'-6"
3-1/2 × 11-7/8	25'-4"	22'-6"	21'-8"	20'-2"	18'-2"
× 14	28'-8"	26'-2"	24'-6"	22'-10"	20'-0"
× 16	31'-10"	28'-10"	27'-2"	25'-4"	20'-2"

[†]Typical loads 40 psf (live), 10 psf (dead); maximum deflection equal to span divided by 480.

4. For structures having trussed roofs or having no bearing partitions on the top floor [Fig. 14.37(a) and (b), respectively], all attic loads are carried by the exterior walls. For nontrussed roofs where a bearing partition exists [Figure 14.37(c)], a portion of the attic load is carried by each exterior wall, and a portion is carried to the lower floor by the partition.

The techniques used to analyze residential structures with the use of tabulated values are demonstrated by the following examples.

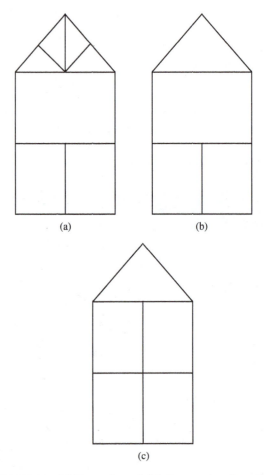

FIGURE 14.37. (a) Trussed roof. (b) Conventional roof, no bearing partition on top floor. (c) Conventional roof with bearing partition on top floor.

Example 14.17   The residence shown in Fig. 14.38 has the following structural details:

- Overall dimensions are 28 ft wide × 42 ft long, and the site consists of loose gravel.
- The roof is trussed and is covered with asphalt shingles.
- Floor joists are 16 in. o.c. (on-center), and the floor area contains curtain partitions.
- Foundation walls are 8-in.-thick concrete block and sit on poured concrete footings that are 2 ft wide and 1 ft thick.
- Columns sit on square footings of poured concrete that are also 1 ft thick.
    a.   Select a suitable floor joist size.
    b.   Check the bearing stress beneath an outside wall footing.
    c.   Select the girder size and specify the number and (equal) spacing of columns.
    d.   Find the load carried by each column and determine the required column size.
    e.   Specify the column footing dimensions.

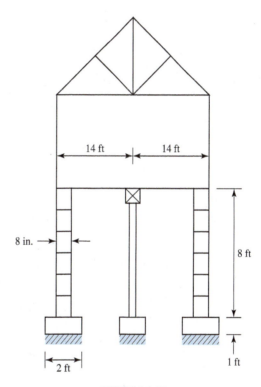

FIGURE 14.38

Solution: a.   Floor joists must span half the structure width, or in this case 14 ft. From Table 14.6, 2 × 10 joists 16 in. o.c. have an allowable span of 15'-6" and are therefore acceptable.

b.   Because the wall footings are continuous along the length of our structure, we may analyze just a 1-ft-long "slice" of the building, as shown in Fig. 14.39. As we will see, this has two advantages: It allows us to use linear dimensions directly from the structure's cross section when computing loads, and it reduces our numerical load values significantly when computing bearing stress at the outside wall.

   The loads on this structure are distributed as shown in Fig. 14.40. Each exterior wall footing carries one-half of the total roof and attic loads, one-fourth of the floor load, and the weights of one foundation wall and one wall footing. The central half of the floor load is transmitted by the girder to columns and their footings, located at discrete intervals along the structure's length.

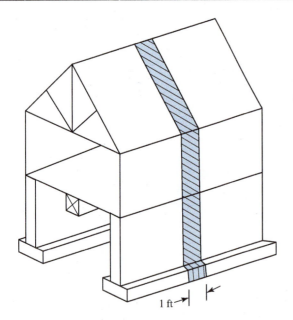

FIGURE 14.39. To determine bearing stress at an outside wall footing, we may analyze just a 1-ft-long "slice" (shaded) of the structure rather than the entire building length.

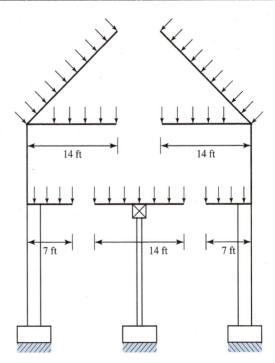

FIGURE 14.40. Live and dead loads applied to each surface are ultimately carried to the ground by the footings. In this case, roof and attic loads, as well as a portion of the floor load, are transmitted to exterior wall footings, while the central half of the floor load is supported by the girder, its columns, and the column footings.

Applying the recommended live and dead loads of Table 14.1 to the appropriate areas in our structure yields the total loads shown in Table (A):

TABLE (A) Example 14.17			
Element	Live + Dead Load (psf)	Area (ft²)	Total Load (lb)
Roof	42 + 12	14	756
Attic	20 + 10	14	420
Floor	40 + (10 + 20)	7	490

Notice that the floor's dead load includes 20 psf for our curtain partitions, and that each area is computed as a product of the appropriate linear dimension from Fig. 14.40 times the 1-ft "slice" length indicated in Fig. 14.39. The total load for each element is then obtained by multiplying its area by the sum of its live and dead loads. Our attic, for instance, has 30 lb/ft² applied to an area of 14 ft² to produce a total load of $30 \times 14 = 420$ lb. Also, the roof area is computed using one-half of the *horizontal* truss span rather than the actual rafter length. The reason for this is indicated in Fig. 14.41(a), where we see that the combined live and dead load, W, acts perpendicular to the truss member of length R. However, if that member is at an angle, A, from the horizontal, then the vertical component of W is ($W \cos A$), as shown in Fig. 14.41(b). (The horizontal components of these loads, $W \sin A$, on both halves of the truss simply push against each other and do not affect bearing stress at the wall footings.) The total vertical force on member R is then

$$(W \cos A) \times R$$

Rearranging this expression, the total force may be written as

$$W \times (R \cos A) = W \times S$$

where S is the horizontal span of R, as shown in the figure.

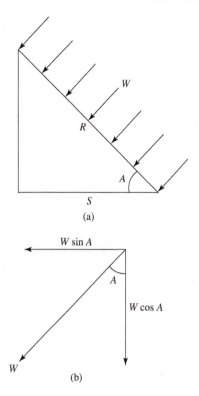

FIGURE 14.41. (a) Combined live and dead load, W, is distributed uniformly across the roof in a direction perpendicular to the roof rafters. (b) Only the vertical component, W cos A, of this load contributes to the bearing stress at the footings.

In computing bearing stress at the exterior wall footing, we must also take into account the weights of foundation wall and wall footing using the material weights of Table 14.2. The results are shown in Table (B).

TABLE (B) Example 14.17

Element	Element Volume (ft^3)	Unit Weight (lb/ft^3)	Total Weight (lb)
Wall	(8/12 × 8 × 1)	85	453
Footing	(2 × 1 × 1)	150	300

All loads from Tables (A) and (B) are supported by the 2-ft-wide and 1-ft -long section of wall footing. The bearing stress, σ_b, then becomes

$$\sigma_b = \frac{P}{A} = \frac{(756 + 420 + 490 + 453 + 300)\,\text{lb}}{2\,\text{ft} \times 1\,\text{ft}}$$

$$\sigma_b = 1210\,\text{psf} \qquad\qquad \textbf{Answer}$$

Because this value is well below the allowable stress for loose gravel soil conditions as given in Table 14.3, the wall footing size (width) is more than adequate.

c. As shown in Fig. 14.42(a), a girder (shaded) represents the *backbone* of any residential structure, whereas the floor joists act as *ribs*. For a one-story structure that is 28 ft wide, Table 14.4 lists three choices of girder sizes, as well as the maximum allowable distance between columns for each girder. Any of these combinations of girder size and column spacing would be suitable; the specific choice is generally a matter of personal preference. However, girder size must be specified before the foundation is constructed because "pockets" are cast in both end walls to support the ends of the girder. (These pockets must be sized so the top of the girder is at the correct level to accept the floor joists.)

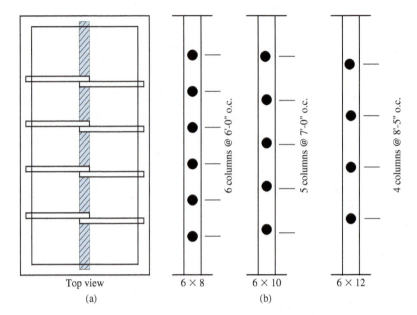

FIGURE 14.42. (a) The girder (shaded) acts as the *backbone* of a residential structure, whereas the floor joists serve as *ribs*. (Drawing not to scale.) (b) Columns are generally spaced along the girder at equal distances that are computed using the overall length of the structure. These distances should not exceed the maximum values given in Table 14.4.

Assuming that equal column spacing is desired, we simply divide the overall structure length by the allowable column spacings to obtain the number of *bays* (openings between columns), rounding any decimal values up to the next highest whole number. For the 6×8 girder whose allowable column spacing is 6'-2" (6.167 ft), the number of bays, N, is

$$N = \frac{42 \text{ ft}}{6.167 \text{ ft}} = 6.81 \text{ (7 bays)}$$

The actual (equal) column spacing, d, then becomes

$$d = \frac{42 \text{ ft}}{N} = \frac{42 \text{ ft}}{7 \text{ days}} = 6'\text{-}0'' \text{ o.c.}$$

The number of columns required, n, is equal to $N - 1$, so six columns at the given spacing are needed for this girder. Similarly for the 6×10 girder (maximum column spacing 7'-9") and the 6×12 girder (maximum column spacing 9'-6"),

6×10

$$N = \frac{42 \text{ ft}}{7.75 \text{ ft}} = 5.42 \text{ (6 bays)}$$

$$d = \frac{42 \text{ ft}}{6 \text{ bays}} = 7'\text{-}0'' \text{ o.c.}$$

$$n = N - 1 = 6 - 1 = 5 \text{ columns}$$

6×12

$$N = \frac{42 \text{ ft}}{9.5 \text{ ft}} = 4.42 \text{ (5 bays)}$$

$$d = \frac{42 \text{ ft}}{5 \text{ bays}} = 8.4' \text{ (or } 8'\text{-}5'')$$

$$n = N - 1 = 5 - 1 = 4 \text{ columns}$$

These three column spacings are depicted in Fig. 14.42(b), and our girder–column choices are summarized in Table (C).

TABLE (C) Example 14.17		
Girder Size	Number of Columns	Column Spacing
6 × 8	6	6'-0"
6 × 10	5	7'-0"
6 × 12	4	8'-5"

d. No matter what girder is selected in part (c), the *total* structural design load carried by the columns remains the same. However, because the smaller 6 × 8 girder requires six columns, while the larger 6 × 12 girder requires only four columns, that portion of the total load carried by *each column* is different for the two girders. To find the load carried by a column, P_c, we must consider the *tributary area* that the column supports. As shown in Fig. 14.43(a), this area (shaded) is defined by one-half the distance in each direction to an adjoining column ($d/2$) and to the side foundation wall (7 ft). This floor area, assigned a combined live and dead load of 70 psf in Table (A), is depicted in Fig. 14.43(b) and is supported by a single column whose total load is the product of tributary area times 70 psf. The column load for each girder is given in Table (D).

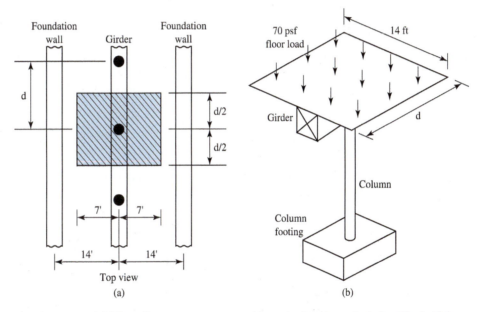

FIGURE 14.43. (a) The *tributary area* supported by a single column is defined by half the distance to an adjoining support (column or side foundation wall) in each direction. (b) The tributary area carries a combined live and dead load of 70 psf, as specified in Table (A).

TABLE (D) Column Loads Corresponding to Girder Choice		
Girder Size	Dimensions of Tributary Area	Column Load (lb)
6 × 8	14 ft × 6'-0"	5880
6 × 10	14 ft × 7'-0"	6860
6 × 12	14 ft × 8'-5"	8250

Should we include the weight of that section of girder being supported by each column? Generally, the effect on column load is small and is neglected. According to Table A.9 of the Appendix, for instance, each 2 × 12 board in our fabricated 6 × 12 girder weighs 4.69 lb per linear foot. (if the particular wood species is known, we may instead use the data from Table 14.2.) The weight of our 8'-5" length of girder is then

$$4.69\frac{lb}{ft} \times 8.4\frac{ft}{board} \times 3\ boards = 118\ lb$$

This is less than 1.5 percent of the 8250-lb column load due to live and dead loads on the floor; such a small increase rarely affects the choice of column size.

From Fig. 14.38, we see that the actual column length may vary between 7 ft and 8 ft, depending on the girder size selected. Table 14.5 tells us that, even at a length of 8 ft, a 3.5-in.-diameter concrete-filled steel column will safely support a compressive load of 19,200 lb, and would therefore be a good selection for any of the three girder sizes and column spacings listed in Table (C).

e. For a square, poured-concrete column footing 1 ft thick and having a side length b (ft), the footing weight, w, is

$$w = (b \times b \times 1)\ ft^3 \times 150\frac{lb}{ft^3} = 150\ b^2\ (lb)$$

Bearing stress beneath the footing becomes

$$\sigma_b = \frac{Footing\ weight + Column\ load}{Contact\ area\ beneath\ footing} = \frac{150b^2 + P_c}{b^2}$$

Solving for b,

$$b = \sqrt{\frac{P_c}{\sigma_b - 150}}$$

Using the values of column load given in Table (D) and an allowable σ_b of 6000 psf (loose gravel, Table 14.3) yields these required footing sizes for our three girder choices:

$$6 \times 8 \qquad b = 1\ ft = 12\ in.$$
$$6 \times 10 \qquad b = 1.08\ ft = 13\ in.$$
$$6 \times 12 \qquad b = 1.19\ ft = 14\ in.$$

Is the footing weight significant or may we neglect it? If we consider only column load, P_c, our formula becomes

$$b = \sqrt{\frac{P_c}{\sigma_b}}$$

The difference between our previous values of b and those obtained with this equation are insignificant for the given structure.

Example 14.18   The two-story residence shown in Fig. 14.44 has the following structural details:

- Building is 30 ft wide, 52 ft long, and has an asphalt shingled roof.
- Roof rafters and ceiling joists are 24 in. o.c.
- Both floors have joists 16 in. o.c. and contain bearing partitions.
- Girder is a fabricated 6 × 12 and is supported on 4-in.-diameter concrete-filled steel columns.
- External walls are faced with 4-in. brick that sits atop foundation walls that are 16-in.-wide poured concrete.
- Wall footings are 30 in. × 1 ft, and column footings are 20 in. diameter × 1 ft thick, all of poured concrete.
  a. Choose appropriate rafter, ceiling, and floor joist sizes.
  b. Find the bearing stress beneath a wall footing.
  c. Specify the number of columns and their (equal) spacing.
  d. Determine the load carried by each column and the column's factor of safety.
  e. Find the bearing stress beneath a column footing.

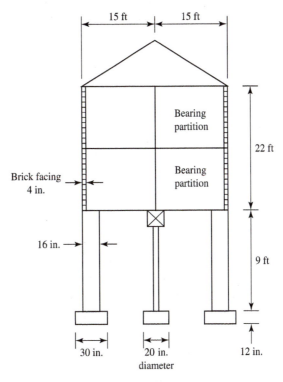

FIGURE 14.44

Solution: a. Rafters and ceiling and floor joists must all span half the structure width (15 ft) at their specified o.c. spacing. The appropriate tables yield these sizes directly:

Rafters (Table 14.8)

$$2 \times 12 \quad @ \quad 24'' \text{ o.c. safely spans } 16'\text{-}9''$$ **Answer**

Ceiling joists (Table 14.7)

$$2 \times 8 \quad @ \quad 24'' \text{ o.c. safely spans } 15'\text{-}6''$$ **Answer**

Floor joists (Table 14.6)

$$2 \times 10 \quad @ \quad 16'' \text{ o.c. safely spans } 15'\text{-}6''$$ **Answer**

b. The loads on this structure are distributed as shown in Fig. 14.45. For an exterior wall footing, the live and dead loads are as follows:

Roof	$54 \text{ psf} \times 15 \text{ ft}^2 = 810 \text{ lb}$
Attic	$30 \text{ psf} \times 7.5 \text{ ft}^2 = 225 \text{ lb}$
2nd floor	$70 \text{ psf} \times 7.5 \text{ ft}^2 = 525 \text{ lb}$
1st floor	$70 \text{ psf} \times 7.5 \text{ ft}^2 = 525 \text{ lb}$

Brick facing:

$$120 \frac{\text{lb}}{\text{ft}^3} \times \left(\frac{4}{12} \times 22 \times 1 \right) \text{ft}^3 = 880 \text{ lb}$$

Foundation wall:

$$150 \frac{\text{lb}}{\text{ft}^3} \times \left(\frac{16}{12} \times 9 \times 1 \right) \text{ft}^3 = 1800 \text{ lb}$$

Wall footing:

$$150 \frac{\text{lb}}{\text{ft}^3} \times \left(\frac{30}{12} \times 1 \times 1 \right) \text{ft}^3 = 375 \text{ lb}$$

$$\text{Total} \quad \overline{5140 \text{ lb}}$$

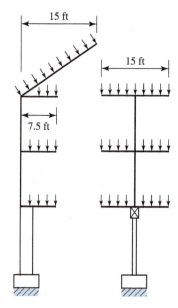

FIGURE 14.45. For this residential structure, the loads carried by an exterior wall footing, and by the girder–columns–column footings, are distributed as shown.

This total load is carried by the 30-in.-wide wall footing, so bearing stress at the footing becomes

$$\sigma_b = \frac{5140\,\text{lb}}{\left(\dfrac{30}{12} \times 1\right)\text{ft}^2} = 2060\,\text{psf}$$ **Answer**

c. From Table 14.4, a 6×12 girder under a 30-ft-wide two-story residence has a maximum column spacing of 6'-8" (6.667 ft). Then,

$$N = \frac{52\,\text{ft}}{6.667\,\text{ft}} = 7.8\,(8\,\text{bays})$$

$$n = N - 1 = 8 - 1 = 7\,\text{columns}$$ **Answer**

$$d = \frac{52\,\text{ft}}{8\,\text{bays}} = 6.5\,\text{ft} = 6'\text{-}6''\,\text{o.c.}$$ **Answer**

d. The tributary areas for the ceiling, second floor, and first floor are each $15\,\text{ft} \times 6'\text{-}6'' = 97.5\,\text{ft}^2$ (see Fig. 14.45). Then, the total load carried by a single column is the sum of these three components:

Attic	$30\,\text{psf} \times 97.5\,\text{ft}^2 = 2925\,\text{lb}$
2nd floor	$70\,\text{psf} \times 97.5\,\text{ft}^2 = 6825\,\text{lb}$
1st floor	$70\,\text{psf} \times 97.5\,\text{ft}^2 = 6825\,\text{lb}$
	Total $\overline{16{,}575\,\text{lb}}$

Load carried by single column: 16,575 lb **Answer**

From Table 14.5, a 4-in.-diameter concrete-filled steel column 9 ft long can safely carry 24,000 lb, so this element has an apparent factor of safety (F.S.) of

$$\text{F.S.} = \frac{24{,}000\,\text{lb}}{16{,}575\,\text{lb}} = 1.45$$ **Answer**

e. Circular column footings are popular because cylindrical forms made of waxed cardboard are inexpensive and are available commercially in many different diameters and lengths. The weight of this footing and the area of contact at its base are

$$\text{Weight} = 150\,\frac{\text{lb}}{\text{ft}^3} \times \left[\frac{\pi}{4}\left(\frac{20}{12}\text{ft}\right)^2 \times 1\,\text{ft}\right]$$

$$= 327\,\text{lb}$$

$$\text{Area} = \frac{\pi}{4} \times \left(\frac{20}{12}\,\text{ft}\right)^2 = 2.18\,\text{ft}^2$$

The column load and footing weight together create a bearing stress of

$$\sigma_b = \frac{16{,}575\,\text{lb} + 327\,\text{lb}}{2.18\,\text{ft}^2} = 7750\,\text{psf}$$ **Answer**

If the existing soil conditions beneath this structure have a maximum allowable bearing stress (see Table 14.3) that is *less* than 7750 psf, then the column footing diameter must be increased.

This same method of load distribution can be combined with our traditional techniques to analyze specific elements in any residential structure. Such analysis is usually required when it is necessary, or desirable, to use construction techniques that are not covered by the tables. These might include replacement of a wooden girder by a steel beam and the use of 2 × 14 boards as floor joists or in fabrication of a 6 × 14 girder. The following example demonstrates how these two design methods (analytical and use of the tables) interface, as well as the typical level of agreement in numerical values obtained with both techniques.

Example 14.19   Table 14.6 indicates that 2 × 12 floor joists placed 24 in. o.c. have a maximum allowable span of 16′-8″. Using the analytical methods presented earlier in this chapter, determine the maximum bending stress created in these joists.

Solution: (Maximum spans for joists are based on a combination of performance criteria: bending stress, deflection, and stiffness [the apparent "springiness" of the floor]. In this example, we are investigating only the requirement for maximum stress. Deflection of this same beam is analyzed in Example 15.11 of the following chapter.)

The tributary area for one joist is shown in Fig. 14.46(a) and has dimensions of 16′-8″ × 24″. Our span values in Table 14.6 are based on a 40-psf live load and assume a dead load of 10 psf. Because each *linear* foot of a joist supports 2 ft² of floor *area* at a combined load of 50 psf [Fig. 14.46(b)], the appropriate loading, shear, and moment diagrams for a joist are as shown in Fig. 14.47. The maximum moment, M, is computed as

$$M = \frac{1}{2} \times 833\,\text{lb} \times 8.33\,\text{ft} = 3470\,\text{ft-lb}$$

$$= 41{,}640\,\text{in.-lb}$$

From Table A.9 of the Appendix, the section modulus for a 2 × 12 board is

$$S = 31.6\,\text{in.}^3$$

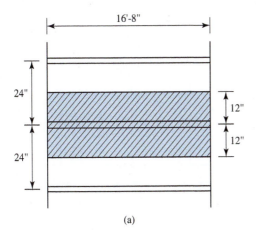

(a)

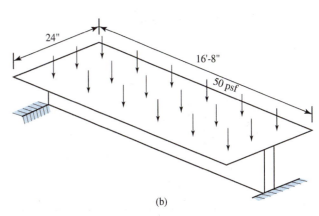

(b)

FIGURE 14.46

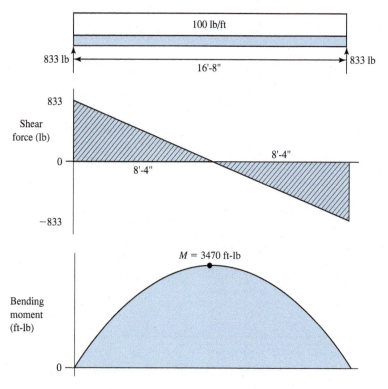

FIGURE 14.47. Loading, shear, and moment diagrams for the floor joist of Example 14.19.

Then, maximum bending stress in the joist becomes

$$\sigma = \frac{M}{S} = \frac{41{,}640 \text{ in.-lb}}{31.6 \text{ in.}^3} = 1320 \text{ psi} \qquad \textbf{Answer}$$

Note that this value is in relative agreement with the maximum bending stress listed in the footnotes of Table 14.6.

PROBLEMS

For probs. 14.66 to 14.71, use tabulated values to select the structural members indicated or to analyze existing conditions at various points in the structure as specified. Note that, for clarity, the figures are not drawn to scale.

14.66 The residence shown
- Is 36 ft long and has an asphalt shingled roof.
- Contains bearing partitions in the living area.
- Has rafters, ceiling joists, and floor joists that are all 24 in. o.c.
- Sits on 10-in.-wide poured foundation walls and footings that are 20 in. wide and 10 in. thick.
- Uses steel columns that sit on square column footings that are also 10 in. thick.
  a. Specify appropriate sizes for rafters, ceiling joists, and floor joists.
  b. Select a suitable girder size, as well as the required number of columns and their (equal) spacing.
  c. Determine the load carried by each column, and select the smallest suitable column size.

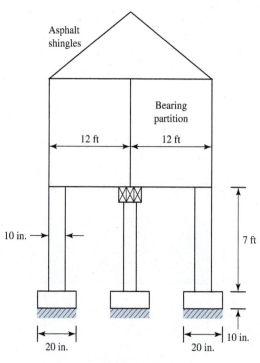

PROB. 14.66

d. Compute the bearing stress at a wall footing.

e. Specify the column-footing dimensions so the bearing stress there does not exceed 2000 psf.

14.67 The structure shown
- Is 54 ft long, uses trusses, and has a slate roof.
- Contains curtain partitions on the second floor and bearing partitions on the first floor.
- Has floor joists that are 12 in. o.c.
- Uses a 6 × 12 girder supported by steel columns that sit on footings 24 in. in diameter and 12 in. thick.
- Has 16-in.-wide foundation walls of poured concrete and 24-in.-wide by 12-in.-thick wall footings.
  a. Select an appropriate floor joist size.
  b. Specify the number of columns required and their (equal) spacing.
  c. Compute the load carried by each column, and select the smallest allowable column size.
  d. Determine the bearing stress at the wall footings and column footings. Are these stresses acceptable if the building site consists of loose gravel?

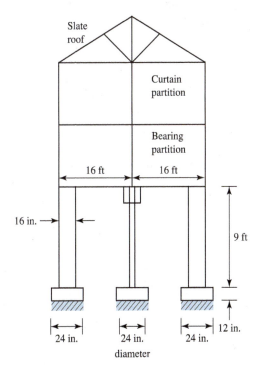

PROB. 14.67

14.68 The garrison-style residence shown
- Is 46 ft long and has an asphalt shingled roof.
- Will be built on a site composed of medium clay.
- Contains bearing partitions on both floors.
- Has 4-in. brick facing on the lower half of the front exterior wall.
- Has rafter and ceiling joist spacing of 24 in. o.c., and a floor–joist spacing of 16 in. o.c.
- Uses wooden columns to support the girder and has 12-in.-thick column footings.

- Has 14-in.-wide poured concrete foundation walls and wall footings that are 28 in. wide and 12 in. thick.
  a. Select appropriate sizes for rafters, ceiling joists, and floor joists.
  b. Check the bearing stress at the front wall footing. Is this an acceptable stress?
  c. Specify a girder size, as well as the number of columns required and their (equal) spacing.
  d. Compute the load carried by each column, and choose an acceptable column size.
  e. Determine the dimensions of square column footings that produce a bearing stress consistent with the site conditions.

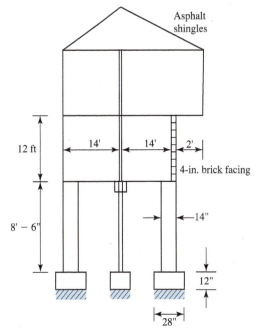

PROB. 14.68

14.69 A barn is to be constructed with a loft apartment as shown. These structural details are known:
- The structure is 50 ft long, has an asphalt shingled roof, and is built on a site containing soft clay.
- The loft is a studio apartment, contains no partitions, has 16-in. o.c. joist spacing, and must support a standard residential floor load.
- The girder is 6 × 12 and is supported by wooden columns on square, 12-in.-thick footings.
- The foundation walls are 10-in. concrete blocks extending from 4 ft below grade (maximum depth of frost penetration) to 5 ft above grade and are supported on 2-ft-wide by 1-ft-thick footings.
  a. Select the most economical floor joist size.
  b. Specify the number of columns and their (equal) center-to-center spacing.
  c. Compute the load carried by each column, and select the smallest appropriate column size.
  d. Determine the smallest column footing that may be used.
  e. Find the bearing stress at a wall footing.

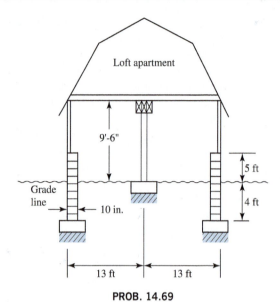

PROB. 14.69

14.70 The A-frame vacation home
- Is 38 ft long and has an asphalt shingled roof.
- Has a rafter spacing of 16 in. o.c. and joist spacing of 24 in. o.c. throughout.
- Contains bearing partitions on the first floor but no partitions in the 16-ft-wide loft living area.
- Has a crawl space rather than a basement and is supported by three walls of 12-in. concrete block (center wall replaces girder and columns) that sit on 20-in.-wide by 12-in.-thick footings.
  a. Specify appropriate sizes for rafters and for floor joists in the loft living area and first floor area.
  b. Compute the bearing stress at an outside wall footing.
  c. Compute the bearing stress at the center wall footing.

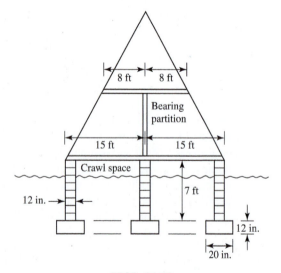

PROB. 14.70

14.71 The residence shown is an architectural style known as New England Saltbox. This particular house has the following structural details:
- The building is 50 ft long and has an asphalt shingled roof.
- The attic allows for limited storage.
- There are bearing partitions on both floors.
- Wooden I-joists are used to support the attic and the floors.
- There are two parallel girders, both 6×10.
- The foundation is 14-in.-thick poured concrete.
- Wall footings are 30 in. wide and 14 in. thick.
- Site conditions consist of loose gravel.
  a. Specify the size and o.c. spacing of floor joists, ceiling joists, and rafters.
  b. Compute the bearing stress at the right-hand wall footing.
  c. For each girder, specify the number of columns required and their equal spacing.
  d. Determine the load carried by one of the columns supporting the right-hand girder.
  e. Find the minimum dimensions of a suitable square column footing. Neglect the footing weight.

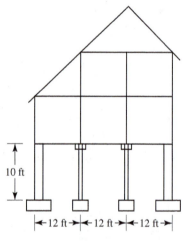

PROB. 14.71

DEFLECTION OF BEAMS DUE TO BENDING

CHAPTER OBJECTIVES

In this chapter, we examine several methods for computing the deflection at any point along a cantilevered or simply supported beam. After completing this material, you should be able to

- Compute beam deflections using the moment-area method.
- Compute beam deflections using the method of superposition.
- Describe the process by which beam deflections may be calculated using computer spreadsheets.
- Use superposition techniques to compute the reactions for beams that are otherwise statically indeterminate.
- (Optional) Compute beam deflections using integration methods and singularity functions.

15.1 INTRODUCTION

The relationship between applied loads and bending and shear stresses in beams was developed in Chapter 14. In addition to limitations on the stresses in the design of beams, deflection in many cases must also be limited.

Although no general rule can be given for limits on deflection, the American Institute of Steel Construction (AISC), in its *Specifications* (1989), limits the maximum load deflection of floor beams that support plastered ceilings to 1/360 of the span, and presumably that prevents cracking of the plaster. The American Association of State Highway and Transportation Officials (AASHTO), in its *Specifications* (1989), limits deflection in steel beams and girders due to live loads and impact to 1/800 of the span and recommends a limitation of 1/1000 for bridges that are used by vehicles and pedestrians. Electrical and mechanical equipment must be supported so the equipment operates properly. Limitations on such supports from 1/1500 to 1/2000 of the span are common. In addition, deflections are required to determine the reactions for statically indeterminate beams, that is, beams for which there are more reactions than can be determined by the equations of static equilibrium.

In this chapter, three methods for calculating beam deflection will be studied: the *moment-area* method, the *superposition* method, and, in a special section marked with an asterisk, the *integration* method. We begin the moment-area method with additional study of the bending-moment diagram.

15.2 BENDING-MOMENT DIAGRAM BY PARTS

When calculating the deflection of a beam by the moment-area method, it is necessary to know the area and location of the centroid of the moment diagram. The calculations are simplified if we imagine that each load and reaction on the beam produces a separate moment diagram. The resulting diagrams are said to be drawn by parts. The sum of all moment diagrams drawn by parts is equivalent to the usual bending-moment diagram for the beam.

Four types of loads will be considered: the couple, the concentrated load, the uniform load, and the triangular load. Figure 15.1 shows the four types of loads and the corresponding free-body diagrams for the beams. Each diagram shows a positive shear and bending moment, and the resultants of the distributed loads are indicated by a superimposed script capital E.

We begin with the free-body diagram for an end couple [Fig. 15.1(b)]. From equilibrium, we sum moments about the centroid of the cross section.

$$\Sigma M_c = M_x + M = 0 \quad \text{or} \quad M_x = -M$$

In a like manner, we sum moments about the centroid of the cross section for the end force in Fig. 15.1(d).

$$\Sigma M_c = M_x + Px = 0 \quad \text{or} \quad M_x = -Px$$

In Fig. 15.1(f), the resultant of the uniformly distributed load is equal in magnitude to the area of the rectangle $R = wx$, and it acts at the geometric center of the rectangle a distance from either end of $x/2$. Summing moments about the centroid of the cross section, we get

$$\Sigma M_c = M_x + wx\frac{x}{2} = 0 \quad \text{or} \quad M_x = -\frac{wx^2}{2}$$

The resultant of the triangular load in Fig. 15.1(h) is equal in magnitude to the area of the triangle $R = (x/2)(w_{max}x/L)$, and it acts at the geometric center of the triangle a distance $2x/3$ from the pointed end and a distance $x/3$ from the flat end

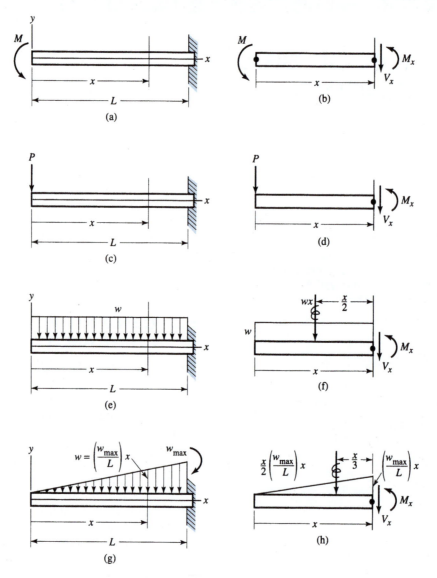

FIGURE 15.1

of the triangle. Summing moments about the centroid of the cross section, we have

$$\Sigma M_c = M_x + \frac{x}{2}\left(\frac{w_{max}x}{L}\right)\frac{x}{3} = 0 \quad \text{or} \quad M_x = -\frac{w_{max}x^3}{6L}$$

The four types of loads, their moment diagrams, and geometric properties of the moment diagrams are summarized in Table 15.1. The table will help us draw moment diagrams by parts.

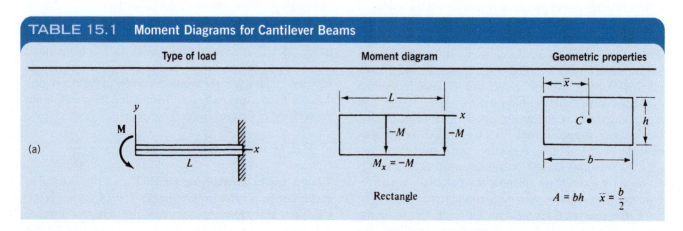

TABLE 15.1	Moment Diagrams for Cantilever Beams		
	Type of load	**Moment diagram**	**Geometric properties**
(a)			

Rectangle

$A = bh \quad \bar{x} = \dfrac{b}{2}$

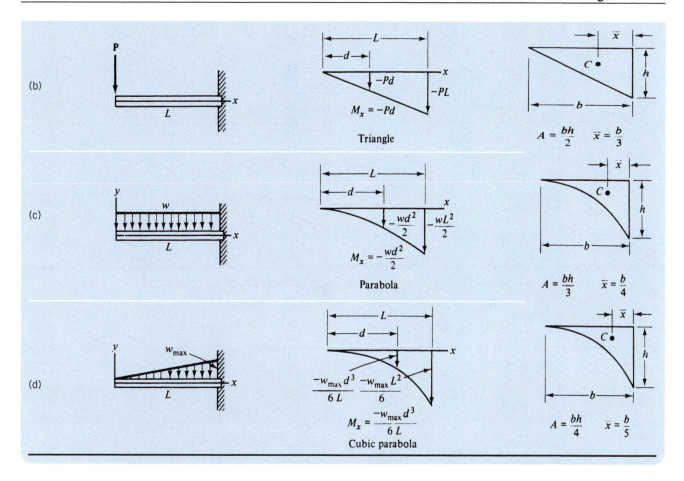

Example 15.1   Draw the moment diagram by parts for the beam shown in Fig. 15.2(a).

Solution: Two concentrated loads act on the cantilever beam. Each concentrated load has a moment diagram that is triangular in shape, as shown in Table 15.1(b). The bending moment is given by

$$M_x = -Pd$$

where x is the location measured from the left end of the beam, P the value of the concentrated load, and d the distance from the load.

Values of the bending moment from $P_1 = 8$ kip are shown in Fig. 15.2(b). At a distance $d = 5$ ft from the load at $x = 5$ ft,

$$M_5 = -(8)(5) = -40 \text{ kip-ft}$$

and at a distance $d = 10$ ft from the load at $x = 10$ ft,

$$M_{10} = -(8)(10) = -80 \text{ kip-ft}$$

Values of the bending moment for $P_2 = 10$ kip are shown in Fig. 15.2(c). At a distance $d = 5$ ft from the load at $x = 10$ ft,

$$M_{10} = -(10)(5) = -50 \text{ kip-ft}$$

The combined moment diagram is shown in Fig. 15.2(d).

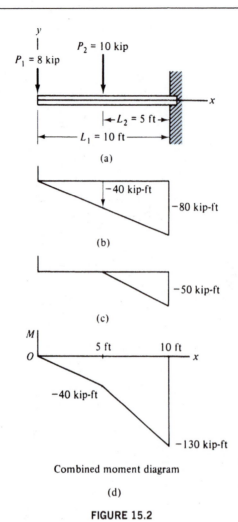

FIGURE 15.2

Example 15.2   Draw the moment diagram by parts for the cantilever beam shown in Fig. 15.3(a).

Solution: A concentrated load and a distributed load act on the cantilever beam. The concentrated load has a moment diagram as given in Table 15.1(b). Values of the bending moment for $P = 36$ kN are shown in Fig. 15.3(b). At a distance $d = 1.5$ m from the load at $x = 1.5$ m,

$$M_{1.5} = -36(1.5) = -54 \text{ kN·m}$$

and at a distance $d = 4.5$ m from the load at $x = 4.5$ m,

$$M_{4.5} = -36(4.5) = -162 \text{ kN·m}$$

The distributed load has a moment diagram that is parabolic in shape, as given in Table 15.1(c). The bending moment is given by

$$M_x = \frac{-wd^2}{2}$$

where x is the location measured from the left end of the beam, w the value of the distributed load, and d the distance along the distributed load. Values of the bending moment for $w = 20$ N/m are shown in Fig. 15.3(c). At a distance of $d = 3$ m along the uniform load at $x = 4.5$ m,

$$M_{4.5} = \frac{-20(3)^2}{2} = -90 \text{ kN·m}$$

The combined moment diagram is shown in Fig. 15.3(d).

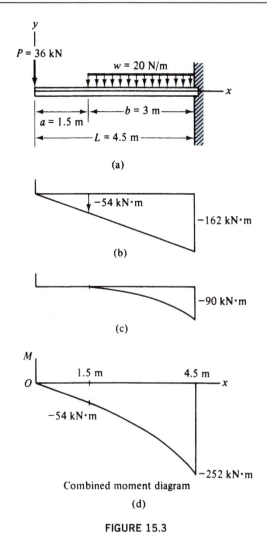

FIGURE 15.3

Example 15.3   Draw the moment diagram by parts for the simply supported beam shown in Fig. 15.4(a).

Solution: The reactions were determined from the equations of equilibrium and are shown in Fig. 15.4(a). The moment diagram for the concentrated upward reaction of 4 kip gives a positive triangular moment diagram [Fig. 15.4(b)] because the reaction is directed upward. The bending moment at a section x at a distance d from the concentrated load P is given by $M_x = Pd$. Therefore,

$$M_4 = 4(4) = 16 \text{ kip-ft} \qquad M_8 = 4(8) = 32 \text{ kip-ft}$$

and $\qquad M_{12} = 4(12) = 48 \text{ kip-ft}$

The moment diagram for the uniformly distributed downward load of 1.5 kip/ft gives a negative parabolic moment diagram [Fig. 15.4(c)]. The bending moment at a section x at a distance d along the distributed load is given by

$$M_x = -\frac{wd^2}{2}$$

Therefore,

$$M_8 = -\frac{1.5(4)^2}{2} = -12 \text{ kip-ft} \quad \text{and} \quad M_{12} = -\frac{1.5(8)^2}{2} = -48 \text{ kip-ft}$$

The combined moment diagram is shown in Fig. 15.4(d).

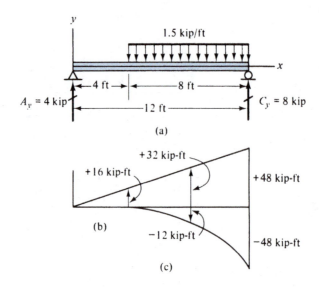

(a)

(b)

(c)

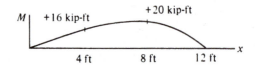

Combined moment diagram

.(d)

FIGURE 15.4

Example 15.4   Draw the moment diagram by parts for the simply supported beam shown in Fig. 15.5(a).

Solution: The reactions were determined. The partial uniform downward load of 15 N/m on the beam from $x = 0$ to $x = 2$ m was replaced by a uniform downward load of 15 N/m from $x = 0$ to $x = 5$ m and a uniform upward load of 15 N/m from $x = 2$ m to $x = 5$ m. The replacement loads when added together give the original load. The replacement loads and reactions are shown in Fig. 15.5(b).

The moment diagram for the concentrated upward reaction or load of 24 N gives a positive triangular moment diagram [Fig. 15.5(d)]. The bending moment at a section x at a distance d from the concentrated load P is given by

$$M_x = Pd$$

Therefore,

$$M_1 = 24(1) = 24 \text{ N·m} \qquad M_2 = 24(2) = 48 \text{ N·m}$$
$$M_{2.5} = 24(2.5) = 60 \text{ N·m} \quad \text{and} \quad M_5 = 24(5) = 120 \text{ N·m}$$

The moment diagram for the distributed upward uniform load of 15 N/m gives a positive parabolic moment diagram [Fig. 15.5(c)]. The bending moment at a section x at a distance d along the distributed load is given by

$$M_x = \frac{wd^2}{2}$$

Therefore,

$$M_{2.5} = \frac{15(0.5)^2}{2} = 1.875 \text{ N·m} \quad \text{and} \quad M_5 = \frac{15(3)^2}{2} = 67.5 \text{ N·m}$$

The moment diagram for the distributed downward uniform load of 15 N/m gives a negative parabolic moment diagram [Fig. 15.5(e)]. The formula for the bending moment is

$$M_x = -\frac{wd^2}{2}$$

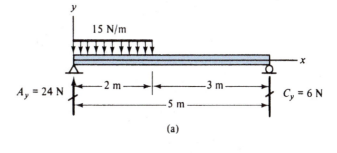

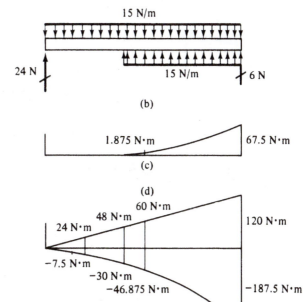

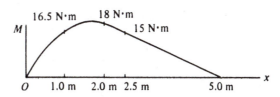

(a)

(b)

(c)

(d)

(e)

(f)

Combined moment diagram (different scale)

FIGURE 15.5

with x and d as defined previously. Therefore,

$$M_1 = -\frac{15(1)^2}{2} = -7.5 \text{ N·m} \qquad\qquad M_2 = -\frac{15(2)^2}{2} = -30 \text{ N·m}$$

$$M_{2.5} = -\frac{15(2.5)^2}{2} = -46.875 \text{ N·m} \quad \text{and} \quad M_5 = -\frac{15(5)^2}{2} = -187.5 \text{ N·m}$$

The combined moment diagram in Fig. 15.5(f) is not drawn to the same scale as the moment diagram by parts.

15.3 MOMENT-AREA METHOD

A beam is bent by loads as shown in Fig. 15.6(a). Before the loads were applied, the neutral surface, line AB, was straight. After the loads are applied, the line is curved. The curved line is called the *deflection curve*. We will now develop relation-ships between the changes in the shape of the deflection curve and the moments that produce the changes. They are based on three assumptions: The beam material follows Hooke's law, the deflections and slopes are small at every point of the deflection curve, and the deflection due to shear forces can be neglected.

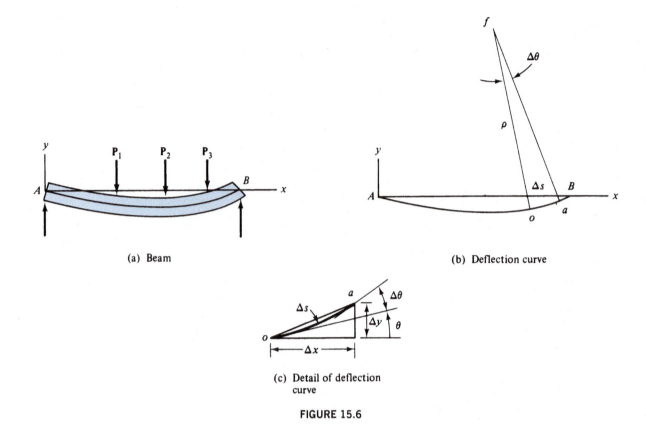

(a) Beam

(b) Deflection curve

(c) Detail of deflection curve

FIGURE 15.6

Consider the deflection curve shown in Fig. 15.6(b). The radius of curvature of the arc Δs is ρ (rho). The angle $\Delta\theta$ in radians is equal to $\Delta s/\rho$. Therefore,

$$\frac{1}{\rho} = \frac{\Delta\theta}{\Delta s} \qquad (a)$$

The slope θ of the deflective curve is assumed at every point to be small [Fig. 15.6(c)]. Consequently, the arc length Δs is approximately equal to Δx. With this substitution in Eq. (a), we have

$$\frac{1}{\rho} = \frac{\Delta\theta}{\Delta x} \qquad (15.1)$$

Eliminating the normal bending stress σ between Eqs. (14.2) and (14.4), that is, $\sigma = Ey/\rho$ and $\sigma = My/I$, another equation for $1/\rho$ is obtained. That is,

$$\frac{1}{\rho} = \frac{M}{EI} \qquad (15.2)$$

Combining Eqs. (15.1) and (15.2), we write

$$\frac{\Delta\theta}{\Delta x} = \frac{M}{EI} \quad \text{or} \quad \Delta\theta = \frac{M}{EI}\Delta x \qquad (15.3)$$

The M/EI Diagram

To draw the M/EI diagram for a beam, the value of the bending moment M at each cross section of the beam is divided by the value of the *flexural stiffness EI* at that cross section. For a beam of constant cross section made of one

material, EI is constant, and the *shape* of the M/EI diagram and the moment diagram is the same.

In Fig. 15.7, we show a loaded beam, the M/EI diagram, and the deflection curve. For illustration, the deflection curve is exaggerated [Fig. 15.7(c)]. The change in slope of the deflection curve from o to a is equal to $\Delta\theta$. From Eq. (15.3), $\Delta\theta$ is equal to the shaded area from o to a of the M/EI diagram. The total change in slope between A and B is obtained by summing up all the changes in slope $\Delta\theta$ from A to B. That is,

$$\theta_B - \theta_A = \Sigma\Delta\theta\,(\text{from } A \text{ to } B) = \Sigma\frac{M}{EI}\Delta x\,(\text{from } A \text{ to } B)$$

$$(15.4)$$

Therefore, *the change in slope between* A *and* B *on a deflection curve is equal to the area of the M/EI diagram between* A *and* B. Equation (15.4) can be restated for numerical applications in a simpler form:

$$\theta_B - \theta_A = A^*(\text{from } A \text{ to } B) \qquad (15.5)$$

where A^* represents the area of the M/EI diagram. This is the *first moment-area theorem*.

The quantity Δt in Fig. 15.7(c) is due to the bending of an element of the beam and is equal to $x_B\,\Delta\theta$. By summing up this effect from A to B, we obtain the vertical distance $B'B$, which is called the tangent deviation $t_{B/A}$. That is,

$$t_{B/A} = \Sigma\,\Delta t\,(\text{from } A \text{ to } B) = \Sigma x_B\,\Delta\theta\,(\text{from } A \text{ to } B)$$

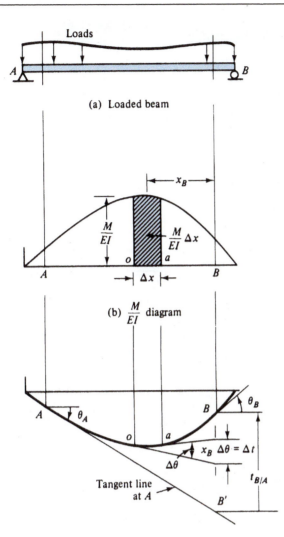

(a) Loaded beam

(b) $\dfrac{M}{EI}$ diagram

(c) Deflection curve

FIGURE 15.7

Substituting for $\Delta\theta$ from Eq. (15.3), we obtain

$$t_{B/A} = \Sigma x_B \frac{M}{EI} \Delta x \text{ (from } A \text{ to } B) \qquad (15.6)$$

Therefore, *the deviation of* B *from a tangent line at* A *is equal to the first moment about a vertical line through* B *of the area of the* M/EI *diagram between* A *and* B. If we use the definition of the centroid, Eq. (15.6) can be restated for numerical application in simpler form:

$$t_{B/A} = \bar{x} \text{ (from } B)A^* \text{(from } A \text{ to } B) \qquad (15.7)$$

where A^* represents the area of the M/EI diagram. This is the *second moment-area theorem.*

Consider the deflection curve shown in Fig. 15.8(a) and the M/EI diagram that has been drawn by parts in Fig. 15.8(b). The deviation of A from a tangent line at B, $t_{A/B}$, is shown in Fig. 15.8(a). The area A^* of the M/EI diagram from A to B and the centroids measured for A, $\bar{x}_A$, are shown in columns 1 and 2 of Table (A). In column 3, the product of areas and centroidal distances is tabulated. The sum of column 3 gives

$$t_{A/B} = 0.990 \text{ in.}$$

Similarly, the deviation of B from a tangent line at A is shown in Fig. 15.8(a). The centroids measured from B, $\bar{x}_B$, are shown in column 4 of the table. In column 5, the product of the areas and the centroidal distances is tabulated. The sum of column 5 gives

$$t_{B/A} = 0.630 \text{ in.}$$

To find the deflection of a simply supported beam requires both the first and the second moment-area theorems. However, the deflection and tangent deviation can be made equal in a cantilever beam problem. Therefore, the deflection can be found from the second moment-area theorem only.

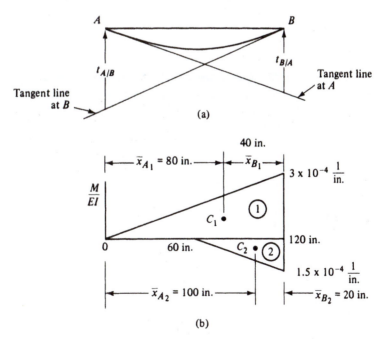

(a)

(b)

FIGURE 15.8

TABLE (A)   Section 15.3

	1	2	3	4	5
	A* (from A to B)	$\bar{x}_A$ (in.)	$\bar{x}_A A^*$ (A to B)	$\bar{x}_B$ (in.)	$\bar{x}_B A^*$ (A to B)
①	$\dfrac{120}{2}(3 \times 10^{-4}) = 0.18$ *0.018*	80	1.440	40	0.720
②	$-\dfrac{60}{2}(1.5 \times 10^{-4}) = -0.0045$ *0.9·1*	100	−0.450	20	−0.090
			$t_{A/B} = 0.990$ in.		$t_{B/A} = 0.630$ in.

15.4 DEFLECTION OF A CANTILEVER BEAM BY THE MOMENT-AREA METHOD

A loaded cantilever beam and the deflection curve for the beam are shown in Fig. 15.9. Because the beam is clamped at B, the slope of the deflection curve is zero at B. Thus, the tangent deviation of any point on the beam from the tangent line at B is directly equal to the deflection of the beam at that point. That is,

$$t_{A/B} = \delta_A \text{ (the deflection at } A)$$

and

$$t_{C/B} = \delta_C \text{ (the deflection at } C)$$

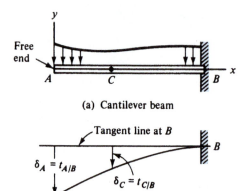

(a) Cantilever beam

(b) Deflection curve

FIGURE 15.9

Example 15.5   Find the maximum deflection for the cantilever beam shown in Fig. 15.10(a).

Solution: The maximum deflection will occur at the free end A. The deviation of A from a tangent line at B, $t_{A/B}$, is equal to the maximum deflection δ_A [Fig. 15.10(b)]. The M/EI diagram between A and B and the centroidal distance measured from a vertical line through A are shown in Fig. 15.10(c). The area A^* of the M/EI diagram from A to B and the centroidal distance measured from A, $\bar{x}_A$, are shown in columns one and two of Table (A). In column three, the product of area and centroidal distance is tabulated. From the second moment-area theorem, the tangent deviation $t_{A/B}$ is equal to the first moment about a vertical line through A of the M/EI diagram between A and B. That is,

$$t_{A/B} = \bar{x}_A A^* \text{ (from } A \text{ to } B)$$

TABLE (A)   Examples 15.5 and 15.6

A*(from A to B)	$\bar{x}_A$	$\bar{x}_A A^*$(A to B)
$\dfrac{L}{2}\left(-\dfrac{PL}{EI}\right) = -\dfrac{PL^2}{EI}$ $\underbrace{\qquad}_{\theta_B - \theta_A}$	$\dfrac{2L}{3}$	$-\dfrac{PL^3}{3EI}$ $\underbrace{\qquad}_{t_{A/B}}$

From column three of the table,

$$t_{A/B} = -\frac{PL^3}{3EI}$$

The maximum deflection $\delta_{max} = t_{A/B}$; therefore,

$$\delta_{max} = -\frac{PL^3}{3EI}$$ **Answer**

The negative sign indicates that the deflection is down.

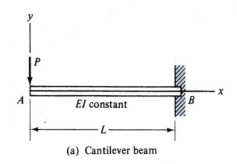

(a) Cantilever beam

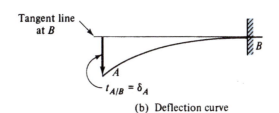

(b) Deflection curve

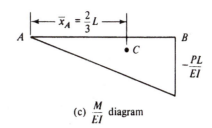

(c) $\dfrac{M}{EI}$ diagram

FIGURE 15.10

Example 15.6   Find the slope at the free end of the cantilever beam in Example 15.5.

Solution: From the first moment-area theorem, the change in slope between A and B is equal to the area of the M/EI diagram between A and B. That is,

$$\theta_B - \theta_A = A^* \text{ (from } A \text{ to } B)$$

From column one of the table,

$$\theta_B - \theta_A = -\frac{PL^2}{2EI}$$

The beam is clamped or fixed at B; therefore, $\theta_B = 0$ and

$$\theta_A = \frac{PL^2}{2EI} \angle \theta_A \qquad\qquad \textbf{Answer}$$

Example 15.7   Find the slope at the free end and the maximum deflection for the W10 × 39 steel cantilever beam shown in Fig. 15.11(a). Use A36 structural steel.

Solution: The modulus $E = 29 \times 10^3$ kip/in.^{2} for A36 steel (Table A.10 of the Appendix), and the moment of iner-tia $I = 209$ in.^{4} for a W10 × 39 beam (Table A.3 of the Appendix). The deflection curve and the M/EI di-agram are shown in Fig. 15.11. The calculations are shown in Table (A). From the first moment-area theorem,

$$\theta_B - \theta_A = A^* \text{ (from } A \text{ to } B)$$

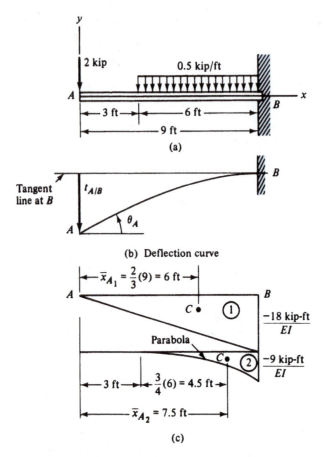

FIGURE 15.11

TABLE (A)   Example 15.7

	A^* (from A to B) $(kip\text{-}ft^2)/EI$	$\bar{x}_A$ (ft)	$\bar{x}_A A^*$ (A to B) $(kip\text{-}ft^3)/EI$
①	$\dfrac{9(-18)}{2EI} = -\dfrac{81}{EI}$	6	$-\dfrac{486}{EI}$
②	$\dfrac{6(-9)}{3EI} = -\dfrac{18}{EI}$	7.5	$-\dfrac{135}{EI}$
	$\Sigma -\dfrac{99}{EI}$ $\theta_B - \theta_A \big)$		$\Sigma -\dfrac{621}{EI}$ $t_{A/B} \big)$

Tabulated in column one, we have

$$\theta_B - \theta_A = -\frac{99\,kip\text{-}ft^2}{EI}$$

Changing ft^2 to $in.^2$ by multiplying by $(12\,in./ft)^2 = 144\,in.^2/ft^2$, we get

$$\theta_B - \theta_A = -\frac{99(144)\,kip\text{-}in.^2}{EI}$$

The flexural stiffness $EI = 29 \times 10^3\,kip/in.^2\,(209\,in.^4) = 29 \times 10^3(209)\,kip\text{-}in.^2$, and the slope $\theta_B = 0$. Therefore,

$$\theta_A = \frac{99(144)\,kip\text{-}in.^2}{29 \times 10^3(209)\,kip\text{-}in.^2} = 2.352 \times 10^{-3}\,rad$$

$$= 2.352 \times 10^{-3}\,rad\,\frac{180°}{\pi\,rad} = 0.1348° \measuredangle\,\theta_A \qquad \textbf{Answer}$$

From the second moment-area theorem,

$$t_{A/B} = \bar{x}_A A^* \text{ (from } A \text{ to } B)$$

From column three of Table (A), we have

$$t_{A/B} = -\frac{621 \text{ kip-ft}^3}{EI}$$

Changing ft^3 to $in.^3$ by multiplying by $(12 \text{ in./ft})^3 = 1728 \text{ in.}^3/ft^3$, we get

$$t_{A/B} = -\frac{621(1728) \text{ kip-in.}^3}{EI}$$

From before the flexural stiffness $EI = 29 \times 10^3(209)$ kip-in.^{2} and the maximum deflection $\delta_{max} = t_{A/B}$; therefore,

$$\delta_{max} = -\frac{621(1728) \text{ kip-in.}^3}{29 \times 10^3(209) \text{ kip-in.}^2} = -0.1770 \text{ in.} \qquad \textbf{Answer}$$

The negative sign indicates that the deflection is down.

PROBLEMS

15.1 through 15.8   Use the moment-area method to find (a) the maximum deflection, and (b) the slope at the free end of the cantilever beam shown in the figure. Draw the M/EI diagram by parts, assume that EI is constant, and neglect the weight of the beam.

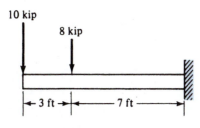

PROB. 15.1

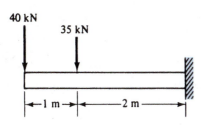

PROB. 15.2

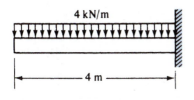

PROB. 15.3

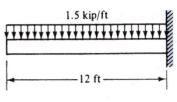

PROB. 15.4

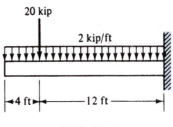

PROB. 15.5

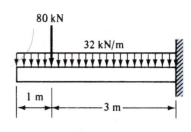

PROB. 15.6

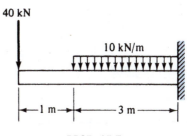

PROB. 15.7

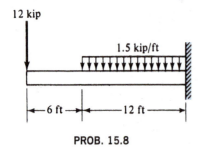

PROB. 15.8

15.9 through 15.11   Use the moment-area method to find (a) the maximum deflection, and (b) the slope at the free end of the cantilever beam shown in the figure. Neglect the weight of the beam. (*Hint:* Draw the M/EI diagram by parts as shown.)

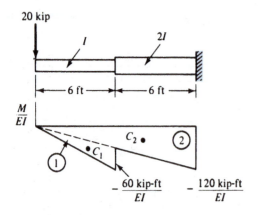

PROB. 15.9

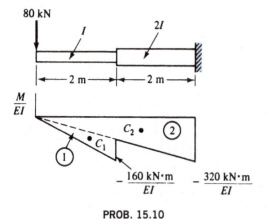

PROB. 15.10

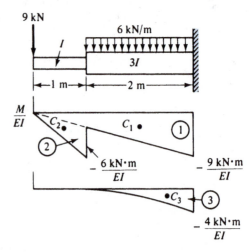

PROB. 15.11

15.12 Determine by the moment-area method (a) the maximum deflection, and (b) the slope at the free end of the cantilever beam shown. Assume that $E = 200 \times 10^6$ kN/m^2 (kPa) and neglect the weight of the beam.

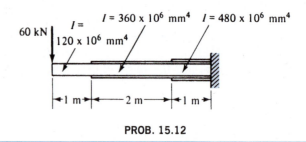

PROB. 15.12

15.5 DEFLECTION OF THE SIMPLY SUPPORTED BEAM BY THE MOMENT-AREA METHOD

For a simply supported beam with the loads all directed either down or up, we will calculate the deflection at the middle of the beam rather than the maximum value of the deflection. In most cases, the deflection at the middle is not substantially different from the maximum deflection and is much easier to calculate.

Consider the loaded simply supported beam [Fig. 15.12(a)] and its deflection curve [Fig. 15.12(b)]. The deviation

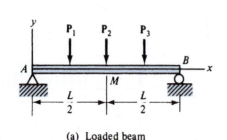

(a) Loaded beam

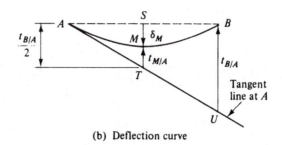

(b) Deflection curve

FIGURE 15.12

of M from a tangent line drawn at A, $t_{M/A}$, and the deviation of B from a tangent line drawn at A, $t_{B/A}$, are shown in the figure. They can both be calculated from the second moment-area theorem. The *magnitude* of the deflection at the middle of the beam

$$|\delta_M| = ST - t_{M/A}$$

From similar triangles ABU and AST, $ST = t_{B/A}/2$; therefore,

$$\delta_M = -\left(\frac{t_{B/A}}{2} - t_{M/A}\right) \tag{15.8}$$

where the negative sign indicates that the deflection is down.

Example 15.8   Calculate the deflection at the middle of the 12-ft span of a simply supported beam with a concentrated load of 10 kip, 9 ft from one of the supports, as shown in Fig. 15.13(a).

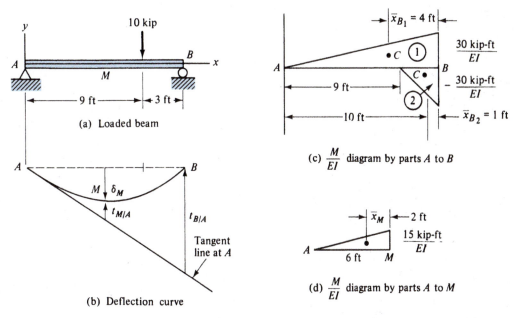

(a) Loaded beam

(b) Deflection curve

(c) $\dfrac{M}{EI}$ diagram by parts A to B

(d) $\dfrac{M}{EI}$ diagram by parts A to M

FIGURE 15.13

Solution: The deflection curve is shown in Fig. 15.13(b). To find the value of $t_{B/A}$, we draw the M/EI diagram by parts in Fig. 15.13(c). The calculations are shown in Table (A). From the sum in column three,

$$t_{B/A} = \frac{675 \text{ kip-ft}^3}{EI}$$

TABLE (A)   Example 15.8

	A*(from A to B) (kip-ft²)/EI	$\bar{x}_B$ (ft)	$\bar{x}_B$A* (A to B) (kip-ft³)/EI
①	$\dfrac{12(30)}{2EI} = \dfrac{180}{EI}$	4	$\dfrac{720}{EI}$
②	$\dfrac{3(-30)}{2EI} = -\dfrac{45}{EI}$	1	$-\dfrac{45}{EI}$
			$\Sigma \; \dfrac{675}{EI}$ $t_{B/A}$

To determine the value of $t_{M/A}$, we draw the M/EI diagram by parts in Fig. 15.13(d). Calculating the value of $t_{M/A}$, we have

$$t_{M/A} = \bar{x}_M A^* \text{ (from } A \text{ to } M) = 2\left(\frac{6}{2}\right)\frac{15}{EI}$$

$$= \frac{90 \text{ kip-ft}^3}{EI}$$

Therefore, from Eq. (15.8),

$$\delta_M = -\left(\frac{t_{B/A}}{2} - t_{M/A}\right) = -\left(\frac{675}{2EI} - \frac{90}{EI}\right)$$

$$= -\frac{247.5 \text{ kip-ft}^3}{EI} = -\frac{4.277(10^8) \text{ lb-in.}^3}{EI}$$

Answer

The formula for the maximum deflection [Table A.14(e) of the Appendix] is

$$\delta_{\max} = -\frac{Pb(L^2 - b^2)^{3/2}}{9\sqrt{3}LEI}$$

For $P = 10,000$ lb, $L = 144$ in., and $b = L/4 = 36$ in., the maximum deflection

$$\delta_{\max} = -\frac{4.348(10^8) \text{lb-in.}^3}{EI}$$

The deflection at the middle of the beam and the maximum deflection differ by approximately 1.6 percent.

Example 15.9   A W10 × 45 simply supported beam with a span of 12 ft has a concentrated load of 10 kip and a distributed load of 1.5 kip/ft, as shown in Fig. 15.14(a). The reactions have been calculated and the moment diagram drawn by parts [Fig. 15.14(b)]. Find the deflection at the middle of the beam.

Solution: The modulus $E = 29 \times 10^3$ ksi for A36 steel (Table A.10 of the Appendix), and the moment of inertia $I = 248$ in.^{4} for the W10 × 45 (Table A.3 of the Appendix). To find the tangent deviation of B from A, we construct the M/EI diagram [Fig. 15.15(a)] and tabulate the calculations in Table (A). The calculations for the tangent deviation of M from A are shown in Fig. 15.15(b) and tabulated in Table (B). The deflection at the middle of the beam is given by Eq. (15.7):

$$\delta_M = -\left(\frac{t_{B/A}}{2} - t_{M/A}\right)$$

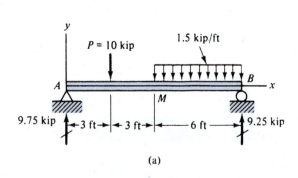

(a)

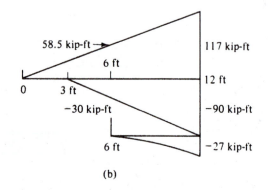

(b)

FIGURE 15.14

TABLE (A) Example 15.9			
	A* (from A to B) (kip-ft²)/EI	$\bar{x}_B$ (ft)	$\bar{x}_B$A*(A to B) (kip-ft³)/EI
①	$\dfrac{12(117)}{2EI} = \dfrac{702}{EI}$	4	$\dfrac{2808}{EI}$
②	$\dfrac{9(-90)}{2EI} = -\dfrac{405}{EI}$	3	$-\dfrac{1215}{EI}$
③	$\dfrac{6(-27)}{3EI} = -\dfrac{54}{EI}$	1.5	$-\dfrac{81}{EI}$
			$\Sigma\ \dfrac{1512}{EI}$, $t_{B/A}$

TABLE (B) Example 15.9			
	A* (from A to M) (kip-ft²)/EI	$\bar{x}_M$ (ft)	$\bar{x}_M$A* (A to M) (kip-ft³)/EI
④	$\dfrac{6(58.5)}{2EI} = \dfrac{175.5}{EI}$	2	$\dfrac{351}{EI}$
⑤	$\dfrac{3(-30)}{2EI} = -\dfrac{45}{EI}$	1	$-\dfrac{45}{EI}$
			$\Sigma\ \dfrac{306}{EI}$, $t_{M/A}$

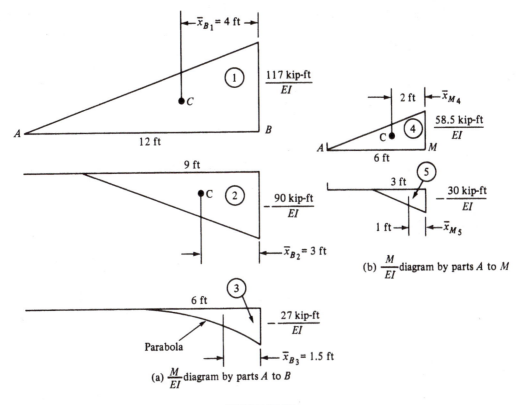

FIGURE 15.15

Substituting for $t_{B/A}$ and $t_{M/A}$, the sums shown in the tables, we have

$$\delta_M = -\left(\frac{1512}{2EI} - \frac{306}{EI}\right) = -\frac{450\text{ kip-ft}^3}{EI}$$

Changing ft^3 to in.^{3} by multiplying by $(12\text{ in./ft})^3 = 1728\text{ in.}^3/\text{ft}^3$, we get

$$\delta_{max} = -\frac{450(1728)\text{ kip-in.}^3}{EI}$$

The flexural stiffness $EI = 29 \times 10^3\text{ kip/in.}^2(248\text{ in.}^4) = 29 \times 10^3(248)\text{ kip-in.}^2$; therefore,

$$\delta_{max} = -\frac{450(1728)\text{ kip-in.}^3}{29 \times 10^3(248)\text{ kip-in.}^2} = -0.1081\text{ in.} \qquad \textbf{Answer}$$

PROBLEMS

15.13 through 15.20   Use the moment-area method to find the deflection at the middle of the simply supported beam shown in the figure. Draw the M/EI diagram by parts; assume that EI is constant, and neglect the weight of the beam.

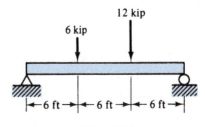

PROB. 15.14

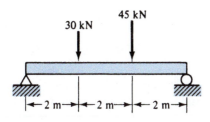

PROB. 15.13

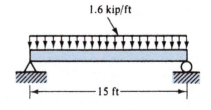

PROB. 15.15

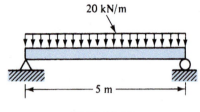

PROB. 15.16

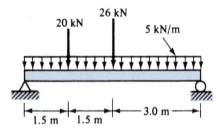

PROB. 15.17

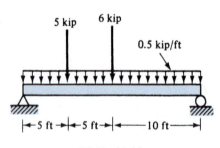

PROB. 15.18

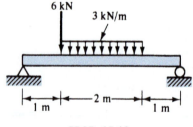

PROB. 15.19

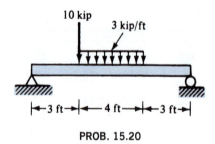

PROB. 15.20

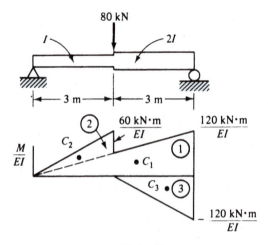

PROB. 15.21

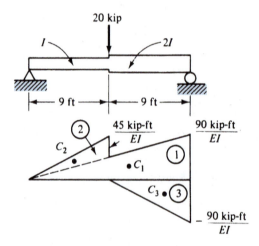

PROB. 15.22

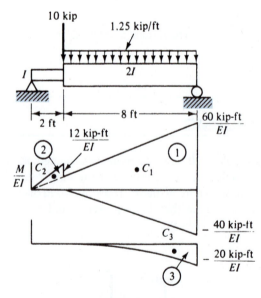

PROB. 15.23

15.21 through 15.23   Use the moment-area method to find the deflection at the middle of the simply supported beam shown in the figure. Neglect the weight of the beam. (*Hint:* Draw the M/EI diagram by parts as shown.)

15.24 Determine by the moment-area method the deflection at the middle of the simply supported beam shown in the figure. Assume that $E = 30 \times 10^6$ psi and neglect the weight of the beam.

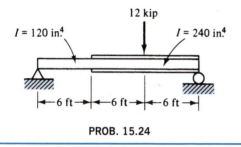

12 kip

$I = 120 \text{ in.}^4$ $I = 240 \text{ in.}^4$

|←6 ft→|←6 ft→|←6 ft→|

PROB. 15.24

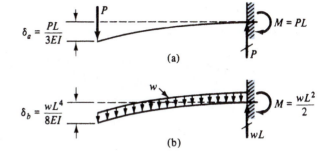

$$\delta_a = \frac{PL}{3EI}$$ $M = PL$

(a)

$$\delta_b = \frac{wL^4}{8EI}$$ $M = \frac{wL^2}{2}$

(b)

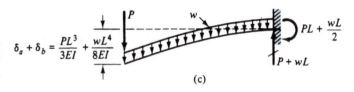

$$\delta_a + \delta_b = \frac{PL^3}{3EI} + \frac{wL^4}{8EI}$$ $PL + \frac{wL}{2}$

$P + wL$

(c)

FIGURE 15.16

15.6 SUPERPOSITION METHOD

Deflections, slopes, and deflection equations for various cantilever and simply supported beams are shown in Table A.14 of the Appendix. All deflections and slopes used in this section were taken from the table.

Consider the two identical cantilever beams shown in Fig. 15.16(a) and (b). One of the beams supports a concentrated load on the free end, and the other beam supports a uniformly distributed load over the entire length. The maximum deflections are from diagrams (b) and (c) of Table A.14 of the Appendix. In the superposition method, we superimpose both loads

on a third identical beam. The reactions and deflections are equal to the sum of the reactions and deflections of the individual beams [Fig. 15.16(c)]. The method is illustrated in the examples that follow.

Example 15.10   Use the methods of superposition to find the deflection at the free end of the cantilever beam shown in Fig. 15.17.

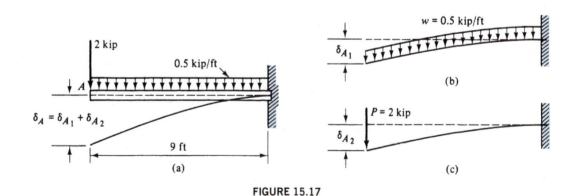

FIGURE 15.17

Solution: Two basic loadings are used as shown in diagrams (b) and (c) of Table A.14 of the Appendix. The deflection at A is equal to the sum of the deflections shown in the table. That is,

$$\delta_A = \frac{PL^3}{3EI} + \frac{wL^4}{8EI}$$

For $P = 2$ kip, $w = 0.5$ kip/ft, and $L = 9$ ft, the deflection

$$\delta_A = \frac{2(9)^3}{3EI} + \frac{0.5(9)^4}{8EI} = \frac{896 \text{ kip-ft}^3}{EI}$$ **Answer**

FLEXIBLE FLOORS

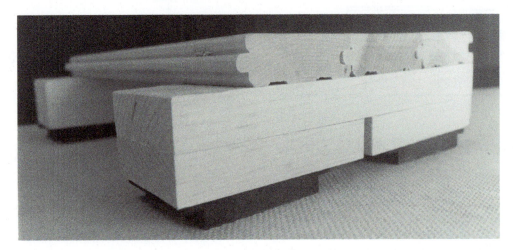

SB.22(a)

SB.22(b)

Beam deflections are not always undesirable. For example, the purpose of a *diving board* (which behaves like a cantilevered beam) is to deflect under the diver's initial "bounce," absorbing energy that it then gives back to the diver as he or she springs into the actual dive. In a similar manner, *flexible floors*, specially designed for athletic activities, are used to cushion the impact of feet on floor, absorbing energy as a foot hits the floor and returning this energy as the foot leaves the floor. Typical construction for such a surface is shown in SB.22(a). The lower two members are fastened together to form a sleeper, which acts like a continuous beam placed across closely spaced rubber pads that serve as supports. Hardwood flooring is laid at right angles to these sleepers. The rubber pads [SB.22(b)] are approximately 2-1/4 in. square, contain a number of air pockets, and behave like mini shock absorbers. Design of the floor, sleeper, and pad combination may be adjusted to control the "springiness" of a floor, matching it to the intended activity and participant weight, such as elementary school gymnastics, women's aerobics, or men's volleyball. These floors are one of the reasons that professional basketball players are able to jump so high and so often. *(Courtesy of Action Floor Systems, LLC, Mercer, Wisconsin, www.actionfloors.com)*

Example 15.11   In Table 14.6, the maximum recommended span for 2 × 12 floor joists at a center-to-center spacing of 24 in. was given as 16′-8″. As specified in the footnotes of that structural table, the joists are assumed to have a modulus of elasticity of 1.4×10^6 psi, support a floor whose distributed load is 40 psf, and have a maximum deflection under load of 1/360 of their span. Using superposition, find the actual midpoint deflection of these joists. (Note that the stress developed in such joists was discussed in Example 14.19.)

Solution: Maximum deflection for a simply supported beam carrying a distributed load is given in Table A.14(g) of the Appendix as

$$\delta_{max} = \frac{5wL^4}{384EI}$$

where $w = 100\,\text{lb/ft}$ (see Example 14.19)
$L = 16'\text{-}8'' = 16.67\,\text{ft}$
$E = 1.4 \times 10^6\,\text{psi}$
$I = 178\,\text{in.}^4$ (see Table A.9 of the Appendix)

Then,

$$\delta_{max} = \frac{5(100\,\text{lb/ft})(16.67\,\text{ft})^4}{384(1.4 \times 10^6\,\text{psi})(178\,\text{in.}^4)} \times \frac{1728\,\text{in.}^3}{1\,\text{ft}^3}$$
$$= 0.697\,\text{in.} \qquad\qquad\qquad \textbf{Answer}$$

However, the joist span is $L = 16'\text{-}8'' = 200\,\text{in.}$, so

$$\frac{L}{360} = \frac{200\,\text{in.}}{360} = 0.556\,\text{in.}$$

Note that the actual deflection exceeds $L/360$ by more than 25 percent. Is the tabulated value wrong? Perhaps not, because no assumed value for *dead load* is specified in the footnotes of the table. If we neglect that load (10 psf), then $w = 80\,\text{lb/ft}$, and the maximum deflection becomes 0.558 in. These apparent contradictions between tabulated and computed values are not uncommon and emphasize the need (and the ability) to spot-check published data.

Example 15.12   A simply supported wooden beam 16 ft long supports a concentrated load of 1000 lb at its midpoint, as shown in Fig. 15.18. If the maximum deflection of this beam is limited to $L/240$, select the lowest suitable grade of machine-graded lumber that may be used when the beam is a nominal:

a. 2 × 10   b. 2 × 12

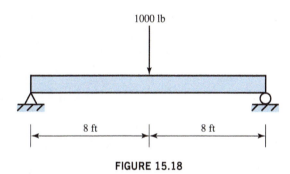

1000 lb

8 ft   8 ft

FIGURE 15.18

Solution: a.   The maximum moment for this beam occurs at midspan and is equal to $M = 4000\,\text{ft-lb} = 48,000\,\text{in.-lb}$. From Table A.9 of the Appendix, the moment of inertia and section modulus for a 2 × 10 are $I = 89.9\,\text{in.}^4$ and $S = 21.4\,\text{in.}^3$, respectively. Checking the maximum bending stress developed,

$$\sigma_{max} = \frac{M}{S} = \frac{48,000\,\text{in.-lb}}{21.4\,\text{in.}^3} = 2240\,\text{psi}$$

Maximum allowable deflection is

$$\delta_{max} = \frac{L}{240} = \frac{192\,\text{in.}}{240} = 0.800\,\text{in.}$$

Rearranging the formula for maximum deflection from Table A.14(f) of the Appendix yields the required modulus of elasticity:

$$E = \frac{PL^3}{48I\delta_{max}} = \frac{(1000\,\text{lb})(192\,\text{in.})^3}{48(89.9\,\text{in.}^4)(0.800\,\text{in.})}$$
$$= 2,050,000\,\text{psi}$$

Therefore, any grade of machine-graded lumber having an allowable bending stress of at least 2240 psi and a modulus of elasticity of at least 2.05×10^6 psi would be suitable for this application. From Table A.11 of the Appendix, select

$$\text{MSR} \quad 2550\text{f-}2.1E \qquad \textbf{Answer}$$

b. For a 2×12 with $I = 178 \text{ in.}^4$ and $S = 31.6 \text{ in.}^3$,

$$\sigma_{\max} = \frac{48{,}000 \text{ in.-lb}}{31.6 \text{ in.}^3} = 1520 \text{ psi}$$

$$E = \frac{(1000 \text{ lb})(192 \text{ in.})^3}{48(178 \text{ in.}^4)(0.800 \text{ in.})} = 1{,}036{,}000 \text{ psi}$$

We therefore select one of the following lumber grades:

$$\text{MSR} \quad 1650\text{f-}1.5E \quad \text{or} \quad \text{MEL} \quad \text{M-11} \qquad \textbf{Answer}$$

Example 15.13   Use the method of superposition to find the deflection at the middle of the simply supported beam shown in Fig. 15.19.

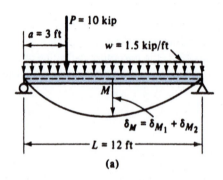

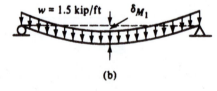

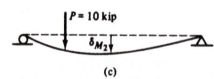

FIGURE 15.19

Solution: Two basic loadings are used. See diagrams (e) and (g) of Table A.14 of the Appendix. The deflection at M is equal to the sum of the deflections shown in the table. That is,

$$\delta_M = \frac{Pa}{48EI}(3L^2 - 4a^2) + \frac{5wL^4}{384EI}$$

For $P = 10$ kip, $w = 1.5$ kip/ft, $a = 3$ ft, and $L = 12$ ft, the deflection

$$\delta_M = \frac{10(3)}{48EI}[3(12)^2 - 4(3)^2] + \frac{5(1.5)(12)^4}{384EI}$$

$$= \frac{652 \text{ kip-ft}^3}{EI} \qquad \textbf{Answer}$$

Example 15.14   Use the method of superposition to find the deflection at the middle of the simply supported beam shown in Fig. 15.20.

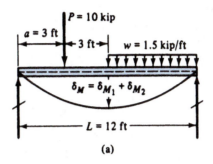

(a)

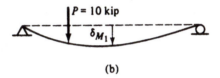

(b)

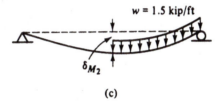

(c)

FIGURE 15.20

Solution: Two basic loadings are used. The concentrated load shown in diagram (e) of Table A.14 of the Appendix can be used directly. However, the uniform load over half the beam will produce half the deflection of the uniform load over the entire beam shown in diagram (g) of Table A.14 of the Appendix. Therefore,

$$\delta_M = \frac{Pa}{48EI}(3L^2 - 4a^2) + \frac{5wL^4}{2(384)EI}$$

For $P = 10$ kip, $w = 1.5$ kip/ft, $a = 3$ ft, and $L = 12$ ft, the deflection

$$\delta_M = \frac{10(3)}{48EI}[3(12)^2 - 4(3)^2] + \frac{5(1.5)(12)^4}{2(384)EI}$$

$$= \frac{450 \text{ kip-ft}^3}{EI}$$ **Answer**

This agrees with the results obtained in Example 15.9 by the moment-area method.

Example 15.15   Use the method of superposition to find the deflection at the free end of the cantilever beam shown in Fig. 15.21(a).

Solution: The concentrated load at the free end of the cantilever beam produces a deflection at A as shown in Fig. 15.21(b). The uniform loading produces a deflection at C, and because the beam is continuous and

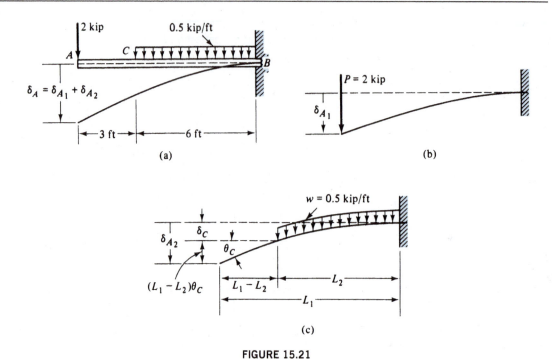

FIGURE 15.21

has a slope at C, an additional deflection occurs at A equal to $(L_1 - L_2)\theta_C$, as shown in Fig. 15.21(c). Therefore, the deflection

$$\delta_A = \delta_{A_1} + \delta_C + (L_1 - L_2)\theta_C \qquad \text{(a)}$$

Deflections and slopes are given in Table A.14 of the Appendix. The deflection δ_A is shown in diagram (b), and the deflection δ_C and slope θ_C are shown in diagram (c). Substituting values from the diagrams into Eq. (a), we have

$$\delta_A = \frac{PL_1^3}{3EI} + \frac{wL_2^4}{8EI} + (L_1 - L_2)\frac{wL_2^3}{6EI}$$

For $P = 2$ kip, $w = 0.5$ kip/ft, $L_2 = 6$ ft, $L_1 = 9$ ft, and $(L_1 - L_2) = 3$ ft, the deflection

$$\delta_A = \frac{2(9)^3}{3EI} + \frac{0.5(6)^4}{8EI} + \frac{3(0.5)(6)^3}{6EI}$$

$$= \frac{621 \text{ kip-ft}^3}{EI} \qquad \qquad \textbf{Answer}$$

This answer agrees with the results obtained in Example 15.7 by the moment-area method.

PROBLEMS

15.25 A cantilevered beam is 5 m long and consists of a W200 × 100 steel member. If a 90 kN concentrated load is applied at a point 1.5 m from the beam's free end, use superposition techniques to find (a) deflection of the free end caused by the beam's own weight, and (b) deflection of the free end caused by the concentrated load.

15.26 When the attic above a ceiling is used for storage, nominal 2 × 8 simply supported ceiling joists spaced 24-in. apart are each assumed to carry a uniformly distributed load of 80 lb/ft. If the ceiling is plastered, maximum deflection of the joists is limited to 1/360th

of their length to prevent cracking of the plastered surface. Using superposition, find the maximum allowable span for Southern pine joists based on this deflection requirement.

15.27 One nominal 2 × 6 wooden member 10 ft long acts as a simply supported beam and carries a distributed load of 80 lb/ft along its entire length. Maximum deflection of the beam must be limited to 1/240 of its span.
   a. Select the lowest grade of MSR or MEL lumber that is suitable for this loading.
   b. If the beam is modified to consist of *two* 2 × 6s placed side by side, does your choice of grade from part (a) change?

15.28 A nominal 2 × 8 wooden beam 12 ft long carries a 900-lb concentrated load placed 4 ft from the right-hand support. If the maximum deflection of this beam must not exceed 1/240 of its span, specify the lowest grade of MSR or MEL lumber that may be used.

15.29 A simply supported 89 × 356 wooden I-joist (see Table A.13 of the Appendix) is 6.5 m long and carries a uniformly distributed load of 2.0 kN/m along its entire length. If the modulus of elasticity for the wood in this joist is 14.0 GPa, find the deflection at the joist's midpoint.

15.30 The seating in a bus station consists of two vertical posts (A–B and C–D) bolted to the floor and a chair rail (B–C) supporting three thermoplastic seats as shown. For easy removal to clean or repair this seat assembly, the rail is pinned to its posts at B and C. All supporting members are 2-in.-square steel tubes with a 1/8-in. wall thickness. If each of the three seats is occupied by a 200-lb person, find (a) the compressive stress in post A–B and C–D, (b) the maximum bending stress in chair rail B–C, and (c) the deflection of rail B–C at its midpoint.

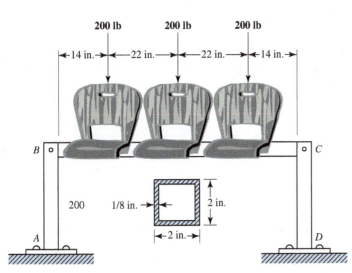

PROB. 15.30

In Probs. 15.31 through 15.38, use Table A.14 of the Appendix and the method of superposition to determine (a) the maximum deflection, and (b) the slope at the free end of the cantilever beam shown in the figure. Assume that EI is constant and neglect the weight of the beam.

15.31 Use the figure for Prob. 15.1.
15.32 Use the figure for Prob. 15.2.
15.33 Use the figure for Prob. 15.3.
15.34 Use the figure for Prob. 15.4.
15.35 Use the figure for Prob. 15.5.
15.36 Use the figure for Prob. 15.6.
15.37 Use the figure for Prob. 15.7.
15.38 Use the figure for Prob. 15.8.

In Probs. 15.39 through 15.44, use Table A.14 of the Appendix and the method of superposition to determine the deflection at the middle of the simply supported beam shown in the figure. Assume that EI is constant and neglect the weight of the beam.

15.39 Use the figure for Prob. 15.13.
15.40 Use the figure for Prob. 15.14.
15.41 Use the figure for Prob. 15.15.
15.42 Use the figure for Prob. 15.16.
15.43 Use the figure for Prob. 15.17.
15.44 Use the figure for Prob. 15.18.

15.7 BEAM DEFLECTIONS USING COMPUTER SOFTWARE

A glance at the superposition formulas in Table A.14 of the Appendix should convince you that, except for relatively simple beam loadings, it can be a formidable task either to compute the deflection at a specific point or to locate the point of maximum deflection on a beam. One powerful tool that can help with these calculations is a type of computer program known as a *spreadsheet*. Such prepackaged software is available for all types of personal computers, and most versions sold under various brand names differ only slightly in their program commands and operations.

As it appears on a computer screen, a spreadsheet has two important elements: an array of boxes (called *cells*) arranged in rows and columns such as those found in the business ledgers used by accountants and bookkeepers, and some type of *editing bar* (usually a separate line or box located at the top of the computer screen) by which the entries on a spreadsheet may be specified or defined. As shown in the abbreviated spreadsheet of Fig. 15.22, columns are commonly *lettered* from left to right, while rows are *numbered*

Editing Bar $---\rightarrow$

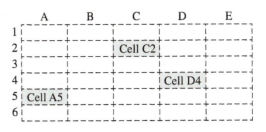

FIGURE 15.22. The screen displayed for a typical spreadsheet contains two important elements: an *editing bar* and *cells*, or boxes, arranged in lettered columns and numbered rows. As indicated by the three entries on the partial spreadsheet shown here, the location, or *address*, of any cell can be designated using its column letter and row number.

from top to bottom. Thus, the location, or *address,* of any cell may be indicated by its column letter and row number; the addresses A5, C2, and D4 are shown within the cells to which they correspond.

Manipulating spreadsheet data—such as specifying numerical values and mathematical operations for individual cells or groups of cells—is usually achieved by a combination of *keyboard commands* and *cursor movements.* The cursor is simply a movable indicator, often in the shape of an arrowhead or cross, that appears on the computer screen and is controlled by the *mouse* or keyboard commands. The mouse, a small, hand-controlled device containing a ball or wheel, is wired to the computer. When rolled over a flat surface, the mouse generates electrical signals that allow the cursor to be moved to any position on the computer screen.

One great advantage of spreadsheet software is that it allows the user to perform repetitive numerical calculations quickly and accurately without having to program an extensive set of directions into the computer. Typical applications include the creation of lists or tables of data for analysis. In addition, most spreadsheet software also contains various graphing capabilities, which can be used to display the relationships between physical quantities in visual form. A general procedure for using spreadsheets is shown in Example 15.16; this is followed by the beam deflection problem of Example 15.17.

Example 15.16   Use spreadsheets to generate a table of coordinates for the equation $y = 2x^2 - 14x + 5$ for $-1 \le x \le 8$.

Solution: A good first step is to label the columns for the x and y values. To do so, we would follow the *typical* sequence given here, again noting that *specific* procedures may vary among different software:

1.   Use the mouse to move the cursor to cell A1.
2.   Click the mouse once to highlight this cell. [See Fig. 15.23(a).]
3.   Use the mouse to move the cursor up to the editing bar.
4.   Use the keyboard to type "= x" into the editing bar space.
5.   Click the mouse once; the heading "x" should now appear in cell A1.
6.   Repeat steps 1 through 5 to label column B1 as "y": highlight cell B1; move to the editing bar and type "= y" [Fig. 15.23(b)], then click the mouse once. At this point, the spreadsheet should appear as shown in Fig. 15.23(c).

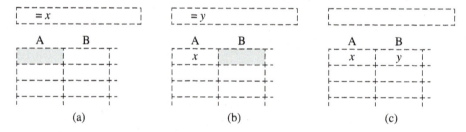

(a)   (b)   (c)

FIGURE 15.23.   To label a column, use the mouse to highlight the top cell in that column, then move the cursor to the editing bar and type in the column heading using the computer keyboard.

Our next step is to establish the set of x values in column A. Instead of typing in each value, we can follow these steps:

7.   Move the cursor to cell A3, highlight the cell, type an initial x value of -1, and click the mouse once.
8.   Move the cursor to cell A4, highlight the cell, and then move to the editing bar. Type "= A3 + 1" and click the mouse. This step will generate our second x value (and all subsequent x values) by adding 1 to the preceding x value. (If x values at intervals of 0.5 or 2 are desired, the typed entries in the editing bar should now read "= A3 + 0.5" or "= A3 + 2," respectively.)
9.   Filling in the remaining cells in the x column can generally be accomplished by highlighting cell A4, holding down the left mouse button, and using the mouse to "drag" the cursor down column A until the final desired x value has been reached. Depending on the software used, this step may require use of either "fill down" or "copy and paste" options available from *pull-down menus* located at the top on the computer screen. Whatever procedure is required, the machine fills in its remaining x cells according to the directions in step 8; this is shown literally in Fig. 15.24(a), with the resulting numerical values listed in Fig. 15.24(b). Note that the entire set of x values can be changed at any time simply by changing the initial value specified in cell A3 and/or the interval established in cell A4.

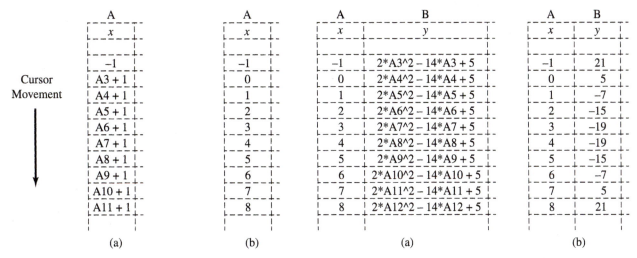

Cursor Movement

A
x
-1
A3 + 1
A4 + 1
A5 + 1
A6 + 1
A7 + 1
A8 + 1
A9 + 1
A10 + 1
A11 + 1

(a)

A
x
-1
0
1
2
3
4
5
6
7
8

(b)

FIGURE 15.24. Once a numerical x value has been installed in cell A3, the value for cell A4 can be specified *literally* in terms of this initial value. If the cursor is now dragged down the column, or the entire column is selected, the computer duplicates this literal relationship of (a) *numerically* to produce the values shown in (b).

A	B
x	y
-1	2*A3^2 – 14*A3 + 5
0	2*A4^2 – 14*A4 + 5
1	2*A5^2 – 14*A5 + 5
2	2*A6^2 – 14*A6 + 5
3	2*A7^2 – 14*A7 + 5
4	2*A8^2 – 14*A8 + 5
5	2*A9^2 – 14*A9 + 5
6	2*A10^2 – 14*A10 + 5
7	2*A11^2 – 14*A11 + 5
8	2*A12^2 – 14*A12 + 5

(a)

A	B
x	y
-1	21
0	5
1	-7
2	-15
3	-19
4	-19
5	-15
6	-7
7	5
8	21

(b)

FIGURE 15.25. (a) To generate y values, first specify the value of cell B3 in terms of the cell address (A3) in which the initial x value is located. (b) Dragging the cursor down column B or selecting the entire column B repeats this literal relationship in the remaining cells to produce a completed table of values.

Finally, to have the computer generate a y value for each given x value, we can follow these steps:

10. Highlight cell B3, then move to the editing bar.

11. Type the given algebraic equation into the editing bar space, replacing variable x by the cell address of the initial x value. A typical representation of this equation might be "= 2*A3*A3 – 14*A3 + 5" or "2*A3^2 – 14*A3 + 5," where the (*) symbol indicates multiplication and the caret (^) denotes exponentiation.

12. Now highlight cell B3, hold down the left mouse button, and drag the cursor down column B. As the cursor passes over each cell, it highlights the cell and duplicates the pattern established in step 11. This process, depicted literally in Fig. 15.25(a), yields the completed table of values listed in Fig. 15.25(b).

From the previous example, you should note that all cell entries on a spreadsheet may be defined in terms of a single initial value. This means that if an entire data set is constructed around that value, then changing (a) the initial value, (b) the incremental value, or (c) the functional relationship between variables will cause the software to *automatically* generate a complete new data set corresponding to these specified conditions. For that reason, it is relatively simple to investigate the effects caused by any changes in the physical constants of an equation or formula and to locate specific points of interest on a graph. This becomes a powerful tool in many technical problems that would otherwise be difficult to analyze or solve.

Example 15.17   A nominal 2 × 8 board used as a floor joist rests on supports that are 192 in. apart. The wood has a modulus of elasticity of 1,760,000 psi. If a concentrated load of 500 lb is applied to this beam at a point 48 in. from the left-hand support,

a. Use a spreadsheet to construct a table of values listing the deflections produced at 6-in. intervals along the entire beam length.

b. Graph your results from part (a).

c. Find the distance, to the nearest half-inch, from the left-hand end to the point of maximum deflection.

Solution: a. The beam is shown in Fig. 15.26. From Table A.14(e) of the Appendix, we see that two formulas are required to compute the deflections (shown as y_1 and y_2 in Fig. 15.26) along the beam:

For $0 \leq x \leq a$,

$$y_1 = \frac{Pbx}{6LEI}(L^2 - x^2 - b^2)$$

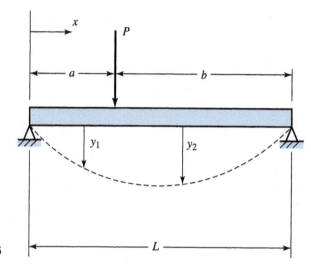

FIGURE 15.26

For $a \le x \le L$,

$$y_2 = \frac{Pb}{6LEI}\left[\frac{L}{b}(x-a)^3 + (L^2 - b^2)x - x^3\right]$$

For this problem,

$P = 500$ lb	$L = 192$ in.
$E = 1{,}760{,}000$ lb/in.^{2}	$a = 48$ in.
$I = 47.6$ in.^{4}	$b = 144$ in.

Substituting these values into our two formulas and simplifying yields the following two algebraic equations.

For $0 \le x \le 48$ in.,

$$y_1 = -0.0000007456x^3 + 0.01202x$$

For 48 in. $\le x \le 192$ in.,

$$y_2 = 0.0000002485x^3 - 0.0001432x^2 + 0.01890x - 0.1099$$

where x, y_1, and y_2 are all in inches.

As shown in Fig. 15.27, only three steps are required to construct a spreadsheet that yields the data in Table (A). Inspection of the table indicates that, because of round-off in the physical

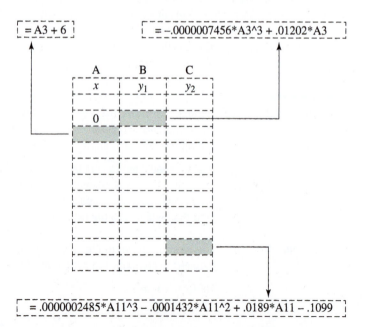

FIGURE 15.27. Construction of
a spreadsheet for Example 15.17

TABLE (A)   Example 15.17

x	y_1	y_2
0	0	
6	0.07196	
12	0.14295	
18	0.21201	
24	0.27817	
30	0.34047	
36	0.39793	
42	0.44960	
48	0.49450	0.49485
54		0.53226
60		0.56226
66		0.58516
72		0.60130
78		0.61100
84		0.61457
90		0.61234
96		0.60463
102		0.59176
108		0.57405
114		0.55184
120		0.52543
126		0.49515
132		0.46133
138		0.42428
144		0.38432
150		0.34179
156		0.29699
162		0.25026
168		0.20192
174		0.15228
180		0.10167
186		0.05041
192		0.00117

constants, the deflections are accurate to only three significant digits. (For example, at $x = 48$ in., deflections y_1 and y_2 should be equal, and at the right-hand support [$x = 192$ in.], the deflection should be exactly zero.)

b. Using the spreadsheet software's graphing option, you should be able to obtain results similar to Fig. 15.28. The graph shown there was plotted in two sections, as denoted by the legend to the right of the graph. Series 1 represents deflection y_1, while Series 2 corresponds to y_2.

c. From the results of part (a), it appears that the maximum deflection occurs near $x = 84$ in. To locate this point more accurately, we may simply replace the x value of 78 in cell A15 with the value 83 and specify the entry for cell A16 as "= A15 + 0.25." Dragging the cursor down column B (or selecting the entire column B) then yields the set of values shown in Table (B). From the data, maximum deflection occurs somewhere between $x = 84$ in. and $x = 85$ in.; to the required accuracy, then, the maximum deflection of this beam occurs at

$$x = 84.5 \text{ in.} \pm 0.5 \text{ in.}$$ **Answer**

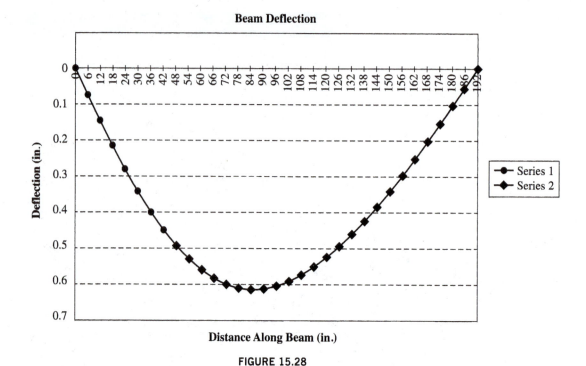

FIGURE 15.28

x	y₂
TABLE (B)	Example 15.17

x	y_2
83	0.61438
83.25	0.61445
83.5	0.61450
83.75	0.61454
84	0.61457
84.25	0.61459
84.5	0.61460
84.75	0.61460
85	0.61459

PROBLEMS

In Probs. 15.45, through 15.47, use a spreadsheet to locate the point of maximum deflection for:

15.45 The same beam and load as Example 15.17 but with a concentrated load of 800 lb added at the beam's midpoint.

15.46 The same beam and load as Example 15.17 but with an additional uniformly distributed load of 80 lb/ft along the entire beam length.

15.47 The same loading as Prob. 15.18 for a W6 × 9 steel member. (Neglect the beam's own weight.)

15.8 STATICALLY INDETERMINATE BEAMS BY THE SUPERPOSITION METHOD

Statically indeterminate beams, beams for which there are more reactions than can be determined by the equations of equilibrium, can be solved by the superposition method. The following examples will illustrate the method.

Example 15.18 Determine the reactions for the statically indeterminate beam shown in Fig. 15.29(a).

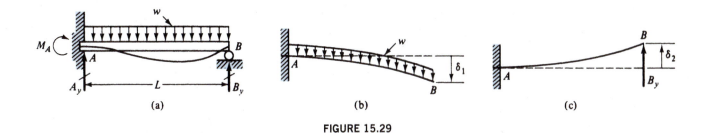

FIGURE 15.29

Solution: The loads on the statically indeterminate beam can be thought of as involving the uniform load w on a cantilever beam and the concentrated reaction or load B_y on the end of an identical cantilever beam [Fig. 15.29(b) and (c)]. The downward deflection δ_1 at B due to the uniform load must be equal to the upward deflection δ_2 at B due to the concentrated load. That is,

$$\delta_1 = \delta_2$$

From diagrams (b) and (c) of Table A.14 of the Appendix,

$$\frac{wL^4}{8EI} = \frac{B_y L^3}{3EI}$$

or
$$B_y = \frac{3}{8}wL \qquad \textbf{Answer}$$

With one reaction known, the other two can be found from the equilibrium equations as follows:

$$\Sigma M_A = 0 \quad -M_A - wL\left(\frac{L}{2}\right) + B_y L = 0$$

$$M_A = -\frac{wL^2}{2} + \frac{3wL^2}{8} = -\frac{wL^2}{8} \qquad \textbf{Answer}$$

$$\Sigma F_y = 0 \quad A_y - wL + B_y = 0$$

$$A_y = \frac{5}{8}wL \qquad \textbf{Answer}$$

Check:

$$\Sigma M_B = 0 \quad -M_A + wL\frac{L}{2} - A_y L = 0$$

$$\frac{wL^2}{8} + \frac{wL^2}{2} - \frac{5wL^2}{8} = 0 \quad \text{OK}$$

Example 15.19 Determine the reaction for the statically indeterminate beam shown in Fig. 15.30(a)

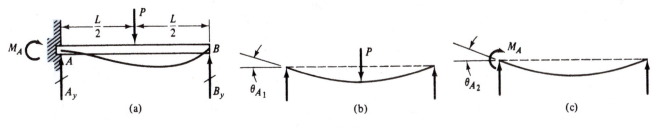

FIGURE 15.30

Solution: The loads can be thought of as made up of a concentrated load at the middle of a simply supported beam and a couple on the left end of an identical simply supported beam, as shown in Fig. 15.30(b) and (c). The slope must be zero at A; therefore,

$$\theta_{A_1} + \theta_{A_2} = 0$$

From diagrams (f) and (h) of Table A.14 of the Appendix,

$$\frac{PL^2}{16EI} + \frac{M_A L}{3EI} = 0$$

$$M_A = -\frac{3}{16}PL \qquad \text{**Answer**}$$

With one of the reactions known, the other two can be found from the equilibrium equations as follows:

$$\Sigma M_A = 0 \qquad -M_A - P\frac{L}{2} + B_y L = 0$$

$$B_y = -\frac{3}{16}P + \frac{P}{2} = \frac{5P}{16} \qquad \text{**Answer**}$$

$$\Sigma M_B = 0 \qquad -M_A - A_y L + P\frac{L}{2} = 0$$

$$A_y = \frac{3}{16}P + \frac{P}{2} = \frac{11P}{16} \qquad \text{**Answer**}$$

Check:

$$\Sigma F_y = 0 \qquad A_y - P + B_y = 0$$

$$\frac{5}{16}P - P + \frac{11}{16}P = 0 \qquad \text{OK}$$

An alternative solution, consisting of two cantilever beams that are fixed at A with the deflection at B made equal to zero, as in Example 15.18, is also possible.

Example 15.20   Determine the reaction for the statically indeterminate beam shown in Fig. 15.31(a).

Solution: The beam fixed on both ends has a zero slope at A and B. By adding the three simple beams shown in Fig. 15.31(b)–(d) and requiring that

$$\theta_{A_1} + \theta_{A_2} + \theta_{A_3} = 0 \qquad \text{(a)}$$

and

$$\theta_{B_1} + \theta_{B_2} + \theta_{B_3} = 0 \qquad \text{(b)}$$

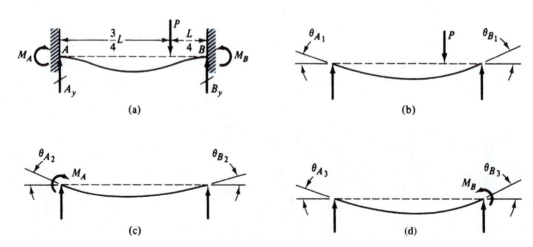

FIGURE 15.31

we can determine the value of the reactions M_A and M_B. From diagram (e) of Table A.14 of the Appendix with $a = 3L/4$ and $b = L/4$,

$$\theta_{A_1} = \frac{5PL^2}{128EI} \quad \text{and} \quad \theta_{B_1} = \frac{7PL^2}{128EI}$$

The other angles can be found directly from diagram (h) of Table A.14 of the Appendix. Substituting values of θ_A into Eq. (a),

$$\frac{5PL^2}{128EI} + \frac{M_AL}{3EI} + \frac{M_BL}{6EI} = 0 \tag{c}$$

and the values of θ_B into Eq. (b), we obtain

$$\frac{7PL^2}{128EI} + \frac{M_AL}{6EI} + \frac{M_BL}{3EI} = 0 \tag{d}$$

Multiplying Eqs. (c) and (d) by $384EI/L$, they reduce to

$$15PL + 128M_A + 64M_B = 0$$
$$21PL + 64M_A + 128M_B = 0$$

Solving for M_B and M_A, we have

$$M_B = -\frac{9}{64}PL \quad \text{and} \quad M_A = -\frac{3}{64}PL \qquad \textbf{Answer}$$

From equilibrium,

$$\Sigma M_A = 0 \qquad -M_A - \frac{3PL}{4} + B_yL + M_B = 0$$
$$B_y = \frac{M_A}{L} + \frac{3PL}{4L} - \frac{M_B}{L}$$

Substituting for M_A and M_B, we obtain

$$B_y = -\frac{3P}{64} + \frac{3P}{4} + \frac{9P}{64} = \frac{54}{64}P \qquad \textbf{Answer}$$

From equilibrium,

$$\Sigma M_B = 0 \qquad -M_A - A_yL + \frac{PL}{4} + M_B = 0$$
$$A_y = -\frac{M_A}{L} + \frac{P}{4} + \frac{M_B}{L}$$

Substituting for M_A and M_B, we obtain

$$A_y = \frac{3P}{64} + \frac{P}{4} - \frac{9P}{64} = \frac{10}{64}P \qquad \textbf{Answer}$$

Check:

$$\Sigma F_y = 0 \qquad A_y - P + B_y = 0$$
$$\frac{10}{64}P - P + \frac{54}{64}P = 0 \quad \text{OK}$$

PROBLEMS

In Probs. 15.48 through 15.65, use the equations of equilibrium and Table A.14 of the Appendix together with superposition to find the reactions for the statically indeterminate beam shown in the figure. Assume EI constant, and neglect the weight of the beam.

15.48 Find the reactions for the beam shown. Use (a) diagrams (b) and (c) of the table, and (b) diagrams (g) and (h) of the table.

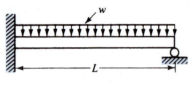

PROB. 15.48

15.49 Find the reactions for Prob. 15.48 if $w = 1.5$ kip/ft and $L = 12$ ft.

15.50 Find the reactions for Prob. 15.48 if $w = 20$kN/m and $L = 4$m.

15.51 Find the reactions for the beam shown if $a = 2L/3$ and $b = L/3$. Use (a) diagrams (a) and (b) of the table, and (b) diagrams (e) and (h) of the table.

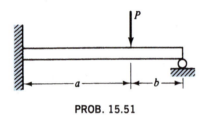

PROB. 15.51

15.52 Find the reactions for Prob. 15.51 if $P = 80$ kN, $a = 2$ m, and $b = 1$ m.

15.53 Find the reactions for Prob. 15.51 if $P = 20$ kip, $a = 12$ ft, and $b = 6$ ft.

15.54 Find the reactions for the beam shown if $a = 3L/4$ and $b = L/4$. Use (a) diagrams (a), (b), and (c) of the table, and (b) diagrams (c), (g), and (h) of the table.

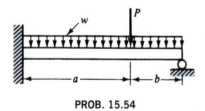

PROB. 15.54

15.55 Find the reactions for Prob. 15.54 if $P = 20$ kip, $w = 2$ kip/ft, $a = 12$ ft, and $b = 4$ ft.

15.56 Find the reactions for Prob. 15.54 if $P = 80$ kN, $w = 32$ kN/m, $a = 3$ m, and $b = 1$ m.

15.57 Find the reactions for the beam shown if $a = 3L/5$ and $b = 2L/5$. Use (a) diagrams (e) and (h) of the table, and (b) diagrams (a), (b), and (d) of the table.

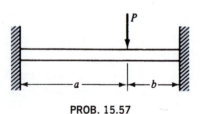

PROB. 15.57

15.58 Find the reactions for Prob. 15.57 if $P = 60$ kN, $a = 1.5$ m, and $b = 1$ m.

15.59 Find the reactions for Prob. 15.57 if $P = 25$ kip, $a = 6$ ft, and $b = 4$ ft.

15.60 Find the reactions for the beam shown. Use (a) diagrams (b), (c), and (d) of the table, and (b) diagrams (g) and (h) of the table.

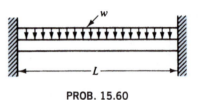

PROB. 15.60

15.61 Find the reactions for Prob. 15.60 if $w = 4$ kip/ft and $L = 12$ ft.

15.62 Find the reactions for Prob. 15.60 if $w = 30$ kN/m and $L = 4$ m.

15.63 Find the reactions for the beam shown if $a = L/4$ and $b = 3L/4$. Use (a) diagrams (e), (g), and (h) of the table, and (b) diagrams (a), (b), (c), and (d) of the table.

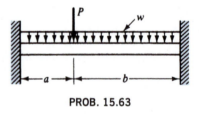

PROB. 15.63

15.64 Find the reactions for Prob. 15.63 if $P = 30$ kip, $w = 2$ kip/ft, $a = 3$ ft, and $b = 9$ ft.

15.65 Find the reactions for Prob. 15.63 if $P = 50$ kN, $w = 15$ kN/m, $a = 1$ m, and $b = 3$ m.

*15.9 DEFLECTION OF BEAMS BY INTEGRATION

In Sec. 15.3, we developed several fundamental relationships for the deflection of beams. They were based on three assumptions: The beam material follows Hooke's law, the deflection and the slope are small at every point of the deflection curve, and the deflection due to shear forces can be neglected.

Returning to Sec. 15.3, we take the limit as Δx approaches zero of the ratio $\Delta\theta/\Delta x$ in Eq. (15.1) and obtain the following equation in differential form.

$$\frac{1}{\rho} = \frac{d\theta}{dx} \tag{15.9}$$

*An asterisk denotes optional material that may be omitted without loss of continuity.

Combining Eqs. (15.9) and (15.2), we write

$$\frac{d\theta}{dx} = \frac{M}{EI} \qquad (15.10)$$

The slope of the deflection curve at every point of the curve is small, and the slope measured in radians can be replaced by the tangent of the slope. In differential form, the tangent of the slope is given by

$$\theta = \frac{d\delta}{dx} \qquad (15.11)$$

Combining Eqs. (15.10) and (15.11), we have

$$\frac{d\theta}{dx} = \frac{d}{dx}\left(\frac{d\delta}{dx}\right) = \frac{d^2\delta}{dx^2} = \frac{M}{EI} \qquad (15.12)$$

Equation (15.12) is known as the differential equation for bending or the Bernoulli-Euler equation to honor its codiscoverers. The equation is integrated once to find the slope equation and twice to find the deflection equation. The process is sometimes called the *double-integration method.* The following examples are used to illustrate the method.

Example 15.21   The uniform cantilever beam shown in Fig. 15.32(a) supports a concentrated load P at its free end. Determine the equation of the deflection curve and the deflection and slope at the free end of the beam.

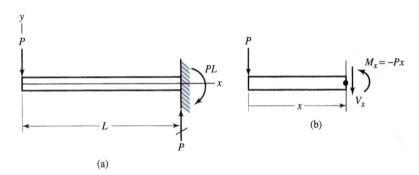

FIGURE 15.32

Solution: Using a part of the beam as a free body [Fig. 15.32(b)], we sum moments about the centroid of a cross section at a distance x from the free end to determine the bending moment.

$$\Sigma M_c = Px + M_x = 0 \quad \text{or} \quad M_x = -Px$$

Substituting for M in Eq. (15.12) and multiplying both sides of the equation by EI, we write

$$EI\frac{d}{dx}\left(\frac{d\delta}{dx}\right) = -Px \qquad (a)$$

Multiplying both sides of the equation by dx and integrating with respect to x, we have

$$EI\int \frac{d}{dx}\left(\frac{d\delta}{dx}\right)dx = -\int Px\,dx$$

or

$$EI\frac{d\delta}{dx} = -\frac{Px^2}{2} + C_1 \qquad (b)$$

To determine the constant C_1, we note that the slope $\theta = d\delta/dx$ is equal to zero at the fixed end $x = L$. Substituting those values into Eq. (b), we obtain the constant of integration.

$$0 = -\frac{PL^2}{2} + C_1 \quad \text{or} \quad C_1 = \frac{PL^2}{2}$$

Substituting the value of C_1 into Eq. (b), we write

or

$$EI\frac{d\delta}{dx} = -\frac{Px^2}{2} + \frac{PL^2}{2} \qquad (c)$$

Again, multiplying both sides of the equation by dx and integrating with respect to x, we have

$$EI\int \frac{d\delta}{dx}dx = -\int \frac{Px^2}{2}dx + \int \frac{PL^2}{2}dx$$

or

$$EI\delta = -\frac{Px^3}{6} + \frac{PL^2x}{2} + C_2 \qquad (d)$$

The constant of integration C_2 is determined from the boundary condition at $x = L$. Substituting $\delta = 0$ at $x = L$ in Eq. (d), we have

$$0 = -\frac{PL^3}{6} + \frac{PL^3}{2} + C_2 \quad \text{or} \quad C_2 = -\frac{PL^3}{3}$$

Substituting C_2 into Eq. (d), we write

$$EI\delta = -\frac{Px^3}{6} + \frac{PL^2x}{2} - \frac{PL^3}{3} \tag{e}$$

Or dividing both sides of the equation by EI and factoring $P/6$ from the right-hand side, we obtain

$$\delta = \frac{P}{6EI}(-x^3 + 3L^2x - 2L^3) \qquad \textbf{Answer}$$

The maximum deflection and maximum slope are at the free end of the beam. Substituting $x = 0$ into Eqs. (c) and (e), we have

$$\delta_{max} = -\frac{PL^3}{3EI} \quad \text{and} \quad \theta_{max} = \frac{d\delta}{dx} = \frac{PL^2}{2EI} \qquad \textbf{Answer}$$

With the load P expressed in kip, the length L in inches, the modulus E in kip/in.^{2}, and the moment of inertia I in in.^{4}, the deflection is in inches and the slope in radians. If the load P is in kN, the length L in mm, the modulus E in kN/mm^2, and the moment of inertia I in mm^4, the deflection is in mm and the slope in radians.

Example 15.22   A uniform simply supported beam supports a uniform load of w force per unit length as shown in Fig. 15.33(a). Determine the equation of the deflection curve and the deflection at the center of the beam.

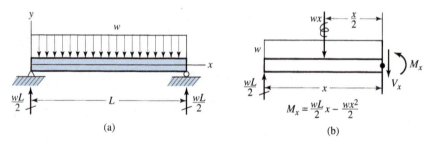

FIGURE 15.33

Solution: Using a part of the beam as a free body [Fig. 15.33(b)], we sum moments about the centroid of the cross section a distance x from the left support to determine the bending moment.

$$\Sigma M_c = -\frac{wL}{2}x + wx\frac{x}{2} + M_x = 0 \quad \text{or} \quad M_x = \frac{wL}{2}x - \frac{wx^2}{2}$$

Substituting for the moment M in Eq. (15.12) and multiplying by EI, we write

$$EI\frac{d^2\delta}{dx^2} = \frac{wL}{2}x - \frac{wx^2}{2} \tag{a}$$

Multiplying by dx and integrating with respect to x, we have

$$EI\frac{d\delta}{dx} = \frac{wLx^2}{4} - \frac{wx^3}{6} + C_1 \tag{b}$$

The deflection is symmetric about the center of the beam; therefore, the slope at the center of the beam where $x = L/2$ is zero. From Eq. (b),

$$0 = \frac{wL^3}{16} - \frac{wL^3}{48} + C_1 \quad \text{or} \quad C_1 = -\frac{wL^3}{24}$$

Substituting the value for C_1 in Eq. (b), we write

$$EI\frac{d\delta}{dx} = \frac{wL}{4}x^2 - \frac{w}{6}x^3 - \frac{wL^3}{24} \qquad \text{(c)}$$

Multiplying Eq. (c) by dx and integrating with respect to x, we have

$$EI\delta = \frac{wLx^3}{12} - \frac{wx^4}{24} - \frac{wL^3x}{24} + C_2 \qquad \text{(d)}$$

At $x = 0$, the deflection $\delta = 0$; therefore, from Eq. (d), $0 = 0 - 0 - 0 + C_2$ or $C_2 = 0$. Substituting $C_2 = 0$ in Eq. (d), we write

$$EI\delta = \frac{wLx^3}{12} - \frac{wx^4}{24} - \frac{wL^3x}{24} \qquad \text{(e)}$$

Dividing both sides of Eq. (e) by EI and factoring $wx/24$ from the right-hand side of the equation, we have

$$\delta = \frac{wx}{24EI}(2Lx^2 - x^3 - L^3) \qquad \textbf{Answer}$$

The maximum deflection is at the center of the beam. From Eq. (e) for $x = L/2$, we have

$$\delta_{max} = -\frac{5wL^4}{384EI} \qquad \textbf{Answer}$$

Example 15.23   Determine the deflection curves for the simply supported uniform beam that supports a concentrated load P as shown in Fig. 15.34(a).

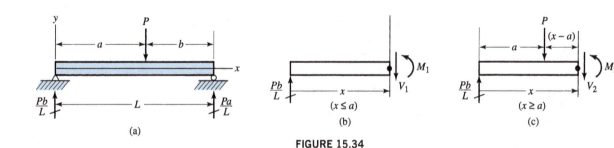

FIGURE 15.34

Solution: The beam is divided at the concentrated load into two sections. Each section has a separate bending-moment equation, slope equation, and deflection equation. Because the beam is continuous, the two sections must join with the same slope and deflection at $x = a$. For the section of the beam [Fig. 15.34(b)] from $x = 0$ to $x = a$, we calculate the bending moment as follows:

$$\Sigma M_c = -\frac{Pb}{L}x + M_1 = 0 \quad \text{or} \quad M_1 = \frac{Pb}{L}x$$

Substituting for M in Eq. (15.12) and multiplying the equation by EI, we obtain

$$EI\frac{d^2\delta_1}{dx^2} = \frac{Pb}{L}x \qquad \text{(a)}$$

Multiplying by dx and integrating Eq. (a) with respect to x, we have

$$EI\frac{d\delta_1}{dx} = \frac{Pbx^2}{2L} + C_1 \qquad \text{(b)}$$

$$EI\delta_1 = \frac{Pbx^3}{6L} + C_1x + C_2 \qquad \text{(c)}$$

For the section of the beam [Fig. 15.34(c)] from $x = a$ to $x = L$, we calculate the bending moment as follows:

$$\Sigma M_c = -\frac{Pb}{L}x + P(x - a) + M_2 = 0$$

or

$$M_2 = \frac{Pb}{L}x - P(x - a)$$

Substituting for M in Eq. (15.12) and multiplying the equation by EI, we obtain

$$EI\frac{d^2\delta_2}{dx^2} = \frac{Pb}{L}x - P(x - a) \tag{d}$$

Multiplying by dx and integrating Eq. (d) with respect to x, we have

$$EI\frac{d\delta_2}{dx} = \frac{Pbx^2}{2L} + \frac{P(x - a)^2}{2} + C_3 \tag{e}$$

$$EI\delta_2 = \frac{Pbx^3}{6L} + \frac{P(x - a)^3}{6} + C_3x + C_4 \tag{f}$$

Because the deflection is continuous, the slopes from Eqs. (b) and (e) are equal at $x = a$. Setting them equal to each other, we write

$$\frac{Pba^2}{2L} + C_1 = \frac{Pba^2}{2L} + \frac{P(a - a)^2}{2} + C_3 \quad \text{or} \quad C_1 = C_3$$

In a similar manner, the deflection from Eq. (c) is equal to the deflection from Eq. (f); therefore,

$$\frac{Pba^3}{6L} + C_1a + C_2 = \frac{Pba^3}{6L} + \frac{P(a - a)^3}{6} + C_3a + C_4$$

The constants C_1 and C_3 are equal; therefore, $C_2 = C_4$.

By not expanding terms of the form $(x - a)^n$ in the moment equations before integrating, equality of the constants, $C_1 = C_3$ and $C_2 = C_4$, is ensured. The term $(x - a)^n$ is always equal to zero at $x = a$, the point where the deflections and slopes join. As a result, the constants have been reduced from four to two.

At the left support, $x = 0$ and $\delta_1 = 0$. From Eq. (c), we have $0 = 0 + 0 + C_2$ or $C_2 = 0$. At the right support, $x = L$ and $\delta_2 = 0$, and from Eq. (f), we get

$$0 = \frac{PbL^3}{6L} - \frac{P(L - a)^3}{6} + C_3L \quad (C_4 = C_2 = 0)$$

or

$$C_3 = \frac{Pb^3}{6L} - \frac{PbL}{6} \quad \text{where } (L - a) = b$$

Substituting the values of the constants into Eqs. (c) and (e), we write

$$EI\delta_1 = \frac{Pbx^3}{6L} + \frac{Pb^3x}{6L} - \frac{PbLx}{6}$$

$$\delta_1 = \frac{Pbx}{6LEI}(x^2 + b^2 - L^2) \qquad \text{for } x \leq a \qquad \textbf{Answer}$$

$$EI\delta_2 = \frac{Pbx^3}{6L} - \frac{P(x - a)^3}{6} + \frac{Pb^3x}{6L} - \frac{PbLx}{6}$$

$$\delta_2 = \frac{Pb}{6LEI}\left[x^3 + (b^2 - L^2)x - \frac{L}{b}(x - a)^3\right] \qquad \text{for } x \geq a \qquad \textbf{Answer}$$

Example 15.24   Determine the deflection equations and the deflection at the middle of the simply supported beam shown in Fig. 15.35(a)

Solution: The beam is divided into three sections. In Fig. 15.35(b)–(d), we draw the free-body diagrams for sections from 0 to 3 ft, 3 to 6 ft, and 6 to 12 ft. From the free-body diagrams and equilibrium equations, we obtain the following bending-moment equations:

$$M_1 = 9.75x$$

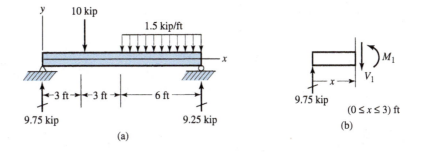

(a)

(b)

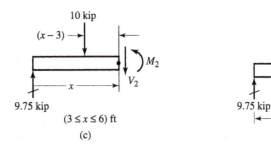

(c)

(d)

FIGURE 15.35

$$M_2 = 9.75x - 10(x - 3)$$

$$M_3 = 9.75x - 10(x - 3) - \frac{1.5(x - 6)^2}{2}$$

Substituting M_1 into Eq. (15.12), we write

$$EI\frac{d^2\delta_1}{dx^2} = 9.75x$$

Multiplying by dx and integrating with respect to x, we have

$$EI\frac{d\delta_1}{dx} = \frac{9.75x^2}{2} + C_1 \tag{a}$$

$$EI\delta_1 = \frac{9.75x^3}{6} + C_1x + C_2 \tag{b}$$

Substituting M_2 into Eq. (15.12), we write

$$EI\frac{d^2\delta_2}{dx^2} = 9.75x - 10(x - 3)$$

Multiplying by dx and integrating with respect to x, we have

$$EI\frac{d\delta_2}{dx} = \frac{9.75x^2}{2} - \frac{10(x - 3)^2}{2} + C_3 \tag{c}$$

$$EI\delta_2 = \frac{9.75x^3}{6} - \frac{10(x - 3)^3}{6} + C_3x + C_4 \tag{d}$$

Substituting M_3 into Eq. (15.12), we write

$$EI\frac{d^2\delta_3}{dx^2} = 9.75x - 10(x - 3) - \frac{1.5(x - 6)^2}{2}$$

Multiplying by dx and integrating with respect to x, we have

$$EI\frac{d\delta_3}{dx} = \frac{9.75x^2}{2} - \frac{10(x - 3)^2}{2} - \frac{1.5(x - 6)^3}{6} + C_5 \tag{e}$$

$$EI\delta_3 = \frac{9.75x^3}{6} - \frac{10(x - 3)^3}{6} - \frac{1.5(x - 6)^4}{24} + C_3x + C_4 \tag{f}$$

The slope is continuous, and Eqs. (a) and (c) are equal at $x = 3$ ft and $C_1 = C_3$. Also, Eqs. (c) and (e) are equal at $x = 6$ ft and $C_3 = C_5$. Therefore,

$$C_1 = C_3 = C_5. \tag{g}$$

Similarly, the deflection is continuous, and Eqs. (b) and (d) are equal at $x = 3$ ft and $C_2 = C_4$. Also, Eqs. (d) and (f) are equal at $x = 6$ ft and $C_4 = C_6$. Therefore,

$$C_2 = C_4 = C_6 \tag{h}$$

By not expanding terms of the form $(x - a)^n$ in the moment equations before integrating, equality of the constants, $C_1 = C_3 = C_5$ and $C_2 = C_4 = C_6$, is ensured. The term $(x - a)^n$ is always equal to zero at the points of the deflection curve where the deflections and slopes join. As a result, the constants are reduced from six to two.

At $x = 0$, the deflection $\delta_1 = 0$. From Eq. (b), $0 = 0 + 0 + C_2$ or $C_2 = 0$, and from Eq. (h), $C_4 = C_6 = 0$. The deflection $\delta_3 = 0$ at $x = 12$ ft, and from Eq. (f), we write

$$0 = \frac{9.75(12)^3}{6} - \frac{10(12-3)^3}{6} - \frac{1.5(12-6)^4}{24} + 12C_5$$

Thus, $C_5 = -126$ kip-ft², and from Eq. (g), $C_1 = C_3 = -126$ kip-ft². Dividing by EI and substituting the value of the constants into Eqs. (b), (d), and (f), we have the following three deflection equations:

$$\delta_1 = \frac{1}{EI}\left[\frac{9.75}{6}x^3 - 126x\right] \qquad \text{for } x \leq 3 \text{ ft}$$

$$\delta_2 = \frac{1}{EI}\left[\frac{9.75}{6}x^3 - \frac{10(x-3)^3}{6} - 126x\right] \qquad \text{for } 3 \leq x \leq 6 \text{ ft} \qquad \textbf{Answer}$$

$$\delta_3 = \frac{1}{EI}\left[\frac{9.75}{6}x^3 - \frac{10(x-3)^3}{6} - \frac{1.5(x-6)^4}{24} - 126x\right] \qquad \text{for } 6 \leq x \leq 12 \text{ ft}$$

At $x = 6$ ft, the middle of the beam, we have the following deflection:

$$\delta_M = \frac{1}{EI}\left[\frac{9.75(6)^3}{6} - \frac{11(6-3)^3}{6} - 126(6)\right] = -\frac{450}{EI} \text{ kip-ft}^3 \qquad \textbf{Answer}$$

---

*15.10 SINGULARITY FUNCTIONS

The integration method provides a convenient method for determining the slope and deflection equations if the moment is a continuous function of x. However, if a separate moment equation is required for each section of a beam with complex loading, the method requires lengthy calculations. To write a single moment equation for a beam with complex loading, we introduce a family of *singularity* or *half-wave functions* designed for that purpose. Although introduced by A. Clebsch in 1862, they were first used to solve beam problems in 1919 by W. H. Macauley.

The functions are shown in Table 15.2. The four types of loads considered—the couple, the concentrated load, the uniform load, and the triangular load—all produce bending moments as described in Table 15.2. The loads are all equal to zero in the interval from $x = 0$ to $x = a$. The bending moment for each load in Table 15.2(a) to (d) is described for the interval from $x = a$ to $x = x$ in the following:

a. *The couple:* the bending moment is a constant $M_x = -M_o$. Notice that the expression shown with the

moment diagram $M_x = -M_o(x - a)^0 = -M_o$ because $(x - a)^0 = 1$.

b. *The concentrated load:* the bending moment is triangular in shape $M_x = -P_o(x - a)^1$.

c. *The uniform load:* the bending moment is parabolic in shape $M_x = -w_o(x - a)^2/2$.

d. *The triangular load:* the bending moment is cubic in shape $M_x = -w_o(x - a)^3/6b$.

The singularity functions for the bending moment must mirror the bending moments previously described. They are listed in Table 15.2 and have the following meaning:

$$\langle x - a\rangle^n = \begin{cases} (x - a)^n & \text{when } x \geq a \\ 0 & \text{when } x < a \end{cases}$$

Thus, we see that the pointed brackets ($\langle \rangle$), sometimes called Macauley brackets, are like ordinary brackets except that they eliminate negative quantities. The usual integration laws apply to the singularity functions as follows:

$$\int \langle x - a\rangle^n dx = \frac{\langle x - a\rangle^{n+1}}{n + 1} + C$$

The following examples illustrate the application of singularity functions for the determination of deflection equations.

---

*An asterisk denotes optional material that may be omitted without loss of continuity.

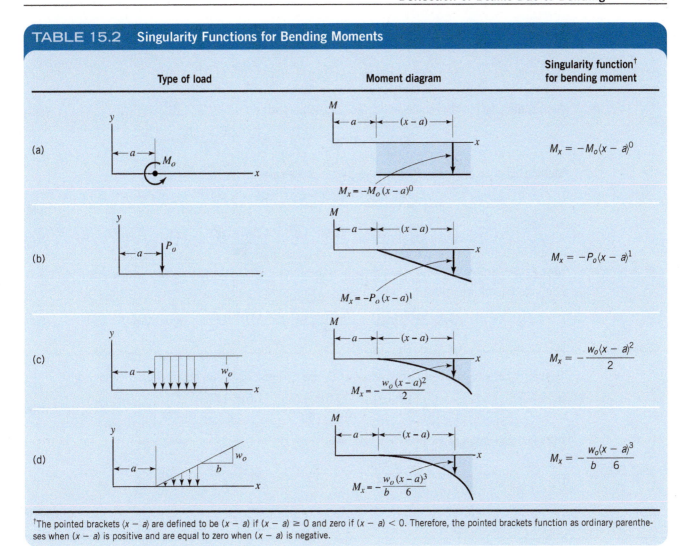

TABLE 15.2 Singularity Functions for Bending Moments

Type of load	Moment diagram	Singularity function[†] for bending moment
(a)	$M_x = -M_o(x-a)^0$	$M_x = -M_o\langle x-a\rangle^0$
(b)	$M_x = -P_o(x-a)^1$	$M_x = -P_o\langle x-a\rangle^1$
(c)	$M_x = -\dfrac{w_o(x-a)^2}{2}$	$M_x = -\dfrac{w_o\langle x-a\rangle^2}{2}$
(d)	$M_x = -\dfrac{w_o}{b}\dfrac{(x-a)^3}{6}$	$M_x = -\dfrac{w_o}{b}\dfrac{\langle x-a\rangle^3}{6}$

[†]The pointed brackets $\langle x-a\rangle$ are defined to be $(x-a)$ if $(x-a) \geq 0$ and zero if $(x-a) < 0$. Therefore, the pointed brackets function as ordinary parentheses when $(x-a)$ is positive and are equal to zero when $(x-a)$ is negative.

Example 15.25   Determine the deflection equation and the deflection at the middle of the simply supported beam shown in Fig. 15.36.

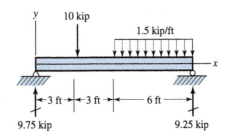

FIGURE 15.36

Solution: There are three sections of the beam, and three singularity functions from Table 15.2 are used to write a single moment equation for the beam.

The reaction $R = 9.75$ kip is *directed up* and produces a positive moment. From (b) of the table, the singularity function $M_x = P_o\langle x-a\rangle^1 = 9.75\langle x\rangle^1$. However, the term $\langle x\rangle$ is replaced by x because the value of x is always positive. The load $P = 10$ kip at $x = a = 3$ ft is *directed down* and produces a negative moment. The singularity function for the moment $M_x = -P_o\langle x-a\rangle^1 = -10\langle x-3\rangle^1$. The uniform load $w = 1.5$ kip/ft starts at $x = a = 6$ ft and is *directed down*, producing a negative moment. From (c) of the table, the singularity function for the moment $M_x = -w_o\langle x - a\rangle^2/2 = -1.5\langle x - 6\rangle^2/2$. For the bending-moment equation, we have

$$M_x = 9.75x - 10\langle x - 3\rangle^1 - \frac{1.5\langle x - 6\rangle^2}{2}$$

Substituting the moment equation into Eq. (15.12), we write

$$EI\frac{d^2\delta}{dx^2} = 9.75x - 10\langle x - 3\rangle^1 - \frac{1.5\langle x - 6\rangle^2}{2}$$

Multiplying by dx and integrating with respect to x, we have

$$EI\frac{d\delta}{dx} = \frac{9.75x^2}{2} - \frac{10\langle x - 3\rangle^2}{2} - \frac{1.5\langle x - 6\rangle^3}{6} + C_1 \tag{a}$$

$$EI\delta = \frac{9.75x^3}{6} - \frac{10\langle x - 3\rangle^3}{6} - \frac{1.5\langle x - 6\rangle^4}{24} + C_1x + C_2 \tag{b}$$

When $x = 0, \delta = 0$; therefore, from Eq. (b), $0 = 0 - 0 - 0 + 0 + C_2$ or $C_2 = 0$. When $x = 12$ ft, $\delta = 0$, and from Eq. (b), we write

$$0 = \frac{9.75(12)^3}{6} - \frac{10(12 - 3)^3}{6} - \frac{1.5(12 - 6)^4}{24} + 12C_1 + 0$$

or $\qquad\qquad C_1 = -126$ kip-ft^2

Dividing by EI and substituting the values of C_1 and C_2 into Eq. (b), we obtain the deflection equation.

$$\delta = \frac{1}{EI}\left[\frac{9.75x^3}{6} - \frac{10\langle x - 3\rangle^3}{6} - \frac{1.5\langle x - 6\rangle^4}{24} - 126x\right] \qquad \textbf{Answer}$$

At the middle of the beam at $x = 6$ ft, we have

$$\delta_M = \frac{1}{EI}\left[\frac{9.75(6)^3}{6} - \frac{10(6 - 3)^3}{6} - \frac{1.5(6 - 6)^4}{24} - 126(6)\right]$$

$$\delta_M = -\frac{450}{EI} \text{ kip-ft}^3 \qquad \textbf{Answer}$$

Example 15.26 Determine the deflection equation and the deflection at the middle of the simply supported beam shown in Fig. 15.37(a).

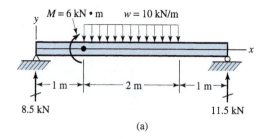

(a)

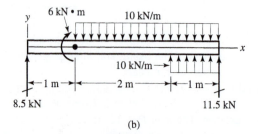

(b)

FIGURE 15.37

Solution: We replace the partial downward load of 10 kN/m from $x = 1$ to $x = 3$ m by a downward load of 10 kN/m from $x = 1$ to $x = 4$ m and an upward load of 10 kN/m from $x = 3$ to $x = 4$ m. The replacement loads added together give the original load. The loads and reactions are shown in Fig. 15.37(b).

We use three singularity functions from Table 15.2. From (b) of the table, the singularity function for the moment of the reaction $R = 8.5$ kN is $M_x = P_o\langle x - a\rangle^1 = 8.5\langle x\rangle^1 = 8.5x$, and for the couple $M = 6$ kN·m from (a) of the table is $M_x = 6\langle x - a\rangle^0 = 6\langle x - 1\rangle^0$. The singularity functions for the bending moment of the uniform loads at $x = 1$ m and 3 m of -10kN/m and $+10$ kN/m from (c) of the table are $M_x = -w_0\langle x - a\rangle^2/2 = -10\langle x - 1\rangle^2/2$ and $+10\langle x - 3\rangle^2/2$. Therefore, for the bending-moment equation in terms of singularity functions, we have

$$M_x = 8.5x + 6\langle x - 1\rangle^0 - \frac{10\langle x - 1\rangle^2}{2} + \frac{10\langle x - 3\rangle^3}{2}$$

Substituting the moment equation into Eq. (15.12), we write

$$EI\frac{d^2\delta}{dx^2} = 8.5x + 6\langle x - 1\rangle^0 - \frac{10\langle x - 1\rangle^2}{2} + \frac{10\langle x - 3\rangle^2}{2}$$

Multiplying by dx and integrating with respect to x, we have

$$EI\frac{d\delta}{dx} = \frac{8.5x^2}{2} + \frac{6\langle x - 1\rangle^1}{1} - \frac{10\langle x - 1\rangle^3}{6} + \frac{10\langle x - 3\rangle^3}{6} + C_1 \qquad \text{(a)}$$

$$EI\delta = \frac{8.5x^3}{6} + \frac{6\langle x - 1\rangle^2}{2} - \frac{10\langle x - 1\rangle^4}{24} + \frac{10\langle x - 3\rangle^4}{24} + C_1 x + C_2 \qquad \text{(b)}$$

When $x = 0, \delta = 0$; therefore, from Eq. (b), $0 = 0 - 0 - 0 + 0 + C_2$ or $C_2 = 0$. When $x = 4$ m, $\delta = 0$, and from Eq. (b), we write

$$0 = \frac{8.5(4)^3}{6} + \frac{6(4 - 1)^2}{2} - \frac{10(4 - 1)^4}{24} - \frac{10(4 - 3)^4}{24} + 4C_1 + 0$$

or $C_1 = -21.08$ kN·m^2

Dividing by EI and substituting the values of C_1 and C_2 into Eq. (b), we have the following deflection equation.

$$\delta = \frac{1}{EI}\left[\frac{8.5x^3}{6} + \frac{6\langle x - 1\rangle^2}{2} - \frac{10\langle x - 1\rangle^4}{24} + \frac{10\langle x - 3\rangle^4}{24} - 21.08x\right] \qquad \textbf{Answer}$$

The maximum deflection occurs where the slope of the deflection curve is zero. The slope equation is a cubic equation and can be solved most easily by trial and error. A trial-and-error solution to find the location of a zero slope yields a value of $x = 1.986$ m.[†] The deflection at the middle of the beam at $x = 2$ m differs from the maximum deflection by less than 0.01 percent. The deflection at the middle of the beam is as follows:

$$\delta_M = \frac{1}{EI}\left[\frac{8.5(2)^3}{6} + \frac{6(2 - 1)^2}{2} - \frac{10(2 - 1)^3}{24} - \frac{10(2 - 2)^3}{24} - 21.08(2)\right]$$

or $$\delta_M = -\frac{28.2}{EI} \text{ kN·m}^3 \qquad \textbf{Answer}$$

[†]A trial-and-error solution is illustrated in Example 15.27.

Example 15.27   For the structural steel beam with an overhang as shown in Fig. 15.38(a), determine (a) the deflection equation, (b) the deflection at the free end A of the beam, (c) the maximum deflection at some point D between B and C, and (d) the minimum required moment of inertia I if the maximum allowable deflection is 1/2000 of the 9-ft span or 0.054 in. (Use a modulus of elasticity for steel $E = 29 \times 10^3$ ksi.)

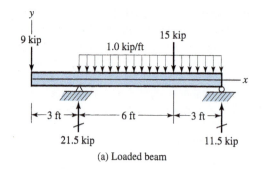

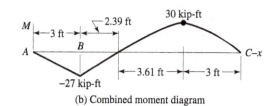

(a) Loaded beam

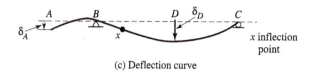

(b) Combined moment diagram

(c) Deflection curve

FIGURE 15.38

Solution: There are four sections of the beam, and four singularity functions are required from Table 15.2 to write a single bending-moment equation. From (b) of the table, the singularity function for a concentrated load or reaction $M_x = \pm P_o \langle x - a \rangle^1$. For the 9-kip load, we have $M_x = -9 \langle x \rangle^1 = -9x$, for the 21.5-kip reaction $M_x = 21.5 \langle x - 3 \rangle^1$, and for the 15-kip load $M_x = -15 \langle x - 9 \rangle^1$. From (c) of the table, the singularity function for the 1-kip/ft uniform load is $M_x = -w_o \langle x - a \rangle^2/2 = -1 \langle x - 3 \rangle^2/2$. Therefore, we have the following bending-moment equation in terms of singularity functions.

$$M_x = -9x + 21.5 \langle x - 3 \rangle^1 - \frac{\langle x - 3 \rangle^2}{2} - 15 \langle x - 9 \rangle^1$$

Substituting the moment equation into Eq. (15.12), we write

$$EI\frac{d^2\delta}{dx^2} = -9x + 21.5 \langle x - 3 \rangle^1 - \frac{\langle x - 3 \rangle^2}{2} - 15 \langle x - 9 \rangle^1$$

Multiplying by dx and integrating with respect to x, we have

$$EI\frac{d\delta}{dx} = -\frac{9x^2}{2} + \frac{21.5 \langle x - 3 \rangle^2}{2} - \frac{\langle x - 3 \rangle^3}{6} - \frac{15 \langle x - 9 \rangle^2}{2} + C_1 \qquad \text{(a)}$$

$$EI\delta = -\frac{9x^3}{6} + \frac{21.5 \langle x - 3 \rangle^3}{6} - \frac{\langle x - 3 \rangle^4}{24} - \frac{15 \langle x - 9 \rangle^3}{6} + C_1 x + C_2 \qquad \text{(b)}$$

When $x = 3$ ft, $\delta = 0$; therefore, from Eq. (b),

$$0 = -9(3)^3/6 + 0 - 0 - 0 + 3C_1 + C_2$$

or $\qquad\qquad 3C_1 + C_2 = 40.5 \qquad\qquad\qquad\qquad\qquad\qquad\qquad\qquad\qquad \text{(c)}$

When $x = 12$ ft, $\delta = 0$; therefore, from Eq. (b),

$$0 = -\frac{9(12)^3}{6} + \frac{21.5(12 - 3)^3}{6} - \frac{(12 - 3)^4}{24} - \frac{15(12 - 9)^3}{6} + 12C_1 + C_2$$

or $\qquad\qquad\qquad 12C_1 + C_2 = 320.625 \qquad\qquad\qquad\qquad\qquad\qquad\qquad \text{(d)}$

TABLE (A)	Example 15.27	
No.	Trial x (ft)	R (kip-ft^2) EI
1	8.3	−1.725
2	8.4	+0.832
3	8.36	−0.200
4	8.37	+0.057
5	8.367	−0.021
6	8.368	+0.006
7	8.3677	−0.003
8	8.3678	+0.001

Solving Eqs. (c) and (d) by elimination, we obtain

$$C_1 = +31.125 \text{ kip-ft}^2; \quad C_2 = -52.875 \text{ kip-ft}^3$$

At $x = 0$, the deflection

$$EI\delta_A = -52.875 \quad \text{or} \quad \delta_A = -\frac{52.9}{EI} \text{ kip-ft}^3 \qquad \textbf{Answer}$$

To find the maximum deflection for the span from B to C, we first determine the location of a zero slope from the slope equation by trial and error. Setting the slope equation equal to R and simplifying the constants, we have

$$R = -4.2x^2 + 10.75\langle x - 3\rangle^2 - \frac{\langle x - 3\rangle^3}{6} - 7.5\langle x - 9\rangle^2 + 31.125$$

From an inspection of the moment diagram in Fig. 15.38(b) and the resulting deflection curve in Fig. 15.38(c), we let $x = 8.3$ ft in the equation for R as a first trial. We then proceed as shown in Table (A). The slope is negative for trial no. 1 and positive for trial no. 2. Therefore, the slope is zero between $x = 8.3$ and 8.4 ft. We continue as in the table until we obtain the desired accuracy. Each new pair of trials, no. 3 and no. 4, etc., bracket the zero slope. Rounding off trial no. 8 to three decimal places, we have $x = 8.368$ ft. From the deflection equation for $x = 8.368$ ft, we get

$$EI\delta_D = -\frac{9(8.368)^3}{6} + \frac{21.5(8.368 - 3)^3}{6} - \frac{(8.368 - 3)^4}{24} + 31.125(8.368) - 52.875$$

$$\delta_D = -\frac{151.7}{EI} \text{ kip-ft}^3 \qquad \textbf{Answer}$$

There is a second point of zero slope for a value of x less than $x = 3$ ft. Setting the slope equation equal to zero and simplifying the constants, we write

$$0 = -4.5x^2 + 31.125$$
$$x^2 = 6.917 \quad \text{or} \quad x = 2.630 \text{ ft}$$

Substituting $x = 2.630$ ft into the deflection equation, we have

$$EI\delta = -1.5(2.630)^3 + 31.125(2.630) - 52.875$$

$$\delta = +\frac{1.697}{EI} \text{ kip-ft}^3$$

The largest deflection occurs at D, and the maximum allowable deflection is 0.054 in.

    Changing ft^3 to in.^{3} by multiplying by $(12 \text{ in./ft})^3 = 1728 \text{ in.}^3/\text{ft}^3$ and solving for the moment of inertia I from the equation for δ_D, we get

$$I = \frac{151.7(1728) \text{ kip-in.}^3}{E\delta_D}$$

The product $E\delta_D = 29 \times 10^3 \text{ kip/in.}^2(0.054 \text{ in.}) = 29 \times 10^3(0.054) \text{ kip/in.}$; therefore,

$$I = \frac{151.7(1728)\text{ kip-in.}^3}{29 \times 10^3(0.054)\text{ kip/in.}} = 167.4 \text{ in.}^4 \qquad \textbf{Answer}$$

From Table A.3 of the Appendix, we select a W14 × 22 beam with a moment of inertia $I = 199$ in.^{4} and a section modulus $S = 29.0$ in.^{3} From Fig. 15.38(b), the maximum moment for the beam is 30 kip-ft at a distance of 3 ft from C. For the bending stress, we have

$$\sigma = \frac{M}{S} = \frac{30 \text{ kip-ft}(12 \text{ in./ft})}{29.0 \text{ in.}^3} = 12.41 \text{ ksi} < 24 \text{ ksi} \qquad \text{OK}$$

The allowable stress for A36 structural steel is 24 ksi; therefore, the W14 × 22 beam is also safe in bending.

PROBLEMS

In Probs. 15.66 through 15.68, use integration to determine (a) the deflection equation, and (b) the slope at the free end.

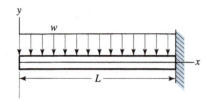

PROB. 15.66

PROB. 15.67

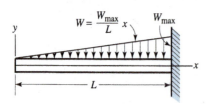

PROB. 15.68

In Probs. 15.69 through 15.71, use integration to determine (a) the deflection equation, and (b) the deflection at the middle of the beam.

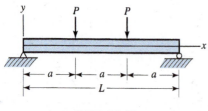

PROB. 15.69

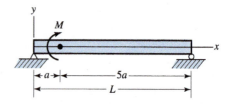

PROB. 15.70

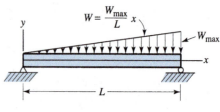

PROB. 15.71

15.72 The overhanging steel beam is loaded by a single concentrated load as shown. Find (a) the deflection equation, (b) the deflection at the free end A, (c) the maximum deflection at some interior point between B and C, and (d) the minimum required moment of inertia I if the maximum allowable deflection is limited to $1/360$ of the 4-m span. The modulus for steel $E = 200 \times 10^6 \text{ kN/m}^2$. (*Hint:* The maximum deflection between B and C is directed upward and occurs at a distance from C of 2.309 m.)

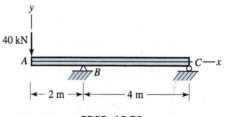

PROB. 15.72

15.73 The beam with an overhang is loaded by a uniform load on the overhanging end from A to B and a concentrated load in the middle of the span BC as shown. Determine (a) the deflection equation, and (b) the deflection at the free end A of the beam.

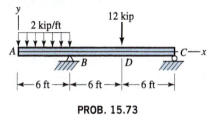

PROB. 15.73

COMBINED STRESSES AND MOHR'S CIRCLE

TABLE 16.1   Basic Formulas

Internal Reactions	Stress	Equation No.
1. Axial force (P)	$\sigma = \dfrac{P}{A}$	(10.3)
2. Bending moment (M)	$\sigma = \dfrac{My}{I}$	(14.5)
	$\sigma_{max} = \dfrac{Mc}{I}$	(14.4)
3. Torsion (T) (For circular bar)	$\tau = \dfrac{Tr}{J}$	(12.10)
(For circular bar)	$\tau_{max} = \dfrac{Tc}{J}$	(12.11)
4. Shear force (V)	$\tau = \dfrac{VQ}{Ib}$	(14.8)
(For rectangular bar)	$\tau_{max} = \dfrac{3V}{2A}$	(14.9)

16.1   INTRODUCTION

The basic formulas for calculating the stress in a member were derived in previous chapters. In deriving the formulas, it was assumed that a single internal reaction was acting at a cross section of the member and that the member was made from material for which stress was proportional to strain. The basic formulas are listed for reference in Table 16.1.

In this chapter, we consider problems in which two or more internal reactions act at a cross section of the member. The stress from each internal reaction will be superimposed or added together to find the resulting stresses. The use of the superposition method is valid if the presence of one internal reaction does not affect the stresses due to another. Only problems where superposition is valid will be considered in the following sections.

16.2   AXIAL FORCES AND BENDING MOMENTS

Axial forces and bending moments [Fig. 16.1(a)] produce normal stresses on the transverse cross section of a bar. These stresses are

$$\sigma_1 = \pm \frac{P}{A} \quad \text{and} \quad \sigma_2 = \pm \frac{My}{I}$$

The stresses are both perpendicular to the cross section and can be added algebraically. Therefore, we write

$$\sigma = \sigma_1 + \sigma_2 = \pm \frac{P}{A} \pm \frac{My}{I} \tag{16.1}$$

The stresses produced by a tensile force and a positive moment are shown in Fig. 16.1(b). The stresses at the top T and bottom B of the beam are given by

$$\sigma_T = \frac{P}{A} - \frac{Mc_T}{I} \quad \text{and} \quad \sigma_B = \frac{P}{A} + \frac{Mc_B}{I}$$

The positive sign indicates tension, and the negative sign indicates compression. The combined neutral axis has been shifted to point N above the centroid of the cross section. The maximum stress in this case occurs at the bottom of the bar.

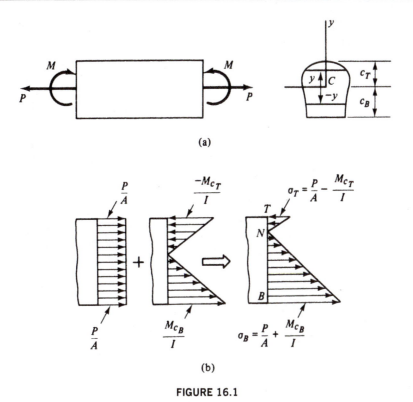

(a)

(b)

FIGURE 16.1

Example 16.1   A rectangular bar with a cross section 2 in. by 8 in. is subjected to an axial tensile force of 20 kip and a bending moment of 60 kip-in., as shown in Fig. 16.2(a). Determine the stress distribution across the bar.

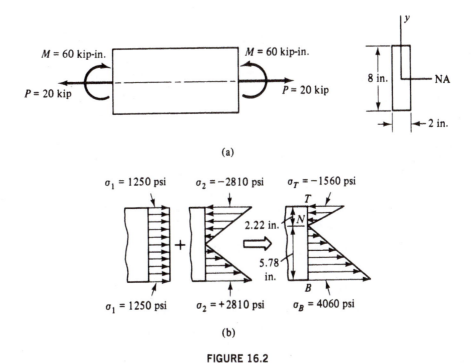

(a)

(b)

FIGURE 16.2

Solution: The cross-sectional area of the bar $A = bh = 2(8) = 16$ in.^{2} and the moment of inertia with respect to the bending neutral axis $I = bh^3/12 = 2(8)^3/12 = 85.33$ in.^{4} The axial stress $\sigma_1 = P/A = 20,000/16 = 1250$ psi and the maximum bending stress $\sigma_2 = Mc/I = 60,000(4)/85.3 = 2810$ psi. Because the bending moment is positive, the stress at the top of the bar $\sigma = \sigma_1 - \sigma_2 = 1250 - 2810 = -1560$ psi and at the bottom of the bar $\sigma = \sigma_1 + \sigma_2 = 1250 + 2810 = 4060$ psi. The combined neutral axis is located above the centroid of the cross section, where the axial stress and bending stress are equal. That is,

$$\frac{P}{A} = \frac{My}{I}$$

$$y = \frac{PI}{AM} = \frac{20,000(85.3)}{16(60,000)} = 1.78 \text{ in.}$$

Thus, the neutral axis is $4.0 - 1.78 = 2.22$ in. from the top of the bar. The distribution of stresses and location of the combined neutral axis are shown in Fig. 16.2(b).

Example 16.2   A simply supported I-beam with span, cross section, and loads is shown in Fig. 16.3(a) and (b). Find the stress distribution in the beam at the cross section of maximum tensile and compressive stress.

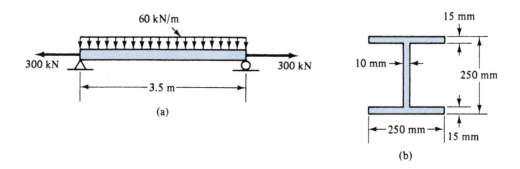

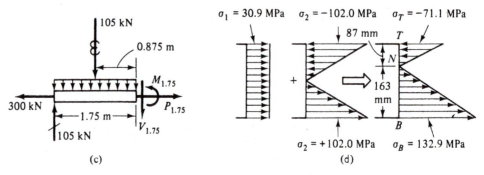

FIGURE 16.3

Solution: The I-shaped cross section can be thought of as made up of a positive rectangle with base $b = 0.250$ m and height $h = 0.250$ m and two negative rectangles with bases $b = 0.120$ m and heights $h = 0.220$ m. Thus, the area $A = 0.250(0.250) - 2(0.120)(0.220) = 9.7 \times 10^{-3}$ m^2 and the moment of inertia with respect to the bending neutral axis,

$$I_{NA} = \frac{0.250(0.250)^3}{12} - \frac{2(0.120)(0.220)^3}{12} = 112.6 \times 10^{-6} \text{ m}^4$$

The axial stress

$$\sigma_1 = \frac{P}{A} = \frac{300}{9.7 \times 10^{-3}} = 30.9 \times 10^3 \text{ kN/m}^2$$

$$= 30.9 \text{ MN/m}^2 (\text{MPa})$$

To find the bending stress, we must determine the maximum bending moment. The maximum bending moment acts on a cross section at the middle of the beam. From the free-body diagram shown in Fig. 16.3(c), we write

$$M_{1.75} = 105(1.75) - 105(0.875)$$

$$= 91.9 \text{ kM·m}$$

Therefore, the bending stress

$$\sigma_2 = \frac{Mc}{I} = \frac{91.9(0.125)}{112.6 \times 10^{-6}} = 102.0 \times 10^3 \, \text{kN/m}^2$$

$$= 102.0 \, \text{MN/m}^2 \, (\text{MPa})$$

Because the bending moment is positive, the stress at the top of the beam $\sigma = \sigma_1 - \sigma_2 = 30.9 - 102.0 = -71.1$ MPa and at the bottom of the beam $\sigma = \sigma_1 + \sigma_2 = 30.9 + 102.0 = 132.9$ MPa. The combined neutral axis is located above the centroid of the cross section, where the axial stress and bending stress are equal. That is,

$$\frac{P}{A} = \frac{My}{I}$$

$$y = \frac{PI}{AM} = \frac{300(112 \times 10^{-6})}{(9.7 \times 10^{-3})(91.9)} = 0.0377 \, \text{m}$$

Thus, the neutral axis is $125 - 38 = 87$ mm from the top of the beam. The distribution of stresses and location of the combined neutral axis are shown in Fig. 16.3(d).

Example 16.3   Find the normal stress distribution on section A–A of the short post with an eccentric load of $P = 15$ kip and cross section as shown in Fig. 16.4(a) and (b).

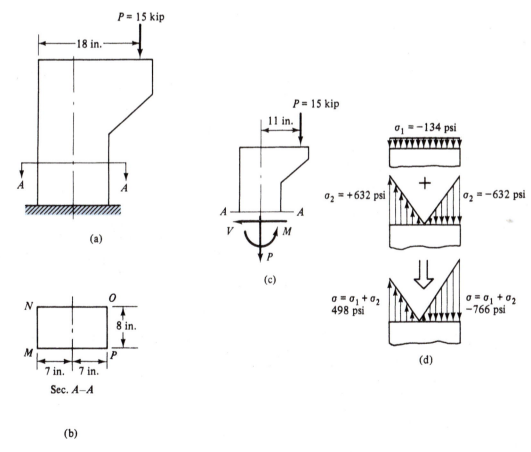

FIGURE 16.4

Solution: The cross-sectional area $A = 8(14) = 112 \, \text{in.}^2$ and the moment of inertia $I_{NA} = 8(14)^3/12 = 1829 \, \text{in.}^4$ The internal reactions at section A–A are found from the free-body diagram [Fig. 16.4(c)] and the equations of equilibrium. The shear force $V = 0$, the axial force $P = -15$ kip, and the bending moment $M = 15(11) = 165$ kip-in. The axial stress

$$\sigma_1 = \frac{P}{A} = \frac{-15{,}000}{112} = -134 \text{ psi} \quad \text{(compression)}$$

The maximum bending stresses

$$\sigma_2 = \pm\frac{Mc}{I} = \pm\frac{165{,}000(7)}{1829} = \pm 632 \text{ psi} \quad \text{(tension or compression)}$$

The axial and bending stresses are added together [Fig. 16.4(d)] to produce a tensile stress $\sigma = -134 + 632 = 498$ psi on side MN, and a compressive stress $\sigma = -134 - 632 = -766$ psi on side OP of the cross section.

Example 16.4   The C-clamp shown in Fig. 16.5(a) is tightened until a force $P = 2.5$ kN is exerted on an object in the clamp. Find the stress distribution in the clamp at the cross section of maximum tensile and compressive stress.

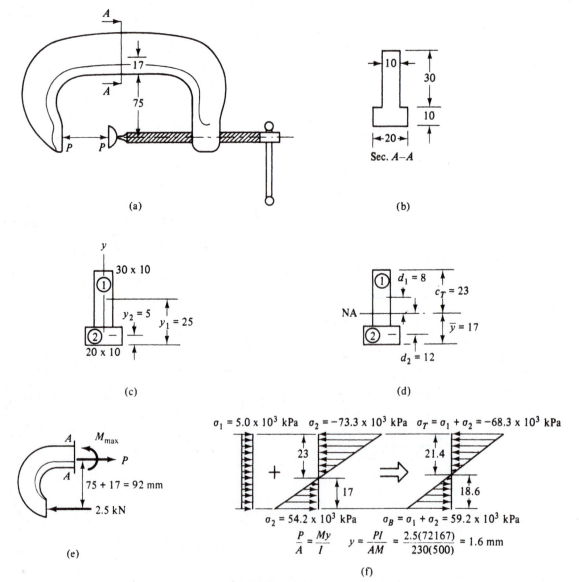

(a)

(b)

(c)

(d)

(e)

(f)

(All dimensions in mm)

FIGURE 16.5

Solution: The maximum bending moment will occur on Section *A–A*. To find the maximum bending moment, we begin by locating the neutral axis of the cross section.

Location of Neutral Axis: The T-shaped cross section [Fig. 16.5(b)] can be divided into area ①, the stem of the T, and area ②, the flange of the T, as shown in Fig. 16.5(c). The calculations for the location of the centroid are tabulated in Table (A). From the sums in the table,

$$\bar{y} = \frac{\Sigma Ay}{\Sigma A} = \frac{8500}{500} = 17 \text{ mm}$$

TABLE (A)   Example 16.4

	b	h	A	y	Ay
①	10	30	300	25	7500
②	20	10	200	5	1000
			$\Sigma A = 500 \text{ mm}^2$		$\Sigma Ay = 8500 \text{ mm}^3$

Moment of Inertia: The moment of inertia with respect to the neutral axis for each area is found from the formula for the centroid moment of inertia, $I_c = bh^3/12$, and the parallel-axis theorem $I_{NA} = I_c + Ad^2$. The composite moment of inertia with respect to the neutral axis is the sum of the individual moments of inertia. The transfer distances are shown in Fig. 16.5(d) and are tabulated in Table (B). From the sums in the table,

$$I_{NA} = \Sigma I_c + \Sigma(Ad^2) = 24.2 \times 10^3 + 48.0 \times 10^3 = 72.2 \times 10^3 \text{ mm}^4$$

TABLE (B)   Example 16.4

	A	d	I_c	Ad^2
①	300	8	22.5×10^3	19.2×10^3
②	200	−12	1.67×10^3	28.8×10^3
			$\Sigma I_c = 24.17 \times 10^3 \text{ mm}^4$	$\Sigma(Ad^2) = 48.0 \times 10^3 \text{ mm}^4$

Internal Reactions: From the free-body diagram of Fig. 16.5(e) and the equations of equilibrium, we write $\Sigma F_x = -2.5 + P = 0$ or $P = 2.5$ kN (tension) and

$$\Sigma M_c = -2.5(92) + M_{max} = 0 \quad \text{or} \quad M_{max} = 230 \text{ kN·mm}$$

Axial and Bending Stresses: The axial stress

$$\sigma_1 = \frac{P}{A} = \frac{2.5}{500} = 0.005 \frac{\text{kN}}{\text{mm}^2}$$

or

$$\sigma_1 = 5.0 \times 10^3 \frac{\text{kN}}{\text{m}^2} \text{(kPa)} \quad \text{(tension)}$$

The bending stress at the top of the cross section

$$\sigma_2 = \frac{-Mc_T}{I} = \frac{-230(23)}{72.2 \times 10^3} = -73.3 \times 10^{-3} \frac{\text{kN}}{\text{mm}^2}$$

or

$$\sigma_2 = -73.3 \times 10^3 \frac{\text{kN}}{\text{m}^2} \text{(kPa)} \quad \text{(compression)}$$

The bending stress at the bottom of the cross section

$$\sigma_2 = \frac{Mc_B}{I} = \frac{230(17)}{72.2 \times 10^3} = 54.2 \times 10^{-3} \text{ kN/mm}^2$$

or

$$\sigma_2 = 54.2 \times 10^3 \frac{\text{kN}}{\text{m}^2} \text{(kPa)} \quad \text{(tension)}$$

Stress Distribution: The stress distribution is shown in Fig. 16.5(f), where the stress at the top of the cross section

$$\sigma_T = \sigma_1 - \sigma_2 = (5.0 - 73.3) \times 10^3 \text{ kN/m}^2$$

or

$$\sigma_T = -68.3 \times 10^3 \text{ kPa} \quad \text{(compression)}$$

and the stress at the bottom of the cross section

$$\sigma_B = \sigma_1 + \sigma_2 = (5.0 + 54.2) \times 10^3 \text{ kN/m}^2$$

or

$$\sigma_B = 59.2 \times 10^3 \text{ kPa} \quad \text{(tension)}$$

PROBLEMS

16.1 A steel bracket is loaded as shown. (a) Determine the maximum tensile and compressive stress, and (b) plot the distribution of normal stress along A–B.

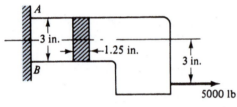

PROB. 16.1

16.2 A C-shaped machine part has dimensions and loads as shown in the figure. For cross section A–A, (a) calculate the maximum tensile and compressive stress, and (b) plot the distribution of normal stress.

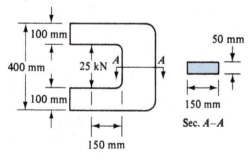

PROB. 16.2

16.3 A bracket has dimensions and is loaded as shown. (a) Determine the maximum tensile and compressive stress, and (b) plot the distribution of normal stress along A–B.

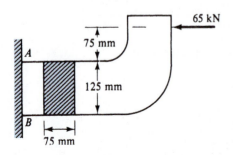

PROB. 16.3

16.4 A horizontal beam consisting of two 7 × 9.8 channels back to back is loaded and supported as shown in the figure. Calculate the maximum tensile and compressive stress on a cross section (a) to the right of B and (b) to the left of B. See Table A.5 in the Appendix for properties of the C7 × 9.8.

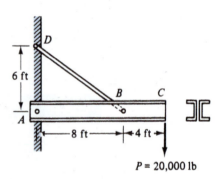

PROB. 16.4

16.5 An L-shaped bracket is supported and loaded as shown. For cross section B–B, (a) determine the maximum tensile and compressive stress, and (b) plot the distribution of normal stress.

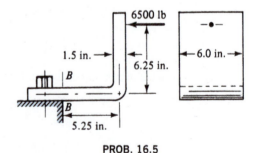

PROB. 16.5

16.6 A davit supports a load $P = 48 \text{ kN}$, as shown in the figure. For cross section A–A, (a) calculate the maximum tensile and compressive stress, and (b) plot the distribution of normal stress.

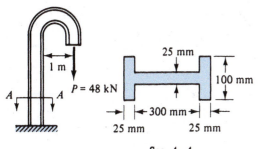

Sec. A–A

PROB. 16.6

16.7 The C-clamp shown in the figure is tightened until the force $P = 5500$ N. For cross section A–A, (a) determine the maximum tensile and compressive stress, and (b) plot the distribution of normal stress.

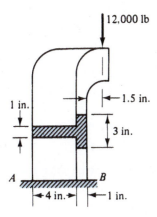

PROB. 16.7

16.8 A machine member has dimensions and a load as shown. For cross section A–B, (a) determine the maximum tensile and compressive stress, and (b) plot the distribution of normal stress.

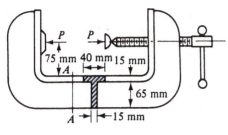

PROB. 16.8

16.9 Two forces act on the open link of a chain as shown in the figure. For cross section A–A, (a) calculate the maximum tensile and compressive stress, and (b) plot the distribution of normal stress.

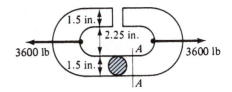

PROB. 16.9

16.10 A 75-mm-diameter round bar is formed into a machine part with a shape as shown. For cross section B–B, (a) determine the maximum tensile and compressive stress, and (b) plot the distribution of normal stress.

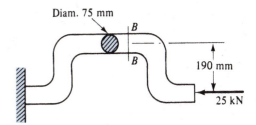

PROB. 16.10

16.11 A machined member has an axial load applied as shown in the figure. For cross section A–A, (a) determine the maximum tensile and compressive stress, and (b) plot the distribution of normal stress.

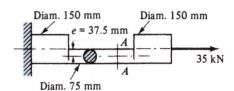

PROB. 16.11

16.12 An offset link has dimensions and loads as shown. For cross section B–B, (a) find the maximum tensile and compressive stress, and (b) plot the distribution of normal stress.

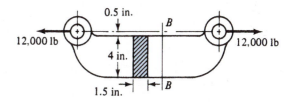

PROB. 16.12

16.13 Solve Prob. 16.12 with cross section B–B modified as shown in the figure.

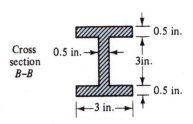

PROB. 16.13

16.14 A 2-in.-diameter bar is bent and loaded as shown. Determine the maximum tensile and compressive stress on a cross section at support A.

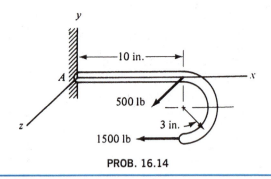

PROB. 16.14

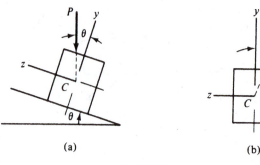

(a)   (b)

FIGURE 16.6

16.3 UNSYMMETRIC BENDING

Problems arise in which the axis of symmetry of the cross section of a beam is inclined at an angle with the vertical [Fig. 16.6(a)] or the load acts at an angle with the axis of symmetry [Fig. 16.6(b)]. These problems can be solved by general bending stress formulas. However, such formulas are complicated and can be avoided by the method of superposition.

In the superposition method, the load is resolved into components that are parallel and perpendicular to the y or symmetric axis. The component of the load parallel to the y axis produces bending about the z axis, M_z. The bending stresses produced are proportional to the distance from the z axis, y, and inversely proportional to the centroidal moment of inertia of the cross section with respect to the z axis. That is,

$$\sigma_1 = \pm \frac{M_z y}{I_z}$$

The component of the load perpendicular to the y axis produces bending about the y axis, M_y. The bending stresses produced are proportional to the distance from the y axis, z, and inversely proportional to the centroidal moment of inertia of the cross section with respect to the y axis. That is,

$$\sigma_2 = \pm \frac{M_y z}{I_y}$$

By superposition, the stress at any point of the cross section is equal to the sum of the stresses; that is,

$$\sigma = \sigma_1 + \sigma_2 = \pm \frac{M_z y}{I_z} \pm \frac{M_y z}{I_y} \qquad (16.2)$$

The following example will illustrate the method.

Example 16.5 A simply supported wooden beam with a cross section 4 in. by 6 in. and a span of 10 ft is used to support a uniformly distributed load of 0.1 kip/ft [Fig. 16.7(a)]. The beam cross section is inclined at an angle of 30° with the vertical plane [Fig. 16.7(b)]. Find the maximum bending stress and the neutral axis.

Solution: The maximum moment occurs at the middle of the beam and can be found from the free-body diagram [Fig. 16.7(c)] and the rotational equation of equilibrium. That is,

$$\Sigma M_c = -w\left(\frac{L}{2}\right)\left(\frac{L}{2}\right) + w\left(\frac{L}{2}\right)\left(\frac{L}{4}\right) + M_{max} = 0$$

or $M_{max} = wL^2/8$. From Fig. 16.7(d), we see that the components of the load parallel and perpendicular to the y axis are

$$w_y = w\cos\theta = 0.1 \cos 30° = 0.0866 \text{ kip/ft}$$
$$w_z = w\sin\theta = 0.1 \sin 30° = 0.0500 \text{ kip/ft}$$

Bending About the z Axis: The maximum bending moment and moment of inertia about the z axis are given by

$$M_z = \frac{w_y L^2}{8} = \frac{0.0866(10)^2}{8} = 1.082 \text{ kip-ft}$$

$$I_z = \frac{bh^3}{12} = \frac{4(6)^3}{12} = 72 \text{ in.}^4$$

Therefore, the maximum bending stress

$$\sigma_1 = \pm \frac{M_z y_{max}}{I_z} = \pm \frac{1082(12)(3)}{72} = \pm 541 \text{ psi}$$

At A and B, the stress is compression, and at C and D, the stress is tension, as shown in Fig. 16.7(e).

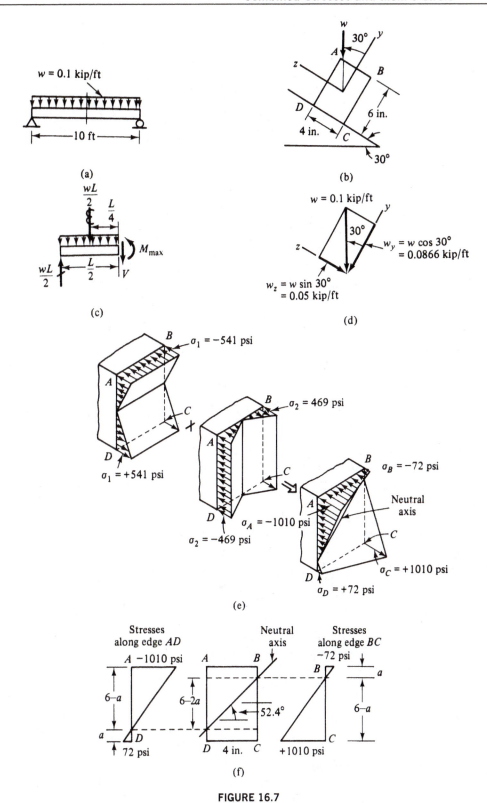

FIGURE 16.7

Bending Moment About the y Axis: The maximum bending moment and moment of inertia about the y axis are given by

$$M_y = \frac{w_z L^2}{8} = \frac{0.05(10)^2}{8} \quad \text{or} \quad M_y = 0.625 \text{ kip-ft}$$

$$I_y = \frac{bh^3}{12} = \frac{6(4)^3}{12} = 32 \text{ in.}^4$$

Therefore, the bending stress

$$\sigma_2 = \pm \frac{M_y z_{max}}{I_y} = \pm \frac{625(12)(2)}{32} = \pm 469 \text{ psi}$$

At A and D, the stress is compression, and at B and C, the stress is tension, as shown in Fig. 16.7(e).

Combined Stresses: From superposition, the stresses at A, B, C, and D are given by

$$\sigma_A = -541 - 469 = -1010 \text{ psi}$$
$$\sigma_B = -541 + 469 = -72 \text{ psi}$$
$$\sigma_C = +541 + 469 = 1010 \text{ psi}$$
and
$$\sigma_D = +541 - 469 = +72 \text{ psi}$$

The combined stresses are shown in Fig. 16.7(e).

Neutral Axis: The neutral axis can be located by finding the points of zero stress on the sides AD and BC of the beam. The stress distribution on the two sides of the beam is shown in Fig. 16.7(f). From similar triangles, we write $a/72 = (6 - a)/1010$ or $a = 0.399$ in. The slope of the neutral axis $\theta = \arctan(6 - 2a)/4 = 52.4°$, as shown in Fig. 16.7(f).

PROBLEMS

16.15 A horizontal W18 × 55 rolled steel beam 15 ft long supports a uniform load of 2 kip/ft. The cross section of the beam is inclined at a slope of 15°. Determine the maximum tensile and compressive stress in the beam if the ends are simply supported. See Table A.3 in the Appendix for properties of steel beams.

16.16 A horizontal simply supported W460 × 82 steel beam 4.5 m long supports a uniform load of 25 kN/m. The cross section of the beam is inclined at a slope of 20°. Determine the maximum tensile and compressive stress in the beam. See Table A.3 (SI Units) in the Appendix for properties of steel beams.

16.17 A simply supported W8 × 35 rolled steel beam 15 ft long supports a concentrated load at midspan. The load is applied through the centroid of the cross section of the beam. If the cross section of the beam is inclined at a slope of 15° and the allowable bending stress is 24 ksi, what load can the beam support? Neglect the weight of the beam. See Table A.3 in the Appendix for properties of steel beams.

16.18 A W200 × 52 rolled steel cantilever beam 1.5 m long supports a concentrated load that is applied at its free end. If the cross section of the beam is inclined at a slope of 15° and the allowable bending stress is 165 MPa, what load can the beam support? Neglect the weight of the beam. See Table A.3 (SI Units) in the Appendix for properties of steel beams.

16.4 ECCENTRICALLY LOADED MEMBERS

The eccentrically loaded member (Fig. 16.8) is a special case where the load produces both axial and bending stresses. The normal stress at any point on a cross section $ABCD$ of the member is found by superposition. Adding the axial stress and the bending stresses about both the y and z axes, we write

$$\sigma_x = \pm \frac{P}{A} \pm \frac{M_y z}{I_y} \pm \frac{M_z y}{I_z} \qquad \text{(a)}$$

The moments of the load P about the y and z axes are $M_y = Pe_z$ and $M_z = Pe_y$. Substituting the moments into Eq. (a), we have

$$\sigma_x = \pm \frac{P}{A} \pm \frac{Pe_z z}{I_y} \pm \frac{Pe_y y}{I_z} \qquad \text{(b)}$$

Notice that the distribution of stress is the same for any cross section of the member.

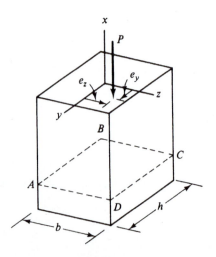

FIGURE 16.8

Example 16.6   A concrete pier (Fig. 16.8) with cross section $b = 24$ in. by $h = 20$ in. supports an axial compressive load $P = 120$ kip. The load is located at $e_y = 6$ in. and $e_z = 5$ in. on top of the member. Determine the normal stresses at the four corners of the pier.

Solution: **Axial Load:** The axial stress is compressive and is given by

$$\sigma_1 = -\frac{P}{A} = \frac{-120}{480} = -0.25 \text{ ksi}$$

Bending About the y Axis: The moment of inertia about the y axis is given by $I_y = hb^3/12 = 20(24)^3/12 = 23{,}040$ in.[4] Therefore, the maximum bending stress

$$\sigma_2 = \pm \frac{Pe_z z_{max}}{I_y} = \pm \frac{120(5)12}{23{,}040}$$

$$= \pm 0.3125 \text{ ksi}$$

The stresses at D and C are compressive and at A and B tensile.

Bending About the z Axis: The moment of inertia about the z axis is given by $I_z = bh^3/12 = 24(20)^3/12 = 16{,}000$ in.[4] Therefore, the maximum bending stress

$$\sigma_3 = \pm \frac{Pe_y y_{max}}{I_z} = \pm \frac{120(6)(10)}{16{,}000}$$

$$= \pm 0.45 \text{ ksi}$$

The stresses at A and D are compressive and at B and C tensile.

Combined Stresses: From superposition, the stresses at A, B, C, and D are added to give

$\sigma_A = -0.25 + 0.3125 - 0.45 = -0.3875$ ksi	**Answer**
$\sigma_B = -0.25 + 0.3125 + 0.45 = +0.5125$ ksi	**Answer**
$\sigma_C = -0.25 - 0.3125 + 0.45 = -0.1125$ ksi	**Answer**
$\sigma_D = -0.25 - 0.3125 - 0.45 = -1.0125$ ksi	**Answer**

PROBLEMS

16.19 A concrete pier 0.55 m × 0.6 m supports a load of 440 kN at point E as shown. Determine the stresses at the four corners of the pier and locate the neutral axis if point E is located (a) distances $e_y = 100$ mm and $e_z = 0$, and (b) distances $e_y = 100$ mm and $e_z = 125$ mm from the axes of symmetry of the cross section. (*Hint:* The neutral axis can be found by locating points of zero stress on two sides of the pier.)

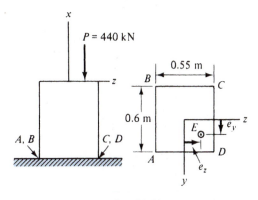

PROB. 16.19

16.20 A concrete pier 20 in. × 24 in. supports a load of 100 kip at point E as shown. Determine the stresses at the four corners of the pier and locate the neutral axis if point E is located at (a) distances $e_y = 0$ and $e_z = 4$ in., and (b) distances $e_y = 6$ in. and $e_z = 4$ in. from the axes of symmetry for the cross section. (*Hint:* The neutral axis can be found by locating points of zero stress on two sides of the pier.)

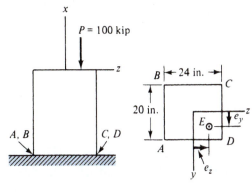

PROB. 16.20

16.21 A W10 × 112 rolled steel section is used as a short compression member to support a load of $P = 125$ kip.

Plates are welded to each end of the member. The load is applied at distances $e_y = 2$ in. and $e_z = 0.7$ in. from the axes of symmetry as shown. Determine the stresses at corners A, B, C, and D of the flanges.

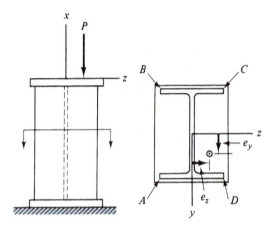

PROB. 16.21 and PROB. 16.22

16.22 A W250 $\times$ 167 rolled steel section is used as a short compression member to support a load $P = 550$ kN. Plates are welded to each end of the member. The load is applied at distances of $e_y = 50$ mm and $e_z = 15$ mm from the axes of symmetry as shown. Determine the stresses at corners A, B, C, and D of the flanges.

16.5 PLANE STRESS

An element of volume that has been cut from a body is shown in Fig. 16.9(a). The edges are parallel to the x, y, and z axes. The sides will be identified in terms of the coordinate axis normal to the side. Thus, side $bcgf$ is called an x plane, and side $aefb$ is called a y plane.

Figure 16.9(b) shows the stress components acting on the same element in a state of plane stress. The normal stresses are indicated by σ_x and σ_y. (The subscript denotes the plane on which the stress acts.) They may be caused by direct tension or compression or bending, or any combination of these. The shear stresses are indicated by τ_{xy} and τ_{yx}. (The first subscript denotes the plane on which the stress acts, and the second subscript denotes the direction of the stress.) Recall that the shear stress on perpendicular planes must be equal (Sec. 12.2). Therefore, $\tau_{xy} = \tau_{yx}$. The shear stresses may be caused by direct shear or torsion, or a combination of the two. The stresses are all positive for the element shown in Fig. 16.9(b). Because many problems involve a state of plane stress, we will consider plane stress in some detail.

16.6 STRESS COMPONENTS ON AN OBLIQUE PLANE

To find the stress components on a plane that is oblique with the x and y plane and normal to the z plane, we cut a section through the element of volume along the desired plane, as shown in Fig. 16.10(a). Stresses on the oblique

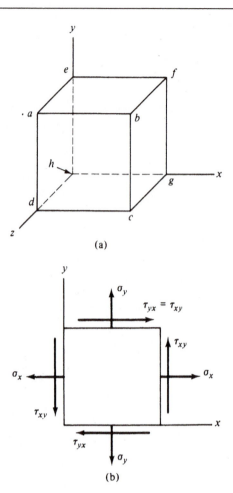

FIGURE 16.9

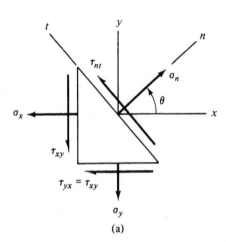

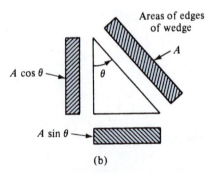

FIGURE 16.10

plane are indicated by σ_n and τ_{nt}, where n and t indicate the normal and tangential axes. The normal axis forms an angle θ with the x axis and the tangential axis forms an angle θ with the y axis.

To draw the free-body diagram of the wedge, each stress must be multiplied by the area on which the stress acts to change it to force. Areas of the edges of the wedge are shown in Fig. 16.10(b). The resulting free-body diagram is shown in Fig. 16.11(a). For convenience, the forces are drawn outward from a common point in Fig. 16.11(b). For equilibrium in the normal and tangential directions,

$$\angle\Sigma F_n = 0 \qquad \sigma_n A - \tau_{xy}A\sin\theta\cos\theta - \sigma_x A\cos\theta\cos\theta$$
$$-\tau_{xy}A\cos\theta\sin\theta - \sigma_y A\sin\theta\sin\theta = 0$$
$$\sigma_n = \sigma_x\cos^2\theta + \sigma_y\sin^2\theta + 2\tau_{xy}\sin\theta\cos\theta \quad \textbf{(16.3)}$$

$$\searrow\Sigma F_t = 0 \qquad \tau_{nt}A + \tau_{xy}A\sin\theta\sin\theta + \sigma_x A\cos\theta\sin\theta$$
$$-\tau_{xy}A\cos\theta\cos\theta - \sigma_y A\sin\theta\cos\theta = 0$$
$$\tau_{nt} = (\sigma_y - \sigma_x)\sin\theta\cos\theta + \tau_{xy}(\cos^2\theta - \sin^2\theta) \quad \textbf{(16.4)}$$

Equations (16.3) and (16.4) can be expressed in a more convenient form. By introducing the trigonometric identities

$$\cos^2\theta = \frac{1 + \cos 2\theta}{2} \qquad \sin^2\theta = \frac{1 - \cos 2\theta}{2}$$
$$2\sin\theta\cos\theta = \sin 2\theta$$

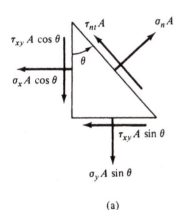

(a)

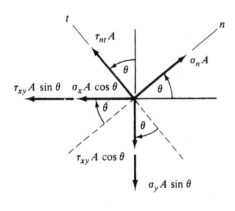

(b)

FIGURE 16.11

we obtain

$$\sigma_n = \frac{\sigma_x + \sigma_y}{2} + \frac{\sigma_x - \sigma_y}{2}\cos 2\theta + \tau_{xy}\sin 2\theta \quad \textbf{(16.5)}$$

$$\tau_{nt} = -\frac{\sigma_x - \sigma_y}{2}\sin 2\theta + \tau_{xy}\cos 2\theta \qquad \textbf{(16.6)}$$

These equations are called *stress transformation equations*. They can be used to find the stresses for any oblique plane. However, we discuss a simpler method based on Mohr's circle in the following section.

16.7 MOHR'S CIRCLE FOR PLANE STRESS

Equations (16.5) and (16.6) can be combined to form the equation of a circle with σ_n and τ_{nt} as the coordinates. To effect the combination, we transpose the term $(\sigma_x + \sigma_y)/2$ in Eq. (16.5) to the left-hand side of the equation and square both sides of the resulting equation. We then square both sides of Eq. (16.6). Adding both squared equations and using the trigonometric identify $\sin^2 2\theta + \cos^2 2\theta = 1$, we have

$$\left(\sigma_m - \frac{\sigma_x + \sigma_y}{2}\right)^2 + \tau_{nt}^2 = \left(\frac{\sigma_x - \sigma_y}{2}\right)^2 + \tau_{xy}^2 \quad \textbf{(16.7)}$$

If we let

$$\sigma_{av} = \frac{\sigma_x + \sigma_y}{2} \quad \text{and} \quad R = \sqrt{\left(\frac{\sigma_x - \sigma_y}{2}\right)^2 + \tau_{xy}^2} \quad \textbf{(16.8)}$$

we obtain

$$(\sigma_n - \sigma_{av})^2 + \tau_{nt}^2 = R^2 \qquad \textbf{(16.9)}$$

the equation of a circle in a plane with σ_n and τ_{nt} as coordinates. The circle is called *Mohr's circle of stress*. In Mohr's circle, normal stresses are plotted along a horizontal axis. Tension is positive and plotted to the right, and compression is negative and plotted to the left. Shear stresses are plotted along a vertical axis. A special sign convention for shear stress is required for construction of Mohr's circle. If the shear stresses tend to rotate the element *clockwise*, they are *positive* and are *plotted up* on the vertical axis. If the shear stresses tend to rotate the element *counterclockwise*, they are *negative* and are *plotted down* on the vertical axis (see Fig. 16.12). The method for constructing Mohr's circle [Fig. 16.13(b)] for the

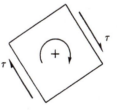

(a) Clockwise: $+\tau$

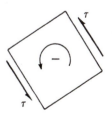

(b) Counterclockwise: $-\tau$

FIGURE 16.12

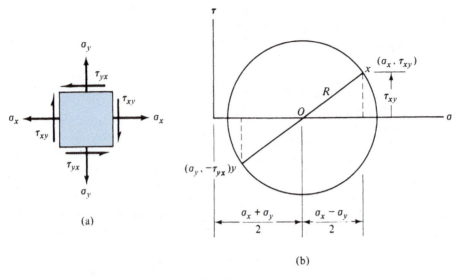

FIGURE 16.13

stresses shown on the element in Fig. 16.13(a) is summarized in the following step-by-step list.

1. On σ and τ axes, plot point x having coordinates $+\sigma_x$ and $+\tau_{xy}$ and point y having coordinates $+\sigma_y$ and $-\tau_{yx}$. (Shear stresses τ_{xy} tend to rotate the element clockwise and are positive, while shear stresses τ_{yx} tend to rotate the element counterclockwise and are negative. Remember that they are always numerically equal.)

2. Draw a line between points x and y. Label the intersection of the line with the σ axis point O. The coordinates of point O are $\sigma = (\sigma_x + \sigma_y)/2$ and $\tau = 0$.

3. With O as the center, draw Mohr's circle through points x and y. The radius of the circle is

$$R = \left[\left(\frac{\sigma_x - \sigma_y}{2} \right)^2 + \tau_{xy}^2 \right]^{1/2}$$

The method for finding stresses on planes of the element that form angles of θ with respect to the x and y axes, as shown in Fig. 16.14(a), are summarized as follows. Rotate the diameter xOy by an angle of 2θ in the *same* direction as for the element [Fig. 16.14(b)]. Notice that we have rotated the diameter xOy of the circle through *twice* the angle that we rotated the element. Coordinate x becomes coordinate n, and coordinate y becomes coordinate t. The coordinates of point n give values of the stresses σ_n and τ_{nt}, and the coordinates of point t give values of the stresses σ_t and $-\tau_{tn}$.

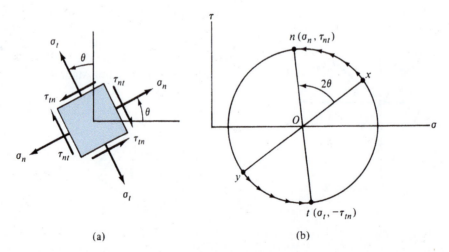

FIGURE 16.14

Example 16.7   The state of stress of an element is shown in Fig. 16.15(a). Use Mohr's circle to find the stresses on planes forming angles of 30° with the x and y axes, as shown in Fig. 16.15(b).

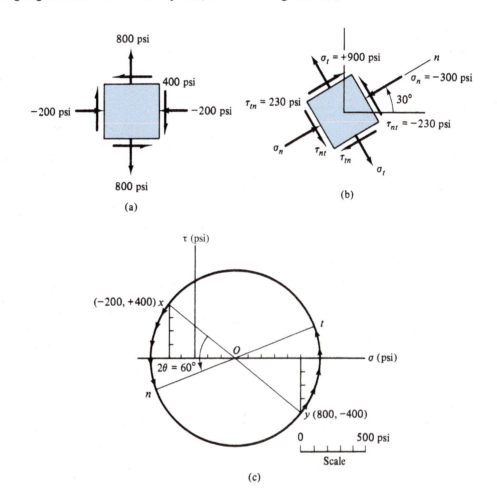

(a)

(b)

(c)

FIGURE 16.15

Solution: The coordinates of point x are $\sigma_x = -200$ psi and $\tau_{xy} = 400$ psi. The coordinates of point y are $\sigma_y = 800$ psi and $\tau_{xy} = -400$ psi. A scale is selected, and Mohr's circle is drawn in Fig. 16.15(c). The diameter xOy is rotated counterclockwise by $2\theta = 60°$. Coordinate x becomes coordinate n, and coordinate y becomes coordinate t. Scaling values from the circle, the coordinates of n and t are

$$\sigma_n = -300 \text{ psi} \qquad \tau_{nt} = -230 \text{ psi} \qquad \textbf{Answer}$$

$$\sigma_t = +900 \text{ psi} \qquad \tau_{tn} = +230 \text{ psi} \qquad \textbf{Answer}$$

16.8 PRINCIPAL STRESSES

On planes where the maximum and minimum normal stresses occur, there are no shearing stresses. These planes are the *principal planes* of stress, and the stresses acting on the principal planes are called the *principal stresses*. From Fig. 16.13, we see that the principal stresses

$$\sigma_{\text{max, min}} = \frac{\sigma_x + \sigma_y}{2} \pm \sqrt{\left(\frac{\sigma_x - \sigma_y}{2}\right)^2 + \tau_{xy}^2} \quad \textbf{(16.10)}$$

where the negative sign is associated with the minimum value.

The angles the principal planes form with the x and y axes are found from the angle the radius Ox forms with the

σ axis in Fig. 16.13. That is,

$$\tan 2\theta = \frac{2\tau_{xy}}{\sigma_x - \sigma_y} \quad \textbf{(16.11)}$$

This equation gives two values of 2θ that are 180° apart and thus two values of θ 90° apart. Accordingly, we establish the direction of the two principal planes (90° apart). It may be necessary to establish the plane on which the stress is a maximum from Eq. (16.5) if the maximum plane is not evident from an inspection of the stresses.

Mohr's circle can also be used to find the principal planes and the principal stresses. The following example will illustrate the method.

Example 16.8   The state of stress of an element is shown in Fig. 16.16(a). Use Mohr's circle to find the principal stresses and the planes of stress. Show your answer on a sketch of the element.

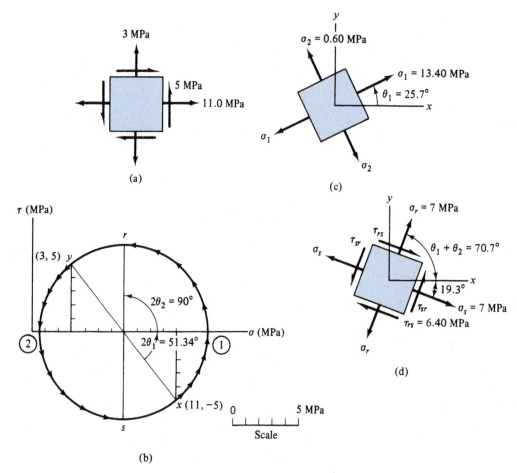

FIGURE 16.16

Solution: The coordinates of point x are $\sigma_x = 11$ MPa and $\tau_{xy} = -5$ MPa. The coordinates of point y are $\sigma_y = 3$ MPa and $\tau_{yx} = 5$ MPa. A scale is selected, and Mohr's circle is drawn in Fig. 16.16(b). The maximum and minimum normal stress must lie along the σ axis. Therefore, the diameter must be rotated counterclockwise by an angle of $2\theta_1$. Coordinate x becomes coordinate ①, and coordinate y becomes ②. The results can be scaled from the circle or determined from the geometry of the circle. The center O of the circle is at

$$\sigma = \frac{\sigma_x + \sigma_y}{2} = \frac{11 + 3}{2} = 7 \text{ MPa}$$

The radius of the circle

$$R = \left[\left(\frac{\sigma_x - \sigma_y}{2} \right)^2 + \tau_{xy}^2 \right]^{1/2} = [(4)^2 + (5)^2]^{1/2}$$

$$= 6.40 \text{ MPa}$$

Therefore, the maximum and minimum stresses are

$$\sigma_{max} = \sigma_1 = 7 + 6.40 = 13.40 \text{ MPa} \qquad \textbf{Answer}$$

$$\sigma_{min} = \sigma_2 = 7 - 6.40 = 0.60 \text{ MPa} \qquad \textbf{Answer}$$

The principal planes are found from

$$\tan 2\theta_1 = \frac{|\tau_{xy}|}{(\sigma_x - \sigma_y)/2} = \frac{5}{4}$$

$$2\theta_1 = 51.34° \qquad \theta_1 = 25.7° \, \measuredangle \qquad \textbf{Answer}$$

The answers are shown in Fig. 16.16(c).

16.9 MAXIMUM SHEAR STRESS

It should be clear from a study of Mohr's circle that the shear stress has a maximum value when the diameter of the circle is in a vertical position. In that position, the magnitude of the shear stress is equal to the radius of the circle, the planes of maximum shear stress form angles of 45° with the planes of principal stress, and the normal stress is equal to the value of σ at the center of the circle.

Example 16.9   Determine the maximum shear stresses and their planes and the normal stresses on the planes for Example 16.8. Show your answers on a sketch of the element.

Solution: In Fig. 16.16(b), we must rotate the horizontal diameter counterclockwise by $2\theta_2 = 90°$ or $\theta_2 = 45°$. Coordinate ① becomes coordinate r, and coordinate ② becomes s. The results can be scaled from the circle or determined from the geometry of the circle.

$$\sigma_r = \sigma_s = \frac{\sigma_x + \sigma_y}{2} = 7 \text{ MPa} \qquad \textbf{Answer}$$

$$\tau_{max_1} = \tau_{rs} = R = 6.40 \text{ MPa} \qquad \textbf{Answer}$$

$$\tau_{max_2} = \tau_{sr} = -R = -6.40 \text{ MPa} \qquad \textbf{Answer}$$

$$\theta = \theta_1 + \theta_2 = 25.7° + 45° = 70.7° \measuredangle \qquad \textbf{Answer}$$

The answers are shown in Fig. 16.16(d).

PROBLEMS

16.23 Sketch Mohr's circle for the state of stress shown in the figures.

16.24 through 16.31   The state of stress is shown in the figure. Use Mohr's circle to find the stresses on planes forming angles of θ with the x and y axes as shown.

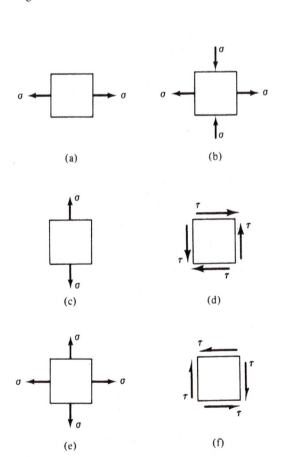

(a)   (b)

(c)   (d)

(e)   (f)

PROB. 16.23

PROB. 16.24

PROB. 16.25

PROB. 16.26

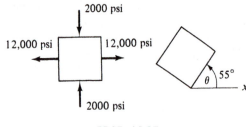

PROB. 16.27

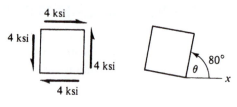

PROB. 16.31

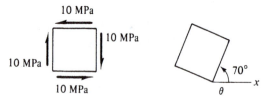

PROB. 16.28

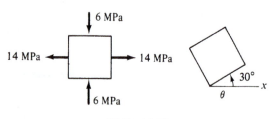

PROB. 16.29

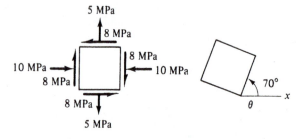

PROB. 16.30

In Probs. 16.32 through 16.39, the state of stress is shown in the figure. Use Mohr's circle to find (a) the principal stresses and planes of principal stress, and (b) the maximum shear stress and the planes of maximum shear stress. Show your results on a sketch.

16.32 Use the figure for Prob. 16.24.

16.33 Use the figure for Prob. 16.25.

16.34 Use the figure for Prob. 16.26.

16.35 Use the figure for Prob. 16.27.

16.36 Use the figure for Prob. 16.28.

16.37 Use the figure for Prob. 16.29.

16.38 Use the figure for Prob. 16.30.

16.39 Use the figure for Prob. 16.31.

16.10 AXIAL STRESS

A condition of axial stress exists when normal stress occurs in one direction only. Such is the case for an axially loaded member, as discussed in Chapter 10. From equilibrium, we determined the normal and shear stresses for an oblique section. The same results can be obtained by Mohr's circle. The following example illustrates the method.

Example 16.10   The state of stress is shown in Fig. 16.17(a). Use Mohr's circle to find the shear stress and normal stress on a plane forming an angle θ with the x axis.

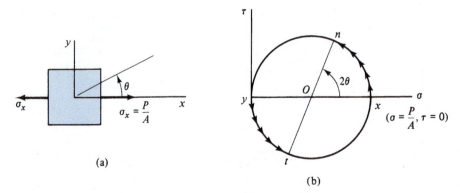

FIGURE 16.17

Solution: The coordinates of point x are $\sigma_x = P/A$ and $\tau_{xy} = 0$. The coordinates of point y are $\sigma_y = 0$ and $\tau_{yx} = 0$. Mohr's circle is drawn in Fig. 16.17(b). The center of the circle is at $\sigma = \sigma_x/2 = P/2A$ and $\tau = 0$, and the radius of the circle $R = \sigma_x/2 = P/2A$. Rotate the diameter yOx counterclockwise through an angle of 2θ. Coordinate x becomes coordinate n, and coordinate y becomes coordinate t. Therefore,

$$\sigma_n = \frac{P}{2A} + \frac{P}{2A}\cos 2\theta = \frac{P}{2A}(1 + \cos 2\theta)$$

From the trigonometric identity,

$$\frac{1 + \cos 2\theta}{2} = \cos^2\theta$$

$$\sigma_n = \frac{P}{A}\cos^2\theta \quad \text{and} \quad \tau_{nt} = \frac{P}{2A}\sin 2\theta$$

The same results were obtained in Sec. 10.9 as Eqs. (10.11) and (10.12) on page 236.

16.11 BIAXIAL STRESS: THIN-WALLED PRESSURE VESSEL

A condition of biaxial stress exists when normal stresses occur in two mutually perpendicular directions. A good example of biaxial stress occurs on the outside of a pressure vessel.

Let Fig. 16.18(a) represent a closed thin-walled pressure vessel such as a boiler. A pressure vessel can be classified as thin-walled if the ratio of the wall thickness t to the radius R of the cylinder is 0.1 or less. Pressure above atmospheric pressure (gauge pressure) on the inside of the curved walls causes the tensile stress σ_C, called *circumferential stress* or *hoop stress*. Pressure on the ends of the cylinder causes tensile stress σ_L, called *longitudinal stress*. The stresses are shown on an element of the wall in Fig. 16.18(b). The wall thickness t is measured in the radial direction R perpendicular to the circumferential and longitudinal directions, C and L, respectively.

The stresses may be calculated by free-body diagrams of selected parts of the pressure vessel together with the enclosed fluid under a uniform internal gauge pressure p. The free-body diagram of Fig. 16.19(a) can be used to determine the circumferential stress. Summing forces—stress multiplied by wall area and pressure multiplied by projected inside area—parallel to the stress σ_C, we obtain

$$\Sigma F_C = 0 \qquad 2Q - P = 0$$
$$2\sigma_C th - p2Rh = 0$$
$$\sigma_C = \frac{pR}{t} \qquad (16.12)$$

The free-body diagram of Fig. 16.19(b) can be used to determine the longitudinal stress. Summing forces—stress multiplied by a ring-shaped wall area and pressure multiplied by a projected circular area—parallel to the stress σ_L

$$\Sigma F_L = 0 \qquad Q - P = 0$$
$$\pi 2Rt\sigma_L - p\pi R^2 = 0$$
$$\sigma_L = \frac{pR}{2t} \qquad (16.13)$$

Notice from Eqs. (16.12) and (16.13) that the longitudinal stress is one-half the circumferential stress.

It can be shown that the normal stress in the walls of the pressure vessel, the radial direction [Fig. 16.18(b)], varies from $\sigma_R = -p$ on the inside of the vessel to $\sigma_R = 0$ on the outside of the vessel. The tensile stress in a *spherical pressure vessel* with a uniform pressure is given by Eq. (16.13).

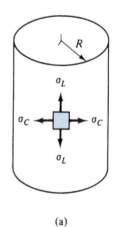

(a)

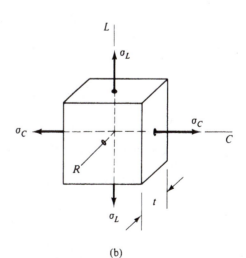

(b)

FIGURE 16.18

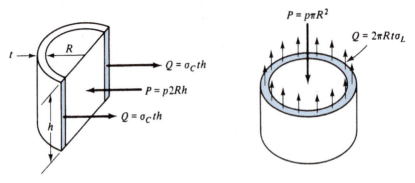

FIGURE 16.19

Example 16.11   A steel boiler 2 m in diameter has a wall thickness of 20 mm. The internal pressure is 1.4 MPa. On the outside surface of the cylindrical shell, determine (a) the maximum shearing stress in a plane parallel to the radial direction, (b) the maximum shearing stress for any plane, and (c) on the inside surface of the cylindrical shell, determine the maximum shearing stress.

Solution: The circumferential stress is given in Eq. (16.12):

$$\sigma_C = \frac{pR}{t} = \frac{1.4(1)}{0.020} = 70 \text{ MPa}$$

and the longitudinal stress is given by Eq. (16.13):

$$\sigma_L = \frac{pR}{2t} = 35 \text{ MPa}$$

The radial stress on the inside surface of the shell is

$$\sigma_R = -p = -1.4 \text{ MPa}$$

and on the outside surface of the shell

$$\sigma_R = 0$$

Outside Surface of Shell: The stresses on the outside surface of the shell are $\sigma_C = 70$ MPa, $\sigma_L = 35$ MPa, and $\sigma_R = 0$. Thus, we have a case of biaxial stress. However, a small volume element on the surface of the shell is a three-dimensional body, and the maximum shear stress may not occur in a plane perpendicular to the circumferential and longitudinal axes. Therefore, the elements must be viewed along each of the three axes to find the maximum shear stress. Such views are shown in Fig. 16.20(a)–(c). The corresponding Mohr's circles are drawn in Fig. 16.20(d). The maximum shear stress can be determined from the circles:

a.   The maximum shearing stress in a plane parallel to the radial direction is found from the circle for an element viewed along the radial axis $\tau_{max} = 17.5$ MPa. The stresses act on planes parallel to the radial axis, which make angles of 45° and 135° with the circumferential axis.

b.   The maximum shearing stress for any plane is found from the circle for an element viewed along the longitudinal axis $\tau_{max} = 35$ MPa. The stresses act on planes parallel to the longitudinal axis, which make angles of 45° and 135° with the circumferential axis.

Inside Surface of Shell: (c) The state of stress in part (c) is not biaxial. However, the method for parts (a) and (b) can be used here with the addition of $\sigma_R = -1.4$ MPa. Mohr's circles are shown for the three views in Fig. 16.21. The maximum shearing stress for any plane

$$\tau_{max} = \frac{\sigma_C - \sigma_R}{2} = \frac{70 + 1.4}{2} = 35.7 \text{ MPa}$$

The stresses act on planes parallel to the longitudinal axis, which make angles of 45° and 135° with the circumferential axis. The stresses determined in part (c) differ from those in part (b) by 2 percent. Therefore, the radial stress is usually neglected in a thin-walled pressure vessel.

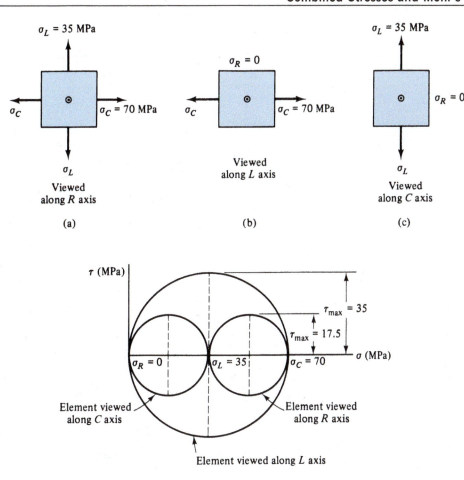

FIGURE 16.20

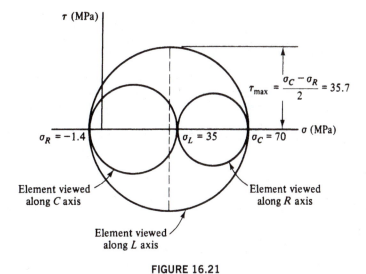

FIGURE 16.21

16.12 PURE SHEAR

A condition of pure shear exists when only shearing stresses are present in two mutually perpendicular directions, as shown in Fig. 16.22(a). Construct Mohr's circle as shown in

Fig. 16.22(b). The maximum tensile and compressive stresses (principal stresses) act on planes that form angles of 45° and 135° with the x axis and their values, $\sigma_1 = -\sigma_2 = \tau_{xy}$. The principal stresses are shown in Fig. 16.22(c).

AIR-SUPPORTED STRUCTURES

SB.23(a)

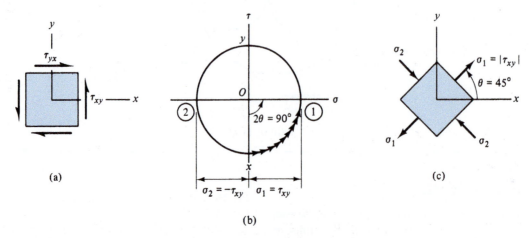

SB.23(b)

Pressurized thin-walled buildings known as *air-supported structures* are finding increased usage in a variety of applications. Although many different types of fabrics may be used, vinyl-coated woven polyester is common. Pieces of fabric are thermally joined at their edges to form the desired structural shape, which is then inflated using air at pressures in the range of 5 to 10 psf. (By comparison, typical inflation pressures for automobile tires range from 4300 to 5700 psf.) These structures are designed to withstand wind and snow loads and can cover or enclose large areas (20 acres is not considered unusual) without support posts. Such buildings have been used as industrial warehouses, aircraft hangars, construction site covers, greenhouses, emergency shelters, portable classrooms, and temporary concert halls, as well as in housing a variety of recreational sites. The structure in SB.23(a), for instance, is located in upstate New York and contains an indoor (golf) driving range, while a similar structure in SB.23(b) covers four tennis courts for year-round use.

FIGURE 16.22

Example 16.12   A hollow circular shaft has an outside diameter of 6 in. and an inside diameter of 4 in. Determine the maximum torque that can be applied if the tensile and compressive stresses are limited to 10,000 psi and 30,000 psi, respectively, and the shear stress is limited to 12,500 psi.

Solution: An element on the surface of the circular shaft is acted on by shear stresses as shown in Fig. 16.23(a). Therefore, the shaft is in a state of pure shear. In pure shear, the principal stresses in tension and compression are equal to the maximum shear stress [see Fig. 16.22(c)]. Because the smallest allowable stress is in tension, the tensile stress will control. From the torsion formula, Eq. (12.11), and the condition that the maximum tensile stress is equal to the maximum shear stress,

$$\sigma_{max} = \tau_{max} = \frac{Tc}{J} \qquad \text{(a)}$$

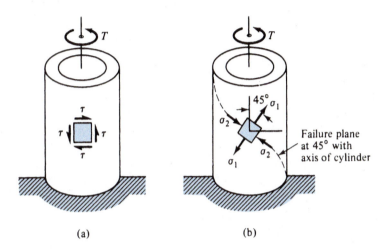

(a)   (b)

FIGURE 16.23

The polar moment of inertia

$$J = \frac{\pi}{32}(d_0^4 - d_i^4) = \frac{\pi}{32}[(6)^4 - (4)^4] = 102.1 \text{ in.}^4$$

$$c = \frac{d_o}{2} = 3 \text{ in.}$$

From Eq. (a),

$$T = \frac{\sigma_{max}J}{c} = \frac{10,000(102.1)}{3} = 340 \times 10^3 \text{ lb-in.}$$

$$= 28.4 \times 10^3 \text{ lb-ft} \qquad \qquad \textbf{Answer}$$

If the torque is increased until failure occurs in tension, the shaft will fail along a helix, as shown in Fig. 16.23(b). The helix forms an angle of 45° with the axis of the shaft.

16.13 COMBINED STRESS PROBLEMS

Various combinations of stress may occur in structural and machine members. One possible combination will be discussed in the following example.

Example 16.13   The machine member shown in Fig. 16.24 is acted on by an axial force of 11 kN and a vertical force of 2.4 kN. Determine the maximum tensile, compressive, and shear stresses. Maximum stresses will occur at points A and B.

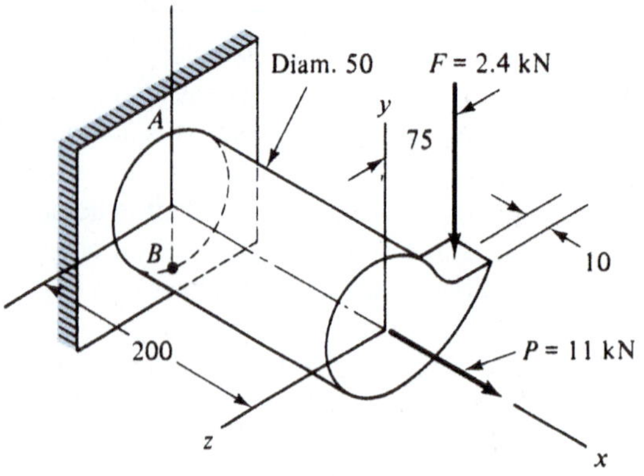

(All dimensions in mm)

FIGURE 16.24

Solution: This problem involves axial, bending, and torsion loads.

Axial Load: The cross section of the bar through A and B has an area $A = \pi d^2/4 = \pi(50)^2/4 = 1.964 \times 10^3 \, mm^2$. Therefore, the axial stress, given by Eq. (10.3),

$$\sigma_x = \frac{P}{A} = \frac{11}{1.964 \times 10^3} = 5.60 \times 10^{-3} \frac{kN}{mm^2}$$

$$= 5.60 \, MPa \qquad \qquad \text{(a)}$$

Bending Loads: The moment of inertia of the bar about the z axis,

$$I_z = \frac{\pi d^4}{64} = \frac{\pi(50)^4}{64} = 306.8 \times 10^3 \, mm^4$$

and the bending moment at the cross section is $M_z = -Fd = -2.4(190) = 456 \, kN \cdot mm$. Therefore, the bending stress, given by Eq. (14.4),

$$\sigma_x = \pm\frac{M_z c}{I_z} = \pm\frac{456(25)}{306.8 \times 10^3} = \pm 37.2 \times 10^{-3} \frac{kN}{mm^2}$$

$$= \pm 37.2 \, MPa \qquad \qquad \text{(b)}$$

The stress is tensile for point A and compressive for point B.

Torsion Load: The polar moments of inertia of the bar about the x axis, $J = 2I_z = 613.6 \times 10^3 \, mm^4$, and the torque about the x axis, $T = Fr = 2.4(75) = 180 \, kN \cdot mm$. Therefore, the torsional stress, given by Eq. (12.11),

$$\tau_{xy} = \frac{Tc}{J} = \frac{180(25)}{613.6 \times 10^3} = 7.33 \times 10^{-3} \frac{kN}{mm^2}$$

$$= 7.3 \, MPa \qquad \qquad \text{(c)}$$

Point A: At point A, the normal stresses from Eqs. (a) and (b) add to give $\sigma_x = 5.6 + 37.2 = 42.8 \, MPa$. The shear stress from Eq. (c) is $\tau_{xy} = 7.3 \, MPa$. An element at point A is shown in Fig. 16.25(a) and the corresponding Mohr's circle in Fig. 16.25(b). From the Mohr's circle,

$$\sigma_{max} = \sigma_1 = \frac{42.8}{2} + \left[\left(\frac{42.8}{2}\right)^2 + (7.3)^2\right]^{1/2}$$

$$\sigma_{max} = \sigma_1 = 21.4 + 22.6 = +40.0 \, MPa \qquad \qquad \text{Answer}$$

$$\sigma_{min} = \sigma_2 = 21.4 - 22.6 = -1.2 \, MPa \qquad \qquad \text{Answer}$$

$$\tau_{max} = 22.6 \, MPa \qquad \qquad \text{Answer}$$

Point B: At point B, the normal stresses from Eqs. (a) and (b) subtract to give $\sigma_x = 5.6 - 37.2 = -31.6 \, MPa$ and the shear stress from Eq. (c) is $\tau_{xy} = 7.3 \, MPa$. An element at point B is shown in

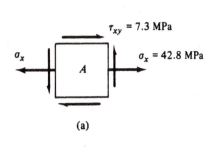

(a)

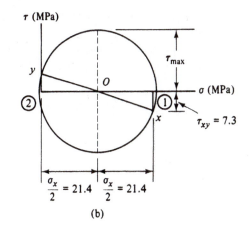

(b)

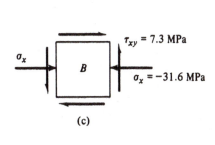

(c)

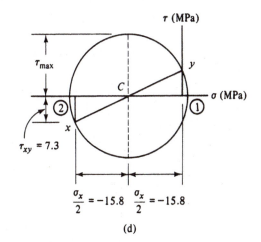

(d)

FIGURE 16.25

Fig. 16.25(c) and the corresponding Mohr's circle in Fig. 16.25(d). From the Mohr's circle,

$$\sigma_{max} = \sigma_1 = \frac{-31.6}{2} + \left[\left(\frac{31.6}{2}\right)^2 + (7.3)^2\right]^{1/2}$$

$$\sigma_1 = -15.8 + 17.4 = +1.6 \text{ MPa} \qquad \text{Answer}$$

$$\sigma_{min} = \sigma_2 = -15.8 - 17.4 = -33.2 \text{ MPa} \qquad \text{Answer}$$

$$\tau_{max} = 17.4 \text{ MPa} \qquad \text{Answer}$$

PROBLEMS

16.40 For the cantilever beam shown, use Mohr's circle to determine the principal stresses and maximum shear stress at A, B, and C for the section in the figure.

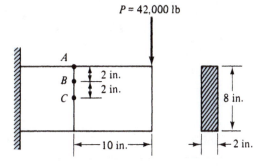

PROB. 16.40

16.41 For the simply supported beam shown, use Mohr's circle to find the principal stresses and planes of principal stress at A, B, and C for the section in the figure.

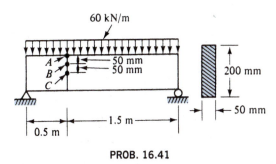

PROB. 16.41

16.42 A cantilever beam is loaded as shown in the figure. Use Mohr's circle to find the principal stresses at A, B, and C of the cross section.

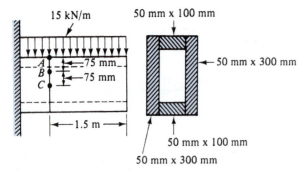

PROB. 16.42

16.43 A beam with overhang, load, and cross section is shown in the figure. Use Mohr's circle to determine the principal stresses at A, B, and C of the cross section.

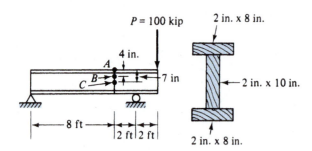

PROB. 16.43

16.44 A simply supported beam with loads and cross section is shown in the figure. The moment of inertia with respect to the neutral axis $I_{NA} = 260.8$ in.⁴ Use Mohr's circle to determine the principal stresses at A, B, and C of the cross section.

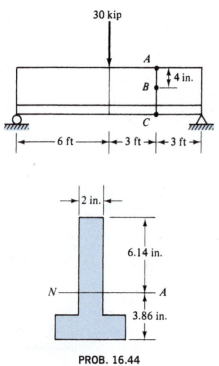

PROB. 16.44

16.45 Two bars measuring 2 in. × 2 in. are glued together to form the member shown. If the allowable stresses in the glue are 900 psi in tension and 500 psi in shear, what is the allowable axial load on the member? Assume that the load is based on the strength of the glue.

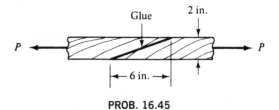

PROB. 16.45

16.46 A concrete test cylinder 6 in. in diameter and 12 in. high failed when subjected to a compressive load of 105,000 lb. The surface of failure formed a right circular cone whose angle was 82°. Determine (a) the compressive and shearing stresses on the plane of failure, and (b) the maximum compressive and shearing stresses in the cylinder at failure. (*Hint:* The plane of failure forms an angle of 41° with the axis of the cylinder.)

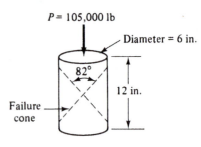

PROB. 16.46

16.47 Two blocks are joined along a diagonal plane by three splines (keys) as shown. (a) If an axial load of $P = 40$ kN is applied to the blocks, what is the shear force on each spline? Neglect friction. (b) If each spline measures 5 mm × 20 mm × 150 mm, what is the shear stress in the splines?

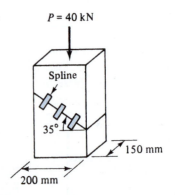

PROB. 16.47

16.48 A steel boiler 2.5 m in diameter has a wall thickness of 25 mm. The internal pressure is 1.6 MPa. Determine (a) the circumferential stress and (b) the longitudinal stress. Use Mohr's circle to find (c) the maximum shearing stress in a plane parallel to the radial direction and (d) the maximum shearing stress for any plane at a point on the outside of the cylinder.

16.49 A spherical steel boiler 8 ft in diameter has a wall thickness of 1 in. The internal pressure is 250 psi. Determine (a) the circumferential stress and (b) the longitudinal stress in the boiler due to the internal pressure. Use Mohr's circle to find (c) the maximum shearing stress in a plane parallel to the radial direction and (d) the maximum shearing stress for any plane at a point on the outside of the sphere.

16.50 For the machine part shown, the load $P = 1500$ lb, the diameter $d = 2$ in., $L = 4$ in., and $w = 1$ in. Determine the following stresses on the cross section at D, E, F, and G: (a) the normal stress due to bending, (b) the shear stress due to torsion, and (c) the shear stress due to the vertical shear force. Use Mohr's circle to find (d) the principal stresses and the maximum shear stresses at D, E, F, and G from the results in parts (a), (b), and (c).

section at D, E, F, and G: (a) the normal stress due to the bending produced by P, (b) the shear stresses due to the torque produced by P, and (c) the shear stresses due to the vertical shear force $V = P$. Use Mohr's circle to find (d) the principal stresses and maximum shear stresses at D, E, F, and G from the results in parts (a), (b), and (c).

16.52 The bracket shown is loaded with a force $P = 1$ kip. The dimensions of the bracket are $L = 6$ in., $R = 3$ in., $w = 4$ in., and $d = 1.5$ in. Determine the following stresses on the cross section at A, B, C, and D: (a) the normal stress due to bending, (b) the shear stress due to torsion, and (c) the shear stress due to the vertical shear force. Use Mohr's circle to find (d) the principal stresses and maximum shear stresses at A, B, C, and D from the results in parts (a), (b), and (c).

16.53 The bracket shown is loaded with a force $P = 4.4$ kN. The dimensions of the bracket are $L = 150$ mm, $R = 75$ mm, $w = 100$ mm, and $d = 40$ mm. Determine the following stresses on the cross section at A, B, C, and D: (a) the normal stress due to bending, (b) the shear stress due to torsion, and (c) the shear stress due to the vertical shear force. Use Mohr's circle to find (d) the principal stresses and maximum shear stresses at A, B, C, and D from the results in parts (a), (b), and (c).

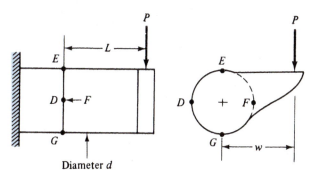

PROB. 16.50 and PROB. 16.51

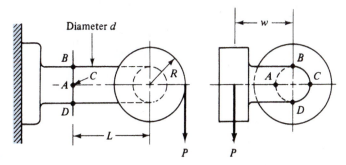

PROB. 16.52 and PROB. 16.53

16.51 The load $P = 6.6$ kN, the diameter $d = 50$ mm, $L = 100$ mm, and $w = 25$ mm for the machine part shown. Determine the following stresses on the cross

COLUMNS

17.1 INTRODUCTION

In previous chapters, we have studied the relationship between the loads applied to a member and the resultant stresses, strains, and deflections. In each problem, the member was in *stable* equilibrium.

Let us now study the column—a long slender member subject to an axial load. For small values of the axial load, the column will remain essentially straight, and the stresses can be determined from the usual formulas for axial loads.

However, as the axial load increases, the column bends and is subject to *small* lateral deflections. The stresses now depend not only on the axial loads, but also on the deflection. Continued small increases in the load after it is close to a certain critical value results in sudden *large* deflections and buckling or collapse without warning. This collapse is not due to failure of the column material but to passage of the column from *stable* to *unstable* equilibrium.

Stable and Unstable Equilibrium

As a first example, consider the three smooth surfaces of Fig. 17.1. The surfaces are all horizontal at point B. A sphere will be in equilibrium at B because the weight W and normal reaction N_R will both be vertical, equal in magnitude, and opposite in direction. The sphere is now moved to C. Three possibilities exist, depending on the unbalanced force. In Fig. 17.1(a), the unbalanced force moves the sphere back to its original position B. This is called *stable* equilibrium. In Fig. 17.1(b), no unbalanced force develops—the sphere is still in equilibrium—and therefore no tendency for it to move away from the new position. This is called *neutral* equilibrium. In Fig. 17.1(c), the unbalanced force moves the sphere away from its original position B. This is called *unstable* equilibrium.

For the second example, consider the hinged bar in Fig. 17.2. The load directed up in Fig. 17.2(a) produces stable equilibrium because a small rotation produces a moment that restores the bar to the vertical equilibrium position. The downward load in Fig. 17.2(b) produces unstable equilibrium because a small rotation produces a moment that moves the bar away from the vertical equilibrium position.

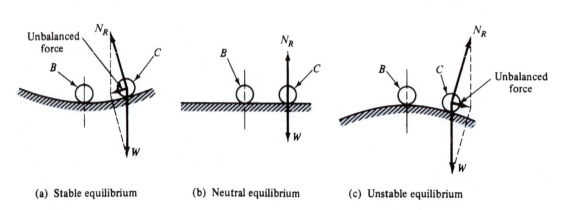

(a) Stable equilibrium   (b) Neutral equilibrium   (c) Unstable equilibrium

FIGURE 17.1

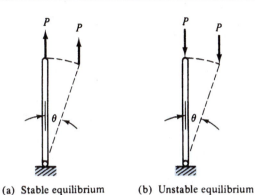

(a) Stable equilibrium   (b) Unstable equilibrium

FIGURE 17.2

Consider a system in equilibrium. Any possible small displacement away from the equilibrium position produces forces. If the forces move the system back to the equilibrium position, the system is stable. If the forces do not, the system is unstable.

17.2 EULER COLUMN FORMULA

The long member or column shown in Fig. 17.3(a) has pin supports on both ends and is acted on by a compressive load. The smallest column load that produces buckling or failure is called the *critical* or *Euler load*. For a load less than the critical load, the column remains straight and in stable equilibrium. For a load equal to or greater than the critical load, the column is unstable and fails. The bent shape shown represents buckling or failure. Our problem is to find the smallest column load that will produce buckling or failure.

From the free-body diagram for a section at a distance x from the top of the column [Fig. 17.3(b)] and the rotational equation of equilibrium, the bending moment $M = -P\delta$.

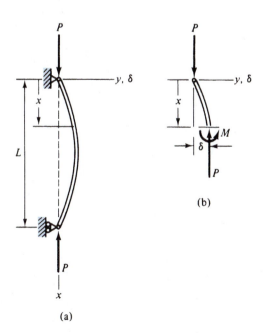

(b)

(a)

FIGURE 17.3

The slope of the deflection curve $\theta = \Delta\delta/\Delta x$. The slope and bending moment are related by Eq. (15.3), $\Delta\theta/\Delta x = M/EI$. Therefore, we write

$$\frac{\Delta(\Delta\delta/\Delta x)}{\Delta x} = \frac{M}{EI} \tag{17.1}$$

Substituting for the bending moment $M = -P\delta$, we have

$$\frac{\Delta(\Delta\delta/\Delta x)}{\Delta x} + \frac{P\delta}{EI} = 0 \tag{17.2}$$

The solution of this equation expresses the deflection δ as a function of x.

If we let the element of length along the column Δx decrease until it approaches a point, Eq. (17.2) becomes a differential equation. The differential equation has known solutions. Guided by these solutions, the deflective curve shown in Fig. 17.3(a) looks like a sine curve from $\theta = 0$ to $\theta = \pi$. We modify the "scale" of the sine curve to make the curve from $\theta = 0$ to $\theta = \pi$ fit the deflection curve from $x = 0$ to $x = L$. To modify the sine curve, we let

$$\delta = A \sin \frac{\pi x}{L} \tag{a}$$

as shown in Fig. 17.4(a). The slope of the deflection curve $\Delta\delta/\Delta x$, where the length Δx decreases until it approaches a point, is given by

$$\frac{\Delta\delta}{\Delta x} = A\frac{\pi}{L}\cos\frac{\pi x}{L} \tag{b}$$

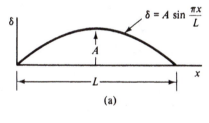

(a)

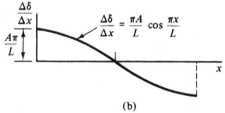

(b)

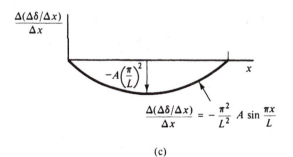

(c)

FIGURE 17.4

as shown in Fig. 17.4(b). The slope of the deflection curve $\Delta(\Delta\delta/\Delta x)/\Delta x$, when the length Δx decreases until it approaches a point, is given by

$$\frac{\Delta(\Delta\delta/\Delta x)}{\Delta x} = -A\frac{\pi^2}{L^2}\sin\frac{\pi x}{L} \qquad \textbf{(c)}$$

and is shown in Fig. 17.4(c). Substituting Eq. (a) and (c) into Eq. (17.2), we write

$$-A\frac{\pi^2}{L^2}\sin\frac{\pi x}{L} + \frac{P}{EI}A\sin\frac{\pi x}{L} = 0 \qquad \textbf{(d)}$$

Equation (d) is satisfied if $A = 0$. This represents no buckling—zero deflection—as can be seen from Eq. (a). The other possible solution requires that $\pi^2/L^2 = P/EI$. The value of the load P that makes the curved shape possible, the critical or Euler load, is, therefore, given by

$$P_C = \frac{\pi^2 EI}{L^2} \qquad \textbf{(17.3)}$$

This is the *Euler column formula* for a column that has pin supports at both ends.

17.3 EFFECTIVE LENGTH OF COLUMNS

For end conditions other than pin supports, the shapes of the deflection curves can be used to modify the Euler column formula. In Fig. 17.5, we show the deflection curve for different end conditions. In each case, the length of a single loop of a sine curve represents the effective length L_e of the column. The extent of a single loop is marked by a dot on the deflection curves. The effective length L_e can be expressed in the form

$$L_e = kL$$

Substituting the effective length into Eq. (17.3), the critical load becomes

$$P_C = \frac{\pi^2 EI}{(kL)^2} \qquad \textbf{(17.4)}$$

where the value of k depends on the end conditions. For example, with both ends fixed as shown in Fig. 17.5(c), $k = 0.5$. The critical load $P_C = 4\pi^2 EI/L^2$. Thus, we see that the critical load is four times the value for an identical column with both ends pinned.

17.4 FURTHER COMMENTS ON THE EULER COLUMN FORMULA

The Euler column formula is usually expressed in terms of the radius of gyration. From Sec. 9.11, the radius of gyration $r = \sqrt{I/A}$ or $I = Ar^2$, where A is the cross-sectional area of the column and I is the minimum moment of inertia of the cross-sectional area. Substituting for the moment of inertia in Eq. (17.4), we have

$$P_C = \frac{\pi^2 EAr^2}{(kL)^2}$$

or in terms of critical stress, $\sigma_C = P_C/A$, we write

$$\sigma_C = \frac{\pi^2 E}{(kL/r)^2} \qquad \textbf{(17.5)}$$

The critical stress σ_C is the average value of the stress over the cross-sectional area A at the critical load P_C. The effective length of the column is given by kL. Buckling will take place about the axis of minimum moment of inertia or equivalently about the axis with a minimum radius of gyration unless prevented by braces. Therefore, r is the minimum radius of gyration. The ratio of the effective length kL to the minimum radius of gyration r is called the *slenderness ratio*.

In Fig. 17.6, we show the relation between critical stress and the slenderness ratio for various materials [Eq. (17.5)]. Because the Euler column formula is based on Hooke's law, the curves are not valid for stresses above the proportional limit of the material.

We are now able to define what is meant by a long column. *A long column is one in which the critical stress σ_C is less than the proportional limit σ_P for the material.*

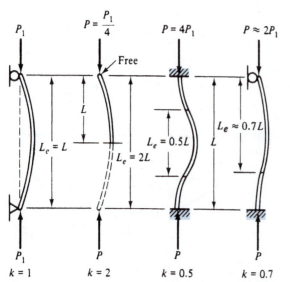

(a) Pinned ends (b) Free-fixed (c) Fixed ends (d) Pinned-fixed

FIGURE 17.5. Various column end constraints.

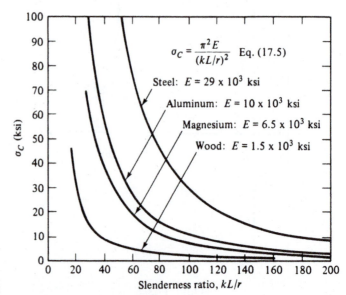

FIGURE 17.6. Euler curves.

Example 17.1   A W14 × 74 steel column has a length of 35 ft. The steel has a proportional limit of 33,000 psi. Use a modulus $E = 29 \times 10^6$ psi and a factor of safety, F.S. = 2.0. Determine the critical or Euler load and the allowable or working load if (a) the ends of the column are pinned, (b) one end is pinned and the other is fixed, and (c) both ends are fixed.

Solution: The minimum radius of gyration for the W14 × 74 column is $r = 2.48$ in., and the cross-sectional area $A = 21.8$ in.^{2}

a.   Ends pinned: for pinned ends, $k = 1$ [Fig. 17.5(a)] and the slenderness ratio $kL/r = 1(35)(12)/2.48 = 169$. The critical Euler stress

$$\sigma_C = \frac{P_C}{A} = \frac{\pi^2 E}{(kL/r)^2} = \frac{\pi^2 29 \times 10^6}{(169)^2} = 10,020 \text{ psi}$$

Because the critical stress is less than the proportional limit stress, Euler's formula applies, and we write

$$P_C = \sigma_C A = 10,020(21.8) = 218,400 \text{ lb}$$

Therefore, the allowable load

$$P_a = \frac{P_C}{\text{F.S.}} = \frac{218,400}{2.0} = 109,200 \text{ lb} \qquad \textbf{Answer}$$

b.   One end pinned, the other fixed: for the end conditions, $k = 0.7$ [Fig. 17.5(d)] and the slenderness ratio $kL/r = 0.7(35)(12)/2.48 = 119$. The critical or Euler stress

$$\sigma_C = \frac{P_C}{A} = \frac{\pi^2 E}{(kL/r)^2} = \frac{\pi^2 29 \times 10^6}{(119)^2} = 20,200 \text{ psi}$$

The critical stress is less than the proportional limit and Euler's formula applies.

$$P_C = \sigma_C A = 20,200(21.8) = 440,000 \text{ lb}$$

Therefore, the allowable load

$$P_a = \frac{P_C}{\text{F.S.}} = \frac{440,000}{2.0} = 220,000 \text{ lb} \qquad \textbf{Answer}$$

c.   Ends fixed: for both ends fixed, $k = 0.5$ [Fig. 17.5(c)] and the slenderness ratio $kL/r = 0.5(35)(12)/2.48 = 85$. The critical or Euler stress

$$\sigma_C = \frac{P_C}{A} = \frac{\pi^2 E}{(kL/r)^2} = \frac{\pi^2 29 \times 10^6}{(85)^2} = 39,600 \text{ psi}$$

Because the critical stress is greater than the proportional limit, Euler's column formula does not apply. An allowable load based on the Euler formula would be meaningless.

PROBLEMS

17.1 Compute the minimum column length for which Euler's formula applies if the column is pinned at both ends and consists of

a.   A 4 × 4 American Standard Timber made of Southern pine (Tables A.9 and A.10 of the Appendix) for which the proportional limit is 6100 psi.

b.   Nominal 4-in. standard weight pipe (Table A.8 of the Appendix) for a proportional limit of 33,000 psi.

17.2 A W14 × 74 steel column has a length of 30 ft. The steel has a modulus $E = 29 \times 10^3$ ksi and a proportional limit $\sigma_P = 33$ ksi. Use a factor of safety of 2.

Determine the critical or Euler load and the allowable load if (a) the ends are pinned, (b) one end is pinned and the other is fixed, and (c) both ends are fixed. For section properties, see Table A.3 of the Appendix.

17.3 A W360 × 110 rolled steel column has a length of 9 m. The steel has a modulus of $E = 200 \times 10^6$ kPa and a proportional limit $\sigma_P = 230 \times 10^3$ kPa. Use a factor of safety of 2. Determine the critical or Euler load and the allowable load if (a) the ends are pinned, (b) one end is pinned and the other is fixed, and (c) both ends are fixed. For section properties, see Table A.3 (SI Units) of the Appendix.

17.4 A boom consists of a 100-mm-diameter steel pipe AB and a cable BC as shown in the figure. The modulus of

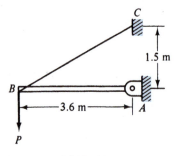

PROB. 17.4

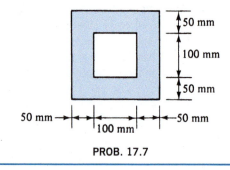

PROB. 17.7

steel $E = 200 \times 10^6$ kPa. Determine the capacity P of the boom. Use the Euler formula with a factor of safety of 3. Neglect the weight of the pipe. Section properties of pipe: $A = 2.05 \times 10^3$ mm^2 and the radius of gyration $r = 38.4$ mm.

17.5 A boom consists of a 4-in.-diameter steel pipe AB and a cable as shown. The modulus of steel $E = 29 \times 10^3$ ksi. Determine the capacity P of the boom. Use the Euler formula with a factor of safety of 2.5. Neglect the weight of the pipe. Section properties of pipe: $A = 3.174$ in.^{2} and the radius of gyration $r = 1.51$ in.

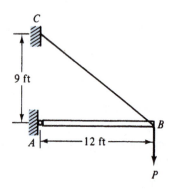

PROB. 17.5

17.6 The cross section of an 18-ft-long column is shown in the figure. Determine the slenderness ratio if (a) the ends are pinned, and (b) one end is fixed and the other end is free.

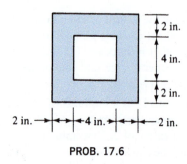

PROB. 17.6

17.7 The cross section of a 5.5-m-long column is shown in the figure. Determine the slenderness ratio if (a) the ends are pinned, and (b) one end is fixed and the other end is free.

17.5 TANGENT MODULUS THEORY

The range of the Euler column formula can be extended if we define a new modulus called the *tangent modulus* E_T. The tangent modulus is the slope of the stress–strain curve at any point. It represents the instantaneous relationship between stress and strain for a particular value of stress. The stress–strain curve for a ductile material, together with the stress versus the tangent modulus E_T curve for the same material, is shown in Fig. 17.7. The slope of the stress–strain curve is a constant up to the proportional limit. Therefore, the tangent modulus is also a constant—shown in the figure as E_{T_P}. Above the proportional limit, the slope drops rapidly as the stress and strain increase. This is shown in the figure for one value as E_{T_Q}. Substituting the tangent modulus E_T for the elastic modulus E in Eq. (17.5), we write

$$\sigma_C = \frac{\pi^2 E_T}{(kL/r)^2} \qquad (17.6)$$

This equation is the *tangent modulus* or *Engesser formula*. Engesser first proposed the formula in 1889.

The Euler and tangent modulus formulas, Eqs. (17.5) and (17.6), have been plotted for the same ductile material in Fig. 17.8. Because the tangent modulus and the elastic modulus are identical below the proportional limit, the two curves for critical stress versus the slenderness ratio coincide. Extensive test results indicate close agreement with the

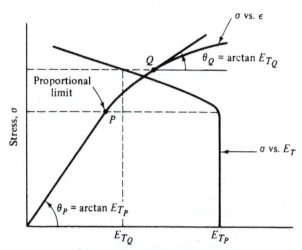

Strain and tangent modulus, ϵ and E_T

FIGURE 17.7

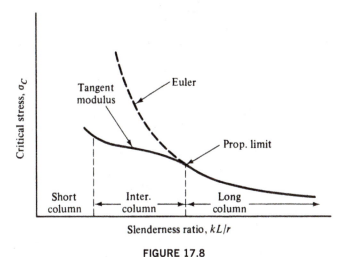

FIGURE 17.8

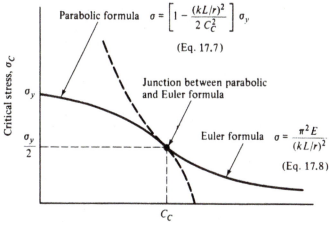

FIGURE 17.9. AISC column formulas (formulas do not include factors of safety).

Euler column formula for long columns in which the critical stress falls below the proportional limit of the material and close agreement with the tangent modulus formula for intermediate columns in which the stress falls above the proportional limit.

Long columns buckle elastically, and intermediate columns buckle inelastically (see Fig. 17.8). Short columns or compression blocks do not buckle. They fail by yielding for ductile materials and fracturing for brittle materials.

The tangent modulus formula requires a trial-and-error solution. Therefore, as is customary in practice, we will use empirical column formulas to find the allowable load or design load.

17.6 EMPIRICAL COLUMN FORMULAS: DESIGN FORMULAS

Many empirical formulas have been proposed, and some are widely used. In general, these formulas are used to determine the allowable or working stress, and the allowable or working load is found by multiplying the stress by the cross-sectional area. Two types of empirical formulas, the parabolic formula and the straight-line formula, together with Euler's formula, are used to analyze columns for structural steel, machine design steel, and structural aluminum.

Structural Steel Columns

The American Institute of Steel Construction, *Manual of Steel Construction—Allowable Stress Design*, 9th ed., Chicago: AISC, 1989, recommends a parabolic formula for short and intermediate columns and the Euler formula for long columns, as shown in Fig. 17.9. Factors of safety are applied to the formulas to find the allowable load. The parabola extends from a stress equal to the yield stresses σ_y when $kL/r = 0$ to a stress equal to one-half the yield stress when $kL/r = C_c$. Because it joins the Euler curve at $kL/r = C_c$, we can evaluate C_c by equating the stress—

one-half the yield stress—to the Euler formula, where $kL/r = C_c$. That is,

$$\frac{\sigma_y}{2} = \frac{\pi^2 E}{(C_c)^2} \quad \text{or} \quad C_c = \sqrt{\frac{2\pi^2 E}{\sigma_y}} \qquad \textbf{(a)}$$

The parabolic equation is given by

$$\sigma_C = \frac{P}{A} = \left[1 - \frac{(kL/r)^2}{2C_c^2}\right]\sigma_y \quad \text{where } 0 \le \frac{kL}{r} \le C_c \quad \textbf{(17.7)}$$

and the Euler equation is given by

$$\sigma_C = \frac{P}{A} = \frac{\pi^2 E}{(kL/r)^2} \quad \text{where } C_c \le \frac{kL}{r} \le 200 \qquad \textbf{(17.8)}$$

The modulus of elasticity for structural steel is taken as 29×10^3 ksi. Equation (17.7) is divided by a variable factor of safety, defined by

$$\text{F.S.} = \frac{5}{3} + \frac{3kL/r}{8C_c} - \frac{(kL/r)^3}{8C_c^3} \qquad \textbf{(17.9)}$$

to give the allowable stress. The factor of safety has a value of 1.67 at $kL/r = 0$ and a value of 1.92 at $kL/r = C_c$. Equation (17.8) is divided by a factor of safety of $23/12 \approx 1.92$ to give the allowable stress. Therefore, the design equation becomes

$$\sigma_a = \frac{\left[1 - \frac{1}{2}\left(\frac{kL/r}{C_c}\right)^2\right]\sigma_y}{\text{F.S.}} \quad \text{where } 0 \le \frac{kL}{r} \le C_c \quad \textbf{(17.10)}$$

and

$$\sigma_a = \frac{149,000}{(kL/r)^2}\,(\text{ksi}) \quad \text{where } C_c \le \frac{kL}{r} \le 200 \quad \textbf{(17.11)}$$

The parabolic equation [Eq. (17.7)] is also used in machine design and is called the *J. B. Johnson formula*.

Machine Design: Steel Columns

In machine design, there is no universally recognized column formula for short or intermediate columns. However, a parabolic formula widely used in column design is the J. B. Johnson formula. Although it is identical to the AISC formula, except for factors of safety, it is usually expressed in a different form. Substituting $C_c^2 = 2\pi^2 E/\sigma_y$ in Eq. (17.7), we have the J. B. Johnson parabolic formula for short and intermediate columns.

$$\sigma_c = \left[1 - \frac{\sigma_y(kL/r)^2}{4\pi^2 E} \right]\sigma_y \quad \text{where } 0 \le \frac{kL}{r} \le C_c \quad \textbf{(17.12)}$$

For long columns we use the Euler column formula [Eq. (17.5)].

$$\sigma_c = \frac{\pi^2 E}{(kL/r)^2} \quad \text{where } \frac{kL}{r} \ge C_c \quad \textbf{(17.13)}$$

Equations (17.12) and (17.13) are divided by a factor of safety to determine the allowable stress.

$$\sigma_a = \frac{\left[1 - \dfrac{\sigma_y(kL/r)^2}{4\pi^2 E} \right]\sigma_y}{\text{F.S.}} \quad \text{where } 0 \le \frac{kL}{r} \le C_c \quad \textbf{(17.14)}$$

$$\sigma_a = \frac{\dfrac{\pi^2 E}{(kL/r)^2}}{\text{F.S.}} \quad \text{where } \frac{kL}{r} \ge C_c \quad \textbf{(17.15)}$$

In structural steel column design, factors of safety range from 1.67 to 1.92. However, in machine design, a *minimum* factor of safety of 2.0 is indicated. In addition, the modulus of elasticity is taken as 30×10^3 ksi.

Aluminum Structural Columns

The Aluminum Association, Inc., *Specifications for Aluminum Structures,* 5th ed., Washington, DC, 1986, gives a constant straight-line formula for short columns, a straight-line formula for intermediate columns, and the Euler formula for long columns. The formulas are expressed in terms of four constants and are shown in Fig. 17.10. The constants are listed in Table (A). The constants depend on the type of aluminum alloy used, and their values *include* a factor of

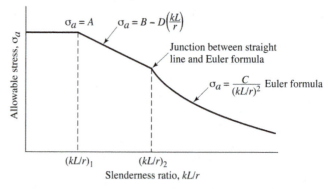

FIGURE 17.10. Aluminum Association column formulas.[†] (Formulas include factors of safety of 1.67 for $kL/r < (kL/r)_1$ and 1.95 for $kL/r \ge (kL/r)_1$.)

[†]Aluminum Association, Inc., *Specifications for Aluminum Structures,* 5th ed. (Washington, DC: Aluminum Association, Inc., 1986). Also see J. R. Kissell, and R. L. Ferry, *Aluminum Structures: A Guide to Their Specifications and Design* (New York: John Wiley & Sons, Inc., 1995).

safety. The *recommended* safety factor for buildings and like structures is 1.67 for short columns (the safety factor on yield) and 1.95 for intermediate and long columns (the safety factor on ultimate).

For the short column, we have

$$\sigma_a = A \quad \text{where } \frac{kL}{r} \le \left(\frac{kL}{r} \right)_1 \quad \textbf{(17.16)}$$

$$\sigma_a = B - D\frac{kL}{r} \quad \text{where} \left(\frac{kL}{r} \right)_1 \le \frac{kL}{r} \le \left(\frac{kL}{r} \right)_2 \quad \textbf{(17.17)}$$

The Euler column formula is given by

$$\sigma_a = \frac{C}{(kL/r)^2} \quad \text{where } \frac{kL}{r} \ge \left(\frac{kL}{r} \right)_2 \quad \textbf{(17.18)}$$

TABLE (A)	Aluminum Column Constants					
	A	**B**	**D**	**C**		
Aluminum Alloy	ksi (MPa)	ksi (MPa)	ksi (MPa)	ksi (MPa)	$\left(\dfrac{kL}{r}\right)_1$	$\left(\dfrac{kL}{r}\right)_2$
6061-T6	19 (131)	20.2 (139)	0.126 (0.869)	51,000 (352×10^3)	9.5	66
2014-T6	28 (193)	30.7 (212)	0.23 (1.586)	54,000 (372×10^3)	12	55

Example 17.2   Using the AISC column formulas, determine the allowable axial load for a W12 × 72 pin-ended column (a) 20 ft long and (b) 40 ft long. The A36 structural steel used has a yield stress of 36 ksi.

Solution: The minimum radius of gyration for the W12 × 72 is $r = 3.04$ in., and the cross-sectional area $A = 21.1 \text{ in.}^2$

a.   The slenderness ratio for a column 20 ft long is $\dfrac{kL}{r} = \dfrac{1(20)(12)}{3.04} = 78.9$. Because the slenderness ratio is less than

$$C_c = \sqrt{\frac{2\pi^2 E}{\sigma_y}} = \sqrt{\frac{2\pi^2(29)(10)^3}{36}} = 126.1$$

we use the parabolic formula for stress. The ratio of the slenderness ratio to C_c,

$$\frac{kL/r}{C_c} = \frac{78.9}{126.1} = 0.626$$

From Eq. (17.9), the factor of safety

$$\text{F.S.} = \frac{5}{3} + \frac{3}{8}(0.626) - \frac{(0.626)^3}{8} = 1.87$$

The allowable stress is given by Eq. (17.10):

$$\sigma_a = \frac{\left[1 - \dfrac{1}{2}\left(\dfrac{kL/r}{C_c}\right)^2\right]\sigma_y}{\text{F.S.}} = \frac{[1 - (1/2)(0.626)^2]36}{1.87} = 15.5 \text{ ksi}$$

The allowable load

$$P_a = \sigma_a A = 15.5(21.1) = 327 \text{ kip} \qquad\qquad \textbf{Answer}$$

b.   The slenderness ratio for a column 40 ft long is

$$\frac{kL}{r} = \frac{1(40)(12)}{3.04} = 157.9$$

Because the slenderness ratio is greater than $C_c = 126.1$, we use the Euler formula for stress given by Eq. (17.11).

$$\sigma_a = \frac{149,000}{(kL/r)^2} = \frac{149,000}{(157.9)^2} = 5.98 \text{ ksi}$$

The allowable load

$$P_a = \sigma_a A = 5.98(21.1) = 126.2 \text{ kip} \qquad\qquad \textbf{Answer}$$

Example 17.3   Find the allowable load for a pin-ended round steel compression member that has a diameter of 30 mm and a length of 0.75 m. Use a factor of safety of 2.5, a yield strength of $\sigma_y = 290 \times 10^3 \text{ kN/m}^2$ and a modulus of $E = 209 \times 10^6 \text{ kN/m}^2$. Use Eq. (17.14) or (17.15).

Solution: To determine the appropriate equation we calculate C_c.

$$C_c = \sqrt{\frac{2\pi^2 E}{\sigma_y}} = \sqrt{\frac{2\pi^2 209 \times 10^6 \text{ kN/m}^2}{290 \times 10^3 \text{ kN/m}^2}} = 119.3$$

The area of a round bar of radius R is πR^2 and the moment of inertia is $\pi R^4/4$ from Table A.2 of the Appendix. Therefore, the radius of gyration

$$r = \sqrt{\frac{I}{A}} = \sqrt{\frac{\pi R^4}{4\pi R^2}} = \frac{R}{2} = \frac{15}{2} = 7.5 \text{ mm}$$

For a pin-ended column, $k = 1$, and the slenderness ratio

$$\frac{kL}{r} = \frac{1(750 \text{ mm})}{7.5 \text{ mm}} = 100$$

The slenderness ratio $kL/r = 100 < C_c = 119.3$, so we use the J. B. Johnson formula, Eq. (17.14).

$$\sigma_a = \frac{\left[1 - \dfrac{\sigma_y(kL/r)^2}{4\pi^2 E} \right]\sigma_y}{\text{F.S.}}$$

$$\sigma_a = \frac{\left[1 - \dfrac{290 \times 10^3\,\text{kN/m}^2(100)^2}{4\pi^2 209 \times 10^6\,\text{kN/m}^2} \right]}{2.5}(290 \times 10^3\,\text{kN/m}^2)$$

$$\sigma_a = 75{,}230\,\text{kN/m}^2 = 75.23\,\text{MN/m}^2$$

The cross-sectional area of the column is $A = \pi R^2 = \pi(15\,\text{mm})^2 = 703.9\,\text{mm}^2$, and the allowable load

$$P_a = \sigma_a A = \left[\frac{75{,}230\,\text{kN}\,(703.9\,\text{mm}^2)}{\text{m}^2} \right]\left(\frac{\text{m}^2}{1 \times 10^6\,\text{mm}^2} \right)$$

$$P_a = 52.9\,\text{kN} \qquad \textbf{Answer}$$

Example 17.4   Determine the allowable axial load for a pin-ended aluminum alloy 2014-T6 angle L3 × 3 × 1/4 (a) 4 ft long, and (b) 2.5 ft long.

Solution: We will assume that the section properties of aluminum angles are the same as those of steel. The minimum radius of gyration for the angle is $r = 0.592$ in., and the cross-sectional area $A = 1.44$ in.^{2}

a.   The slenderness ratio for a column 4 ft long is

$$\frac{kL}{r} = \frac{1(4)(12)}{0.592} = 81.1$$

Because the slenderness ratio is greater than 55, we use the Euler column formula given by Eq. (17.18) with 2014-T6 aluminum alloy to find the allowable stress.

$$\sigma_a = \frac{54{,}000}{(kL/r)^2} = \frac{54{,}000}{(81.1)^2} = 8.21\,\text{ksi}$$

The allowable load

$$P_a = \sigma_a A = 8.21(1.44) = 11.82\,\text{kip} \qquad \textbf{Answer}$$

b.   The slenderness ratio for a column 2.5 ft long is

$$\frac{kL}{r} = \frac{1(2.5)(12)}{0.592} = 50.7$$

Because the slenderness ratio is less than 55, we use the straight-line formula given by Eq. (17.17) with 2014-T6 aluminum alloy to find the allowable stress.

$$\sigma_a = B - D\frac{kL}{r} = 30.7 - 0.23(50.7) = 19.04\,\text{ksi}$$

The allowable load

$$P_a = \sigma_a A = 19.04\,\text{kip/in.}^2\,(1.44\,\text{in.}^2) = 27.4\,\text{kip} \qquad \textbf{Answer}$$

PROBLEMS

17.8 The bracket shown supports a force P. Assume that both members have their ends pinned in the plane of the bracket ABC and fixed in a perpendicular plane. The modulus $E = 207 \times 10^6\,\text{kN/m}^2$, and the yield stress $\sigma_y = 375 \times 10^3\,\text{kN/m}^2$. Determine the allowable load if the factor of safety is 2. Use Eq. (17.14) or (17.15).

17.9 In the mechanism shown, called a toggle joint, a small force can be used to exert a much larger force F. Assume that both members have their ends pinned in the plane of the mechanism ABC and fixed in a perpendicular plane. The modulus $E = 30 \times 10^3\,\text{ksi}$, and the yield stress $\sigma_y = 50\,\text{ksi}$. Determine the allowable load if the factor of safety is 2.5. Use Eq. (17.14) or (17.15).

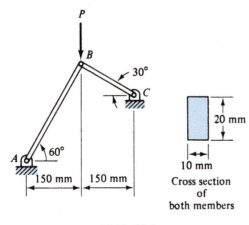

PROB. 17.8

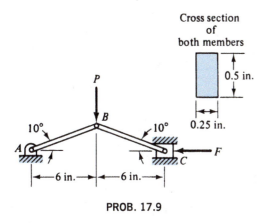

PROB. 17.9

17.10 A rectangular steel bar has a cross section 15 mm by 25 mm and a length of 300 mm. The bar has hinged ends. The modulus $E = 207 \times 10^6$ kN/m^2, and the yield strength $\sigma_y = 290 \times 10^3$ kN/m^2. Determine the allowable axial compressive load if the safety factor is 2.5. Use Eq. (17.14) or (17.15).

17.11 Determine the allowable axial compressive load for a pin-ended steel tube with an outside diameter of 3 in., a wall thickness of 0.3 in., and a length of 4 ft. The steel has a yield strength $\sigma_y = 42$ ksi and a modulus $E = 30 \times 10^3$ ksi. Use Eq. (17.14) or (17.15) with a safety factor of 3.0.

17.12 What is the allowable axial load for a W10 × 45 steel section used as a column 14 ft long if the member has pin ends and is braced at the midpoint in the weak direction? $E = 29 \times 10^3$ ksi and $\sigma_y = 36$ ksi. Use AISC formulas—Eq. (17.10) or (17.11). For section properties, see Table A.3 of the Appendix.

17.13 What is the allowable load for a W250 × 67 section used as a column 4.5 m long if the member has pin ends and is braced at the midpoint in the weak direction? $E = 200 \times 10^6$ kN/m^2 and $\sigma_y = 250 \times 10^3$ kN/m^2. Use AISC formulas—Eq. (17.10) or (17.11). For section properties, see Table A.3 of the Appendix.

17.14 Determine the allowable axial compressive load that can be supported by a steel column 10 m long with the

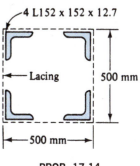

PROB. 17.14

cross section shown. $E = 200 \times 10^6$ kN/m^2, and $\sigma_y = 250 \times 10^3$ kN/m^2. Use AISC formulas—Eq. (17.10) or (17.11). For section properties, see Table A.6 of the Appendix.

17.15 Solve Prob. 17.14 if the column is aluminum alloy 6061-T6. Use Eq. (17.16), (17.17), or (17.18). For section properties, see Table A.6 of the Appendix.

17.16 Member BC of the pin-connected truss has a cross section as shown. $E = 200 \times 10^6$ kN/m^2, and $\sigma_y = 250 \times 10^3$ kN/m^2. Determine the load P that will produce an allowable stress in the member. Use AISC formulas—Eq. (17.10) or (17.11).

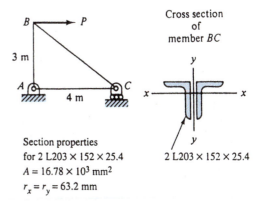

Section properties
for 2 L203 × 152 × 25.4
$A = 16.78 \times 10^3$ mm^2
$r_x = r_y = 63.2$ mm

PROB. 17.16

17.17 Member BC of the pin-connected truss has a cross section as shown. $E = 29 \times 10^3$ ksi, and $\sigma_y = 42$ ksi. Determine the load P that will produce an allowable stress in the member. Use AISC formulas—Eq. (17.10) or (17.11).

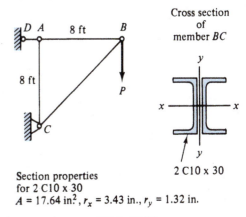

Section properties
for 2 C10 × 30
$A = 17.64$ in.^{2}, $r_x = 3.43$ in., $r_y = 1.32$ in.

PROB. 17.17

17.18 An aluminum tube BD is loaded by cables as shown. Determine the allowable load P based on buckling of the bar BD if it is a 6061-T6 aluminum structural tube with outside dimensions of 75 mm by 65 mm and a wall thickness of 10 mm. Use Eq. (17.16), (17.17), or (17.18).

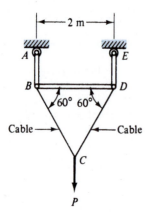

PROB. 17.18

17.19 A pin-ended aluminum boom BC is supported by a cable as shown. The boom is made of a 2014-T6 aluminum pipe with a 2.75-in. outside diameter and a 0.25-in. wall thickness. Determine the allowable load based on the boom. Use Eq. (17.16), (17.17), or (17.18).

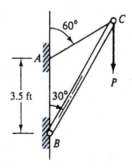

PROB. 17.19

BOLTED, RIVETED, AND WELDED STRUCTURAL CONNECTIONS

CHAPTER OBJECTIVES

In this chapter, we analyze the behavior of connections formed between steel members using bolts, rivets, or welds. After completing this material, you should be able to

- Describe the four most common modes of failure for axially loaded bolted connections.

- Compute the maximum allowable axial load for a bolted connection.

- Determine the maximum allowable loads for shear connections used in the fabrication of building frames.

- Analyze existing weld patterns to determine their maximum axial-load capacities.

- Design weld patterns that will safely carry specified axial loads.

18.1 INTRODUCTION

Steel structures are made up of members formed from rolled plates and shapes that are connected with welds or fasteners such as bolts or rivets. Welds fuse the members together, while bolts or rivets form mechanical connections between the members. For many years, rivets were the principal method used for making structural connections. However, because of the economic advantages that high-strength bolts and welds offer, they have largely replaced rivets and unfinished bolts as connectors. In this chapter, we limit ourselves mainly to the analysis and design of high-strength bolts and welds. For completeness, rivets and ordinary unfinished bolts are also discussed. Where space limitations permit, we will follow the American Institute of Steel Construction (AISC) *Manual of Steel Construction—Allowable Stress Design*, 9th ed., Chicago, 1989.

18.2 RIVETS AND BOLTS

Rivets are formed in place while hot. As the rivet cools, it contracts. This sets up clamping forces on the joined plates and normal forces between the plates. The normal forces are

H. W. Morrow, Elements of Steel Design, 1st, © 1987. Electronically reproduced by permission of Pearson Education, Inc., Upper Saddle River, New Jersey.

unpredictable, and it is customary in design calculations to neglect the resulting frictional resistance of the connection.

Unfinished bolts—also known as ordinary or common bolts—are used in much the same way as rivets. The clamping force of the bolt is unpredictable, and the resulting frictional resistance of the connection is neglected in design. Therefore, the design of connections with unfinished bolts is essentially the same as for rivets, except for substantially lower allowable shear stresses.

High-strength bolts have tensile strengths several times larger than those for unfinished bolts. They are tightened until they develop approximately 70 percent of the minimum ultimate tensile strength of the bolt. The connected plates are clamped together tightly so that most of the load transfer between plates takes place by friction.

The three types of connections made with high-strength bolts are the Friction-type connection (F), the bearing-type connection with threads iNcluded in the shear plane (N), and the bearing-type connection with threads eXcluded from the shear plane (X). The design of bearing-type N and X connections is essentially the same as for rivets, except for different allowable stresses.

Allowable shear stresses for various rivets and bolts are given in Table 18.1.

To transfer loads from one structural member to another requires that the connections designed are safe, are economical, and can be fabricated. Two typical bolted or riveted shear connections are shown in Fig. 18.1.

18.3 METHODS OF FAILURE FOR BOLTED JOINTS

If the load on a connection is large enough, the connection can fail in any one of the four following ways.

1. *Tension failure* of the connecting plates can occur when the plates, weakened by holes, fracture on a cross section including the holes or when general yielding of the plates occurs [Fig. 18.2(a)].

   The cross-sectional area not including the holes is called the gross area A_g, and the cross-sectional area including the holes is called the net area A_n. The gross area and net area of the plate

$$A_g = wt$$

and $$A_n = [w - n(d + 1/8)]t$$

TABLE 18.1   Allowable Shear Stresses on Bolts and Rivets (ksi)[a]

Fastener	Friction-Type Connection	Bearing-Type Connection
Rivets		
A502 Grade 1		17.5
Grade 2		22.0
Bolts		
A307 (Unfinished)		10.0
A325-N[b]		21.0
A325-X[c]		30.0
A490-N[b]		28.0
A490-X[c]		40.0
A325-F	17.5	
A490-F	22.0	

[a]Stresses in this table to be used with a cross-sectional area are based on *nominal* diameter of connector.

[b]Threads iNcluded in shear plane.

[c]Threads eXcluded from shear plane.

where w = width of the plate, in.
   t = thickness of the plate, in.
   d = rivet diameter, in.
   n = number of bolt holes in the critical section

The 1/8 in. added to the diameter in the net section accounts for the hole that is 1/16 in. larger than the bolt plus an additional 1/16 in. for damage due to the punching of the hole. The allowable tensile stress on the gross area

$$\sigma_t = 0.6\sigma_y \qquad (18.1)$$

and on the net area

$$\sigma_t = 0.5\sigma_u \qquad (18.2)$$

where σ_y is the yield strength of the plate and σ_u is the tensile or ultimate strength of plate.

2. *Shear failure* of the bolt can take place in either single or double shear, depending on the number of shear planes involved as shown in Fig. 18.2(b). (See Sec. 10.5 for a discussion of shear stress in bolts.) The allowable shear stress for various bolts and rivets is listed in Table 18.1.

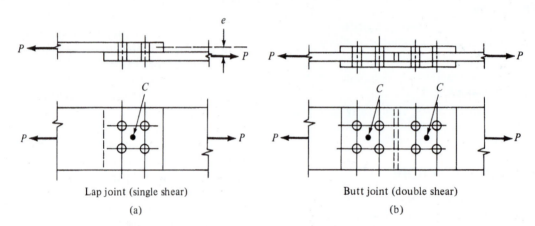

Lap joint (single shear)   Butt joint (double shear)
(a)   (b)

FIGURE 18.1. Bolted or riveted shear connections.

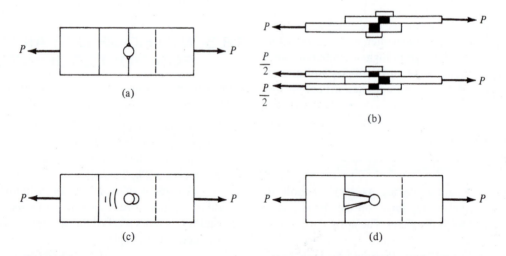

FIGURE 18.2. Methods of failure of bolted or riveted joints: (a) plate tension failure, (b) bolt single or double shear, (c) plate bearing failure, and (d) plate tear-out.

INTELLIGENT BOLTS

SB.24(a)

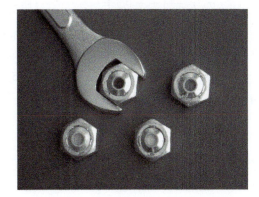

SB.24(b)

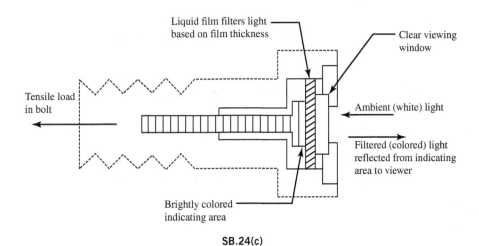

Liquid film filters light based on film thickness

Clear viewing window

Tensile load in bolt

Ambient (white) light

Filtered (colored) light reflected from indicating area to viewer

Brightly colored indicating area

SB.24(c)

In Application Sidebar 3, we saw that *torque* may be used as an indication of how tightly a bolted connection is assembled. For some critical applications, however, this method is not sufficiently accurate. Instead, direct measurement of the *strain* produced by a tensile load in the bolt is preferred. The "intelligent" bolts shown in SB.24(a) and (b) each contain an *optical* strain indicator in the bolt head itself, so that the tensile load clamping the connection together may be determined visually at a glance. These indicators function as *variable-density filters*, are constructed as shown in SB.24(c), and operate as follows: Incoming ambient light passes through a layer of liquid that is encapsulated in a circular film envelope within the bolt head, hits a colored indicating area behind the liquid, and is reflected back through this layer, appearing as colored light to the viewer. Under a tensile load, the bolt deforms internally, allowing the liquid layer to increase in thickness. This changes the liquid's light-filtering properties and thus alters the color of reflected light seen by the viewer. From no-load conditions, to 50 percent load, to 100 percent of the design load, the color of these indicators changes from red-orange, to maroon, to black respectively. *(Courtesy of Stress Indicators, Inc., Bethesda, Maryland, www.smartbolts.com)*

3. *Bearing failure* of the connected plates occurs when the plates crush due to the bearing force of the bolt on the plate [Fig. 18.2(c)]. (See Sec. 10.6 for a discussion of bearing stress for bolts.)

Tests have shown that the critical or failure bearing stress when divided by a factor of safety of 2.0 to obtain an allowable bearing stress σ_b is given by

$$\sigma_b = \frac{\sigma_u l_e}{2d} \qquad (18.3)$$

where σ_u = tensile or ultimate strength of plate, ksi
l_e = distance from center of a fastener to the nearest edge of an adjacent fastener or to the free edge of a connecting part in the direction of the stress, in.
d = diameter of the fastener, in.
s = distance from center of a fastener to the center of an adjacent fastener, in.

The center-to-edge distance l_e in terms of the center-to-center distance s is given by $l_e = s - 0.5d$. Rewriting

Eq. (18.3) in terms of the center-to-center distance s (spacing of the connectors), we have

$$\sigma_b = \frac{\sigma_u}{2}\left(\frac{s}{d} - 0.5\right) \qquad (18.4)$$

In addition, the AISC *Specification* limits the bearing stress to a value determined from the equation

$$\sigma_b \leq 1.5\sigma_u \qquad (18.5)$$

4. *Tear-out* or shear of the connected plates can take place when the distance from the bolt to the edge of the plate

in the direction of stress is too small [Fig. 18.2(d)]. Tear-out is avoided by providing an adequate edge distance.

18.4 AXIALLY LOADED BOLTED AND RIVETED CONNECTIONS

In the following examples, we determine the tensile capacity of various shear connections.

Example 18.1   Two $5/8 \times 10$-in. plates made of A36 steel are joined by bolts or rivets as shown in Fig. 18.3. (A36 steel has an ultimate strength of $\sigma_u = 58$ ksi and a yield strength of $\sigma_y = 36$ ksi.) Determine the maximum allowable load on the plates if the bolts or rivets are as follows:

a.   7/8-in. A502 Grade 1 rivets
b.   7/8-in. A325-N bolts

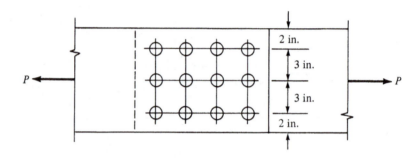

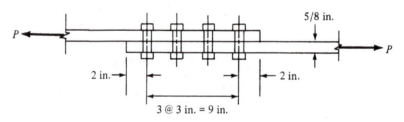

FIGURE 18.3

Solution: a.   7/8-in.-diameter A502 Grade 1 rivets

Tensile Capacity: *Based on the gross area:* The allowable tensile stress $\sigma_t = 0.6\sigma_y = 0.6(36) = 21.6$ ksi and the gross area $A_g = (5/8)(10) = 6.25$ in.^{2}; thus

$$P_{all} = \sigma_t A_g = 21.6(6.25) = 135.0 \text{ kip}$$

Based on the net area: The allowable tensile stress $\sigma_t = 0.5\sigma_u = 29.0$ ksi and the net area

$$A_n = [10 - 3(7/8 + 1/8)](5/8) = 4.375 \text{ in.}^2$$

Thus, $$P_{all} = \sigma_t A_n = 29.0(4.375) = 126.9 \text{ kip}$$

Shear Capacity: One 7/8-in. A502 Grade 1 rivet in single shear has an allowable shear stress $\tau = 17.5$ ksi (Table 18.1) and a shear area $A_{rivet} = \pi d^2/4 = \pi(7/8)^2/4 = 0.601$ in.^{2}; therefore, the allowable load for twelve rivets in single shear

$$P_{all} = \tau A_{rivet} N = 17.5(0.601)(12) = 126.2 \text{ kip}$$

Bearing Capacity: From Eqs. (18.3), (18.4), and (18.5) the bearing stress for one 7/8-in. rivet is given by the smallest of the following values:

$$\sigma_b = \frac{\sigma_u l_e}{2d} = \frac{58(2)}{2(0.875)} = 66.3 \text{ ksi}$$

$$\sigma_b = \frac{\sigma_u}{2}\left(\frac{s}{d} - 0.5\right) = \frac{58}{2}\left(\frac{3}{0.875} - 0.5\right) = 84.9 \text{ ksi}$$

$$\sigma_b \le 1.5\sigma_u = 1.5(58) = 87 \text{ ksi}$$

The area in bearing for one 7/8-in. rivet $A_b = dt = (7/8)(5/8) = 0.547$ in.^{2}; therefore, the allowable load for 12 rivets in bearing

$$P_{all} = \sigma_b A_b N = 66.3(0.547)(12) = 434.2 \text{ kip}$$

The smallest allowable load is in shear. Accordingly,

$$P_{max} = P_{all}(\text{shear}) = 126.2 \text{ kip} \qquad \qquad \textbf{Answer}$$

b. 7/8-in.-diameter A325-N bolts. The tensile and bearing capacity is unchanged from part (a).

Shear Capacity: One 7/8-in. A325-N bolt in single shear has an allowable shear stress $\tau = 21.0$ ksi (Table 18.1) and the shear area $A_{bolt} = \pi d^2/4 = 0.601$ in.^{2}; therefore, the allowable load for 12 bolts in single shear

$$P_{all} = \tau A_{bolt} N = 21.0(0.601)(12) = 151.4 \text{ kip}$$

The smallest allowable load is in tension. Thus,

$$P_{max} = P_{all}(\text{tension}) = 126.9 \text{ kip} \qquad \qquad \textbf{Answer}$$

Example 18.2   Two 1/2 × 10-in. plates made of A36 steel are joined by a butt splice as shown in Fig. 18.4. (A36 steel has an ultimate strength $\sigma_u = 58$ ksi and yield strength $\sigma_y = 36$ ksi.) Determine the maximum allowable load on the plates if 7/8-in.-diameter A325-F bolts are used for the connection.

Solution: The combined thickness of the two cover plates is greater than the thickness of the main plate; therefore, the main plate will control for tension and bearing.

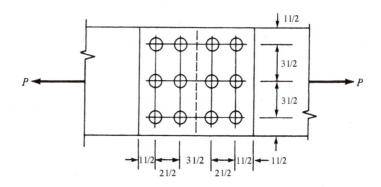

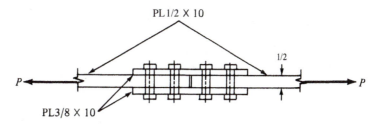

(All dimensions in inches)

FIGURE 18.4

Tensile Capacity: *Based on gross area:* The allowable tensile stress $\sigma_t = 0.6\sigma_y = 0.6(36) = 21.6\,\text{ksi}$ and the gross area $A_g = (1/2)(10) = 5.0\,\text{in.}^2$; thus

$$P_{all} = \sigma_t A_g = 21.6(5.0) = 108\,\text{kip}$$

Based on net area: The allowable tensile stress $\sigma_t = 0.5\sigma_u = 29\,\text{ksi}$ and the net area

$$A_n = [10 - 3(7/8 + 1/8)](1/2) = 3.5\,\text{in.}^2$$

Thus, $P_{all} = \sigma_t A_n = 29.0(3.5) = 101.5\,\text{kip}$

Shear Capacity: One 7/8-in. A325-F bolt in double shear has an allowable shear stress $\tau = 17.5\,\text{ksi}$ (Table 18.1) and the shear area $A_{bolt} = 2\pi d^2/4 = \pi(0.875)^2/2 = 1.203\,\text{in.}^2$; therefore, the allowable load for six bolts in double shear

$$P_{all} = \tau A_{bolt} N = 17.5(1.203)(6) = 126.3\,\text{kip}$$

Bearing Capacity: Although the possibility of a friction-type shear connection slipping into bearing is remote, connections must still meet the requirements guarding against excessive bearing stress. From Eqs. (18.3), (18.4), and (18.5), the bearing stress for one 7/8-in.-diameter bolt is given by the smallest of the following values.

$$\sigma_b = \frac{\sigma_u l_e}{2d} = \frac{58(1.5)}{2(0.875)} = 49.7\,\text{ksi}$$

$$\sigma_b = \frac{\sigma_u}{2}\left(\frac{s}{d} - 0.5\right) = \frac{58}{2}\left(\frac{2.5}{0.875} - 0.5\right) = 68.4\,\text{ksi}$$

$$\sigma_b \le 1.5\sigma_u = 1.5(58) = 87\,\text{ksi}$$

Therefore, the allowable load for six bolts in bearing

$$P_{all} = \sigma_b A_b N = 49.7(7/8)(1/2)(6) = 130.4\,\text{kip}$$

The smallest allowable load is in tension; therefore,

$$P_{max} = P_{all}(\text{tension}) = 101.5\,\text{kip}\qquad\textbf{Answer}$$

Example 18.3   Two L6 × 4 × 3/8 angles made of A36 steel are connected to a 5/8-in. gusset plate as shown in Fig. 18.5. (A36 steel has an ultimate strength of $\sigma_u = 58\,\text{ksi}$ and a yield stress of $\sigma_y = 36\,\text{ksi}$.) Determine the allowable load on the connection if six 3/4-in.-diameter A325-X bolts are used in the connection.

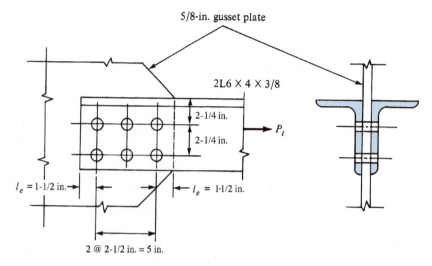

FIGURE 18.5

Solution: **Tensile Capacity:** *Based on the gross area of the two angles:* The allowable tensile stress $\sigma_t = 0.6\sigma_y = 0.6(36) = 21.6\,\text{ksi}$ and the gross area for two angles $A_g = 2(3.61) = 7.22\,\text{in.}^2$ (Table A.7 of the Appendix); thus,

$$P_{all} = \sigma_t A_g = 21.6(7.22) = 156.0\,\text{kip}$$

Based on the net area of the two angles: The allowable tensile stress $\sigma_t = 0.5\sigma_u = 29$ ksi and the net area

$$A_n = A_g - n(d + 1/8)t = 7.22 - 4(3/4 + 1/8)(3/8) = 5.91 \text{ in.}^2$$

According to AISC *Specifications,* because the load is transmitted by bolts through only one leg (side) of the angle and there are three bolts in the direction of the stress, the net area is only 85 percent effective. Therefore,

$$P_{all} = \sigma_t(0.85A_n) = 29(0.85)(5.91) = 145.7 \text{ kip}$$

Shear Capacity: One 3/4-in. A325-X bolt in double shear has an allowable shear stress $\tau = 30.0$ ksi (Table 18.1) and the shear area $A_{bolt} = 2\pi d^2/4 = \pi(0.75)^2/2 = 0.884$ in.^{2}; therefore, the allowable load for six bolts in double shear

$$P_{all} = \tau A_{bolt}N = 30.0(0.884)(6) = 159.0 \text{ kip}$$

Bearing Capacity: From Eqs. (18.3), (18.4), and (18.5), the bearing stress for one bolt is given by the smallest of the following values.

$$\sigma_b = \frac{\sigma_u l_e}{2d} = \frac{58(1.5)}{2(0.75)} = 58 \text{ ksi}$$

$$\sigma_b = \frac{\sigma_u}{2}\left(\frac{s}{d} - 0.5\right) = \frac{58}{2}\left(\frac{2.25}{0.75} - 0.5\right) = 72.5 \text{ ksi}$$

$$\sigma_b \leq 1.5\sigma_u = 1.5(58) = 87 \text{ ksi}$$

Bearing on the gusset plate will control; therefore, the area in bearing $A_b = td = (5/8)(3/4) = 0.467$ in.^{2} and the allowable load for six bolts in bearing

$$P_{all} = \sigma_b A_b N = 58(0.469)(6) = 163.2 \text{ kip}$$

The smallest allowable load is in tension; thus,

$$P_{max} = P_{all}(\text{tension}) = 145.7 \text{ kip} \qquad \textbf{Answer}$$

18.5 SHEAR CONNECTIONS FOR BUILDING FRAMES

The shear or web-framing connection shown in Fig. 18.6 has been the most commonly used method of framing for small buildings where the beams are simply supported. In the shear connection, the web of the beam is connected by two web-framing angles to the web of a girder or the flange or web of a column. The framing angles are made thin so they are as flexible as possible. Most of the flexibility is due to the bending and twisting of the angles. Although the connections do develop moments, the moments are not considered in design. The design procedure for shear connections is illustrated in the following example.

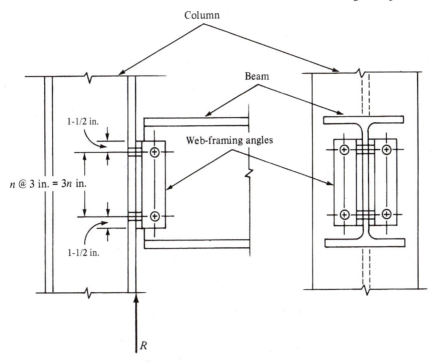

FIGURE 18.6. Double-angle web-framing connection.

Example 18.4 A W24 × 84 beam is connected to a W12 × 58 column with a double-angle web-framing connection as shown in Fig. 18.6. The connection is made with two L3-1/2 × 3-1/2 × 1/4 and 3/4-in. A325-N bolts. It is bolted to the web of the beam by five bolts ($n = 4$ in the figure) in double shear and the flange of the column by ten bolts in single shear. Rolled sections are of A36 steel with a yield stress $\sigma_y = 36$ ksi and an ultimate strength of $\sigma_u = 58$ ksi. What reaction can the connection support?

Solution: **Angle Connected to Web of Beam:** One 3/4-in. A325-N bolt has an allowable shear stress $\tau = 21.0$ ksi (Table 18.1) and a double shear area $A_{bolt} = 2\pi d^2/4 = \pi(0.75)^2/2 = 0.884$ in.^{2} Therefore, the allowable reaction for five bolts in double shear

$$R_{all} = \tau A_{bolt} N = 21.0(0.884)(5) = 92.8 \text{ kip}$$

Bearing on the web of the beam will control. The allowable bearing stress $\sigma_b \leq 1.5\sigma_u = 1.5(58) = 87$ ksi [Eq. (18.5)] and the area in bearing $A_b = t_w d = 0.470(0.75) = 0.352$ in.^{2}; therefore, the allowable reaction for five bolts in bearing

$$R_{all} = \sigma_b A_b N = 87(0.352)(5) = 253.1 \text{ kip}$$

Angles Connected to Flange of Column: For the shear capacity, we notice that ten bolts in single shear have the same capacity as five bolts in double shear; therefore, from the previous calculation

$$R_{all} = 92.8 \text{ kip}$$

Bearing capacity on angles and flange of column will both exceed bearing capacity of web of beam because their bearing areas are all larger. The smallest allowable reaction is based on the shear capacity of the bolts; therefore,

$$R_{max} = R_{all}(\text{shear}) = 92.8 \text{ kip} \qquad \textbf{Answer}$$

However, according to AISC *Specifications,* the shear stress through the angles along the bolt line should not exceed $0.3\sigma_u$. The area in shear along the bolt line $A = 2[L - n(d + 1/8)]t$, where $L =$ length of angles, in.; $n =$ number of bolts in row, in.; $d =$ diameter of bolts, in.; and $t =$ angle thickness, in.

$$A = 2[14.5 - 5(3/4 + 1/8)](5/16) = 6.33 \text{ in.}^2$$

$$\tau = \frac{R}{A} = \frac{92.8}{6.33} = 14.7 \text{ ksi} < 0.3\sigma_u = 0.3(58) = 17.4 \text{ ksi} \qquad \text{OK}$$

PROBLEMS

18.1 Two 1/2 × 10-in. plates of A36 steel are joined by 7/8-in.-diameter bolts or rivets as shown. Determine

PL1/2 × 10

2 in.
3 in.
3 in.
2 in.

1-3/4 in.   1-3/4 in.

2 @ 3 in. = 6 in.

1/2 in.

PROB. 18.1 and PROB. 18.2

the tensile capacity of the plates if the bolts or rivets are as follows:

a. A502 Grade 1 rivets.
b. A325-N bolts.

18.2 Two 1/2 × 10-in. plates of A36 steel are joined by 1-in.-diameter bolts as shown. Determine the tensile capacity of the plates if the bolts are as follows:

a. A325-F bolts.
b. A325-X bolts.

18.3 Two plates are spliced together with 3/4-in.-diameter A325-X bolts by a butt splice as shown. Determine the tensile capacity of the splice if there are three bolts in a cross section and the member is made of A36 steel. The width, thickness, and total number of bolts on each side of the splice are as follows:

a. $w = 12$ in., $t = 3/4$ in., $N = 9$.
b. $w = 8$ in., $t = 5/8$ in., $N = 6$.

18.4 Two 1/2 × 6-in. plates are spliced together by two 1/4 × 6-in. plates and 3/4-in.-diameter bolts as shown. There are two bolts in a cross section and four bolts on each side of the splice (eight bolts total), and the member is made of A36 steel. Determine

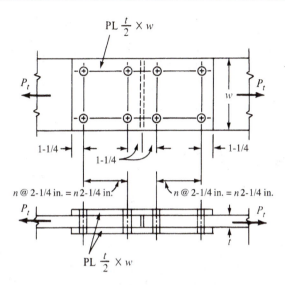

PROB. 18.3 and PROB. 18.4

the tensile capacity of the plates if the bolts are as follows:

a. A325-F bolts.
b. A325-X bolts.

18.5 The long leg of two L8 × 6 × 1/2s are bolted to a gusset plate by 7/8-in.-diameter bolts as shown. There are two rows of bolts and the bolt spacing $s = 2$-5/8 in. The angles are made of A36 steel. Determine the tensile capacity of the angles if the type of bolt, total number of bolts, and edge distance are as follows:

a. A325-F, $N = 14$, $l_e = 1$-1/4 in.
b. A325-X, $N = 8$, $l_e = 2$ in.

Note: Because the load is transmitted by bolts through only one leg (side) and there are at least three bolts in the direction of the stress, the net area is only 85 percent effective (AISC *Specifications*).

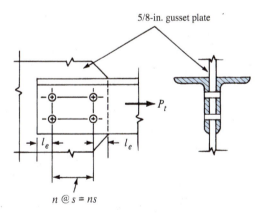

PROB. 18.5 and PROB. 18.6

18.6 The long legs of two L6 × 4 × 3/8s are bolted to a gusset plate by 3/4-in.-diameter bolts as shown. There are two rows of bolts, the bolt spacing $s = 2$-1/4 in., and the edge distance $l_e = 1$-1/4 in. The angles are made of A36 steel. Determine the tensile capacity of the angles if the type of bolt and total number of bolts are as follows:

a. A325-F, $N = 10$.
b. A325-N, $N = 8$.

Note: See Prob. 18.5.

18.7 Two C10 × 20s are bolted to a gusset plate by four rows of 7/8-in.-diameter bolts as shown. The channels are made of A36 steel. Determine their tensile capacity if the type and total number of bolts are as follows:

a. A325-F, $N = 12$.
b. A325-N, $N = 12$.

Note: Because the load is transmitted by bolts through only part of the channel (web) and there are at least three bolts in the direction of stress, the net area is only 85 percent effective (AISC *Specifications*).

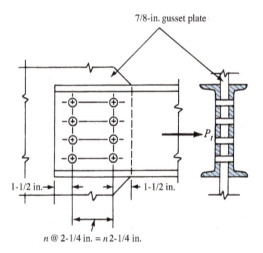

PROB. 18.7 and PROB. 18.8

18.8 Repeat Prob. 18.7 if a total of eight A490-F bolts are used. There will be four rows of bolts with only two bolts in the direction of stress, and according to AISC *Specifications*, the net area is only 75 percent effective.

18.9 A W18 × 55 beam is connected to two W12 × 72 columns by double-angle web-framing connections as shown. Each connection consists of two L4 × 3-1/2 × 5/16 angles connected to the web of the beam by three bolts and the flange of the column by six bolts. If the angles and structural members are made of A36 steel, the connectors are 3/4-in.-diameter A325-N bolts, $l_e = 1$-1/2 in., and $s = 3$ in., determine the maximum uniform load w over a span $L = 20$ ft that the beam can safely support. (*Hint:* Determine the reaction R that the connections can support. Knowing R, determine the maximum uniform load w that the beam can support if the allowable bending stress $\sigma = 0.66\sigma_y$. The beam is assumed to be simply supported.)

18.10 Repeat Prob. 18.9 for a W18 × 40 beam and W12 × 58 columns if the connectors are 7/8-in.-diameter bolts and the span $L = 10$ ft.

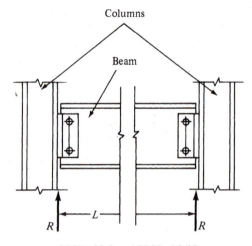

Columns

Beam

R L R

PROB. 18.9 and PROB. 18.10

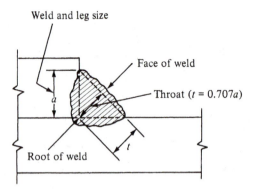

Weld and leg size

Face of weld

Throat ($t = 0.707a$)

a

t

Root of weld

FIGURE 18.8. Fillet weld terminology.

18.6 WELDS

Welding is the joining of two pieces of metal by creating a strong metallurgical bond between them by heat or pressure or both. There are numerous welding processes, but the one we consider here is the manual shielded-metal arc welding (SMAW) process.

The most common types of welds are the *fillet weld* and the *groove weld* (Fig. 18.7). Fillet welds can be used for lap joints and groove welds for butt joints.

Allowable Weld Stresses

Groove Welds   The tension, compression, and shearing stresses permitted through the throat of a groove weld are the

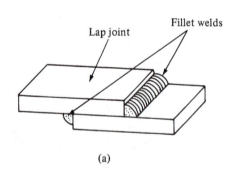

Lap joint   Fillet welds

(a)

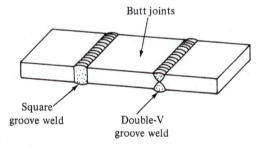

Butt joints

Square groove weld   Double-V groove weld

(b)

FIGURE 18.7. Types of joints and welds.

TABLE 18.2	Shear Strength q (kip per in. of weld) for Fillet Welds, Weld Size a (in.)		
Electrode	E60	E70	E80
σ_u (ksi)	60	70	80
q (kip/in.)	12.73a	14.85a	16.97a

same as those for the base metal. The throat for a complete-penetration weld is the thickness of the thinnest part joined.

Fillet Welds   The strength of the fillet weld is assumed to be the *strength in shear through the throat of the weld* (Fig. 18.8). For the SMAW process, a weld with equal legs of nominal size a has a throat

$$t = 0.707a \qquad (a)$$

and the allowable shear stress on the throat

$$\tau = 0.3\sigma_u \qquad (b)$$

where σ_u is the ultimate strength of the weld material. The shear strength per inch of weld $q = t\tau$; therefore, from Eqs. (a) and (b), the allowable strength for a weld size a (in.) per inch of weld

$$q = 0.2121(\sigma_u)a \qquad (c)$$

For the E60 electrode with a tensile strength $\sigma_u = 60$ ksi, the allowable strength for weld size a (in.) per inch of weld

$$q = 0.2121(60)a = 12.73a$$

The weld strengths for various electrodes are listed in Table 18.2.

18.7 AXIALLY LOADED WELDS

The design of the welded connection for tension or compression members requires the welds to be distributed so the connection does not introduce eccentricity into the loading and to be at least as strong as the members joined.

WELD TESTING

SB.25(a)

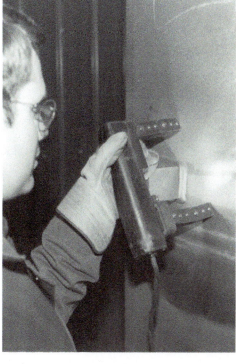

SB.25(b)

Regardless of how carefully a welded connection is designed, discontinuities and defects in the welds (including voids, cracks, and inclusions of foreign matter) can dramatically reduce the strength of the connection. For this reason, some method of *weld testing* is often performed, particularly on critical welds such as those found in pressurized vessels and piping. Several procedures recommended by the American Society for Nondestructive Testing (see Table 11.1) are liquid penetrant testing, magnetic particle testing, ultrasonic testing, eddy-current testing, and radiographic testing. In the magnetic particle test, the handheld sensor's magnetic field is first checked using a test "target," in this case an octagonal plate, the bottom surface of which contains a number of radial grooves [SB.25(a)]. When iron powder is dusted onto the top surface of the plate and a magnetic field is applied, the powder forms a pattern that reveals the existence of these defects (grooves), which would not otherwise be detected through visual inspection. If iron powder is now applied to some portion of an actual weld and the sensor is placed alongside this weld [SB.25(b)], any internal defects readily become apparent. *(Courtesy of RANOR, Inc., Westminster, Massachusetts, www.ranor.com)*

Example 18.5   Two PL-1/2 × 6s of A36 steel are joined by a lap joint as shown in Fig. 18.9. Determine the minimum lap required to develop the full tensile capacity of the plates using E60 electrodes and the required weld size.

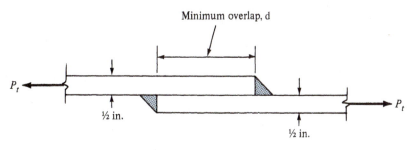

Minimum overlap, d

P_t ½ in.   ½ in. P_t

FIGURE 18.9

Solution: The minimum lap

$$d = 5t \text{(thinnest plate)} = 5(1/2) = 2\text{-}1/2 \text{ in.}$$

(AISC *Specifications*)

The tensile capacity of the plate

$$P_t = \sigma_t A_g = 0.6\sigma_y A_g = 21.6(1/2)(6) = 64.8 \text{ kip}$$

The effective weld length $L = 2(6) = 12$ in.; therefore, the required shear strength of the fillet weld

$$q = \frac{P_t}{L} = \frac{64.8}{12} = 5.5 \text{ kip/in.}$$

For the E60 electrode from Table 18.2, $q = 12.73a$. Thus,

$$q = 5.5 = 12.73a \quad \text{or} \quad a = \frac{5.5}{12.73} = 0.432 \text{ in.}$$

Use a 7/16-in. fillet weld. **Answer**

Example 18.6 A PL $- t \times 7$ of A36 steel is fillet welded to a gusset plate as shown in Fig. 18.10. The plate supports a tensile load of 93 kip. Determine the required plate thickness and weld size if the effective length of the weld is 15 in. Use an E70 electrode.

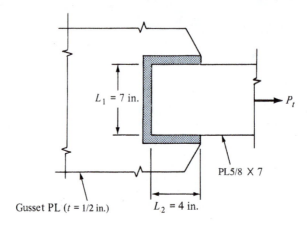

Gusset PL ($t = 1/2$ in.) $L_2 = 4$ in.

$L_1 = 7$ in. P_t

PL5/8 X 7

FIGURE 18.10

Solution: The required plate thickness

$$t = \frac{P_t}{7(0.6\sigma_y)} = \frac{93}{7(21.6)} = 0.603 \text{ in.} \quad \text{Use 5/8 in.} \qquad \textbf{Answer}$$

For an E70 electrode from Table 18.2, $q = 14.85a$. For a 15-in. weld, $P_t = qL = 14.85(15)a = 223a$; therefore,

$$P_t = 93 = 223a \quad \text{or} \quad a = \frac{93}{223} = 0.417 \text{ in.}$$

Use a 7/16-in. fillet weld. **Answer**

Example 18.7 An L6 $\times$ 4 $\times$ 1/2 of A36 steel is connected to a gusset plate as shown in Fig. 18.11(a). Determine the lengths of welds L_1 and L_2 for a balanced connection if the connection develops the full tensile capacity of the angle. Use a 3/8-in. weld and an E70 electrode.

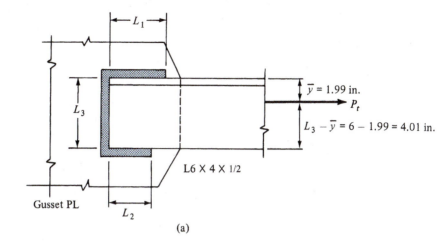

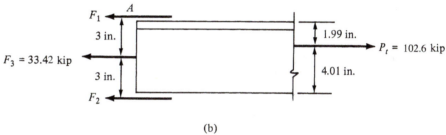

(b)

FIGURE 18.11

Solution: The allowable tensile stress $\sigma_t = 0.6\sigma_y = 0.6(36) = 21.6$ ksi; therefore, the tensile capacity of the angle

$$P_t = \sigma_t A_g = 21.6(4.75) = 102.6 \text{ kip} \quad \text{(Table A.7 of the Appendix)}$$

For a 3/8-in. weld and E70 electrode from Table 18.2, $q = 14.85a = 14.85(3/8) = 5.57$ kip/in. The reactive forces for each of the three welds together with the load on the angle are shown in Fig. 18.11(b). The force $F_3 = qL = 5.57L = 5.57(6) = 33.42$ kip. From equilibrium, we have from Fig. 18.10(b),

$$\Sigma M_A = F_2(6) + 33.42(3) - 102.6(1.99) = 0$$
$$F_2 = 17.32 \text{ kip}$$
and
$$\Sigma F_x = 102.6 - F_2 - F_1 - 33.42 = 0$$
$$F_1 = 51.86 \text{ kip}$$

The lengths of welds required

$$L_1 = \frac{F_1}{q} = \frac{51.86}{5.57} = 9.31 \text{ in.} \quad \text{(use 10 in.)} \qquad \textbf{Answer}$$
and
$$L_2 = \frac{F_2}{q} = \frac{17.32}{5.57} = 3.11 \text{ in.} \quad \text{(use 4 in.)} \qquad \textbf{Answer}$$

PROBLEMS

18.11 A $1/2 \times 6$-in. plate of A36 steel is fillet welded to a gusset plate as shown. Determine the weld length required to develop the full tensile strength of the plate. Use an E60 electrode and welds as follows:

   a.   Weld size $a = 3/8$ in.

   b.   Weld size $a = 1/4$ in.

18.12 Repeat Prob. 18.11 with an E70 electrode.

18.13 A $5/8 \times 8$-in. plate of A36 steel is fillet welded to a gusset plate as shown. Determine the lengths L of weld

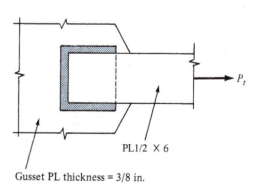

Gusset PL thickness = 3/8 in.

PROB. 18.11 and PROB. 18.12

to develop the full tensile strength of the plate. Use an E70 electrode, end returns on both side welds of 1 in., and welds as follows:

a.   Weld size $a = 3/8$ in.
b.   Weld size $a = 1/2$ in.

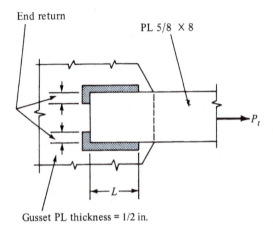

Gusset PL thickness = 1/2 in.

PROB. 18.13

Note: The end return is used to reduce the effect of stress concentrations. AISC *Specifications* require that the end return be at least two times the weld size. The end return is included in the weld length.

18.14 A two-angle $6 \times 6 \times 1/2$ tension member of A36 steel is welded with an E60 electrode to a gusset plate as shown. The weld is balanced. Determine the weld lengths if the welds are as follows:

a.   Weld size $a = 3/8$ in.
b.   Weld size $a = 5/16$ in.

18.15 Repeat Prob. 18.14 with weld size $= 3/8$ in. and lengths as follows:

a.   Use $L_1 = L_2$ (welds not balanced).
b.   Omit the end weld ($L_3 = 0$), but use an end return of 1 in. on each side weld. (See Prob. 18.13.)

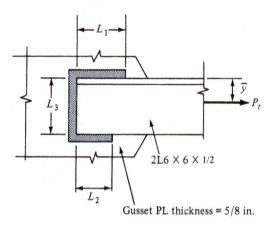

Gusset PL thickness = 5/8 in.

PROB. 18.14 and PROB. 18.15

18.16 An angle $8 \times 6 \times 1/2$ of A36 steel is fillet welded to a gusset plate as shown. Determine the tensile capacity of the angle for a balanced connection and the lengths of 7/16-in. welds required if we use the following electrodes:

a.   E60 electrode.
b.   E70 electrode.

18.17 Repeat Prob. 18.16 with an angle $4 \times 3\text{-}1/2 \times 5/16$ and 1/4-in. weld size.

18.18 Repeat Prob. 18.16 without an end weld ($L_3 = 0$), but use an end return of 1 in. on each side weld. (See Prob. 18.13.)

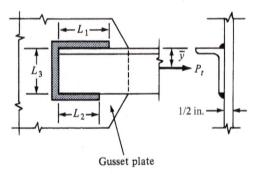

Gusset plate

PROB. 18.16, PROB. 18.17, and PROB. 18.18

APPENDIX

Credits: Tables A.3 through A.8 are taken from or based on data in *AISC Steel Construction Manual,* (New York: AISC, 1980). Copyright © American Institute of Steel Construction, Inc. Reprinted with permission. All rights reserved. Tables in SI units are rounded conversions from customary units by the authors. Data for Table A.11 was provided by the Forest Products Laboratory of the USDA Forest Service. Tables A.12 and A.13 tabulated for common sizes of LVL and I-joists from a number of manufacturers of engineered wood products.

Appendix

TABLE A.1 Areas and Centroids of Simple Shapes

Shapes		Area	$\bar{x}$	$\bar{y}$
Rectangle		bh	$\dfrac{b}{2}$	$\dfrac{h}{2}$
Triangle		$\dfrac{bh}{2}$	$\dfrac{b}{3}$	$\dfrac{h}{3}$
Circle		$\dfrac{\pi d^2}{4}$	$\dfrac{d}{2}$	$\dfrac{d}{2}$
Semicircle		$\dfrac{\pi d^2}{8}$	$\dfrac{d}{2}$	$\dfrac{4r}{3\pi}$
Parabolic spandrel	$y = \dfrac{h}{b^2}x^2$	$\dfrac{bh}{3}$	$\dfrac{3b}{4}$	$\dfrac{3h}{10}$

TABLE A.2 Moments of Inertia of Simple Shapes

Shapes		I_x	I_{x_c}
Rectangle		$\dfrac{bh^3}{3}$	$\dfrac{bh^3}{12}$
Triangle		$\dfrac{bh^3}{12}$	$\dfrac{bh^3}{36}$
Circle		$\dfrac{5\pi r^4}{4}$	$\dfrac{\pi r^4}{4}$
Semicircle		$\dfrac{\pi r^4}{8}$	$0.0349\pi r^4$
Quarter-circle		$\dfrac{\pi r^4}{16}$	$0.01747\pi r^4$

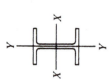

TABLE A.3 (U.S. Customary Units) W Shapes: Properties for Designing (Selected Listing)

Designation	Area in.²	Depth in.	Web Thickness in.	Flange Width in.	Flange Thickness in.	Axis X–X I in.⁴	Axis X–X S in.³	Axis X–X r in.	Axis Y–Y I in.⁴	Axis Y–Y S in.³	Axis Y–Y r in.
W36 × 160	47.0	36.01	0.650	12.000	1.020	9750	542	14.4	295	49.1	2.50
W33 × 221	65.0	33.93	0.775	15.805	1.275	12,800	757	14.1	840	106	3.59
W24 × 84	24.7	24.10	0.470	9.020	0.770	2370	196	9.79	94.4	20.9	1.95
W21 × 62	18.3	20.99	0.400	8.240	0.615	1330	127	8.54	57.5	13.9	1.77
× 44	13.0	20.66	0.350	6.500	0.450	843	81.6	8.06	20.7	6.36	1.26
W18 × 55	16.2	18.11	0.390	7.530	0.630	890	98.3	7.41	44.9	11.9	1.67
× 50	14.7	17.99	0.355	7.495	0.570	800	88.9	7.38	40.1	10.7	1.65
× 40	11.8	17.90	0.315	6.015	0.525	612	68.4	7.21	19.1	6.35	1.27
× 35	10.3	17.70	0.300	6.000	0.425	510	57.6	7.04	15.3	5.12	1.22
W16 × 100	29.4	16.97	0.585	10.425	0.985	1490	175	7.10	186	35.7	2.51
× 67	19.7	16.33	0.395	10.235	0.665	954	117	6.96	119	23.2	2.46
× 50	14.7	16.26	0.380	7.070	0.630	659	81.0	6.68	37.2	10.5	1.59
× 40	11.8	16.01	0.305	6.995	0.505	518	64.7	6.63	28.9	8.25	1.57
× 31	9.12	15.88	0.275	5.525	0.440	375	47.2	6.41	12.4	4.49	1.17
× 26	7.68	15.69	0.250	5.500	0.345	301	38.4	6.26	9.59	3.49	1.12
W14 × 159	46.7	14.98	0.745	15.565	1.190	1900	254	6.38	748	96.2	4.00
× 132	38.8	14.66	0.645	14.725	1.030	1530	209	6.28	548	74.5	3.76

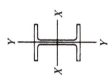

TABLE A.3 (U.S. Customary Units) (cont'd.)

Designation	Area in.²	Depth in.	Web Thickness in.	Flange Width in.	Flange Thickness in.	Axis X–X I in.⁴	Axis X–X S in.³	Axis X–X r in.	Axis Y–Y I in.⁴	Axis Y–Y S in.³	Axis Y–Y r in.
× 99	29.1	14.16	0.485	14.565	0.780	1110	157	6.17	402	55.2	3.71
× 74	21.8	14.17	0.450	10.070	0.785	796	112	6.04	134	26.6	2.48
× 53	15.6	13.92	0.370	8.060	0.660	541	77.8	5.89	57.7	14.3	1.92
× 38	11.2	14.10	0.310	6.770	0.515	385	54.6	5.87	26.7	7.88	1.55
× 34	10.0	13.98	0.285	6.745	0.455	340	48.6	5.83	23.3	6.91	1.53
× 30	8.85	13.84	0.270	6.730	0.385	291	42.0	5.73	19.6	5.82	1.49
× 30	7.69	13.91	0.255	5.025	0.420	245	35.3	5.65	8.91	3.54	1.08
× 22	6.49	13.74	0.230	5.000	0.335	199	29.0	5.54	7.00	2.80	1.04
W12 × 136	39.9	13.41	0.790	12.400	1.250	1240	186	5.58	398	64.2	3.16
× 96	28.2	12.71	0.550	12.160	0.900	833	131	5.44	270	44.4	3.09
× 72	21.1	12.25	0.430	12.040	0.670	597	97.4	5.31	195	32.4	3.04
× 58	17.0	12.19	0.360	10.010	0.640	475	78.0	5.28	107	21.4	2.51
× 45	13.2	12.06	0.335	8.045	0.575	350	58.1	5.15	50.0	12.4	1.94
× 30	8.79	12.34	0.260	6.520	0.440	238	38.6	5.21	20.3	6.24	1.52
× 26	7.65	12.22	0.230	6.490	0.380	204	33.4	5.17	17.3	5.34	1.51
× 22	6.48	12.31	0.260	4.030	0.425	156	25.4	4.91	4.66	2.31	0.847
× 19	5.57	12.16	0.235	4.005	0.350	130	21.3	4.82	3.76	1.88	0.822
× 16	4.71	11.99	0.220	3.990	0.265	103	17.1	4.67	2.82	1.41	0.773

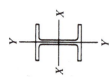

TABLE A.3 (U.S. Customary Units) (cont'd.)

Designation	Area in.²	Depth in.	Web Thickness in.	Flange Width in.	Flange Thickness in.	Axis X-X I in.⁴	Axis X-X S in.³	Axis X-X r in.	Axis Y-Y I in.⁴	Axis Y-Y S in.³	Axis Y-Y r in.
W10 × 112	32.9	11.36	0.755	10.415	1.250	716	126	4.66	236	45.3	2.68
× 100	29.4	11.10	0.680	10.340	1.120	623	112	4.60	207	40.0	2.65
× 77	22.6	10.60	0.530	10.190	0.870	455	85.9	4.49	154	30.1	2.60
× 54	15.8	10.09	0.370	10.030	0.615	303	60.0	4.37	103	20.6	2.56
× 45	13.3	10.10	0.350	8.020	0.620	248	49.1	4.32	53.4	13.3	2.01
× 39	11.5	9.92	0.315	7.985	0.530	209	42.1	4.27	45.0	11.3	1.98
× 22	6.49	10.17	0.240	5.750	0.360	118	23.2	4.27	11.4	3.97	1.33
W8 × 67	19.7	9.00	0.570	8.280	0.935	272	60.4	3.72	88.6	21.4	2.12
× 35	10.3	8.12	0.310	8.020	0.495	127	31.2	3.51	42.6	10.6	2.03
× 24	7.08	7.93	0.245	6.495	0.400	82.8	20.9	3.42	18.3	5.63	1.61
× 15	4.44	8.11	0.245	4.015	0.315	48.0	11.8	3.29	3.41	1.70	0.876
× 10	2.96	7.89	0.170	3.940	0.205	30.8	7.81	3.22	2.09	1.06	0.841
W6 × 25	7.34	6.38	0.320	6.080	0.455	53.4	16.7	2.70	17.1	5.61	1.52
× 12	3.55	6.03	0.230	4.000	0.280	22.1	7.31	2.49	2.99	1.50	0.918
× 9	2.68	5.90	0.170	3.940	0.215	16.4	5.56	2.47	2.19	1.11	0.905

TABLE A.3 (SI Units) W Shapes: Properties for Designing (Selected Listing)

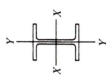

Designation	Area mm²	Depth mm	Web Thickness mm	Flange Width mm	Flange Thickness mm	Axis X-X I 10⁶mm⁴	Axis X-X S 10³mm³	Axis X-X r mm	Axis Y-Y I 10⁶mm⁴	Axis Y-Y S 10³mm³	Axis Y-Y r mm
W920 × 238	30,300	915	16.5	305	25.9	4060	8870	366	122.8	805	63.7
W840 × 329	41,900	862	19.7	401	32.4	5330	12,370	357	350	1746	91.4
W610 × 125	15,900	612	11.9	229	19.6	986	3220	249	39.3	343	49.7
W530 × 92	11,800	533	10.2	209	15.6	554	2080	217	23.9	229	45.0
× 65	8390	525	8.9	165	11.4	351	1337	205	8.62	104.5	32.1
W460 × 82	10,500	460	9.9	191	16.0	370	1609	187.7	18.69	195.7	42.2
× 74	9480	457	9.0	190	14.5	333	1457	187.4	16.69	175.7	42.0
× 60	7610	455	8.0	153	13.3	255	1121	183.1	7.95	103.9	32.3
× 52	6650	450	7.6	152	10.8	212	942	178.5	6.37	83.8	30.9
W410 × 149	19,000	431	14.9	265	25.0	620	2880	180.6	77.4	584	63.8
× 100	12,700	415	10.0	260	16.9	397	1913	176.8	49.5	381	62.4
× 74	9480	413	9.7	180	16.0	274	1327	170.0	15.48	172.0	40.4
× 60	7610	407	7.7	178	12.8	216	1061	168.5	12.03	135.2	39.8
× 46.1	5880	403	7.0	140	11.2	156.1	775	162.9	5.16	73.7	29.6
× 38.7	4950	399	6.4	140	8.8	125.3	628	159.1	3.99	57.0	28.4
W360 × 237	30,100	380	18.9	395	30.2	791	4160	162.1	311	1575	101.6
× 196	25,000	372	16.4	374	26.2	637	3420	159.6	228	1219	95.5
× 147	18,800	360	12.3	370	19.8	462	2570	156.8	167	903	94.2
× 110	14,100	360	11.4	256	19.9	331	1839	153.2	55.8	436	62.9

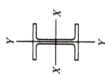

TABLE A.3 (SI Units) (cont'd.)

Designation	Area mm²	Depth mm	Web Thickness mm	Flange Width mm	Flange Thickness mm	Axis X-X I 10⁶mm⁴	Axis X-X S 10³mm³	Axis X-X r mm	Axis Y-Y I 10⁶mm⁴	Axis Y-Y S 10³mm³	Axis Y-Y r mm
× 79	10,100	354	9.4	205	16.8	225	1271	149.3	24.0	234	48.7
× 57	7230	358	7.9	172	13.1	160.2	895	148.9	11.11	129.2	39.2
× 51	6450	355	7.2	171	11.6	141.5	797	148.1	9.70	113.5	38.8
× 44.6	5710	352	6.9	171	9.8	121.1	688	145.6	8.16	95.4	37.8
× 38.7	4960	353	6.5	128	10.7	102.0	578	143.4	3.71	58.0	27.3
× 32.7	4190	349	5.8	127	8.5	82.8	474	140.6	2.91	45.8	26.4
W310 × 202	25,700	341	20.1	315	31.8	516	3026	141.7	165.7	1052	80.3
× 143	18,200	323	14.0	309	22.9	347	2149	138.1	112.4	728	78.6
× 107	13,600	311	10.9	306	17.0	248	1595	135.0	81.2	531	77.3
× 86	11,000	310	9.1	254	16.3	197.7	1275	134.1	44.5	350	63.6
× 67	8520	306	8.5	204	14.6	145.7	952	130.8	20.8	204	49.4
× 44.6	5670	313	6.6	166	11.2	99.1	633	132.2	8.45	101.8	38.6
× 38.7	4940	310	5.8	165	9.7	84.9	548	131.1	7.20	87.3	38.2
× 32.7	4180	313	6.6	102	10.8	64.9	415	124.6	1.940	38.0	21.5
× 28.3	3590	309	6.0	102	8.9	54.1	350	122.8	1.565	30.7	20.9
× 23.8	3040	305	5.6	101	6.7	42.9	281	118.8	1.174	23.2	19.65

TABLE A.3 (SI Units) (cont'd.)

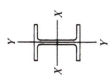

Designation	Area	Depth	Web Thickness	Flange		Elastic Properties					
				Width	Thickness	Axis X–X			Axis Y–Y		
						I	S	r	I	S	r
	mm²	mm	mm	mm	mm	10^6mm⁴	10^3mm³	mm	10^6mm⁴	10^3mm³	mm
W250 × 167	21,200	289	19.2	265	31.8	298	2060	118.6	98.2	741	68.1
× 149	19,000	282	17.3	263	28.4	259	1837	116.8	86.2	656	67.4
× 115	14,600	269	13.5	259	22.1	189.4	1408	113.9	64.1	495	66.3
× 80	10,200	256	9.4	255	15.6	126.1	985	111.2	42.9	336	64.9
× 67	8580	257	8.9	204	15.7	103.2	803	109.7	22.2	218	50.9
× 58	7420	252	8.0	203	13.5	87.0	690	108.3	18.73	184.5	50.2
× 32.7	4190	258	6.1	146	9.1	49.1	381	108.3	4.75	65.1	33.7
W200 × 100	12,710	229	14.5	210	23.7	113.2	989	94.4	36.9	351	53.9
× 52	6650	206	7.9	204	12.6	52.9	514	89.2	17.73	173.8	51.6
× 35.7	4570	201	6.2	165	10.2	34.5	343	86.9	7.62	92.4	40.8
× 22.3	2860	206	6.2	102	8.0	19.98	194.0	83.6	1.419	27.8	22.3
× 14.9	1910	200	4.3	100	5.2	12.82	128.2	81.9	0.870	17.40	21.3
W150 × 37.2	4740	162	8.1	154	11.6	22.2	274	68.4	7.12	92.5	38.8
× 17.9	2290	153	5.8	102	7.1	9.20	120.3	63.4	1.245	24.4	23.3
× 13.4	1730	150	4.3	100	5.5	6.83	91.1	62.8	0.912	18.24	23.0

TABLE A.4 (U.S. Customary Units) S Shapes: Properties for Designing (Selected Listing)

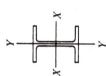

Designation	Area in.²	Depth in.	Web Thickness in.	Flange Width in.	Flange Thickness in.	Axis X–X I in.⁴	Axis X–X S in.³	Axis X–X r in.	Axis Y–Y I in.⁴	Axis Y–Y S in.³	Axis Y–Y r in.
S24 × 100	29.3	24.00	0.745	7.245	0.870	2390	199	9.02	47.7	13.2	1.27
× 90	26.5	24.00	0.625	7.125	0.870	2250	187	9.21	44.9	12.6	1.30
× 80	23.5	24.00	0.500	7.000	0.870	2100	175	9.47	42.2	12.1	1.34
S20 × 96	28.2	20.30	0.800	7.200	0.920	1670	165	7.71	50.2	13.9	1.33
× 86	25.3	20.30	0.660	7.060	0.920	1580	155	7.89	46.8	13.3	1.36
× 75	22.0	20.00	0.635	6.385	0.795	1280	128	7.62	29.8	9.32	1.16
× 66	19.4	20.00	0.505	6.255	0.795	1190	119	7.83	27.7	8.85	1.19
S18 × 70	20.6	18.00	0.711	6.251	0.691	926	103	6.71	24.1	7.72	1.08
× 54.7	16.1	18.00	0.461	6.001	0.691	804	89.4	7.07	20.8	6.94	1.14
S15 × 50	14.7	15.00	0.550	5.640	0.622	486	64.8	5.75	15.7	5.57	1.03
× 42.9	12.6	15.00	0.411	5.501	0.622	447	59.6	5.95	14.4	5.23	1.07
S12 × 50	14.7	12.00	0.687	5.477	0.659	305	50.8	4.55	15.7	5.74	1.03
× 40.8	12.0	12.00	0.462	5.252	0.659	272	45.4	4.77	13.6	5.16	1.06

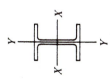

TABLE A.4 (U.S. Customary Units) (cont'd.)

Designation	Area in.²	Depth in.	Web Thickness in.	Flange Width in.	Flange Thickness in.	Axis X-X I in.⁴	Axis X-X S in.³	Axis X-X r in.	Axis Y-Y I in.⁴	Axis Y-Y S in.³	Axis Y-Y r in.
× 35	10.3	12.00	0.428	5.078	0.544	229	38.2	4.72	9.87	3.89	0.980
× 31.8	9.35	12.00	0.350	5.000	0.544	218	36.4	4.83	9.36	3.74	1.00
S10 × 35	10.3	10.00	0.594	4.944	0.491	147	29.4	3.78	8.36	3.38	0.901
× 25.4	7.46	10.00	0.311	4.661	0.491	124	24.7	4.07	6.79	2.91	0.954
S8 × 23	6.77	8.00	0.441	4.171	0.426	64.9	16.2	3.10	4.31	2.07	0.798
× 18.4	5.41	8.00	0.271	4.001	0.426	57.6	14.4	3.26	3.73	1.86	0.831
S7 × 20	5.88	7.00	0.450	3.860	0.392	42.4	12.1	2.69	3.17	1.64	0.734
× 15.3	4.50	7.00	0.252	3.662	0.392	36.7	10.5	2.86	2.64	1.44	0.766
S6 × 17.25	5.07	6.00	0.465	3.565	0.359	26.3	8.77	2.28	2.31	1.30	0.675
× 12.5	3.67	6.00	0.232	3.332	0.359	22.1	7.37	2.45	1.82	1.09	0.705
S5 × 14.75	4.34	5.00	0.494	3.284	0.326	15.2	6.09	1.87	1.67	1.01	0.620
× 10	2.94	5.00	0.214	3.004	0.326	12.3	4.92	2.05	1.22	0.809	0.643

Elastic Properties

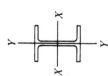

TABLE A.4 (SI Units) S Shapes: Properties for Designing (Selected Listing)

Designation	Area	Depth	Web Thickness	Flange Width	Flange Thickness	Axis X–X			Axis Y–Y		
						I	S	r	I	S	r
	mm²	mm	mm	mm	mm	10^6mm⁴	10^3mm³	mm	10^6mm⁴	10^3mm³	mm
S610 × 149	18,900	610	18.9	184	22.1	995	3260	229	19.85	216	32.4
× 134	17,100	610	15.9	181	22.1	937	3070	234	18.69	207	33.1
× 119	15,160	610	12.7	178	22.1	874	2870	240	17.56	197.3	34.0
S510 × 143	18,190	516	20.3	183	23.4	695	2690	195.5	20.89	228	33.9
× 128	16,320	516	16.8	179	23.4	658	2550	201	19.48	218	34.5
× 112	14,190	508	16.1	162	20.2	533	2100	193.8	12.40	153.1	29.6
× 98	12,520	508	12.8	159	20.2	495	1949	198.8	11.53	145.0	30.3
S460 × 104	13,290	457	18.1	159	17.6	385	1685	170.2	10.03	126.2	27.5
× 81.4	10,390	457	11.7	152	17.6	335	1466	179.6	8.66	113.9	28.9
S380 × 74	9480	381	14.0	143	15.8	202	1060	146.0	6.53	91.3	26.2
× 64	8130	381	10.4	140	15.8	186.1	977	151.3	5.99	85.6	27.1
S310 × 74	9480	305	17.4	139	16.7	127.0	833	115.7	6.53	94.0	26.2

Elastic Properties

TABLE A.4 (SI Units) (cont'd.)

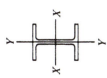

Designation	Area	Depth	Web Thickness	Flange Width	Flange Thickness	Axis X-X I	Axis X-X S	Axis X-X r	Axis Y-Y I	Axis Y-Y S	Axis Y-Y r
	mm²	mm	mm	mm	mm	10⁶mm⁴	10³mm³	mm	10⁶mm⁴	10³mm³	mm
× 60.7	7740	305	11.7	133	16.7	113.2	742	120.9	5.66	85.1	27.0
× 52	6650	305	10.9	129	13.8	95.3	625	119.7	4.11	63.7	24.9
× 47.3	6030	305	8.9	127	13.8	90.7	595	122.6	3.90	61.4	25.4
S250 × 52	6650	254	15.1	126	12.5	61.2	482	95.9	3.48	55.2	22.9
× 37.8	4810	254	7.9	118	12.5	51.6	406	103.6	2.83	48.0	24.3
S200 × 34	4370	203	11.2	106	10.8	27.0	262	78.6	1.794	33.8	20.3
× 27.4	3490	203	6.9	102	10.8	24.0	236	82.9	1.553	30.5	21.1
S180 × 30	3790	178	11.4	98	10.0	17.65	198.3	68.2	1.319	26.9	18.66
× 22.8	2900	178	6.4	93	10.0	15.28	171.7	72.6	1.099	23.6	19.47
S150 × 25.7	3270	152	11.8	91	9.1	10.95	144.1	57.8	0.961	21.1	17.14
× 18.6	2370	152	5.9	85	9.1	9.20	121.1	62.3	0.758	17.84	17.88
S130 × 22.0	2800	127	12.5	83	8.3	6.33	99.7	47.5	0.695	16.75	15.75
× 15	1900	127	5.4	76	8.3	5.12	80.6	51.9	0.508	13.37	16.35

TABLE A.5 (U.S. Customary Units) C Shapes: Properties for Designing (Selected Listing)

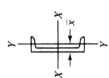

Designation	Area	Depth	Web Thickness	Flange Width	Flange Thickness	Axis X–X				Axis Y–Y			
	in.²	in.	in.	in.	in.	I in.⁴	S in.³	r in.		I in.⁴	S in.³	r in.	x̄ in.
C15 × 40	11.8	15.00	0.520	3.520	0.650	349	46.5	5.44		9.23	3.37	0.886	0.777
× 33.9	9.96	15.00	0.400	3.400	0.650	315	42.0	5.62		8.13	3.11	0.904	0.787
C12 × 30	8.82	12.00	0.510	3.170	0.501	162	27.0	4.29		5.14	2.06	0.763	0.674
× 25	7.35	12.00	0.387	3.047	0.501	144	24.1	4.43		4.47	1.88	0.780	0.674
× 20.7	6.09	12.00	0.282	2.942	0.501	129	21.5	4.61		3.88	1.73	0.799	0.698
C10 × 30	8.82	10.00	0.673	3.033	0.436	103	20.7	3.42		3.94	1.65	0.669	0.649
× 25	7.35	10.00	0.526	2.886	0.436	91.2	18.2	3.52		3.36	1.48	0.676	0.617
× 20	5.88	10.00	0.379	2.739	0.436	78.9	15.8	3.66		2.81	1.32	0.692	0.606
× 15.3	4.49	10.00	0.240	2.600	0.436	67.4	13.5	3.87		2.28	1.16	0.713	0.634
C9 × 20	5.88	9.00	0.448	2.648	0.413	60.9	13.5	3.22		2.42	1.17	0.642	0.583
× 15	4.41	9.00	0.285	2.485	0.413	51.0	11.3	3.40		1.93	1.01	0.661	0.586
× 13.4	3.94	9.00	0.233	2.433	0.413	47.9	10.6	3.48		1.76	0.962	0.669	0.601

Elastic Properties

TABLE A.5 (U.S. Customary Units) (cont'd.)

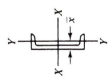

Designation	Area in.²	Depth in.	Web Thickness in.	Flange Width in.	Flange Thickness in.	Axis X–X I in.⁴	Axis X–X S in.³	Axis X–X r in.	Axis Y–Y I in.⁴	Axis Y–Y S in.³	Axis Y–Y r in.	$\bar{x}$ in.
C8 × 18.75	5.51	8.00	0.487	2.527	0.390	44.0	11.0	2.82	1.98	1.01	0.599	0.565
× 13.75	4.04	8.00	0.303	2.343	0.390	36.1	9.03	2.99	1.53	0.854	0.615	0.553
× 11.5	3.38	8.00	0.220	2.260	0.390	32.6	8.14	3.11	1.32	0.781	0.625	0.571
C7 × 14.75	4.33	7.00	0.419	2.299	0.366	27.2	7.78	2.51	1.38	0.779	0.564	0.532
× 12.25	3.60	7.00	0.314	2.194	0.366	24.2	6.93	2.60	1.17	0.703	0.571	0.525
× 9.8	2.87	7.00	0.210	2.090	0.366	21.3	6.08	2.72	0.968	0.625	0.581	0.540
C6 × 13	3.83	6.00	0.437	2.157	0.343	17.4	5.80	2.13	1.05	0.642	0.525	0.514
× 10.5	3.09	6.00	0.314	2.034	0.343	15.2	5.06	2.22	0.866	0.564	0.529	0.499
× 8.2	2.40	6.00	0.200	1.920	0.343	13.1	4.38	2.34	0.693	0.492	0.537	0.511
C5 × 9	2.64	5.00	0.325	1.885	0.320	8.90	3.56	1.83	0.632	0.450	0.489	0.478
× 6.7	1.97	5.00	0.190	1.750	0.320	7.49	3.00	1.95	0.479	0.378	0.493	0.484
C4 × 7.25	2.13	4.00	0.321	1.721	0.296	4.59	2.29	1.47	0.433	0.343	0.450	0.459
× 5.4	1.59	4.00	0.184	1.584	0.296	3.85	1.93	1.56	0.319	0.283	0.449	0.457

Elastic Properties

TABLE A.5 (SI Units) C Shapes: Properties for Designing (Selected Listing)

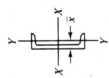

Designation	Area	Depth	Web Thickness	Flange Width	Flange Thickness	Axis X–X			Axis Y–Y			
	mm²	mm	mm	mm	mm	I 10^6mm⁴	S 10^3mm³	r mm	I 10^6mm⁴	S 10^3mm³	r mm	$\bar{x}$ mm
C380 × 60	7610	381	13.2	89.4	16.5	145.3	763	138.2	3.84	55.1	22.5	19.74
× 50.4	6430	381	10.2	86.4	16.5	131.1	688	142.8	3.38	50.9	22.9	19.99
C310 × 45	5690	305	13.0	80.5	12.7	67.4	442	108.8	2.14	33.8	19.39	17.12
× 37	4740	305	9.8	77.4	12.7	59.9	393	112.4	1.861	30.9	19.81	17.12
× 30.8	3930	305	7.2	74.7	12.7	53.7	352	116.9	1.615	28.3	20.27	17.73
C250 × 45	5690	254	17.1	77.0	11.1	42.9	338	86.8	1.640	27.1	16.98	16.48
× 37	4740	254	13.4	73.3	11.1	38.0	299	89.5	1.399	24.3	17.18	15.67
× 30	3790	254	9.6	69.6	11.1	32.8	258	93.0	1.170	21.6	17.57	15.39
× 22.8	2900	254	6.1	66.0	11.1	28.1	221	98.4	0.949	19.02	18.09	16.10
C230 × 30	3790	229	11.4	67.3	10.5	25.3	221	81.7	1.007	19.18	16.30	14.81
× 22	2850	229	7.2	63.1	10.5	21.2	185.2	86.2	0.803	16.65	16.79	14.88
× 19.9	2540	229	5.9	61.8	10.5	19.94	174.1	88.6	0.733	15.75	16.99	15.27

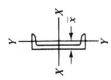

Designation	Area mm²	Depth mm	Web Thickness mm	Flange Width mm	Flange Thickness mm	Axis X–X I 10⁶mm⁴	Axis X–X S 10³mm³	Axis X–X r mm	Axis Y–Y I 10⁶mm⁴	Axis Y–Y S 10³mm³	Axis Y–Y r mm	Axis Y–Y x̄ mm
C200 × 27.9	3550	203	12.4	64.2	9.9	18.31	180.4	71.8	0.824	16.53	15.24	14.35
× 20.5	2610	203	7.7	59.5	9.9	15.03	148.1	75.9	0.637	14.02	15.62	14.05
× 17.1	2180	203	5.6	57.4	9.9	13.57	133.7	78.9	0.549	12.80	15.87	14.50
C180 × 22.0	2790	178	10.6	58.4	9.3	11.32	127.2	63.7	0.574	12.79	14.34	13.51
× 18.2	2320	178	8.0	55.7	9.3	10.07	113.1	65.9	0.487	11.50	14.49	13.34
× 14.6	1852	178	5.3	53.1	9.3	8.87	99.7	69.2	0.403	10.23	14.75	13.72
C150 × 19.3	2470	152	11.1	54.8	8.7	7.24	95.3	54.1	0.437	10.47	13.30	13.06
× 15.6	1994	152	8.0	51.7	8.7	6.33	83.3	56.3	0.360	9.22	13.44	12.67
× 12.2	1548	152	5.1	48.8	8.7	5.45	71.7	59.3	0.288	8.04	13.64	12.98
C130 × 13.4	1703	127	8.3	47.9	8.1	3.70	58.3	46.6	0.263	7.35	12.43	12.14
× 10.0	1271	127	4.8	44.4	8.1	3.12	49.1	49.5	0.199	6.20	12.51	12.29
C100 × 10.8	1374	102	8.2	43.7	7.5	1.911	37.5	37.3	0.180	5.62	11.45	11.66
× 8.0	1026	102	4.7	40.2	7.5	1.602	31.4	39.5	0.133	4.65	11.39	11.61

Elastic Properties

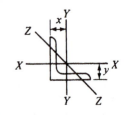

TABLE A.6 (U.S. Customary Units) Angles, Equal Legs: Properties for Designing

Size and Thickness	Weight per Foot	Area	Elastic Properties Axis X–X or Y–Y				Axis Z–Z r
			I	S	r	x or y	
in.	lb	in.^{2}	in.^{4}	in.^{3}	in.	in.	in.
L8 × 8 × 1	51.0	15.0	89.0	15.8	2.44	2.37	1.56
× 3/4	38.9	11.4	69.7	12.2	2.47	2.28	1.58
× 1/2	26.4	7.75	48.6	8.36	2.50	2.19	1.59
L6 × 6 × 1	37.4	11.0	35.5	8.57	1.80	1.86	1.17
× 3/4	28.7	8.44	28.2	6.66	1.83	1.78	1.17
× 5/8	24.2	7.11	24.2	5.66	1.84	1.73	1.18
× 1/2	19.6	5.75	19.9	4.61	1.86	1.68	1.18
× 3/8	14.9	4.36	15.4	3.53	1.88	1.64	1.19
L5 × 5 × 3/4	23.6	6.94	15.7	4.53	1.51	1.52	0.975
× 1/2	16.2	4.75	11.3	3.16	1.54	1.43	0.983
× 3/8	12.3	3.61	8.74	2.42	1.56	1.39	0.990
L4 × 4 × 3/4	18.5	5.44	7.67	2.81	1.19	1.27	0.778
× 5/8	15.7	4.61	6.66	2.40	1.20	1.23	0.779
× 1/2	12.8	3.75	5.56	1.97	1.22	1.18	0.782
× 3/8	9.8	2.86	4.36	1.52	1.23	1.14	0.788
× 1/4	6.6	1.94	3.04	1.05	1.25	1.09	0.795
L3-1/2 × 3-1/2 × 3/8	8.5	2.48	2.87	1.15	1.07	1.01	0.687
× 1/4	5.8	1.69	2.01	0.794	1.09	0.968	0.694
L3 × 3 × 1/2	9.4	2.75	2.22	1.07	0.898	0.932	0.584
× 3/8	7.2	2.11	1.76	0.833	0.913	0.888	0.587
× 1/4	4.9	1.44	1.24	0.577	0.930	0.842	0.592
L2-1/2 × 2-1/2 × 3/8	5.9	1.73	0.984	0.566	0.753	0.762	0.487
× 1/4	4.1	1.19	0.703	0.394	0.769	0.717	0.491
× 3/16	3.07	0.902	0.547	0.303	0.778	0.694	0.495
L2 × 2 × 3/8	4.7	1.36	0.479	0.351	0.594	0.636	0.389
× 1/4	3.19	0.938	0.348	0.247	0.609	0.592	0.391
× 1/8	1.65	0.484	0.190	0.131	0.626	0.546	0.398

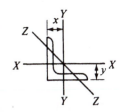

TABLE A.6   (SI Units) Angles, Equal Legs: Properties for Designing

Size and Thickness	Mass per Meter	Area	Elastic Properties				Axis Z–Z r
			Axis X–X or Y–Y				
			I	S	r	x or y	
mm	kg	mm²	× 10⁶mm⁴	× 10³mm³	mm	mm	mm
L203 × 203 × 25.4	75.9	9680	37.0	259	62.0	60.2	39.6
× 19.0	57.9	7350	29.0	200	62.7	57.9	40.1
× 12.7	39.3	5000	20.2	137.0	63.5	55.6	40.4
L152 × 152 × 25.4	55.7	7100	14.78	140.4	45.7	47.2	29.7
× 19.0	42.7	5450	11.74	109.1	46.5	45.2	29.7
× 15.9	36.0	4590	10.07	92.8	46.7	43.9	30.0
× 12.7	29.2	3710	8.28	75.5	47.2	42.7	30.0
× 9.52	22.2	2810	6.41	57.8	47.8	41.7	30.2
L127 × 127 × 19.0	35.1	4480	6.53	74.2	38.4	38.6	24.8
× 12.7	24.1	3060	4.70	51.8	39.1	36.3	25.0
× 9.52	18.3	2330	3.64	39.7	39.6	35.3	25.1
L102 × 102 × 19.0	27.5	3510	3.19	46.0	30.2	32.3	19.8
× 15.9	23.4	2970	2.77	39.3	30.5	31.2	19.8
× 12.7	19.0	2420	2.31	32.3	31.0	30.0	19.9
× 9.52	14.6	1845	1.815	24.9	31.2	29.0	20.0
× 6.35	9.8	1250	1.265	17.21	31.8	27.7	20.2
L89 × 89 × 9.52	12.6	1600	1.195	18.85	27.2	25.7	17.45
× 6.35	8.6	1090	0.837	13.01	27.7	24.6	17.63
L76 × 76 × 12.7	14.0	1774	0.924	17.53	22.8	23.7	14.83
× 9.52	10.7	1360	0.733	13.65	23.2	22.6	14.91
× 6.35	7.3	929	0.516	9.46	23.6	21.4	15.04
L64 × 64 × 9.52	8.9	1116	0.410	9.38	19.13	19.35	12.37
× 6.35	6.1	768	0.293	6.46	19.53	18.21	12.47
× 4.76	4.57	582	0.228	4.97	19.76	17.63	12.57
L51 × 51 × 9.52	6.99	877	0.1994	5.75	15.09	16.15	9.88
× 6.35	4.75	605	0.1448	4.05	15.47	15.04	9.93
× 3.18	2.48	312	0.0791	2.15	15.90	13.92	10.11

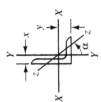

TABLE A.7 (U.S. Customary Units) Angles, Unequal Legs: Properties for Designing

Size and Thickness	Weight per Foot	Area	Axis X–X				Axis Y–Y				Axis Z–Z	
in.	lb	in.²	I in.⁴	S in.³	r in.	y in.	I in.⁴	S in.³	r in.	x in.	r in.	Tan α
L8 × 6 × 1	44.2	13.0	80.8	15.1	2.49	2.65	38.8	8.92	1.73	1.65	1.28	0.543
× 3/4	33.8	9.94	63.4	11.7	2.53	2.56	30.7	6.92	1.76	1.56	1.29	0.551
× 1/2	23.0	6.75	44.3	8.02	2.56	2.47	21.7	4.79	1.79	1.47	1.30	0.558
L7 × 4 × 3/4	26.2	7.69	37.8	8.42	2.22	2.51	9.05	3.03	1.09	1.01	0.860	0.324
× 3/8	13.6	3.98	20.6	4.44	2.27	2.37	5.10	1.63	1.13	0.870	0.880	0.340
L6 × 4 × 5/8	20.0	5.86	21.1	5.31	1.90	2.03	7.52	2.54	1.13	1.03	0.864	0.435
× 1/2	16.2	4.75	17.4	4.33	1.91	1.99	6.27	2.03	1.15	0.987	0.870	0.440
× 3/8	12.3	3.61	13.5	3.32	1.93	1.94	4.90	1.60	1.17	0.941	0.877	0.446
L6 × 3-1/2 × 3/8	11.7	3.42	12.9	3.24	1.94	2.04	3.34	1.23	0.988	0.787	0.767	0.350
× 5/16	9.8	2.87	10.9	2.74	1.95	2.01	2.85	1.04	0.996	0.763	0.772	0.352
L5 × 3 × 1/2	12.8	3.75	9.45	2.91	1.59	1.75	2.58	1.15	0.829	0.750	0.648	0.357
× 3/8	9.8	2.86	7.37	2.24	1.61	1.70	2.04	0.888	0.845	0.704	0.654	0.364
L4 × 3-1/2 × 3/4	9.1	2.67	4.18	1.49	1.25	1.21	2.95	1.17	1.06	0.955	0.727	0.755
× 5/16	7.7	2.25	3.56	1.26	1.26	1.18	2.55	0.994	1.07	0.932	0.730	0.757
L4 × 3 × 1/2	11.1	3.25	5.05	1.89	1.25	1.33	2.42	1.12	0.864	0.827	0.639	0.543
× 3/8	8.5	2.48	3.96	1.46	1.26	1.28	1.92	0.866	0.879	0.782	0.644	0.551
× 1/4	5.8	1.69	2.77	1.00	1.28	1.24	1.36	0.599	0.896	0.736	0.651	0.558
L3-1/2 × 2-1/2 × 3/8	7.2	2.11	2.56	1.09	1.10	1.16	1.09	0.592	0.719	0.660	0.537	0.496
× 1/4	4.9	1.44	1.80	0.755	1.12	1.11	0.777	0.412	0.735	0.614	0.544	0.506
L3 × 2 × 3/8	5.9	1.73	1.53	0.781	0.940	1.04	0.543	0.371	0.559	0.539	0.430	0.428
× 1/4	4.1	1.19	1.06	0.542	0.597	0.993	0.392	0.260	0.574	0.493	0.435	0.440
L2-1/2 × 2 × 3/8	5.3	1.55	0.912	0.547	0.768	0.831	0.514	0.363	0.577	0.581	0.420	0.614
× 1/4	3.62	1.06	0.654	0.381	0.784	0.787	0.372	0.254	0.592	0.537	0.424	0.626

Elastic Properties

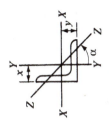

TABLE A.7 (SI Units) Angles, Unequal Legs: Properties for Designing

Size and Thickness	Mass per Meter	Area	Axis X-X				Axis Y-Y				Axis Z-Z	
			I	S	r	y	I	S	r	x	r	Tan α
mm	kg	mm²	×10⁶mm⁴	×10³mm³	mm	mm	×10⁶mm⁴	×10³mm³	mm	mm	mm	
L203 × 152 × 25.4	65.8	8390	33.6	247	63.2	67.3	16.14	146.2	43.9	41.9	32.5	0.543
× 19.0	50.3	6410	26.4	191.7	64.3	65.0	12.78	113.4	44.7	39.6	32.8	0.551
× 12.7	34.2	4350	18.44	131.4	65.0	62.7	9.03	78.5	45.5	37.3	33.0	0.558
L178 × 102 × 19.0	39.0	4960	15.73	138.0	56.4	63.8	3.77	49.7	27.7	25.7	21.8	0.324
× 9.52	20.2	2570	8.57	72.8	57.7	60.2	2.12	26.7	28.7	22.1	22.4	0.340
L152 × 102 × 15.9	29.8	3780	8.78	87.0	48.3	51.6	3.13	41.6	28.7	26.2	21.9	0.435
× 12.7	24.1	3060	7.24	71.0	48.5	50.5	2.61	33.3	29.2	25.1	22.1	0.440
× 9.52	18.3	2330	5.62	54.4	49.0	49.3	2.04	26.2	29.7	23.9	22.3	0.446
L152 × 89 × 9.52	17.4	2210	5.37	53.1	49.3	51.8	1.390	20.2	25.1	19.99	19.48	0.350
× 7.94	14.6	1850	4.54	44.7	39.5	51.1	1.186	17.04	25.3	19.38	19.61	0.352
L127 × 76 × 12.7	19.0	2420	3.93	47.7	40.4	44.4	1.074	18.85	21.1	19.05	16.46	0.357
× 9.52	14.6	1850	3.07	36.7	40.9	43.2	0.849	14.55	21.5	17.88	16.61	0.364
L102 × 89 × 9.52	13.5	1720	1.740	24.4	31.8	30.7	1.228	19.17	26.9	24.3	18.47	0.755
× 7.94	11.5	1450	1.482	20.6	32.0	30.0	1.061	16.29	27.2	23.7	18.54	0.757
L102 × 76 × 12.7	16.5	2100	2.10	31.0	31.8	33.8	1.007	18.35	21.9	21.0	16.23	0.543
× 9.52	12.6	1600	1.648	23.9	32.0	32.5	0.799	14.19	22.3	19.86	16.36	0.551
× 6.35	8.6	1090	1.153	16.38	32.5	31.5	0.566	9.82	22.8	18.69	16.54	0.558
L89 × 64 × 9.52	10.7	1360	1.066	17.86	27.9	29.5	0.454	9.70	18.26	16.76	13.64	0.496
× 6.35	7.3	929	0.749	12.37	28.4	28.2	0.323	6.75	18.67	15.59	13.82	0.506
L76 × 51 × 9.52	8.8	1120	0.637	12.80	23.9	26.4	0.226	6.08	14.20	13.69	10.92	0.428
× 6.35	6.1	768	0.454	8.88	24.3	25.2	0.1632	4.26	14.58	12.52	11.05	0.440
L64 × 51 × 9.52	7.9	1000	0.380	8.96	19.51	21.1	0.214	5.95	14.66	14.76	10.67	0.614
× 6.35	5.39	684	0.272	6.24	19.91	19.99	0.1548	4.16	15.04	13.64	10.77	0.626

TABLE A.8 (U.S. Customary Units) Pipe Dimensions and Properties

Nominal Diameter	Outside Diameter	Inside Diameter	Wall Thickness	Weight per Foot	Area	I	S	r
in.	in.	in.	in.	lb	in.^{2}	in.^{4}	in.^{3}	in.
Standard Weight								
2	2.375	2.067	0.154	3.65	1.07	0.666	0.561	0.787
2-1/2	2.875	2.469	0.203	5.79	1.70	1.53	1.06	0.947
3	3.500	3.068	0.216	7.58	2.23	3.02	1.72	1.16
3-1/2	4.000	3.548	0.226	9.11	2.68	4.79	2.39	1.34
4	4.500	4.026	0.237	10.79	3.17	7.23	3.21	1.51
5	5.563	5.047	0.258	14.62	4.30	15.2	5.45	1.88
6	6.625	6.065	0.280	18.97	5.58	28.1	8.50	2.25
8	8.625	7.981	0.322	28.55	8.40	72.5	16.8	2.94
10	10.750	10.020	0.365	40.48	11.9	161	29.9	3.67
12	12.750	12.000	0.375	49.56	14.6	279	43.8	4.38
Extra Strong								
2	2.375	1.939	0.218	5.02	1.48	0.868	0.731	0.766
2-1/2	2.875	2.323	0.276	7.66	2.25	1.92	1.34	0.924
3	3.500	2.900	0.300	10.25	3.02	3.89	2.23	1.14
3-1/2	4.000	3.364	0.318	12.50	3.68	6.28	3.14	1.31
4	4.500	3.826	0.337	14.98	4.41	9.61	4.27	1.48
5	5.563	4.813	0.375	20.78	6.11	20.7	7.43	1.84
6	6.625	5.761	0.432	28.57	8.40	40.5	12.2	2.19
8	8.625	7.625	0.500	43.39	12.8	106	24.5	2.88
10	10.750	9.750	0.500	54.74	16.1	212	39.4	3.63
12	12.750	11.750	0.500	65.42	19.2	362	56.7	4.33

TABLE A.8 (SI Units) Pipe Dimensions and Properties

Nominal Diameter	Outside Diameter	Inside Diameter	Wall Thickness	Mass per Meter	Area	I	S	r
mm	mm	mm	mm	kg	mm²	10⁶mm⁴	10³mm³	mm
				Standard Weight				
51	60.3	52.5	3.91	5.43	690	0.277	9.19	20.0
64	73.0	62.7	5.16	8.62	1100	0.637	17.4	24.1
76	88.9	77.9	5.49	11.3	1440	1.26	28.2	29.5
90	102	90.1	5.74	13.6	1730	1.99	39.2	34.0
102	114	102	6.02	16.1	2050	3.01	52.6	38.4
127	141	128	6.55	21.8	2770	6.33	89.3	47.8
152	168	154	7.11	28.2	3600	11.7	139	57.2
203	219	203	8.18	42.5	5420	30.2	275	74.7
254	273	255	9.27	60.2	7680	67.0	490	93.2
305	324	305	9.53	73.7	9420	116	718	111
				Extra Strong				
51	60.3	49.3	5.54	7.47	955	0.361	12.0	19.5
64	73.0	59.0	7.01	11.4	1450	0.799	22.0	23.5
76	88.9	73.7	7.62	15.3	1950	1.62	36.5	29.0
90	102	85.4	8.08	18.6	2370	2.61	51.5	33.3
102	114	97.2	8.56	22.3	2850	4.00	70.0	37.6
127	141	122	9.53	30.9	3940	8.62	122	46.7
152	168	146	11.0	42.5	5420	16.9	200	55.6
203	219	194	12.7	64.6	8260	44.1	401	73.2
254	273	248	12.7	81.4	10,400	88.2	646	92.2
305	324	298	12.7	97.3	12,400	151	929	110

TABLE A.9 (U.S. Customary Units) American Standard Timber Sizes: Properties for Designing[†] (Selected Listing)

Nominal Size	American Standard Dressed Size	Area of Section	Weight per Foot	Moment of Inertia	Section Modulus
in.	in.	in.^{2}	lb	in.^{4}	in.^{3}
2 × 4	1.5 × 3.5	5.25	1.46	5.36	3.06
× 6	5.5	8.25	2.29	20.8	7.56
× 8	7.25	10.9	3.02	47.6	13.1
× 10	9.25	13.9	3.85	89.9	21.4
× 12	11.25	16.9	4.69	178	31.6
4 × 4	3.5 × 3.5	12.2	3.40	12.5	7.15
× 6	5.5	19.2	5.35	48.5	17.6
× 8	7.25	25.4	7.05	111	30.7
× 10	9.25	32.4	8.99	231	49.9
× 12	11.25	39.4	10.9	415	73.8
6 × 6	5.5 × 5.5	30.2	8.40	76.3	27.7
× 8	7.5	41.2	11.5	193	51.6
× 10	9.5	52.2	14.5	393	82.7
× 12	11.5	63.2	17.6	697	121
× 14	13.5	74.2	20.6	1128	167
8 × 8	7.5 × 7.5	56.2	15.6	263	70.3
× 10	9.5	71.2	19.8	536	113
× 12	11.5	86.2	24.0	951	165
× 14	13.5	101	28.1	1538	228
× 16	15.5	116	32.3	2327	300
10 × 10	9.5 × 9.5	90.2	25.1	679	143
× 12	11.5	109	30.3	1204	209
× 14	13.5	128	35.6	1948	289
× 16	15.5	147	40.9	2948	380
× 18	17.5	166	46.2	4243	485
× 20	19.5	185	51.5	5870	602
12 × 12	11.5 × 11.5	132	36.7	1458	253
× 14	13.5	155	43.1	2358	349
× 16	15.5	178	49.5	3569	460
× 18	17.5	201	55.9	5136	587
× 20	19.5	224	62.3	7106	729
× 22	21.5	247	68.7	9524	886
× 24	23.5	270	75.1	12,440	1058

[†]All weights and properties calculated for dressed sizes. Weights based on 40 lb/ft^3. Moments of inertia and section modulus calculated from formulas.

TABLE A.9   (SI Units) American Standard Timber Sizes: Properties for Designing[†] (Selected Listing)

Nominal Size	American Standard Dressed Size	Area of Section	Mass per Meter	Moment of Inertia	Section Modulus
mm	mm	$\times 10^3 mm^2$	kg	$\times 10^6 mm^4$	$\times 10^3 mm^3$
50 × 100	38 × 89	3.38	2.03	2.23	50.2
× 150	140	5.32	3.19	8.69	124
× 200	184	6.99	4.20	19.7	214
× 250	235	8.93	5.36	41.1	350
× 300	286	10.87	6.52	74.1	518
100 × 100	89 × 89	7.92	4.75	5.23	117
× 150	140	12.5	7.48	20.4	291
× 200	184	16.4	9.83	46.2	502
× 250	235	20.9	12.5	96.3	819
× 300	286	25.4	15.3	174	1213
150 × 150	140 × 140	19.6	11.8	32.0	457
× 200	191	26.7	16.0	81.3	851
× 250	241	33.7	20.2	163	1355
× 300	292	40.9	24.6	290	1989
× 350	343	48.0	28.8	471	2745
200 × 200	191 × 191	36.5	21.9	111	1161
× 250	241	46.0	27.6	223	1849
× 300	292	55.8	33.5	396	2714
× 350	343	65.5	39.3	642	3745
× 400	394	75.2	45.2	974	4942
250 × 250	241 × 241	58.1	34.8	281	2333
× 300	292	70.4	42.2	500	3425
× 350	343	82.7	49.6	810	4726
× 400	394	95.0	57.0	1228	6235
× 450	445	107	64.3	1770	7954
× 500	495	119	71.6	2436	9842
300 × 300	292 × 292	85.3	51.2	606	4150
× 350	343	100	60.1	982	5726
× 400	394	115	69.0	1488	7555
× 450	445	130	78.0	2144	9637
× 500	495	144	86.7	2951	11,920
× 550	546	159	95.7	3961	14,510
× 600	597	174	105	5178	17,350

[†]All masses and properties calculated for dressed sizes. Masses based on 600 kg/m³. Moments of inertia and section modulus calculated from formulas.

TABLE A.10 (U.S. Customary Units) Typical Mechanical Properties of Selected Engineering Materials

No.	Material	Specific Weight (lb/ft³)	Modulus (10³ ksi)		Tensile Strength (ksi)		Compressive Strength (ksi)		Shear Strength (ksi)		Coefficient of Thermal Expansion (10⁻⁶/°F)	Ductility (% Elongation in 2 in.)
			E	G	Yld.ᵃ	Ult.	Yld.ᵃ,ᵇ	Ult.ᶜ	Yld.	Ult.		
	Steel											
	Carbon—structural											
1	ASTM–A36	490	29	11.5	36	58	—	—	21	—	6.5	23
	High strength, low alloy (HSLC)											
2	A633 Grade A	490	30	12.0	42	68/83	—	—	24	—	6.5	23
3	Grade E	490	30	12.0	50/60	75/100	—	—	29/35	—	6.5	23
	Stainless											
4	(301) Annealed	489	28	9.5	40	110	—	—	24	—	9.6	60
5	Cold worked	489	28	9.5	140	185	—	—	84	—	9.6	9
	Cast iron											
	Gray											
6	Ferritic class 25	450	9.7	4.1	20	25	—	95	—	35	6.0	0.4
	Ductile											
7	Ferritic (60–40–18)	456	22	8.5	40	60	—	108	—	42	6.4	18
	Aluminum alloy											
8	1060 0	169	10.1	3.7	4	10	—	—	2.3	6.2	12.9	42
9	Hard H18	169	10.1	3.7	18	19	—	—	10.3	11.9	12.9	6
10	6061 Annealed	170	10	3.7	8	18	—	—	4.6	11.2	13.1	30
11	Heat treated–T6	170	10	3.7	40	45	—	—	22.9	28.1	13.1	12

No.	Material	Specific Weight (lb/ft³)	Modulus (10³ ksi)		Tensile Strength (ksi)		Compressive Strength (ksi)		Shear Strength (ksi)		Coefficient of Thermal Expansion (10⁻⁶/°F)	Ductility (% Elongation in 2 in.)
			E	G	Yld.[a]	Ult.	Yld.[a,b]	Ult.[c]	Yld.	Ult.		
	Magnesium alloy											
12	AZ 31B–H24	112	6.5	2.4	32	42	—	—	—	18.3	14.0	15
13	AZ 91C–T6	114	6.5	2.4	19	40	—	—	—	17.5	14.0	5
	Copper alloy											
14	C26800 Annealed	529	15	5.6	14	46	—	—	8	25.4	11.2	65
15	Cold worked	529	15	5.6	60	74	—	—	34.2	40.8	11.2	8
16	C71500 Annealed	557	16.5	6.2	20	44	—	—	11.4	24.3	9.0	40
17	Cold worked	557	16.5	6.2	68	75	—	—	38.8	41.3	9.0	12
	Concrete											
18	Low strength	150	2.5	—	—	—	—	1	—	0.2	6.0	—
19	High strength	150	4.5	—	—	—	—	10	—	2	6.0	—
	Wood[d]											
20	Douglas fir (coast)	30	1.9	—	—	—	6.5 (0.9)[d]	7.4	—	1.1	3.0	—
21	Southern pine (longleaf)	36	2.0	—	—	—	6.1 (1.2)[d]	8.4	—	1.5	3.0	—

[a]Yield point or for an offset of 0.2%; proportional limit for wood.
[b]For ductile material, the yield strength in compression may be assumed to be the same as in tension.
[c]For ductile material, the ultimate strength in compression is indefinite; may be assumed to be the same as in tension.
[d]Clear seasoned wood, 12% moisture content; properties parallel to the grain and perpendicular (in parentheses) to the grain.

TABLE A.10 (SI Units) Typical Mechanical Properties of Selected Engineering Materials

No.	Material	Density (kg/m³)	Modulus (GN/m²)		Tensile Strength (MN/m²)		Compressive Strength (MN/m²)		Shear Strength (MN/m²)		Coefficient of Thermal Expansion (10⁻⁶/°C)	Ductility (% Elongation in 2 mm)
			E	G	Yld.[a]	Ult	Yld.[a,b]	Ult.[c]	Yld.	Ult.		
	Steel											
	Carbon—structural											
1	ASTM-A36	7850	200	79	250	400	—	—	145	—	11.7	23
	High strength low alloy (HSLC)											
2	A633 Grade A	7850	207	83	290	470/570	—	—	165	—	11.7	23
3	Grade E	7850	207	83	375/410	520/690	—	—	200/240	—	11.7	23
	Stainless											
4	(301) Annealed	7830	193	65	275	760	—	—	165	—	17.3	60
5	Cold worked	7830	193	65	965	1280	—	—	580	—	17.3	9
	Cast iron											
	Gray											
6	Ferritic class 25	7210	67	28	140	170	—	650	—	240	10.8	0.4
	Ductile											
7	Ferritic (60–40–18)	7300	150	59	275	415	—	745	—	290	11.5	18
	Aluminum alloy											
8	1060 0	2710	70	26	28	70	—	—	16	42.5	23.2	42
9	Hard H18	2710	70	26	125	130	—	—	71	82	23.2	6
10	6061 Annealed	2720	69	26	55	125	—	—	31.5	77	23.6	30
11	Heat treated-T6	2720	69	26	275	310	—	—	158	195	23.6	12

No.	Material	Density (kg/m³)	Modulus (GN/m²)		Tensile Strength (MN/m²)		Compressive Strength (MN/m²)		Shear Strength (MN/m²)		Coefficient of Thermal Expansion (10⁻⁶/°C)	Ductility (% Elongation in 2 mm)
			E	G	Yld.[a]	Ult.	Yld.[a,b]	Ult.[c]	Yld.	Ult.		
	Magnesium alloy											
12	AZ 31B–H24	1790	45	16.5	220	290	—	—	—	125	25.2	25.6
13	AZ 91C–T6	1830	45	16.5	130	275	—	—	—	120	25.2	5
	Copper alloy											
14	C26800 Annealed	8470	105	39	97	320	—	—	55	175	20.2	65
15	Cold worked	8470	105	39	415	510	—	—	235	280	20.2	8
16	C71500 Annealed	8920	115	43	140	305	—	—	79	168	16.2	40
17	Cold worked	8920	115	43	470	520	—	—	268	285	16.2	12
	Concrete											
18	Low strength	2400	17	—	—	—	—	6.9	—	1.4	10.8	—
19	High strength	2400	31	—	—	—	—	68.9	—	13.8	10.8	—
	Wood[d]											
20	Douglas fir (coast)	480	13	—	—	—	45 (6.2)[d]	51	—	7.6	5.4	—
21	Southern pine (longleaf)	576	14	—	—	—	42 (8.2)[d]	58	—	10.3	5.4	—

[a]Yield point or for an offset of 0.2%; proportional limit for wood.

[b]For ductile material, the yield strength in compression may be assumed to be the same as in tension.

[c]For ductile material, the ultimate strength in compression is indefinite; may be assumed to be the same as in tension.

[d]Clear seasoned wood, 12% moisture content; properties parallel to the grain and perpendicular (in parentheses) to the grain.

TABLE A.11   Common Grades for Machine-Graded Lumber[a]

Grade Name	Fb MPa (psi)		E GPa (× 10⁶ psi)		Ft MPa (psi)		Fcll MPa (psi)	
MSR								
1350f–1.3E	9.3	(1350)	9.0	(1.3)	5.2	(750)	11.0	(1600)
1450f–1.3E	10.0	(1450)	9.0	(1.3)	5.5	(800)	11.2	(1625)
1650f–1.5E	11.4	(1650)	10.3	(1.5)	7.0	(1020)	11.7	(1700)
1800f–1.6E	12.4	(1800)	11.0	(1.6)	8.1	(1175)	12.1	(1750)
1950f–1.7E	13.4	(1950)	11.7	(1.7)	9.5	(1375)	12.4	(1800)
2100f–1.8E	14.5	(2100)	12.4	(1.8)	10.9	(1575)	12.9	(1875)
2250f–1.9E	15.5	(2250)	13.1	(1.9)	12.1	(1750)	13.3	(1925)
2400f–2.0E	16.5	(2400)	13.8	(2.0)	13.3	(1925)	13.6	(1975)
2550f–2.1E	17.6	(2550)	14.5	(2.1)	14.1	(2050)	14.0	(2025)
2700f–2.2E	18.6	(2700)	15.2	(2.2)	14.8	(2150)	14.4	(2100)
2850f–2.3E	19.7	(2850)	15.9	(2.3)	15.9	(2300)	14.8	(2150)
MEL								
M–10	9.7	(1400)	8.3	(1.2)	5.5	(800)	11.0	(1600)
M–11	10.7	(1550)	10.3	(1.5)	5.9	(850)	11.5	(1675)
M–14	12.4	(1800)	11.7	(1.7)	6.9	(1000)	12.1	(1750)
M–19	13.8	(2000)	11.0	(1.6)	9.0	(1300)	12.6	(1825)
M–21	15.9	(2300)	13.1	(1.9)	9.7	(1400)	13.4	(1950)
M–23	16.5	(2400)	12.4	(1.8)	13.1	(1900)	13.6	(1975)
M–24	18.6	(2700)	13.1	(1.9)	12.4	(1800)	14.5	(2100)

[a]Other grades are available and permitted.
F_b is the allowable 10-year load duration bending stress parallel to the grain.
E is the modulus of elasticity.
F_t is the allowable 10-year load duration tensile stress parallel to the grain.
F_{cll} is the allowable 10-year load duration compressive stress parallel to the grain.

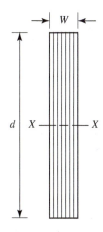

TABLE A.12   (U.S. Customary Units) Properties of LVL Boards

Width × Depth in.	Area in.^{2}	I_{x-x} in.^{4}	S_{x-x} in.^{3}
1-3/4 × 9-1/2	16.6	125	26.3
× 11-7/8	20.8	244	41.1
× 14	24.5	400	57.2
× 16	28.0	597	74.7
× 18	31.5	851	94.5

TABLE A.12   (SI Units) Properties of LVL Boards

Width × Depth mm	Area × 10^{3}mm^2	I_{x-x} × 10^{6}mm^4	S_{x-x} × 10^{4}mm^3
44 × 241	10.7	52.0	43.1
× 302	13.4	102	67.4
× 356	15.8	167	93.7
× 406	18.1	249	122
× 457	20.3	354	155

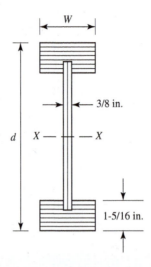

TABLE A.13 (U.S. Customary Units) Properties of Wooden I-Joists			
Width × Depth in.	Area in.^{2}	I_{x-x} in.^{4}	S_{x-x} in.^{3}
1-1/2 × 9-1/2	6.52	76.7	16.1
× 11-7/8	7.41	135	22.8
1-3/4 × 9-1/2	7.17	87.8	18.5
× 14	8.86	232	33.1
× 16	9.61	323	40.4
2 × 9-1/2	7.83	98.9	20.8
× 11-7/8	8.72	172	29.0
× 14	9.52	258	36.9
× 16	10.3	359	44.8
3-1/2 × 11-7/8	12.7	282	47.5
× 14	13.5	417	59.6
× 16	14.2	572	71.4

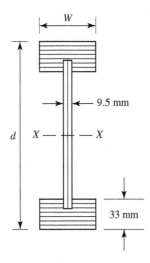

Width × Depth mm	Area × 10^3mm²	I_{x-x} × 10^6mm⁴	S_{x-x} × 10^4mm³
38 × 241	4.21	31.9	26.4
× 302	4.78	56.2	37.4
44 × 241	4.63	36.5	30.3
× 356	5.72	96.6	54.2
× 406	6.20	134	66.2
51 × 241	5.05	41.2	34.1
× 302	5.63	71.6	47.5
× 356	6.14	107	60.5
× 406	6.65	149	73.4
89 × 302	8.19	117	77.8
× 356	8.71	174	97.7
× 406	9.16	238	117

TABLE A.13   (SI Units) Properties of Wooden I-Joists

TABLE A.14 Deflections, Slopes, and Deflection Equations for Beams

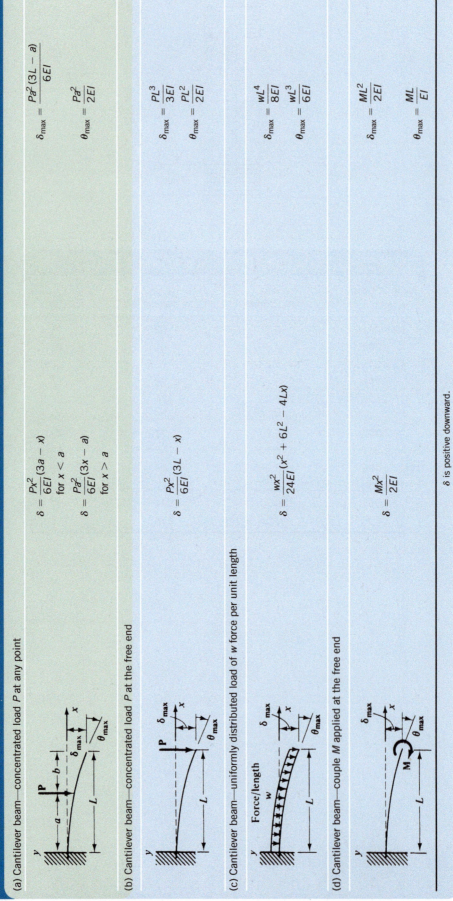

(a) Cantilever beam—concentrated load P at any point

$$\delta = \frac{Px^2}{6EI}(3a - x)$$
for $x < a$

$$\delta = \frac{Pa^2}{6EI}(3x - a)$$
for $x > a$

$$\delta_{max} = \frac{Pa^2(3L - a)}{6EI}$$

$$\theta_{max} = \frac{Pa^2}{2EI}$$

(b) Cantilever beam—concentrated load P at the free end

$$\delta = \frac{Px^2}{6EI}(3L - x)$$

$$\delta_{max} = \frac{PL^3}{3EI}$$

$$\theta_{max} = \frac{PL^2}{2EI}$$

(c) Cantilever beam—uniformly distributed load of w force per unit length

$$\delta = \frac{wx^2}{24EI}(x^2 + 6L^2 - 4Lx)$$

$$\delta_{max} = \frac{wL^4}{8EI}$$

$$\theta_{max} = \frac{wL^3}{6EI}$$

(d) Cantilever beam—couple M applied at the free end

$$\delta = \frac{Mx^2}{2EI}$$

$$\delta_{max} = \frac{ML^2}{2EI}$$

$$\theta_{max} = \frac{ML}{EI}$$

δ is positive downward.

(e) Beam freely supported at the ends—concentrated load P at any point

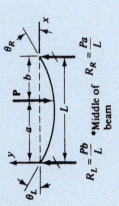

$R_L = \dfrac{Pb}{L}$ *Middle of beam $R_R = \dfrac{Pa}{L}$

$$\delta = \frac{Pbx}{6LEI}(L^2 - x^2 - b^2)$$
for $x < a$

$$\delta = \frac{Pb}{6LEI}\left[\frac{L}{b}(x-a)^3 - x^3 + (L^2 - b^2)x\right]$$

$\delta^*_M = \dfrac{Pa}{48EI}(3L^2 - 4a^2)$ $\delta^*_M = \dfrac{Pb}{48EI}(3L^2 - 4b^2)$
for $b > a$ for $a > b$

$$\delta_{max} = \frac{Pb(L^2 - b^2)^{3/2}}{9\sqrt{3}LEI}$$
at $x = \sqrt{\dfrac{L^2 - b^2}{3}}$

$\theta_L = \dfrac{Pab(2L - a)}{6LEI}$

$\theta_R = \dfrac{Pab(2L - b)}{6LEI}$

(f) Beam freely supported at ends—concentrated load P at the center

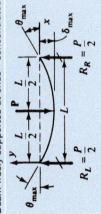

$R_L = \dfrac{P}{2}$ $R_R = \dfrac{P}{2}$

$$\delta = \frac{Px}{48EI}(-4x^2 + 3L^2)$$
for $x < \dfrac{L}{2}$

$$\delta = \frac{Px}{48EI}\left[\frac{8}{x}\left(x - \frac{L}{2}\right)^3 - 4x^2 + 3L^2\right]$$
for $x > \dfrac{L}{2}$

$\delta_{max} = \dfrac{PL^3}{48EI}$
at $x = \dfrac{L}{2}$

$\theta_{max} = \dfrac{PL^2}{16EI}$

(g) Beam freely supported at the ends—uniformly distributed load of w force per unit length

$R_L = \dfrac{wL}{2}$ $R_R = \dfrac{wL}{2}$

$$\delta = \frac{wx}{24EI}(L^3 - 2Lx^2 + x^3)$$

$\delta_{max} = \dfrac{5wL^4}{384EI}$
at $x = \dfrac{L}{2}$

$\theta_{max} = \dfrac{wL^3}{24EI}$

(h) Beam freely supported at the ends—couple M at the right end

$R_L = \dfrac{M}{L}$ $R_R = \dfrac{M}{L}$

$$\delta = \frac{Mx}{6EIL}(L^2 - x^2)$$

$\theta_L = \dfrac{ML}{6EI}$ $\theta_R = \dfrac{ML}{3EI}$

$\delta_{max} = \dfrac{ML^2}{9\sqrt{3}EI}$
at $x = \dfrac{L}{\sqrt{3}}$

δ is positive downward.

ANSWERS TO EVEN-NUMBERED PROBLEMS

Chapter 1

1.2 (a) 33.5 ft, (b) 147.6 ft, (c) 0.669 ft, (d) 15.09 ft

1.4 (a) 0.0805 hp, (b) 156,597 W

1.6 (a) 0.01189 kip, (b) 1.540 kip

1.8 (a) 204 kg, (b) 357 kg, (c) 122.3 kg, (d) 449 kg

1.10 229 lb/ft, 3.34 kN/m, 341 kg/m

1.12 36.3 ksi, 7.98 ksi

1.14 58.0 ksi, 10.15 ksi

1.16 (a) 2.74 m, (b) 7.52 m

1.18 $\theta = 29.5°$

1.20 $b = 713$ ft

1.22 230.7 ft

1.24 Angle $ABC = 44.3°$

1.26 (a) All nonright angles are 45°, (b) 132 ft

1.28 (a) $b = 4.79$ in., $c = 3.72$ in.
(b) $c = 4.57$ m, $A = 29.2°$, $B = 102.8°$
(c) $B = 49.3°$, $C = 70.7°$, $c = 8.72$ ft
(d) $C = 106.6°$, $A = 25.2°$, $B = 48.2°$

1.30 $AB = 5.26$ ft, $\theta = 12.4°$

1.32 $OB = 390$ mm, $OB' = 275$ mm,
Distance moved $= 115.0$ mm

1.34 (a) 85.2°
(b) 59.5°
(c) 94.8°

1.36 $h = 97.4$ ft (will hit house)

1.38 $R = 22.0$ ft, $B = 32.8°$

1.40 $x = 107, y = 77$

1.42 $m = 4.5, n = 4.5$

1.44 $3x - 5y = 0$
$6x + 1y = -4$

1.46 $\$10.10 = F + 12r$
$\$13.95 = F + 19r$
$F = \$3.50, r = \$0.55/lb$

1.48 $x = -2, y = 1, z = 3$

1.50 $x = 2, y = 3, z = -1$

1.52 $H = $ weight of heaviest casting
$L = $ weight of lightest casting
$M = $ weight of middle casting
$H + L + M = 1867$
$H - L = 395$
$2L = (M + H) - 427$
$H = 875$ lb, $L = 480$ lb, $M = 512$ lb

Chapter 2

2.2 100 N @ −58°

2.4 11.60 kN @ 80.1°

2.6 $Q = 40.2$ kN, $R = 53.4$ kN

2.8 $P = 6.53$ kN, $Q = 9.93$ kN

2.10 13 kip @ −35°

2.12 8.5 kN @ 22°

2.14 (a) 1414 lb parallel, 1414 lb perpendicular
(b) 518 lb parallel, 1932 lb perpendicular

2.16 (a) -27.6 kN$-x'$, 4.86 kN$-y'$
(b) 1299 lb$-x'$, -750 lb$-y'$
(c) 5.10 kip$-x'$, 8.83 kip$-y'$
(d) -448 N$-x'$, -39.2 N$-y'$

2.18 perpendicular component $= 948.8$ lb
parallel component $= 1350$ lb

2.20 1905 kN @ 65.7°

2.22 8.05 kip @ −169°

2.24 2790 lb @ −20.3°

2.26 101.0 N @ −57.9°

2.28 11.61 kN @ 80.0°

2.30 13.23 kip @ −35.2°

2.32 8.47 kip @ 22.4°

2.34 14.63 kN @ 160.6°

2.36 9.30 kip @ 143.7°

Chapter 3

3.10 $F_{BA} = 4.28$ kip, $F_{BC} = 16.30$ kip

3.12 $F_{AB} = 4.86$ kip, $F_{AC} = -4.17$ kip

3.14 $N_R = 77.2$ lb, $F = 32.7$ lb

3.16 $W_B = 34.2$ lb, $N_R = 94.0$ lb

3.18 $F_1 = 1660$ lb, $F_2 = 318$ lb

3.20 $F_{BA} = 2770$ N, $m_1 = 350$ kg

3.22 $T_{BA} = 2.66$ kN, $N_R = 0.910$ kN

3.24 $\theta_3 = 35.7°$, $F_4 = 123.8$ lb

Chapter 4

4.2 $M_O = M_R = 86$ kN·m, $M_P = 278$ kN·m,
$M_Q = -106$ kN·m

4.4 $M_A = 16,000$ lb-ft, $M_B = 3000$ lb-ft, $M_C = 10,000$ lb-ft

4.6 $M_P = 148,000$ lb-in., $M_Q = 4940$ lb-in.,
$M_R = 68,900$ lb-in., $M_S = 121,800$ lb-in.

4.8 $R = -3$ kip, $\bar{x}_A = 8$ ft

4.10 $R = -40$ kN, $\bar{x}_A = 6.9$ m

4.12 $R = 9.67$ kN @ −76.5° located 6.29 m to the right of A and
2.71 m to the left of C

4.14 $R = 6.52$ kip @ −101.5° located 3.41 ft to the right of A and
4.59 ft to the left of D

4.16 $R = 38.1$ kip @ −23.2° located 21.4 ft above J and 18.6 ft
below F

4.18 (a) $+60$ lb @ C and -60 lb @ B

(b) $+80$ lb @ E and -80 lb @ D

4.20 $M = 6.44$ kip-ft clockwise

4.22 (a) $F = -80$ lb @ B, $M_C = -960$ lb-in.

(b) $F = -80$ lb @ C, $M_C = -960$ lb-in.

4.24 $F_o = 696$ lb @ $21.0°$, $M_C = +2500$ lb-in.

4.26 $P_A = 12$ kip @ $-100°$, $M_C = -8.75$ kip-in.

4.28 (a) $R = -50$ kip, $\bar{x}_A = 10$ ft to the right of A

(b) $R = -50$ kip @ A, $M_C = -500$ kip-ft

4.30 (a) $R = -43.5$ kN, $\bar{x}_A = 1$ m

(b) $R = -43.5$ kN @ A, $M_C = 43.5$ kN·m

4.32 (a) $R = -8.8$ kN @ 3.89 m to the right of A

(b) $R = -8.8$ kN @ A, $M_C = -34.2$ kN·m

Chapter 5

5.2 (a) $B = 242$ lb, $C = 256$ lb, (b) $F_{min} = 48$ lb

(c) $F_{max} = 966$ lb

5.4 $A_x = 0$, $A_y = 1.65$ kN↑, $C = 1.10$ kN↑

5.6 $B_x = 2696$ lb

$B_y = 1389$ lb

$C = 2704$ lb

5.8 $A_x = 0$, $A_y = 34$ kN, $M_A = -124.8$ kN·m

5.10 $A_x = -35$ kip, $A_y = -75$ kip, $F_{JI} = -90$ kip

5.12 $A_y = 22.5$ kip, $B_x = 0$, $B_y = 2.5$ kip

5.14 $B_x = 0$, $B_y = 15$ kN, $M_B = -45$ kN·m

5.16 $A_x = 0$, $A_y = 2450$ N, $T_{BC} = 2940$ N

5.18 $A_x = 514$ lb, $A_y = 613$ lb, $T = 526$ lb

5.20 $A_x = -739$ N, $A_y = 106.7$ N, $C = 853$ N @ $-30°$

5.22 $A_x = 0$, $A_y = 11.5$ kip, $E_y = 14.5$ kip

5.24 $B_y = -2810$ lb, $C_x = 0$, $C_y = 712$ lb

5.26 $B_x = -60$ N, $B_y = 13.33$ N, $C_y = 86.7$ N

5.28 $A_x = -257$ lb, $A_y = -857$ lb, $F_{GF} = -924$ lb

5.30 $T_{BD} = 466$ lb, $T_{AC} = 413$ lb

5.32 $F_{CB} = -490$ N, $A = 693$ N @ $45°$

5.34 $A = 747$ N @ $171.8°$, $B = 853$ N @ $-30°$

5.36 $B = 61.5$ N @ $167.5°$, $C = 86.7$ N upward

5.38 (b) improperly constrained, statically indeterminate, not in equilibrium

(c) completely constrained, statically determinate
$C = 16.67$ kip, $E_x = -16.67$ kip, $E_y = 15$ kip

(d) completely constrained, statically determinate
$G = 20$ kip slope $4/3$, $E_x = 12$ kip, $E_y = 1.0$ kip

(e) improperly constrained, statically indeterminate, not in equilibrium

(f) completely constrained, statically determinate
$G = 20.8$ kip, $E_x = 16.67$ kip, $E_y = 2.5$ kip

(g) improperly constrained, statically indeterminate

(h) completely constrained, statically indeterminate

(i) completely constrained, statically determinate
$E = -16.67$ kip, $C_x = 16.67$ kip, $C_y = 15$ kip

Chapter 6

6.2 $AB = 2000$ lb (C), $AC = 2154$ lb (T), $BC = 2828$ lb (C)

6.4 $F_{AB} = 3.60$ kN, $F_{AC} = 6.0$ kN, $F_{CB} = -7.0$ kN

6.6 $F_{AB} = 12.11$ kN, $F_{CB} = -28.2$ kN, $F_{AC} = -8.10$ kN

6.8 $F_{BA} = -30$ kN, $F_{DC} = -30$ kN, $F_{CA} = F_{BC} = F_{AD} = 0$

6.10 $F_{AD} = 34.4$ kN, $F_{AB} = 23.6$ kN, $F_{CD} = 34.4$ kN,

$F_{CB} = -47.1$ kN, $F_{DB} = 16.67$ kN

6.12 Member forces in kN: $af = dj = -27.0$, $bg = ci = -21.5$, $ef = ej = 22.5$, $eh = 15.0$,

$fg = ij = -8.3$, $gh = hi = 7.5$

6.14 $A_x = 0$, $A_y = 2.5$ kip, $C_y = 2.5$ kip

$F_{AD} = 5.91$ kip (T), $F_{AB} = 5.36$ kip (C), $F_{BD} = 4.53$ kip (T), $F_{DC} = 3.80$ kip (T), $F_{BC} = 6.52$ kip (C)

6.16 $AB = -13.5$, $AG = +5.4$, $BC = -10.0$,

$BD = -13.5$, $BG = -10.0$, $CD = +11.2$,

$DE = +10.0$, $DG = -3.4$,

$EF = +11.2$, $EG = -15.0$,

$FG = -10.0$, $GH = -21.0$

6.18 JK, JI, LI, HM, NG, PE, RC, BC

6.20 HF, FI, IE, EJ, LB

6.22 $F_{BC} = 2.29$ kip (C), $F_{FE} = 7.08$ kip (T), $F_{BE} = 3.82$ kip (C)

6.24 $A_x = 1414$ lb, $A_y = 354$ lb, $F_{GE} = 354$ lb (T)

$F_{BC} = 500$ lb (T), $F_{AF} = 1060$ lb (C), $F_{BF} = 1000$ lb (T)

6.26 $F_{BC} = 3.0$ kip (T), $F_{GF} = 6.0$ kip (C), $F_{GC} = 5.0$ kip (T),

$F_{CD} = 6.0$ kip (T), $F_{FE} = 9.0$ kip (C), $F_{FD} = 5.0$ kip (T)

6.28 $F_{FE} = 2.33$ kip (T), $F_{BC} = 4.67$ kip (C),

$F_{BE} = 4.21$ kip (T)

6.30 (a) $D = 571$ lb, $F = 2429$ lb

(b) $A = 3964$ lb, $C = 3007$ lb

6.32 Reactions: $E_x = 800$ N, $E_y = 400$ N,

$A = -800$ N, $F_{AB} = 800$ N (C),

$F_{BC} = 1132$ N (C), $F_{BD} = 800$ N (T)

6.34 $B_x = 2833$ lb on AC, $B_y = 3437$ lb on BD, $A_x = -583$ lb on AC, $A_y = 3437$ lb on AC, $D_x = 2833$ lb on BD, $D_y = -2937$ lb on BD

6.36 $A_x = 0$, $A_y = -2000$ N on AC, $E = 3000$ N on ED, $D_x = 0$, $B_x = 0$, $D_y = -1000$ N on DF, $C_x = 0$, $C_y = 2000$ N on DF, $B_y = -4000$ N on DE

6.38 Reactions: $E = 1.625$ kN, $A_y = 2.5$ kN, $A_x = -1.625$ kN, $F_{AE} = 2.5$ kN (T), $C_x = 1.625$ kN on AD, $C_y = 5.0$ kN on AD, $E_x = 1.625$ kN on CE, $E_y = 2.5$ kN on CE

6.40 (a) $DB = 1143$ lb↗

(b) $A_x = 506.4$ lb ←

$A_y = 600.0$ lb↓

6.42 (a) $P = 0.625$ kip, $R_1 = 0.625$ kip,

$R_2 = 1.25$ kip

(b) $P = 0.415$ kip, $R = 1.667$ kip

(c) $P = 0.1562$ kip, $R_1 = 0.625$ kip,

$R_2 = 0.3125$ kip, $R_3 = 0.1562$ kip,

$R_4 = 0.3125$ kip

(d) $P = 0.417$ kip, $R_1 = 0.833$ kip,

$R_2 = 0.833$ kip

6.44 (a) $P = 264$ N, $B = 957$ N

(b) $P = 57.4$ lb, $B = 208$ lb

6.46 $Q = 6.6$ kN; Force on pin A, B, and C: 6.69 kN

6.48 (a) $B = 3800$ N↓

(b) $A = 3580$ N↓

6.50   (a) $F = 6370\,\text{N}\uparrow$

     (b) $B = 21,100\,\text{N}\downarrow$

     (c) $A = 14,900\,\text{N}\downarrow$

6.52 $J = K = 10.3\,\text{N}$

6.54 $B = 55.1\,\text{kN}, E = 38.1\,\text{kN}$

6.56   (a) $T_{AB} = T_{BC} = 465\,\text{N}; T_{CD} = 903\,\text{N},$

     (b) $h = 16.7\,\text{m}$, (c) $L_C = 29.1\,\text{m}$

6.58   (a) $T_{\min} = 2.73\,\text{kN}$, (b) $T_{\max} = 3.79\,\text{kN}$,

     (c) $T_{\max} = 5.16\,\text{kN}$, (d) $L_C = 48.6\,\text{m}$

6.60   (a) $L_C = 209.4\,\text{m}$, (b) $T_{\max} = 10,850\,\text{N}; h = 27.0\,\text{ft}$

Chapter 7

7.2   (a) $P_x = P_y = P_z = 57.7\,\text{lb}, Q_x = -128.6\,\text{lb},$

     $Q_y = 42.9\,\text{lb}, Q_z = 64.3\,\text{lb}$

     (b) $R = 173.3\,\text{lb}, \theta_x = 114.1°, \theta_y = 54.5°,$

     $\theta_z = 45.3°$

7.4   (a) $200\,\text{kN}: (0, -200, 0)\,\text{kN}; 150\,\text{kN}: (-106.1,$

     $84.9, -63.6)\,\text{kN}; 180\,\text{kN}: (-154.4, 0, 92.6)\,\text{kN}$

     (b) $R = 286\,\text{kN}, \theta_x = 155.5°, \theta_y = 113.7°,$

     $\theta_z = 84.2°$

7.6   (a) $25.8\,\text{lb}$, (b) $120.4\,\text{N}$

7.8 $AB = 654\,\text{N}, AC = 427\,\text{N}, AD = 288\,\text{N}$

7.10 $AB = 0, AC = 12.81\,\text{kN}, AD = -10.0\,\text{kN}$

7.12 $AB = -2.30\,\text{kN}, AC = -1.957\,\text{kN}, AD = 3.41\,\text{kN}$

7.14 $M_x = -61.2\,\text{lb-ft}, M_y = -9840\,\text{lb-ft}, M_z = 183.6\,\text{lb-ft}$

7.16 $M_x = -518\,\text{lb-ft}, M_y = 558\,\text{lb-ft},$

     $M_z = 20.4\,\text{lb-ft}$

7.18 $R_x = 30\,\text{lb}, \bar{y} = 12.67\,\text{in.}, \bar{z} = -5.17\,\text{ft}$

7.20 $T_A = 1.462\,\text{kN}, T_B = 3.65\,\text{kN}, T_C = 2.19\,\text{kN}$

7.22 $T_{AC} = 42.1\,\text{kN}, T_{AB} = 36.5\,\text{kN}, D_x = 63.0\,\text{kN},$

     $D_y = 60.1\,\text{kN}, D_z = 0$

7.24 $A_x = -4070\,\text{lb}, A_y = 4440\,\text{lb}, B_x = -926\,\text{lb},$

     $B_y = 5560\,\text{lb}, B_z = 0$

7.26 $T_{CD} = 129.1\,\text{N}, A_x = -53.7\,\text{N}, A_y = 61.2\,\text{N},$

     $B_x = -28.0\,\text{N}, B_y = 255\,\text{N}, B_z = 94.2\,\text{N}$

Chapter 8

8.2   (a) $150\,\text{N}$, (b) $390\,\text{N}$, (c) $294\,\text{N}$

8.4 $290\,\text{N}, 41.3\,\text{N}$

8.6 $135.4\,\text{lb}, 20.9\,\text{lb}$

8.8 $64.3\,\text{in.}$

8.10 0.443

8.12 0.150

8.14 $1.0\,\text{ft}$

8.16 $3120\,\text{N}$

8.18 $596\,\text{lb}$

8.20 $33.4°$

8.22   (a) 0.110, (b) 0.166

8.24 $2930\,\text{N}$

8.26 $0.5\,\text{lb-ft}$

8.28 $1404\,\text{N·m}$

8.30 0.136

8.32 $26.4\,\text{lb-in.}$

8.34 $3560\,\text{lb-in.}$

8.36 $P = 4.91\,\text{N}$

8.38 $a = 0.828\,\text{in.}$

Chapter 9

9.6 $\bar{x} = 150\,\text{mm}, \bar{y} = 120\,\text{mm}$

9.8 $\bar{x} = 2.4\,\text{in.}, \bar{y} = 2.2\,\text{in.}$

9.10 $\bar{x} = 42\,\text{mm}, \bar{y} = 43.6\,\text{mm}$

9.12 $\bar{x} = 75\,\text{mm}, \bar{y} = 94.9\,\text{mm}$

9.14 $\bar{x} = 3.36\,\text{in.}, \bar{y} = 2.4\,\text{in.}$

9.16 $\bar{x} = 0, \bar{y} = 6.82\,\text{in.}$

9.18 $\bar{x} = 1.41\,\text{in.}, \bar{y} = 5.43\,\text{in.}$

9.20   (a) $781 \times 10^6\,\text{mm}^4$, (b) $779 \times 10^6\,\text{mm}^4$,

     (c) -0.25 percent

9.22   (a) $49.1 \times 10^9\,\text{mm}^4$, (b) $50.8 \times 10^9\,\text{mm}^4$,

     (c) 3.4 percent

9.24   (a) $72.9\,\text{in.}^4, 124.9\,\text{in.}^4$, (b) $292\,\text{in.}^4, 572\,\text{in.}^4$

9.26   (a) $6.36 \times 10^9\,\text{mm}^4, 6.36 \times 10^9\,\text{mm}^4$

     (b) $31.8 \times 10^9\,\text{mm}^4, 31.8 \times 10^9\,\text{mm}^4$

9.28 $I = 2.083 \times 10^6\,\text{mm}^4$ without stiffening ribs

     $I = 3.825 \times 10^6\,\text{mm}^4$ with stiffening ribs

9.30 $I_{yc} = 763.3 \times 10^6\,\text{mm}^4, I_{xc} = 1.343 \times 10^9\,\text{mm}^4$

9.32 $I_{yc} = 4.23 \times 10^6\,\text{mm}^4, I_{xc} = 16.9 \times 10^6\,\text{mm}^4$

9.34 $I_{yc} = 21.9 \times 10^6\,\text{mm}^4, I_{xc} = 13.2 \times 10^6\,\text{mm}^4$

9.36 $I_{xc} = 19.5 \times 10^6\,\text{mm}^4, I_{yc} = 11.5 \times 10^6\,\text{mm}^4$

9.38 $I_{xc} = 650\,\text{in.}^4, I_{yc} = 104\,\text{in.}^4$

9.40 $I_{xc} = 70.7 \times 10^6\,\text{mm}^4, I_{yc} = 37.4 \times 10^6\,\text{mm}^4$

9.42 $I_{xc} = 1100\,\text{in.}^4, I_{yc} = 120.4\,\text{in.}^4$

9.44 $I_{xc} = 1.300 \times 10^9\,\text{mm}^4, I_{yc} = 0.792 \times 10^9\,\text{mm}^4$

9.46 $I_{xc} = 180 \times 10^6\,\text{mm}^4$

9.48   (a) $I = 143\,\text{in.}^4$, (b) $I = 469\,\text{in.}^4$, (c) $I = 236.4\,\text{in.}^4$

9.50 $J_c = 52.0 \times 10^6\,\text{mm}^4, J = 156.3 \times 10^6\,\text{mm}^4$

9.52 $J_c = 2.54\,\text{in.}^4, J = 11.93\,\text{in.}^4$

9.54 $23.6\,\text{mm}, 17.56\,\text{mm}$

9.56 $9.76\,\text{in.}, 3.82\,\text{in.}$

9.58 $\bar{x} = 3b/4, \bar{y} = 3h/10$

9.60 $\bar{x} = 4b/5, \bar{y} = 2h/7$

9.62 $bh^3/3$

9.64 $2hb^3/7, 8hb^3/175$

Chapter 10

For Probs. 10.2 through 10.14, answers are given for the bar to the left of a vertical section and above a horizontal section. Axial forces are identified as tension or compression. Shear for vertical sections is identified by the subscript y and for horizontal sections by the subscript x.

10.2   Section M–M: $F = -7.5\,\text{kN}, V_x = 3.0\,\text{kN},$

        $M = -4.2\,\text{kN·m}$

     Section N–N: $F = 3.0\,\text{kN}, V_y = -5.33\,\text{kN},$

        $M = -2.13\,\text{kN·m}$

10.4   Section M–M: $F = 0, V_y = -2800\,\text{lb}, M = 4800\,\text{lb-ft}$

     Section N–N: $F = 0, V_y = 4200\,\text{lb}, M = 10,400\,\text{lb-ft}$

10.6   Section M–M: $F = 0$, $V_y = -900$ lb, $M = 3600$ lb-ft
Section N–N: $F = 0$, $V_y = 300$ lb, $M = 5400$ lb-ft

10.8   Section M–M: $F = +2.0$ kip, $V_x = -0.5$ kip,
$M = 21.0$ kip-ft; Section N–N: $F = 0.5$ kip, $V_y = -2.0$ kip,
$M = +6.0$ kip-ft

10.10   Section R–R: $F = -3.64$ kip, $V = 3.89$ kip
@ $-136.8°$, $M = 31.1$ kip-ft

10.12   Section R–R: $F = 0$, $V_y = 6.25$ kN, $M = +2.5$ kN·m

10.14   Section R–R: $F = 0$, $V_y = -24$ N, $M = 2.4$ N·m,
$T = 0$; Section S–S: $F = 40$ N,
$V_y = 24$ N, $M = -3.0$ N·m, $T = -4.8$ N·m

10.16 $\sigma_{AB} = 6.37$ ksi, $\sigma_{BC} = 4.42$ ksi

10.18   24.4 ksi

10.20   688 MN/m^2 (MPa)

10.22   (a) 19.42 ksi, (b) 9.39 ksi

10.24   (a) 9298 psi, (b) 23,232 lb, (c) $d = 2.47$ in.

10.26   (a) 37,500 lb, (b) 1.154 in.

10.28   (a) 75 psi, (b) 14.92 psi

10.30   8350 psi

10.32   24.2 mm

10.34   3.93 mm

10.36   (a) 97.4 kip, (b) 68.5 kip

10.38   25.6 kip

10.40 $d = 0.924$ in.

10.42 $d = 0.810$ in., 3.95 in. × 3.95 in.

10.44   1.143 kip

10.46   1.343 in.

10.48   (a) $\sigma = 10{,}940$ psi, $\tau = 15{,}630$ psi
(b) $\sigma_{max} = 33{,}300$ psi, $\tau_{max} = 16{,}630$ psi

10.50   (a) $F_N = 819$ lb, $F_V = 358$ lb
(b) $\sigma = 69.9$ psi, $\tau = 30.6$ psi

Chapter 11

11.2   0.0320 mm

11.4   120 mm

11.6   (a) 230 MPa, (b) 207 × 10^3 MPa, (c) 272 MPa,
248 MPa, (d) 279 MPa, (e) 295 MPa, (f) 69.5 percent,
(g) 38.0 percent

11.8   0.400 mm

11.10   (a) 1.36, (b) 2.73 mm

11.12   (a) 202 MPa, (b) 1.24, (c) 8.08 mm

11.14   (a) 118.4 MPa, (b) 2.17 mm

11.16   (a) MSR 2700f-2.2E; MEL M-24 (b) MSR: 0.0895 in.;
MEL: 0.104 in.

11.18   109.1 kN

11.20   219 MPa, 109.9 × 10^3 MPa

11.22   80.0 kip, −160.0 kip; 20.0 ksi, −20.0 ksi;
0.667×10^{-3}, -0.667×10^{-3}

11.24   (a) steel: 7405 psi, concrete: 1149 psi (b) 0.0184 in.

11.26   (a) steel: 16,129 psi, wood: 1057 psi (b) steel: 2.23, wood:
6.15

11.28   deflection (steel and copper): 1.282 mm; stress in steel:
64.1 MPa, stress in copper: 70.5 MPa

11.30   106.7 kN

11.32   bolt stress: 35,710 psi, tube stress: 8930 psi

11.34   0.286

11.36   557 kN, −0.010 mm, −0.035 mm

11.38   (a) 11,310 psi, (b) 63,110 lb

11.40   (a) 945 psi, (b) 136,000 lb

11.42   79.8°C

11.44   4100 psi (tension)

11.46   steel: −275 psi; concrete: 62.5 psi

11.48   193.8 × 10^3 kN/m^2 (kPa), 0.773 mm

11.50   (a) 16.43 ksi, (b) 16.43 ksi, (c) 16.48 ksi

11.52   13.54 kN, 43.8 kN, 69.1 percent

11.54   (a) 24.0 kN, (b) 200 N/mm^2 (MPa), (c) 307 N/mm^2 (MPa)

11.56   44.4 kN

Chapter 12

12.2   82.4 GPa

12.4   0.250

12.6   20,400 psi

12.8   29.3 kN·m

12.10   3.55°

12.12 AB: 2.09 in.; BC: 1.83 in.

12.14   (a) 18.0 lb-in, (b) 5867 psi, (c) 1.45

12.16   (a) AB: 81.5 mm; BC: 102.7 mm, (b) 31.1°

12.18   10.20 mm

12.20   861 rpm

12.22   1.86 in.

12.24   (a) 3242 N·m, (b) 14 mm

12.26   (a) 3-11/16 in., (b) 6650 psi, (c) 0.774 in., (d) 13,808 psi

12.28   (a) 17.48 mm
(b) 127.8 MPa (bearing); 87.2 MPa (shear)
(c) 108.2 hp

12.30   530 hp

Chapter 13

13.2   (a) 29.0 kN, 9.00 kN, −31.0 kN
(b) 0, 29.0 kN·m, 43.4 kN·m, 0

13.4   (a) −8.00 kip, −18.00 kip
(b) 0, −32.0 kip-ft, −176.0 kip-ft

13.6   (a) 0, −10.00 kN;
(b) −4.00 kN·m, −4.00 kN·m

13.8   (a) −8000 lb, 5600 lb, −4400 lb
(b) 0, −16,000 lb-ft, 17,600 lb-ft, 0

13.10   (a) −24.0 kN, −54.0 kN
(b) 0, −24.0 kN·m, −43.2 kN·m, −108.0 kN·m

For Probs. 13.12 through 13.62, the location of a beam cross
section is indicated by the distance measured from the left end
of the beam. A prime (′) is used to indicate left of the section and
a double prime (″) right of the section.

13.12 $V_0 = V_4' = 13.00$ kip, $V_4'' = V_8' = 1.00$ kip,
$V_8'' = V_{16} = -7.00$ kip

13.14 $V_0 = V_4' = -7000$ lb,
$V_4'' = V_{10}' = 10,000$ lb,
$V_{10}'' = V_{18} = -4000$ lb

13.16 $V_0 = V_{1.2}' = -5.20$ kN, $V_{1.2}'' = V_{3.6} = -1.400$ kN

13.18 $V_0 = 16.00$ kip, $V_{5.33} = 0$, $V_8 = V_{12} = -8.00$ kip

13.20 $V_0 = V_1' = 22$ kN, $V_1'' = V_2 = -2$ kN, $V_4' = -18$ kN, $V_4'' = 0$

13.22 $V_0 = 32.1$ kip, $V_8' = 8.1$ kip, $V_8'' = -7.9$ kip,
$V_{20}' = -43.9$ kip, $V_{20}'' = 30.0$ kip, $V_{30} = 0$

13.24 $V_0 = V_8 = -4.0$ kip, $V_{20}' = -10$ kip, $V_{20}'' = 0$

13.26 $V_0 = 4$ kip, $V_{2.86} = 0$, $V_{10}' = -10$ kip, $V_{10}'' = 0$

13.28 $V_0 = 42$ kN, $V_{1.5} = 31.5$ kN, $V_3 = 0$, $V_6' = -42$ kN,
$V_6'' = 0$

13.30 $V_0 = 0$, $V_6 = -9$ kip, $V_{12}' = -18$ kip, $V_{12}'' = 0$

13.32 $V_0 = 0$, $V_3 = -54$ kN, $V_6' = -72$ kN, $V_6'' = 0$

13.34 $V_0 = V_6' = 26.25$ kip, $V_6'' = V_{12}' = -33.75$ kip,
$V_{12}'' = 22.5$ kip, $V_{15} = 0$

13.36 $M_0 = 0$, $M_{1.5} = 2.55$ kN·m, $M_3 = 3.3$ kN·m,
$M_{4.5} = 0$

13.38 $M_0 = 0$, $M_{1.6} = 28.8$ kN·m, $M_{2.29} = 0$,
$M_3 = -30$ kN·m, $M_4 = 0$

13.40 $M_0 = 0$, $M_5 = 34,250$ lb-ft, $M_{12} = 41,600$ lb-ft,
$M_{16} = 39,200$ lb-ft, $M_{20} = 0$

13.42 $M_0 = 0$, $M_4 = -36$ kip-ft, $M_{5.53} = 0$, $M_7 = 34.5$ kip-ft,
$M_{12.31} = 0$, $M_{16} = -24$ kip-ft, $M_{18} = 0$

13.44 $M_0 = 0$, $M_{9.5} = 361$ kN·m, $M_{12} = 336$ kN·m,
$M_{18} = 0$

13.46 $M_0 = 0$, $M_4 = 120$ kip-ft, $M_{14} = 270$ kip-ft,
$M_{16} = 264$ kip-ft, $M_{22} = 228$ kip-ft, $M_{30} = 0$

13.48 $M_0, M_2 = -2.40$ kN·m, $M_{2.55} = 0$, $M_4 = 6.40$ kN·m,
$M_{4.67} = 6.67$ kN·m, $M_8 = 0$

13.50 $M_0 = 0$, $M_6 = -36.0$ kip-ft, $M_{12.16} = 200$ kip-ft,
$M_{18} = -32.0$ kip-ft, $M_{22} = 0$

13.52 $M_0 = -41.6$ kN·m, $M_{1.2} = -20.0$ kN·m,
$M_{2.4} = M_3 = -12.80$ kN·m

13.54 $M_0 = 0$, $M_3 = 9$ kip-ft, $M_4 = 12$ kip-ft,
$M_5 = 9$ kip-ft, $M_8 = 0$

13.56 $M_0 = 0$, $M_{0.634} = 12.99$ kN·m, $M_{1.5} = 0$,
$M_{2.366} = -12.99$ kN·m, $M_3 = 0$

13.58 $M_0 = 0$, $M_6 = -3$ kip-ft, $M_{12} = -24$ kip-ft

13.60 $M_0 = 0$, $M_{2.81} = 197.7$ kN·m, $M_{5.62} = 0$,
$M_6 = -56.8$ kN·m, $M_{7.5} = 0$

13.62 Distributed load of 2.4 kN/m along entire length of beam;
concentrated load of 40 kN downward @ $x = 3$ m;
concentrated support forces of 40 kN upward @ $x = 0$,
and 24 kN upward @ $x = 10$ m.
$M_0 = 0$, $M_3 = 109.2$ kN·m, $M_{10} = 0$

Chapter 14

14.2 11.49 kN·m

14.4 (a) 4520 psi, −4520 psi; (b) 3960 psi

14.6 2.99 kN·m

14.8 (a) 4730 psi, −4730 psi; (b) 1578 psi

14.10 (a) 7.01 ksi, −11.86 ksi, (b) 2.00 ksi

14.12 (a) 50.4 MPa, −89.6 MPa, (b) 12.41 MPa

14.14 (a) 11.11 ksi, −20.79 ksi, (b) 2.12 ksi

14.16 5.90 kip/ft

14.18 14.02 kN

14.20 0.798 kip/ft

14.22 (a) 11.78 ksi, −11.78 ksi, (b) 2.34 kip/ft

14.24 (a) 9.77 ksi; (b) 3.69 kip/ft

14.26 (a) 6.87 ksi, −6.87 ksi, (b) 4.60 kip/ft

14.28 298 lb/ft

14.30 (a) 1160 psi, (b) 25.5 ft

14.32 8.25 MPa

14.34 42.0 kip

14.36 68.4 kPa

14.38 (a) 3.47 ksi, (b) 3.38 ksi, (c) 2.7 percent

14.40 12.75 MPa

14.42 $\tau_{max} = 1.369$ ksi

14.44 8.06 in.

14.46 132.4 mm

14.48 7.04 in.

14.50 (a) $S = 40.0$ in.^{3} (b) W14 × 30

14.52 (a) 2 × 12, (b) 2 × 10

14.54 4 in. × 12 in.

14.56 200 mm × 300 mm

14.58 6 in. × 12 in.

14.60 W18 × 55 (U.S. customary units)

14.62 W310 × 32.7 (SI units)

14.64 W16 × 26 (U.S. customary units)

14.66 (a) 2 × 10 rafters, 2 × 8 ceiling joists, 2 × 10 floor joists
(b) 6 × 12 girder, three columns @ 9′-0″ o.c. spacing
(c) 10,800 lb, select 3.5-in. steel column
(d) 1400 psf
(e) 29 in. × 29 in.

14.68 (a) 2 × 12 rafters, 2 × 8 ceiling joists, 2 × 10 floor joists
(b) 1950 psf, OK for medium clay
(c) 6 × 12 girder, 6 columns @ 6′-7″ o.c. spacing
(d) 15,850 lb, select 6 × 6 spruce–fir–pine column
(e) 28 in. × 28 in.

14.70 (a) 2 × 10 rafters, 2 × 6 floor joists in loft, 2 × 12 floor
joists on first floor, (b) 1425 psf, (c) 1374 psf

Chapter 15

15.2 (a) −523 kN·m^3/EI, (b) 250 kN·m^2/EI

15.4 (a) −6.72 × 10^6 kip-in.^{3}/EI,
(b) 6.22 × 10^4 kip-in.^{2}/EI

15.6 (a) −2.10 MN·m^3/EI, (b) 0.701 MN·m^2/EI

15.8 (a) −5.15 × 10^7 kip-in.^{3}/EI
(b) 3.42 × 10^5 kip-in.^{2}/EI

15.10 (a) −1.920 MN·m^3/EI, (b) 0.800 MN·m^2/EI

15.12 (a) −15.76 mm, (b) 0.376°

15.14 −3.22 × 10^6 kip-in.^{3}/EI

15.16 $-162.8 \text{ kN·m}^3/EI$

15.18. $-4.52 \times 10^6 \text{ kip-in.}^3/EI$

15.20 $-6.86 \times 10^5 \text{ kip-in.}^3/EI$

15.22 $-3.15 \times 10^6 \text{ kip-in.}^3/EI$

15.24 -0.333 in.

15.26 11.93 ft

15.28 MSR 2250f-1.9E or MEL M-21

15.30 (a) 320 psi, (b) 11,600 psi, (c) 0.205 in.

15.32 (a) $-523 \text{ kN·m}^3/EI$, (b) $250 \text{ kN·m}^2/EI$

15.34 (a) $-6.72 \times 10^6 \text{ kip-in.}^3/EI$

       (b) $6.22 \times 10^4 \text{ kip-in.}^2/EI$

15.36 $-2.10 \text{ MN·m}^3/EI$, (b) $0.701 \text{ MN·m}^2/EI$

15.38 (a) $-5.15 \times 10^7 \text{ kip-in.}^3/EI$, (b) $3.42 \times 10^5 \text{ kip-in.}^2/EI$

15.40 $-3.22 \times 10^6 \text{ kip-in.}^3/EI$

15.42 $-162.8 \text{ kN·m}^3/EI$

15.44 $-4.52 \times 10^6 \text{ kip-in.}^3/EI$

15.46 $y_{\max} = 2.02$ in. @ $x = 92.7$ in. from left end of beam

15.48 $R_A = \dfrac{5}{8}wL, R_B = \dfrac{3}{8}wL, M_A = \dfrac{wL^2}{8}$

15.50 $R_A = 50 \text{ kN}, R_B = 30 \text{ kN}, M_A = 40 \text{ kN·m}$

15.52 $R_A = 38.5 \text{ kN}, R_B = 41.5 \text{ kN}, M_B = 35.6 \text{ kN·m}$

15.54 $R_A = \dfrac{5}{8}wL + \dfrac{47}{128}P, R_B = \dfrac{3}{8}wL + \dfrac{81}{128}P,$

       $M_A = \dfrac{15}{128}PL + \dfrac{wL^2}{8}$

15.56 $R_A = 109.4 \text{ kN}, R_B = 98.6 \text{ kN},$
       $M_A = 101.5 \text{ kN·m}$

15.58 $R_A = 21.1 \text{ kN}, R_B = 38.9 \text{ kN}$
       $M_A = 14.4 \text{ kN·m}, M_B = 21.6 \text{ kN·m}$

15.60 $R_A = R_B = \dfrac{wL}{2}, M_A = M_B = \dfrac{wL^2}{12}$

15.62 $R_A = R_B = 60 \text{ kN}, M_A = M_B = 40 \text{ kN·m}$

15.64 $R_A = 37.3 \text{ kip}, R_B = 16.69 \text{ kip}$
       $M_A = 74.6 \text{ kip-ft}, M_B = 40.9 \text{ kip-ft}$

15.66 (a) $\delta = \dfrac{wx^2}{24EI}\left[-x^2 + \dfrac{4L^3}{x} - \dfrac{3L^4}{x^2}\right]$

       (b) $\theta = \dfrac{wL^3}{6EI}$

15.68 (a) $\delta = \dfrac{W_{\max}}{120EIL}(-x^5 + 5L^4x - 4L^5)$

       (b) $\theta = \dfrac{W_{\max}L^3}{24EI}$

15.70 (a) $\delta = \dfrac{M}{72EIL}(-12x^3 + 36L\langle x - a\rangle^2 - 13L^2x)$

       (b) $\delta = -\dfrac{Ma^2}{2EI}$

15.72 (a) $\delta = \dfrac{1}{EI}\left[-\dfrac{20x^3}{3} + 10\langle x - 2\rangle^3 + 186.7x - 320\right]$

       (b) $\delta = -\dfrac{320 \text{ kN·m}^3}{EI}$

       (c) $\delta = \dfrac{82.3 \text{ kN·m}^3}{EI}$

       (d) $I = 144 \times 10^6 \text{ mm}^4$

Chapter 16

16.2   (a) 33.3 MPa, −26.7 MPa

16.4   (a) 78.9 ksi, −78.9 ksi, (b) 72.0 ksi, −85.9 ksi

16.6   (a) 48.5 MPa, −56.2 MPa

16.8   (a) 6040 psi, −6590 psi

16.10 (a) 109.0 MPa, −120.3 MPa

16.12 (a) 9500 psi, −5500 psi

16.14 (a) 10,980 psi, −11,940 psi

16.16 147.9 MPa, −147.9 MPa

16.18 32.7 kN

16.20 (a) A and B: 0 ksi; C and D: −0.417 ksi, neutral axis along
       side AB

       (b) A: −0.375 ksi; B: 0.375 ksi; C: −0.0416 ksi;
       D: −0.792 ksi, neutral axis intersects side AB 10 in.
       from B and side BC 21.6 in. from B.

16.22 A: 28.2 MPa; B: −1.460 MPa; C: −23.7 MPa;
       D: −50.4 MPa

16.24 $\sigma_n = 8200 \text{ psi}, \sigma_t = -4200 \text{ psi}, \tau_{nt} = -1270 \text{ psi}$

16.26 $\sigma_n = 12,160 \text{ psi}, \sigma_t = 14,160 \text{ psi}, \tau_{nt} = -2800 \text{ psi}$

16.28 $\sigma_n = 1.368 \text{ ksi}, \sigma_t = -1.368 \text{ ksi}, \tau_{nt} = 3.76 \text{ ksi}$

16.30 $\sigma_n = 9.00 \text{ MPa}, \sigma_t = -1.000 \text{ MPa}, \tau_{nt} = 8.66 \text{ MPa}$

16.32 (a) 8320 psi, −4320 psi @ 35.8°; (b) 6320 psi @ 80.8°

16.34 (a) 12,450 psi, −14,450 psi @ 66.0°,
       (b) 13,450 psi @ 21.0°

16.36 (a) 4.00 ksi, −4.00 ksi @ 45.0°, (b) 4.00 ksi @ 0°

16.38 (a) 14.00 MPa, −6.00 MPa @ 0°,
       (b) 10.00 MPa @ 45.0°

16.40 A: 19,690 psi, 0,9850 psi; B: 10,600 psi, −818 psi,
       5740 psi; C: 3940 psi, −3940 psi, 3940 psi

16.42 A: 6.60 MPa, 0; B: 3.51 MPa, −0.212 MPa;
       C: 1.027 MPa, −1.027 MPa

16.44 A: 0, −12.71 ksi; B: 0.196 ksi, −4.63 ksi; C: 7.99 ksi, 0

16.46 (a) −1598 psi, 1839 psi; (b) −3710 psi, 1857 psi

16.48 (a) 80.0 MPa, (b) 40.0 MPa, (c) 20.0 MPa, (d) 40.0 MPa

16.50 (a) 0, 7640 psi, 0, −7640 psi, (b) 955 psi, (c) 637 psi, 0,
       637 psi, 0, (d) D: 318 psi, −318 psi, 318 psi;
       E: 7760 psi, −117 psi, 3940 psi; F: 1592 psi, −1592 psi,
       1592 psi; G: 117 psi, −7760 psi, 3940 psi

16.52 (a) 0, 27.2 ksi, 0, −27.2 ksi, (b) 6.04 ksi, (c) 0.755 ksi, 0 ksi,
       0.755 ksi, 0 ksi, (d) A: 6.79 ksi, −6.79 ksi, 6.79 ksi; B:
       28.4 ksi, −1.28 ksi, 14.86 ksi; C: 5.28 ksi, −5.28 ksi,
       5.28 ksi; D: 1.28 ksi, −28.4 ksi, 14.86 ksi

Chapter 17

17.2   (a) $P_c = 296 \text{ kip}, P_a = 148 \text{ kip}$
       (b) $P_c = 604.3 \text{ kip}, P_a = 302.2 \text{ kip}$ (c) $P_a = 360 \text{ kip}$

17.4   64.0 kN

17.6   (a) 83.6, (b) 167.3

17.8   5.46 kN

17.10 26.8 kN

17.12 253 kip

17.14 1.917 MN

17.16 1.438 MN

17.18 396 kN

Chapter 18

18.2 (a) 96.1 kip, (b) 96.1 kip

18.4 (a) 61.6 kip, (b) 61.9 kip

18.6 (a) 145.6 kip, (b) 145.6 kip

18.8 189.8 kip

18.10 10.79 kip/ft

18.12 (a) 11.6 in., (b) 17.5 in.

18.14 (a) L_1 = 15.8 in., L_2 = 4.3 in., L_3 = 6 in. (on both angles)

(b) L_1 = 19.5 in., L_2 = 5.7 in., L_3 = 6 in. (on both angles)

18.16 (a) L_1 = 14.1 in., L_2 = 4.1 in., L_3 = 8 in.

(b) L_1 = 11.5 in., L_2 = 2.9 in., L_3 = 8 in.

18.18 (a) L_1 = 17.1 in., L_2 = 7.1 in., end return: 1 in.

(b) L_1 = 14.5 in., L_2 = 5.9 in., end return: 1 in.

INDEX

A

Acceleration of gravity, 7
Accuracy, numerical, 8–10
Action and reaction, 45
Air-supported structure, 422
Allowable stress
    for aluminum alloy columns, 434
    definition of, 232, 243
    for machine design steel columns, 433–434
    for structural steel columns, 433
Alternate interior angles, 12
Aluminum Association, 245, 434
Ambiguous case, 15
American Airlines, 6
American Association of State Highway and Transportation
    Officials, 353
American Ceramic Society, 245
American Concrete Institute International, 245
American Institute of Steel Construction (AISC), 245, 328,
    353, 433, 439, 442, 445, 446, 447, 452
American Institute of Timber Construction, 245
American Iron and Steel Institute, 245
American National Standards Institute, 245
American Society for Nondestructive Testing, 245, 449
American Society of Civil Engineers, 245
American Society of Mechanical Engineers, 8, 245
American Society for Testing and Materials, 245
American Standard beam, 185–186, 462–465
American Standard channel, 186, 466–469
American Wood Council, 245
Angles, L shapes, 186, 470–473
Angles
    alternate interior, 12
    corresponding, 12
    of depression, 13
    of elevation, 13
    of friction, 164–167
    of repose, 166
    supplementary, 15
    or twist for a circular shaft, 270–272, 274
    vertical, 12
Anticlastic curvature, 306
APA: The Engineered Wood Association, 246
Arch, Gateway, 142
Archimedes, 1, 9
Arm of the couple, 75
Aspdin, Joseph, 338
Associations, professional, 245–246, 353
Average stress, 219–223

Axes, 58
Axial loads
    columns, 428
    combined stress, 423–425
    deformation, 240, 247
    statically indeterminate problems, 251–253, 256–258
    strain due to, 240–241, 247, 251–253
    stresses due to, 220–223, 418–419
    stresses on oblique sections, 236–238, 418–419
    thermal stresses and strains, 256–258
Axial stress
    bending plus, 410–411
    welds, 448–451
Axis of symmetry, centroid, 182–183

B

Babylonians, 1
BASIC programs, 25–27
Beams
    American Standard, 185, 186, 462–465
    axial force in, 284
    bending moment in, 284, 286–290, 294–301
    bending stress formula for, 307–313
    bending stress in, 305–313
    cantilever, 284
    cantilever, moment diagrams, 354–357
    continuous, 285, 372
    deflection of (*see* Deflection of beams)
    distributed loading of, 78–79
    elastic section modulus of, 308
    fixed or built-in, 285
    flexure formula for, 307–313
    hybrid-composite, 4
    lateral buckling or twisting, 328
    maximum bending stress in, 308
    maximum load deflection, 353
    neutral axis of, 306
    overhanging, 284
    propped, 285
    pure bending, 305–306
    reactions for, 81–82, 285–286, 382–385
    shear flow, formula for, 324–326
    shear force in, 284, 286
    shearing stress, formula for, 316–323
    shearing stresses in, 316
    simply supported (simple), 284
    statically indeterminate, 285, 382–385
    steel, structural, 327–328, 332–335
    stresses in, 307–313, 316–323

SI Prefixes

Multiplication Factor	Prefix	Symbol
$1,000,000,000 = 10^9$	giga	G
$1,000,000 = 10^6$	mega	M
$1,000 = 10^3$	kilo	k
$0.1 = 10^{-1}$	deci[†]	d
$0.01 = 10^{-2}$	centi[†]	c
$0.001 = 10^{-3}$	milli	m
$0.000001 = 10^{-6}$	micro	μ (mu)
$0.000000001 = 10^{-9}$	nano	n

[†]These prefixes are not recommended except for decimeter in the measurement of areas and volumes and centimeter for nontechnical length measurements.

Some Common Large Numbers (USA and Modern British)

Name	Value	Exponential Form
million	1,000,000	10^6
billion	1,000,000,000	10^9
trillion	1,000,000,000,000	10^{12}
quadrillion	1,000,000,000,000,000	10^{15}
googol	. . . . . .	10^{100}